Image and Video Processing and Recognition Based on Artificial Intelligence (Volume II)

Image and Video Processing and Recognition Based on Artificial Intelligence (Volume II)

Editors

Kang Ryoung Park
Sangyoun Lee
Euntai Kim

MDPI • Basel • Beijing • Wuhan • Barcelona • Belgrade • Manchester • Tokyo • Cluj • Tianjin

Editors
Kang Ryoung Park
Dongguk University
Seoul
Republic of Korea

Sangyoun Lee
Yonsei University
Seoul
Republic of Korea

Euntai Kim
Yonsei University
Seoul
Republic of Korea

Editorial Office
MDPI
St. Alban-Anlage 66
4052 Basel, Switzerland

This is a reprint of articles from the Special Issue published online in the open access journal *Sensors* (ISSN 1424-8220) (available at: https://www.mdpi.com/journal/sensors/special_issues/IVPRBAI).

For citation purposes, cite each article independently as indicated on the article page online and as indicated below:

LastName, A.A.; LastName, B.B.; LastName, C.C. Article Title. *Journal Name* **Year**, *Volume Number*, Page Range.

ISBN 978-3-0365-7602-2 (Hbk)
ISBN 978-3-0365-7603-9 (PDF)

Contents

About the Editors

Kang Ryoung Park

Kang Ryoung Park received his B.S. and M.S. degrees in electronic engineering from Yonsei University, Seoul, South Korea, in 1994 and 1996, respectively. He received his Ph.D. degree in electrical and computer engineering from Yonsei University in 2000. He has been a professor in the division of electronics and electrical engineering at Dongguk University since March 2013. His research interests include image processing and deep learning.

Sangyoun Lee

Sangyoun Lee received his B.S. and M.S. degrees in electrical and electronic engineering from Yonsei University, Seoul, South Korea, in 1987 and 1989, respectively, and his Ph.D. degree in electrical and computer engineering from the Georgia Institute of Technology, Atlanta, GA, USA, in 1999. He is currently a professor and the head of the Graduate School of Electrical and Electronic Engineering, and is the head of the Image and Video Pattern Recognition Laboratory, Yonsei University. His research interests include all aspects of computer vision, with a special focus on pattern recognition for face detection and recognition, advanced driver-assistance systems, and video codecs.

Euntai Kim

Euntai Kim received his B.S., M.S., and Ph.D. degrees in electronic engineering from Yonsei University, Seoul, South Korea, in 1992, 1994, and 1999, respectively. From 1999 to 2002, he was a full-time lecturer in the Department of Control and Instrumentation Engineering, Hankyong National University, Kyonggi-do, Korea. Since 2002, he has been with the faculty of the School of Electrical and Electronic Engineering, Yonsei University, where he is currently a professor. He was a visiting researcher at the Berkeley Initiative in Soft Computing, University of California, Berkeley, CA, USA, in 2008. He was also a visiting researcher at the Korea Institute of Science and Technology (KIST), Korea, in 2018. His current research interests include computational intelligence, statistical machine learning and deep learning and their application in intelligent robotics, autonomous vehicles, and robot vision.

Preface to "Image and Video Processing and Recognition Based on Artificial Intelligence (Volume II)"

Recent developments have led to the powerful application of artificial intelligence (AI) techniques in image and video processing and recognition. While this state-of-the-art technology has matured, its performance is still affected by various environmental conditions and heterogeneous databases. The purpose of this Special Issue was to bring together high-quality and state-of-the-art academic papers on challenging issues in the field of AI-based image and video processing and recognition. We solicited original papers of unpublished and completed research that were not under review by any other conference, magazine, or journal. Topics of interest included, but were not limited to, the following:

- AI-based image processing, understanding, recognition, compression, and reconstruction;
- AI-based video processing, understanding, recognition, compression, and reconstruction;
- Computer vision based on AI;
- AI-based biometrics;
- AI-based object detection and tracking;
- Approaches that combine AI techniques and conventional methods for image and video processing and recognition;
- Explainable AI (XAI) for image and video processing and recognition;
- Generative adversarial network (GAN)-based image and video processing and recognition;
- Approaches that combine AI techniques and blockchain methods for image and video processing and recognition.

Kang Ryoung Park, Sangyoun Lee, and Euntai Kim
Editors

sensors

Article

Unsupervised Video Summarization Based on Deep Reinforcement Learning with Interpolation

Ui Nyoung Yoon, Myung Duk Hong and Geun-Sik Jo *

Artificial Intelligence Laboratory, Department of Electrical and Computer Engineering, Inha University, Incheon 22212, Republic of Korea
* Correspondence: gsjo@inha.ac.kr; Tel.: +82-032-860-7447

Abstract: Individuals spend time on online video-sharing platforms searching for videos. Video summarization helps search through many videos efficiently and quickly. In this paper, we propose an unsupervised video summarization method based on deep reinforcement learning with an interpolation method. To train the video summarization network efficiently, we used the graph-level features and designed a reinforcement learning-based video summarization framework with a temporal consistency reward function and other reward functions. Our temporal consistency reward function helped to select keyframes uniformly. We present a lightweight video summarization network with transformer and CNN networks to capture the global and local contexts to efficiently predict the keyframe-level importance score of the video in a short length. The output importance score of the network was interpolated to fit the video length. Using the predicted importance score, we calculated the reward based on the reward functions, which helped select interesting keyframes efficiently and uniformly. We evaluated the proposed method on two datasets, SumMe and TVSum. The experimental results illustrate that the proposed method showed a state-of-the-art performance compared to the latest unsupervised video summarization methods, which we demonstrate and analyze experimentally.

Keywords: video summarization; unsupervised learning; reinforcement learning; piecewise linear interpolation

Citation: Yoon, U.N.; Hong, M.D.; Jo, G.-S. Unsupervised Video Summarization Based on Deep Reinforcement Learning with Interpolation. *Sensors* **2023**, *23*, 3384. https://doi.org/10.3390/s23073384

Academic Editors: Euntai Kim, Sangyoun Lee and Kang Ryoung Park

Received: 10 February 2023
Revised: 13 March 2023
Accepted: 22 March 2023
Published: 23 March 2023

1. Introduction

Individuals spend time on online video-sharing platforms such as YouTube to search for videos. To reduce the search time, thumbnails or summary videos are used to efficiently and quickly grasp the video content [1]. Over the past few years, video summarization has become important and has been actively researched to search through video content or produce summary videos from long videos. The video summarization problem is a challenging task in predicting the frame-level or shot-level importance scores of videos [2] and is an abstract and subjective multimodal task without explicit audio-visual patterns or semantic rules. If a frame of the video is interesting or informative, the importance score of the frame should be high. These high-scored frames are selected to create the video summary. Recently, various methods that show high performance using deep learning have been proposed [3–5]. Deep learning-based video summarization methods are divided into supervised and unsupervised learning-based methods. For supervised learning-based methods, creating a labeled dataset is a challenge. Furthermore, it is hard to produce a dataset covering various domains or scenes. For this reason, we focused on developing an unsupervised video summarization method.

The reinforcement learning (RL) based unsupervised video summarization method proposed in [6] demonstrated an improved performance. Specifically, to train the neural network using RL, there is an efficient and explicit evaluation method to select keyframes, which is a reward function. Using the evaluation method, the deep neural network efficiently trains the various features of video such as representativeness, diversity, and uniformity. Using RL, we proposed Interp-SUM in the previous work [3], which uses the piecewise linear interpolation method. With the interpolation method, we mitigated high variance problems and improved performance with a shorter output of the network. However, since we fixed the length of output from the network, Interp-SUM had limitations in increasing the performance for long and short videos. For several videos, the keyframes were selected only in specific scenes, or the interesting keyframes were not adequately selected. Furthermore, previous RL-based video summarization methods have several weaknesses. First, it is difficult to capture the visual and temporal context with their deep neural networks. Second, many methods use the reward or loss function to train the network by calculating the visual difference among the keyframes without considering the temporal distribution of the keyframes. A summary of the video, which keeps the director's storyline by selecting keyframes uniformly, helps people easily understand the video.

In this paper, we propose a new reinforcement learning-based video summarization framework with the interpolation method, which is composed of a new network and a new reward function such as a temporal consistency reward. To increase the performance, we used graph-level features. We also present the transformer and convolutional neural network (CNN)-based video summarization network to accurately predict the importance scores, as shown in Figure 1. The overall contributions are as follows. (i) We present a lightweight video summarization network with a transformer network and 1D convolutional neural network to capture the local and global context feature representations and for efficient interpolation. (ii) We use graph-level features as input features of the video summarization network to efficiently capture long and short context. (iii) We present the temporal consistency reward function to select interesting keyframes efficiently and uniformly.

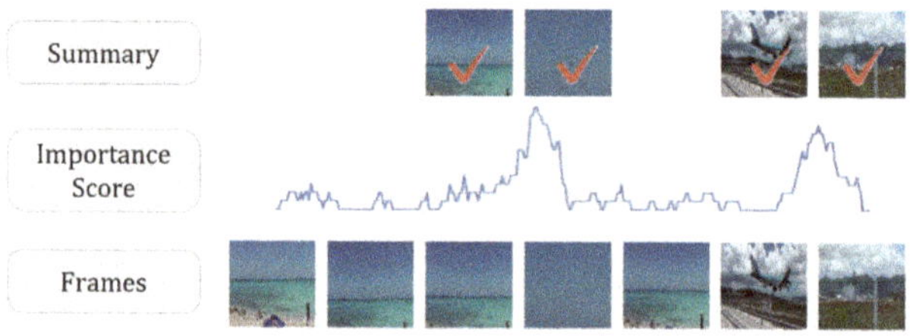

Figure 1. Overview: Our goal was to predict the accurate importance score of the keyframes to produce a summary.

2. Background and Related Work

2.1. Video Summarization

Video summarization methods are divided into supervised and unsupervised learning-based methods. Both methods use the video summarization dataset, which includes the frame-level or shot-level importance scores of the video annotated by several users [2,7]. The supervised learning-based method trains the model with the frame-level or shot-level features of the video as the input to predict the importance scores. With the dataset, this method calculates the cost with the difference between the predicted importance score and annotated importance score. The method minimizes the cost of finding the best model. Many supervised learning-based methods have been proposed.

In [8], the memory-augmented video summarizer was proposed. The memory network provides supporting knowledge extracted from the whole video efficiently. The global attention mechanism was used to predict the importance score of a specific shot by adjusting the score with a holistic understanding of the raw video. In [9], they presented an LSTM-based network with a determinantal point process (DPP) that encodes the probability to sample frames to learn representativeness and diversity. In [10], a dilated temporal relational (DTR) unit in the generator was presented to enhance the temporal context representation among video frames. To train the network to obtain the best summary of the video, the adversarial learning method was used with three-player loss functions. In [11], to predict the importance score to select key shots, an attention-based encoder–decoder network was proposed. This network used an encoder with a bidirectional LSTM network and a decoder with an attention mechanism to train the video representation.

However, the issue with supervised learning-based methods is that it is very difficult to make a human-labeled video summarization dataset including videos of various categories. However, the unsupervised learning-based method does not need a human-labeled dataset. Many unsupervised learning-based methods have been proposed.

In [4], the attention autoencoder (AAE) network replaced the VAE in the SUM-GAN to improve the training efficiency and performance from the adversarial autoencoder (AAE) proposed in SUM-GAN. The interesting frames to summarize the video were weighted while training the networks. In [5], the proposed CSNet (chunk and stride network) was based on a variational autoencoder (VAE) and generative adversarial network (GAN) architecture to efficiently train the local and global contexts of the video to predict the video summary well. In [12], an adversarial autoencoder (AAE) based video summarization model was proposed. The selector LSTM selected the frames using the input frame-level features of the video. Then, the variational autoencoder (VAE) generated a reconstructed video using the selected frames. To train the entire network, the discriminator distinguishes between the original input video and the reconstructed video. In particular, four loss functions were used to train the model. In [13], the Cycle-SUM, a SUM-GAN variant, was proposed. However, the model used a cycle generative adversarial network with two VAE-based generators and two discriminators to preserve the information in the original video in the summary video. In [14], the proposed tessellation approach was a video summarization method that finds visually similar clips and selects the clips that maintain temporal coherence using the Viterbi algorithm, which is a graph-based method. Rochan et al. proposed an unsupervised learning-based SUM-FCN [15]. The method presented a new FCN architecture with the temporal convolution converted from spatial convolution to handle the video sequence. The method selects frames using the output score of the decoder and calculates the loss function with a repelling regularizer to enforce the diversity of the frames in the summary video.

2.2. *Policy Gradient Method*

Deep reinforcement learning combines a deep neural network with a reinforcement learning method [16]. The policy gradient method is one of the model-free reinforcement learning methods. The policy gradient method parametrizes the policy to the deep neural network model and optimizes the model by maximizing reward over the state distribution defined by the policy using gradient descent methods such as stochastic gradient descent (SGD). The method calculates and minimizes the cost with the objective function to train the neural network. However, policy gradient methods have several problems such as the low sample efficiency problem [17] and the high variance problem. The reason for the low sample efficiency problem is that the agent requires more samples such as human experience to learn actions in the environment (states) compared to humans as the agent is not as intelligent as a human. The high variance problem of the estimated gradient is caused by the long-horizon problem and the high-dimensional action space [18]. The long-horizon problem arises from the hugely delayed reward for a long sequence of decisions to achieve a goal. In this paper, we used a policy gradient with a baseline to reduce the variance and increase the number of episodes to mitigate the sample efficiency problem. We also used the piecewise linear interpolation to mitigate the high variance problem.

3. Method

The video summarization problem is formulated as a keyframe selection problem using the importance score predicted by the video summarization network. As illustrated in Figure 2, to predict accurate importance scores, we developed a video summarization network that consists of a transformer encoder network and the Pointwise Conv 1D network. The transformer encoder network encodes the input graph-level features, and Pointwise

Conv 1D network decodes the features to efficiently generate importance score candidates. The importance score candidates are interpolated to the importance scores with piecewise linear interpolation and are converted to a frame-selection action to select keyframes as a summary using the Bernoulli distribution. Furthermore, the proposed temporal consistency reward function and adopted diversity and representativeness reward functions were used to measure how good the generated summary with the importance scores was. Then, we trained the video summarization network with the policy gradient-based training method using the calculated reward.

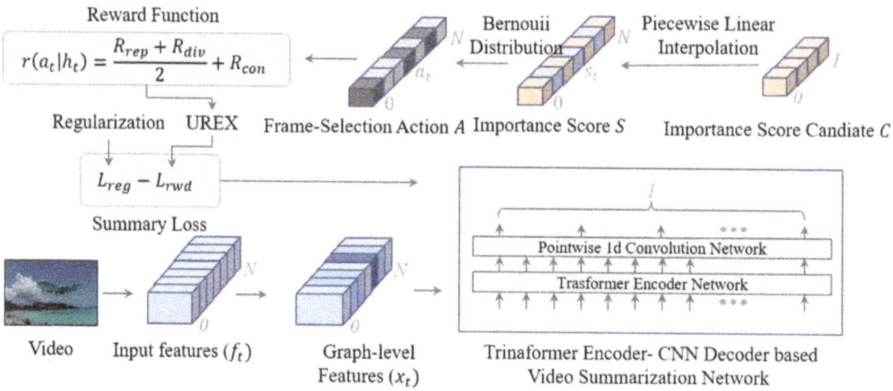

Figure 2. Unsupervised video summarization framework-based on deep reinforcement learning with piecewise linear interpolation.

3.1. *Video Summarization Network*

First, suppose the sequence length is N and the frame number is t, the frame-level visual features $\{f_t\}_{t=1}^{N}$ are extracted from the input video using GoogleNet [19], which is a powerful CNN for image classification trained with the ImageNet dataset. Let the embedding size of the frame-level visual features is M, and the shape of the frame-level visual features is like (N, M). Feature extraction is important for capturing the visual characteristics of the frame image as a low-dimensional feature vector. Therefore, the extracted features efficiently calculate the visual differences among frames in the video.

In recent years, graph neural networks (GNNs) have been extensively studied and it has shown state-of-the-art performance in various deep learning applications [20]. In this paper, we considered the keyframe selection problem as a graph-based anchor node finding or graph-based pathfinding problem. Because graph-level features have a relationship and the structural information of nodes, it is very useful to capture the temporal dependency among keyframes, specifically, to capture the relationship between scenes. In GNN, the graph-level features and graph representation mean that the low-dimensional node embeddings are encoded by a neural network. However, in this paper, we simply defined the graph-level features (1), that is, the features made of multiplying the node-level features F (N × M) and adjacency matrix A (N × N) without trainable embeddings, as shown in Figure 3. We built an adjacency matrix that represents edge information with node-level features.

$$\{x_t\}_{t=1}^{N} = FA \tag{1}$$

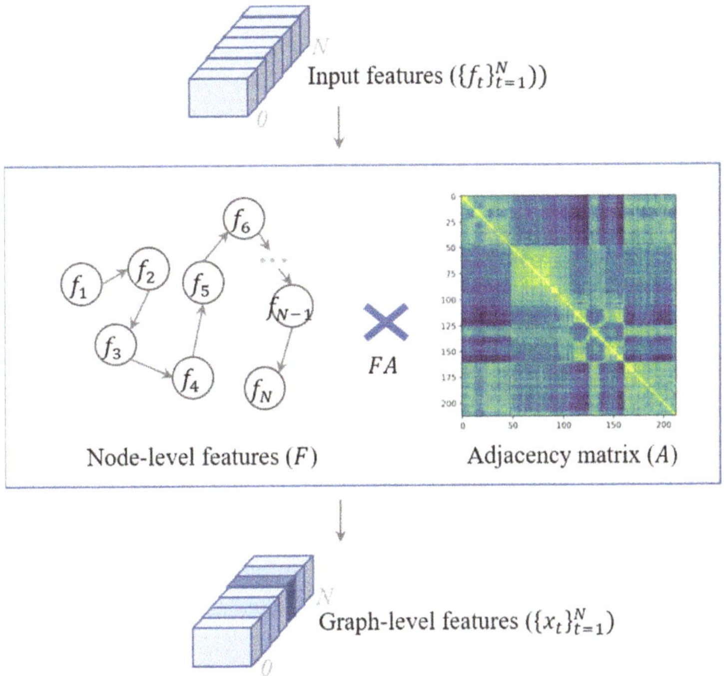

Figure 3. Convert the input features to the graph-level features.

As illustrated in Figure 4a, the proposed video summarization network had a transformer and CNN to predict the keyframe-level importance score candidates with graph-level features. Transformer networks are widely used in sequence learning such as natural language understanding and video understanding because the network learns spatio-temporal context very efficiently. The original transformer network is composed of the encoder and decoder network and we only used the encoder network. The transformer encoder network in our network was made of four layers and eight heads. After the transformer encoder network, we encoded the features from 1024 to 512, which is an embedding size M using a fully connected layer. We used the layer normalization (LayerNorm) layer, which can efficiently train and distribute importance score candidate values uniformly to prevent the importance scores of each frame that do not have the same values after the sigmoid function because the sigmoid function minimizes the difference in the importance scores among frames. As illustrated in Figure 4b, the Pointwise 1D convolutional (Conv 1D) network is a very lightweight network to train temporal dependencies. It summarizes the frame-level features into the short length I of features to make importance score candidates using the convolution filter as a sliding window. In the previous work, Interp-SUM [3], since the method fixed the length I of the importance score candidate, had limitations in terms of increasing the performance for the long and short videos. On the other hand, the Pointwise Conv 1D network provides flexibility for various lengths of videos. For example, if the sequence length of a video is 213, kernel size is 15, and stride size is 3, then the length of the importance score candidates is 65 as shown in Figure 4a. After the network, we reduced the embedding size to 1 and used the sigmoid function to obtain the importance score candidates $C = \{c_t\}_{t=1}^{N}$.

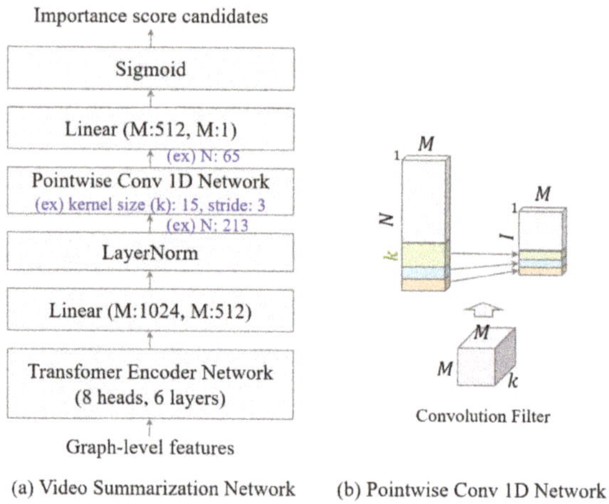

Figure 4. The overall architecture of our network.

3.2. Piecewise Linear Interpolation

Interpolation is an estimation method that finds new data points based on the range of certain known data points. Specifically, piecewise linear interpolation connects the data points with the linear line and calculates intermediate data points on the line [3]. We first aligned the importance score candidates C to fit the sequence length N of the input features with the same intervals. We also interpolated the importance score candidates to the importance score S using piecewise linear interpolation. Finally, we obtained the importance scores of each frame as frame-selection probabilities to select the keyframes. The policy gradient-based reinforcement learning method has a high variance problem. The high variance problem occurs in the high-dimensional action space, for example, the frame-selection action space for video summarization. Moreover, since we used the Bernoulli distribution to select frames based on an exploration strategy, the reward of the frame-selection action changed in every step in the case of high-dimensional action space. Next, the variance of the gradient estimate calculated with a cumulated reward increased, and a high variance problem caused lower training efficiency and performance. To mitigate the high variance problem, we proposed an interpolation method. When the frames were selected using interpolated importance scores, adjacent frames had similar importance scores and were selected together. This had the effect of reducing the action space and mitigated the high variance problem. The interpolation method facilitated the generation of a natural sequence of summary frames because the near keyframes that had high importance scores were selected by each other. Moreover, the interpolation method reduced the computational complexity of the video summarization network. This makes the video summarization network learn faster because the network needs to predict only the I length of the important candidates, and not all.

To select the keyframes as a summary, the Bernoulli distribution (2) was used, which is a discrete probability distribution to convert the importance score S to the frame-selection action $A = \{a_t | a_t \in \{0, 1\}, t = 1, .., N\}$. If the frame-selection action of a frame is equal to 1, this keyframe is selected as a summary. The Bernoulli distribution promotes the exploration of the various summaries of the video as it randomly creates variants of the frame-selection action.

$$A \sim Bernoulli(a_t; s_t) = \begin{cases} s_t, & for\ a_t = 1 \\ 1 - s_t, & for\ a_t = 0 \end{cases} \quad (2)$$

3.3. Reward Functions

To train the video summarization network efficiently, we adopted the diversity reward function and the representativeness reward function [6], and we proposed a temporal consistency reward function. The temporal consistency reward function and representativeness reward function considered the visual similarity distance and temporal similarity distance among keyframes.

The diversity reward function R_{div} (3) computes the dissimilarity among the keyframes selected by the frame-selection action with the extracted features. With this reward function, the network is trained to predict the importance score to selecting diverse frames as keyframes of the summary. The summary, consisting of these keyframes, allows individuals to easily grasp the content of the video. To maintain the storyline of the video and reduce the computational complexity, we limited the temporal distance to 20 to calculate the dissimilarity among the selected keyframes. Without this limitation, even when the flashback scenes or similar scenes were far from the selected keyframe, they were ignored when selecting diverse frames.

Let the indices of the selected keyframes be $\mathcal{J} = \{i_k | a_{i_k} = 1,\ k = 1,2,\ldots,\ |\mathcal{J}|\}$, and the diversity reward function is

$$R_{div} = \frac{1}{\mathcal{J}(\mathcal{J}-1)} \sum_{t\in \mathcal{J}} \sum_{\substack{t'\in\mathcal{J} \\ t\neq t'}} \left(1 - \frac{x_t^T x_{t'}}{\|x_t\|_2 \|x_{t'}\|_2}\right) \quad (3)$$

The representativeness reward function R_{rep} (4) computes the similarity between the selected keyframes and all the keyframes of the video using the extracted features. With this reward function, the network is trained to predict the importance score to select keyframes of the summary that represent the video. $\mathrm{D_t}$ is the distance between the selected keyframes and $\mathrm{s_t}$ is the importance score of the selected keyframe.

$$R_{rep} = exp\left(-\frac{1}{N}\sum_{t=1}^{N} D_t \times s_t\right) \quad (4)$$

$$D_t = \min_{t'\in\mathcal{J}} \|x_t - x_{t'}\|_2 \quad (5)$$

To create a superior summary of the video and increase performance, we applied the importance score S to the representativeness reward function, for example, $D \times s_t$. In particular, if the video summarization network predicts a high importance score for the representative keyframes that are selected, the distance D is decreased and the reward function returns a high reward. If the importance score of other keyframes except the representative keyframes is low, the reward function also returns a high reward for the selected keyframes. Therefore, the video summarization network is trained to minimize the average distance of the selected keyframes and to maximize the importance scores of the representative keyframes by minimizing the importance scores of other keyframes. This means that the reward is dynamically updated by applying the importance score to efficiently select more representative keyframes.

$$R_{con} = \frac{1}{\log\left(\sum_{k=1}^{\mathcal{J}} (i_k - j_k)^2 / |\mathcal{J}|\right)} \quad (6)$$

$$j_k = (\underset{k\in|\mathcal{J}|}{\mathrm{argmin}} \|x_{i_k} - x\,\|) \quad (7)$$

The temporal consistency reward function R_{con} (6) is proposed to efficiently and uniformly select representative shot-level keyframes. To calculate the reward function, we repeated the process to find the nearest neighbor of the selected keyframes j_k (7) in all keyframes $\{\mathrm{x_t}\}_{\mathrm{t=1}}^{\mathrm{N}}$ until the number of all keyframes $|\mathcal{J}|$, and we calculated the distance between j_k and j_k. To normalize the reward, the distance was divided by $|\mathcal{J}|$, and the log probability was used. In the training process, the temporal consistency reward was increased by minimizing the distance.

To explain the temporal consistency reward in detail, as described in Figure 5, we calculated the similarities among the selected keyframes such as $V_i^{4'}$ of $V_i^{summary}$ and the other keyframes of V_i^{all}. We selected the keyframes such as V_i^3, which were most similar to the keyframes of $V_i^{4'}$, except for itself. Then, we defined the K_A as a group of keyframes that had similar scenes around the neighbor keyframes and K_B as a group of keyframes that had no similar scenes near the selected keyframes or had no similar scenes in the video, as described in Figure 5. The keyframes in K_A are the representative keyframes of the surrounding keyframes. To figure out the number of keyframes included in K_A, we calculated the distance between the selected keyframe and the nearest neighbor of the selected keyframe and counted the selected keyframes that were shorter than the specified threshold. All of the keyframes were counted, except K_A, as the K_B. Depending on the storyline intended by the director, scenes similar to the keyframes included in K_B often appear sparsely or are only shown once. This means that the summary created with the keyframes included in K_B limits the users' understanding of the video content because these keyframes are typically less interesting. In other words, these keyframes are not representative. Therefore, these keyframes need to be removed from the summary to increase the users' understanding. Another advantage is that the temporal consistency reward function mitigates the problem of the excessive selection of specific scenes in the video and helps to select keyframes uniformly because the reward function helps to find local representative keyframes among neighboring keyframes, and not global representative keyframes. However, in the case of videos with large parts of static content, the proposed reward function increases the information redundancy in summary. However, the diversity reward function mitigates the redundancy problem while training.

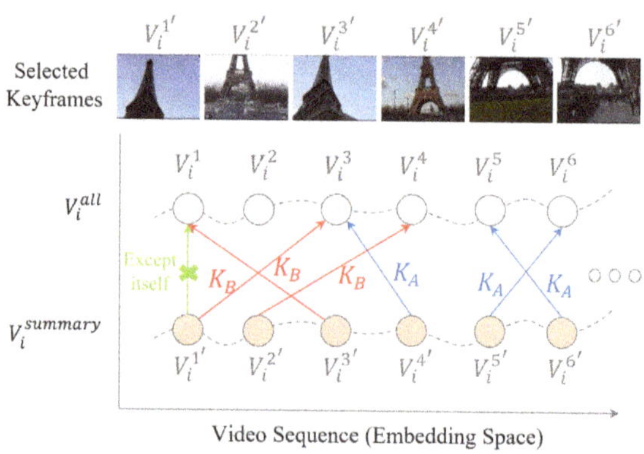

Figure 5. Example of the temporal consistency reward function.

3.4. *Training with Policy Gradient*

The quality of the generated summary was evaluated with the sum of the rewards. With the reward, the proposed video summarization network was trained as a parameterized policy π_θ using the policy gradient method. The method is a reinforcement learning method that explores a better action strategy to obtain a better summary using the gradient descent algorithm. To explore the variety of action strategies, the objective function of the exploration strategy of exploring the under-appreciated reward (UREX) method was used [21]. If the log probability log $\pi_\theta(a_t \mid h_t)$ of the action $\pi_\theta(a_t \mid h_t)$ under the policy underestimates its reward $(a_t \mid h_t) = (R_{rep} + R_{div})/2 + R_{con}$, the action will be further explored by the exploration strategy.

To calculate the objective function of UREX $\mathcal{O}_{UREX}$, the log probabilities of the action and the rewards in episode $\mathcal{J}$ were used. $\mathcal{O}_{UREX}$ is the expectation of the reward $R(a_t \mid h_t)$, which is the sum of the rewards and the reward-augmented maximum likelihood (RAML) objective function. The set of normalized importance weights of the rewards for each episode j is computed using the softmax function to approximate the RAML objective function.

$$\mathcal{O}_{UREX}(\theta;\tau) = \mathbb{E}_{h\sim p(h_t)}\left\{\sum_{a\in\mathcal{A}} R(a_t \mid h_t)\right\} \tag{8}$$

The baseline, an essential technique for the policy gradient to reduce the variance of the gradient estimate and improve computational efficiency, was applied. The baseline is calculated as the moving average of rewards experienced thus far. To reduce the variance efficiently, we calculated the baselines per input video. As illustrated in the equation below, baseline $\mathcal{B}$ was computed as the sum of baseline b_1 for each input video and the baseline b_2, which was the average of all baselines for all videos. Finally, the L_{rwd} is maximized as a cost to train the network.

$$\mathcal{B} = 0.7 \times b_1 + 0.3 \times b_2 \tag{9}$$

$$L_{rwd} = \mathcal{O}_{UREX}(\theta;\tau) - \mathcal{B} \tag{10}$$

3.5. Regularization

The regularization term L_{reg} proposed in [6] was used to control the probability of selecting the keyframe using the importance score. If most of the importance scores are close to 1 or 0, the probability of selecting the wrong keyframes as a summary can be increased. For example, if the importance scores of all keyframes are 1, the video summarization network selects all keyframes as the summary. Consequently, this term was used to make the importance score closer to 0.5 while training. The number (0.5) means that this term helps to select keyframes as summaries evenly based on the exploration strategy of reinforcement learning. To avoid the rapid convergence of the importance score to 0.5 while training, 0.01 was multiplied as below.

$$L_{reg} = 0.01 \times \left(\frac{1}{N} \times \sum_{1}^{N} s_t - 0.5\right)^2 \tag{11}$$

After all of the loss functions were computed, the final loss for video summarization $\mathrm{L}_{\mathrm{summary}}$ was calculated, and the backpropagation was conducted.

$$L_{summary} = L_{reg} - L_{rwd} \tag{12}$$

Algorithm 1 applies to the training procedure of the proposed video summarization network with the policy gradient method.

Algorithm 1. Training Video Summarization Network

```
Input: Graph-level features of the video (x_t)
Output: Proposed network's parameters (θ)
for the number of iterations do
    C ← Network(x_t) % Generate importance score candidate
    S ← Piecewise linear interpolation of C
    A ← Bernoulli Distribution(S)% Action A from the score S
    % Calculate Reward Functions and A Loss Function using A and S
    {θ} ←+ −∇(L_reg − L_rwd)% Minimization
    % Update the network using the policy gradient method:
end for
```

3.6. Generating a Video Summary

In the SumMe and TVSum datasets, we used the shot-level importance score to compare with other methods. To detect the shots of the video, the kernel temporal segmentation (KTS) method, which detects change points such as shot boundaries, is used [22]. The shot-level importance scores are calculated by averaging the frame-level importance scores in a shot. To generate the video summary, key shots were selected over the top 15% of the video length and sorted by the score. This step applies the same concept as the 0–1 Knapsack problem to maximize the importance of the summary video as described in [6].

4. Experiments

4.1. Dataset

Our proposed video summarization method was evaluated on two datasets: SumMe [2] and TVSum [7]. The SumMe dataset consists of 25 videos covering various topics such as extreme sports or airplane landings. Each video length was about 1 to 6.5 min long, and the average number of frames was 4692.8. The average number of shot changes was 29.76. The frame-level importance scores for each video were annotated. The TVSum dataset consists of 50 videos of various topics such as vlogs, news, and documentary. The shot-level importance scores for each video were annotated and the videos varied from 2 to 10 min, and the average number of frames was 7047.06. The average number of shot changes was 47.46.

4.2. Evaluation Setup

For a fair comparison with other methods on two datasets, the evaluation method used in [6] to compute the F-measure as a performance metric was applied. First, the precision and recall were calculated based on the result. Then, the F-measure was computed. Let G be the generated shot-level summary by our proposed method and A be the user-annotated summary in the dataset. Both the precision and recall were calculated based on the amount of temporal overlap between G and A, as seen below.

$$\text{Precision} = \frac{\text{Duration of overlap between G and A}}{\text{Duration of G}} \tag{13}$$

$$\text{Recall} = \frac{\text{Duration of overlap between G and A}}{\text{Duration of A}} \tag{14}$$

$$\text{F}-\text{measure} = \frac{2 \times \text{ Precision } \times \text{ Recall}}{\text{Precision } + \text{ Recall}} \times 100\% \tag{15}$$

The 5-fold cross-validation was used to find the performance of our method for a fair comparison. Our method was tested for five different random splits and the result of the average performance was taken. To create random splits, the videos were split into training and validation datasets. Moreover, the F-measure was computed using the validation dataset.

4.3. Implementation Details

The proposed video summarization network was developed using Pytorch 1.7.1 and consisted of a transformer encoder network and a Pointwise 1D Conv network. The transformer encoder network consisted of six transformer encoder layers with eight heads and 512 hidden units. Moreover, the Pointwise 1D Conv network was based on 15 kernel sizes and three stride sizes. The Adam optimizer was used to train the network with a learning rate of 0.00001 for 1000 epochs.

4.4. Performance Evaluation

4.4.1. Quantitative Evaluation

As illustrated in Table 1, the proposed methods with different kernel sizes and stride sizes were compared to analyze the performance of a Pointwise 1D Conv network for interpolation. By changing the receptive field size using the kernel size, the importance score candidates were produced to capture the short-term or long-term context information in a video efficiently. By changing the stride size, we reduced the size of important score candidates to be interpolated to make a lightweight video summarization network and mitigate the high variance problems, as explained in the previous section. For the SumMe dataset, the method with 15 kernel sizes and three stride sizes showed the highest performance. For the TVSum dataset, the method with five kernel sizes and three stride sizes showed the highest performance.

Table 1. Results (F-measure, %) of the comparison among our methods with different kernel and stride sizes.

Kernel Size	Stride Size	SumMe	TVSum
3	3	50.78	59.76
5	3	51.58	**60.08**
10	3	49.74	59.54
15	3	**51.66**	59.86
15	5	50.70	59.62
15	10	47.88	56.54
20	3	49.16	59.06
25	3	46.22	56.46

We noted that the performance was decreased because of the lack of context information caused by the shorter importance score candidates as the kernel size and stride size increased. In addition, the performance decreased rapidly as the stride size increased. We observed that the performance degradation was caused by the missed context information between scenes due to the short length of the importance score candidates because the length of the importance score candidates is mainly influenced by the stride size. The result of the TVSum dataset showed better performance than the result of the SumMe dataset at a smaller kernel size. This is because the training efficiency of the video summarization network was low since the length of the importance score candidates rapidly shortened when the kernel size was large, and the video length was long like the videos in the TVSum dataset. In other words, the training efficiency of the video summarization network was decreased by the shortened importance score candidates. We chose the proposed method as the best-performing method showing the best F-measure (51.66) on the SumMe dataset and the second best F-measure (59.86) on the TVSum dataset. We chose the best-performing method because increasing the performance of a small number of videos such as the SumMe dataset is more complicated than the TVSum dataset.

As illustrated in Table 2, variants of the proposed method were compared. Our method without the Pointwise 1D Conv network and interpolation showed lower performance on both datasets than the proposed method, which means that the proposed convolutional neural network and interpolation method are useful for improving the performance. Our method without the temporal consistency reward function showed a lower performance on the SumMe dataset, but showed a higher performance on the TVSum dataset. In the case of the SumMe dataset, the average length of the videos was short and the number of keyframes was small. The temporal consistency reward function helped to improve the performance in the dataset, which had many similar scenes among the keyframes. In the case of the TVSum dataset, the temporal consistency reward function was less effective. Because the video was long and the number of keyframes was large, similar scenes among keyframes were small. Our method without graph-level features showed a lower performance result on both datasets. This means that the graph-level features are very effective in improving performance.

Table 2. Results (F-measure, %) of the comparison among variants with 15 kernel sizes and three stride sizes.

Method	SumMe	TVSum
Ours w/o 1D Conv Network and w/o Interpolation	50.02	58.70
Ours w/o Temporal Consistency Reward	49.88	**60.58**
Ours w/o Graph-level Features	50.46	57.26
Ours	**51.66**	59.86

Table 3 illustrates the difference between our proposed method and the existing unsupervised-based state-of-the-art methods. The results demonstrate that our proposed method showed a state-of-the-art performance on the SumMe dataset and high performance on the TVSum dataset. However, AC-SUM-GAN was 1.24% better than our proposed method on the TVSum dataset and DSR-RL-GRU was 0.57% better on the TVSum dataset. Although AC-SUM-GAN and DSR-RL-GRU performed better on the TVSum dataset, AC-SUM-GAN was more complicated than the proposed method and both methods could not make a natural sequence of summary like the proposed method without the interpolation method. However, these methods showed a lower performance than the proposed method on the SumMe dataset. Additionally, our proposed method demonstrated significantly improved results compared with the experimental results of our previous proposal (Interp-SUM). Specifically, the results showed that the videos in the SumMe dataset were close to the average keyframe length (293) such as 'Car over camera' (293, 73.2%), 'Air force one' (300, 60.0%), and 'Kids playing in leaves' (213, 50.2%), which showed the highest F-measure (%). In addition, the summary results of the short videos such as 'Fire Domino'(108, 60.2%), and 'Paluma jump' (172, 61.9%) showed a high F-measure (%). However, the result of long videos such as 'Cockpit landing' (604, 28.8%), and 'Uncut evening flight' (645, 19.4%) showed a low F-measure (%). We noted from our analysis that it was difficult to find representative keyframes in the case of long videos, as selecting keyframes from the long video caused a high variance problem. However, there were cases where the video length was short, but the F-measure (%) was high such as 'Fire Domino' (108, 60.2%). We noted that the proposed method selected representative keyframes well when the visual dissimilarity among scenes was high.

Table 3. Results (F-measure, %) of the comparison among the unsupervised-based methods tested on SumMe and TVSum. +/− indicates better/worse performance than ours.

Method	SumMe	TVSum
SUM-GAN [12]	39.1 (−)	51.7 (−)
SUM-FCN [15]	41.5 (−)	52.7 (−)
DR-DSN [6]	41.4 (−)	57.6 (−)
Cycle-SUM [13]	41.9 (−)	57.6 (−)
CSNet [5]	51.3 (−)	58.8 (−)
UnpairedVSN [23]	47.5 (−)	55.6 (−)
SUM-GAN-AAE [4]	48.9 (−)	58.3 (−)
CSNet+GL+RPE [24]	50.2 (−)	59.1 (−)
AC-SUM-GAN [25]	50.8 (−)	**60.6 (+)**
DSR-RL-GRU [26]	50.3 (−)	**60.2 (+)**
AuDSN-SD [27]	47.7 (−)	59.8 (−)
Interp-SUM [3]	47.68 (−)	59.14 (−)
Ours	**51.66**	**59.86**

4.4.2. Qualitative Evaluation

Figure 6 is an example of the predicted importance scores by the proposed video summarization method and the video thumbnails of the keyframes, as presented in Figure 6a,c. The proposed video summarization network with the interpolation method generated a more natural sequence of a summary than the network without the interpolation method and the network selected the keyframes in the main content of the video well and uniformly. With the interpolation method, the network predicted the importance scores of the main content as being similar to the importance score of the highest important keyframe. As presented in Figure 6b,c, the method without linear interpolation did not properly select all keyframes of the main content as a summary because it is hard to predict the importance score accurately by the unsupervised learning-based method.

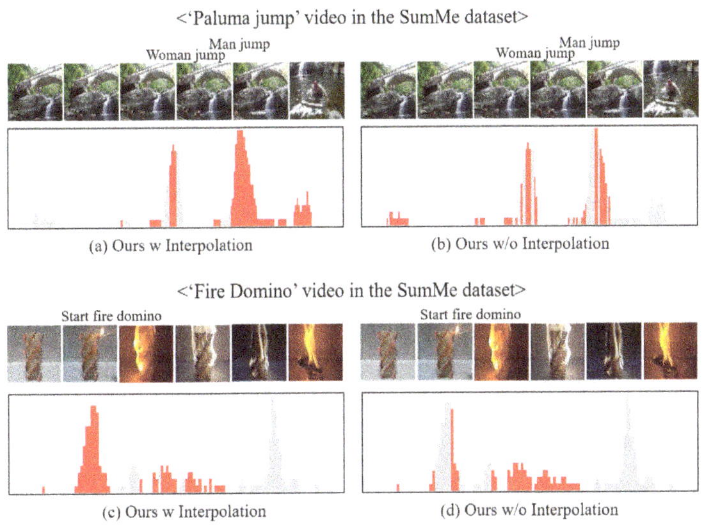

Figure 6. Example of the predicted importance scores by the proposed method and video thumbnails of the keyframes (The gray color means the importance score and the red color means the selected keyframes by the proposed method).

5. Conclusions

In this paper, we proposed an unsupervised video summarization method based on deep reinforcement learning with an interpolation method. We designed a lightweight video summarization network to predict the accurate importance score candidates of keyframes and we interpolated the importance score candidates with a piecewise linear interpolation method to generate a natural sequence of summary and to mitigate the high variance problem. To train the video summarization network by the reinforcement learning method efficiently, we used graph-level features and proposed a temporal consistency reward function to select keyframes uniformly and adopted the representativeness and diversity reward functions. The experimental results illustrate that the proposed method showed state-of-the-art performance compared to the latest unsupervised video summarization methods, which we demonstrated and analyzed experimentally. Moreover, the proposed method robustly summarized various types of videos in the two datasets. However, based on the experimental results, the proposed method showed a lower performance on long videos due to the high variance problem. Conversely, the proposed method showed the best performance on the videos that were shorter than about 300 keyframes. Therefore, the proposed method is very useful for the video summarization of short-form videos.

Author Contributions: Conceptualization, U.N.Y., M.D.H. and G.-S.J.; Methodology, U.N.Y.; Software, U.N.Y.; Validation, U.N.Y. and M.D.H.; Formal analysis, U.N.Y.; Investigation, U.N.Y.; Resources, U.N.Y.; Data curation, U.N.Y.; Writing—original draft preparation, U.N.Y.; Writing—review and editing U.N.Y., M.D.H. and G.-S.J.; Visualization, U.N.Y.; Supervision, G.-S.J.; Project administration, G.-S.J.; Funding acquisition, G.-S.J. All authors have read and agreed to the published version of the manuscript.

Funding: This research was funded in part by the National Research Foundation of Korea (NRF) and in part by an INHA UNIVERSITY Research Grant.

Institutional Review Board Statement: Not applicable.

Informed Consent Statement: Not applicable.

Data Availability Statement: Not applicable.

Acknowledgments: This work was supported by the Ministry of Education of the Republic of Korea and the National Research Foundation of Korea (NRF-2020S1A5C2A02093223).

Conflicts of Interest: The authors declare no conflict of interest.

References

1. Ejaz, N.; Mehmood, I.; Baik, S.W. Efficient visual attention based framework for extracting key frames from videos. *J. Image Commun.* **2013**, *28*, 34–44. [CrossRef]
2. Gygli, M.; Grabner, H.; Riemenschneider, H.; Gool, L.V. Creating summaries from user videos. In Proceedings of the European Conference on Computer Vision (ECCV), Santiago, Chile, 7–13 December 2015; pp. 505–520.
3. Yoon, U.N.; Hong, M.D.; Jo, G.S. Interp-SUM: Unsupervised Video Summarization with Piecewise Linear Interpolation. *Sensors* **2021**, *21*, 4562. [CrossRef] [PubMed]
4. Apostolidis, E.; Adamantidou, E.; Metsai, A.; Mezaris, V.; Patras, I. Unsupervised Video Summarization via Attention-Driven Adversarial Learning. In Proceedings of the International Conference on Multimedia Modeling (MMM), Daejeon, Korea, 5–8 January 2020; pp. 492–504.
5. Jung, Y.J.; Cho, D.Y.; Kim, D.H.; Woo, S.H.; Kweon, I.S. Discriminative feature learning for unsupervised video summarization. In Proceedings of the AAAI Conference on Artificial Intelligence, Honolulu, HI, USA, 27 January–1 February 2019; pp. 8537–8544.
6. Zhou, K.; Qiao, Y.; Xiang, T. Deep Reinforcement Learning for Unsupervised Video Summarization with Diversity-Representativeness Reward. In Proceedings of the AAAI Conference on Artificial Intelligence, New Orleans, LA, USA, 2–7 February 2018; Volume 32, pp. 7582–7589.
7. Song, Y.; Vallmitjana, J.; Stent, A.; Jaimes, A. Tvsum: Summarizing web videos using titles. In Proceedings of the IEEE Conference on Computer Vision and Pattern Recognition (CVPR), Boston, MA, USA, 7–12 June 2015; pp. 5179–5187.
8. Feng, L.; Li, Z.; Kuang, Z.; Zhang, W. Extractive Video Summarizer with Memory Augmented Neural Networks. In Proceedings of the 26th ACM International Conference on Multimedia, Seoul, Republic of Korea, 22–26 October 2018; pp. 976–983.
9. Zhang, K.; Chao, W.L.; Sha, F.; Grauman, K. Video Summarization with Long Short-term Memory. In Proceedings of the European Conference on Computer Vision (ECCV), Amsterdam, The Netherlands, 8–16 October 2016; pp. 766–782.
10. Zhang, Y.; Kampffmeyer, M.; Zhao, X.; Tan, M. DTR-GAN: Dilated Temporal Relational Adversarial Network for Video Summarization. In Proceedings of the ACM Turing Celebration Conference (ACM TURC), Shanghai, China, 19–20 May 2018; pp. 1–6.
11. Ji, J.; Xiong, K.; Pang, Y.; Li, X. Video Summarization with Attention-Based Encoder-Decoder Networks. *IEEE Trans. Circuits Syst. Video Technol.* **2020**, *30*, 1709–1717. [CrossRef]
12. Mahasseni, B.; Lam, M.; Todorovic, S. Unsupervised Video Summarization with Adversarial LSTM Networks. In Proceedings of the IEEE Conference on Computer Vision and Pattern Recognition, Honolulu, HI, USA, 22–25 July 2017; pp. 202–211.
13. Yuan, L.; Tay, F.E.; Li, P.; Zhou, L.; Feng, F. Cycle-SUM: Cycle-consistent Adversarial LSTM Networks for Unsupervised Video Summarization. In Proceedings of the Thirty-Third AAAI Conference on Artificial Intelligence, Honolulu, HI, USA, 27 January–1 February 2019; Volume 33, pp. 9143–9150.
14. Kaufman, D.; Levi, G.; Hassner, T.; Wolf, L. Temporal Tessellation: A Unified Approach for Video Analysis. In Proceedings of the IEEE International Conference on Computer Vision (ICCV), Venice, Italy, 22–29 October 2017; pp. 94–104.
15. Rochan, M.; Ye, L.; Wang, Y. Video Summarization Using Fully Convolutional Sequence Networks. In Proceedings of the European Conference on Computer Vision (ECCV), Munich, Germany, 8–14 September 2018; pp. 347–363.
16. Silver, D.; Lever, G.; Heess, N.; Degris, T.; Wierstra, D.; Riedmiller, M. Deterministic Policy Gradient Algorithms. In Proceedings of the 31st International Conference on International Conference on Machine Learning (ICML), Beijing, China, 21–26 June 2014; pp. 387–395.
17. Yu, Y. Towards Sample Efficient Reinforcement Learning. In Proceedings of the Twenty-Seventh International Joint Conference on Artificial Intelligence (IJCAI), Stockholm, Sweden, 13–19 July 2018; pp. 5739–5743.
18. Lehnert, L.; Laroche, R.; Seijen, H.V. On Value Function Representation of Long Horizon Problems. In Proceedings of the Thirty-Second AAAI Conference on Artificial Intelligence, New Orleans, LA, USA, 2–7 February 2018; pp. 3457–3465.
19. Szegedy, C.; Liu, W.; Jia, Y.; Sermanet, P.; Reed, S.; Anguelov, D.; Erhan, D.; Vanhoucke, B.; Rabinovich, A. Going deeper with convolutions. In Proceedings of the IEEE Conference on Computer Vision and Pattern Recognition (CVPR), Boston, MA, USA, 7–12 June 2015; pp. 1–9.
20. Wu, y.; Omar, B.F.E.; Xi, L.; Fei, W. Adaptive Graph Representation Learning for Video Person Re-Identification. *IEEE Trans. Image* **2020**, *29*, 8821–8830. [CrossRef] [PubMed]

21. Nachum, O.; Norouzi, M.; Schuurmans, D. Improving Policy Gradient by Exploring Under-Appreciated Rewards. *arXiv* **2016**, arXiv:1611.09321.
22. Potapov, D.; Douze, M.; Harchaoui, Z.; Schmid, C. Category-specifc video summarization. In Proceedings of the European Conference on Computer Vision (ECCV), Zurich, Switzerland, 6–12 September 2014; pp. 540–555.
23. Rochan, M.; Wang, Y. Video Summarization by Learning from Unpaired Data. In Proceedings of the IEEE Conference on Computer Vision and Pattern Recognition (CVPR), Long Beach, CA, USA, 16–20 June 2019; pp. 7902–7911.
24. Yunjae, J.; Donghyeon, C.; Sanghyun, W.; Inso, K. Global-and-Local Relative Position Embedding for Unsupervised Video Summarization. In Proceedings of the European Conference on Computer Vision (ECCV), Virtual, 23–28 August 2020; Volume 12370, pp. 167–183.
25. Evlampios, A.; Eleni, A.; Alexandros, M.I.; Vasileios, M.; Ioannis, P. AC-SUM-GAN: Connecting Actor-Critic and Generative Adversarial Networks for Unsupervised Video Summarization. *IEEE Trans. Circuits Syst. Video Technol.* **2021**, *31*, 3278–3292.
26. Aniwat, P.; Yi, G.; Fangli, Y.; Wentian, X.; Zheng, Z. Self-Attention Recurrent Summarization Network with Reinforcement Learning for Video Summarization Task. In Proceedings of the IEEE International Conference on Multimedia and Expo (ICME), Virtual, 5–9 July 2021; pp. 1–6.
27. Xu, W.; Yujie, L.; Haoyu, W.; Longzhao, H.; Shuxue, D. A Video Summarization Model Based on Deep Reinforcement Learning with Long-Term Dependency. *Sensors* **2022**, *22*, 7689.

sensors

MDPI

Article

YOLO Series for Human Hand Action Detection and Classification from Egocentric Videos

Hung-Cuong Nguyen [1], Thi-Hao Nguyen [1], Rafał Scherer [2] and Van-Hung Le [3,*]

1 Faculty of Engineering Technology, Hung Vuong University, Viet Tri City 35100, Vietnam
2 Department of Intelligent Computer Systems, Czestochowa University of Technology, 42-218 Czestochowa, Poland
3 Faculty of Basic Science, Tan Trao University, Tuyen Quang City 22000, Vietnam
* Correspondence: van-hung.le@mica.edu.vn; Tel.: +84-973512275

Abstract: Hand detection and classification is a very important pre-processing step in building applications based on three-dimensional (3D) hand pose estimation and hand activity recognition. To automatically limit the hand data area on egocentric vision (EV) datasets, especially to see the development and performance of the "You Only Live Once" (YOLO) network over the past seven years, we propose a study comparing the efficiency of hand detection and classification based on the YOLO-family networks. This study is based on the following problems: (1) systematizing all architectures, advantages, and disadvantages of YOLO-family networks from version (v)1 to v7; (2) preparing ground-truth data for pre-trained models and evaluation models of hand detection and classification on EV datasets (FPHAB, HOI4D, RehabHand); (3) fine-tuning the hand detection and classification model based on the YOLO-family networks, hand detection, and classification evaluation on the EV datasets. Hand detection and classification results on the YOLOv7 network and its variations were the best across all three datasets. The results of the YOLOv7-w6 network are as follows: FPHAB is $P = 97\%$ with $Thesh_{IOU} = 0.5$; HOI4D is $P = 95\%$ with $Thesh_{IOU} = 0.5$; RehabHand is larger than 95% with $Thesh_{IOU} = 0.5$; the processing speed of YOLOv7-w6 is 60 fps with a resolution of 1280×1280 pixels and that of YOLOv7 is 133 fps with a resolution of 640×640 pixels.

Keywords: hand detection; hand classification; YOLO-family networks; convolutional neural networks (CNNs); egocentric vision

Citation: Nguyen, H.-C.; Nguyen, T.-H.; Scherer, R.; Le, V.-H. YOLO Series for Human Hand Action Detection and Classification from Egocentric Videos. *Sensors* **2023**, *23*, 3255. https://doi.org/10.3390/s23063255

Academic Editors: Sylvain Girard and Nizar Bouguila

Received: 6 February 2023
Revised: 20 February 2023
Accepted: 12 March 2023
Published: 20 March 2023

1. Introduction

Building an application to support the rehabilitation of the hand after surgery is an issue of interest in artificial intelligence, machine learning, deep learning, and computer vision. The quantification of the patient's hand function after surgery was previously only based on the subjectivity of the doctors. To have objective and accurate assessments and exercise orientation for patients, it is necessary to support an assessment system. Through the research process, we propose a model to build a help system, as illustrated in Figure 1. Figure 1 includes three steps: Input—image sequence from EV; hand tracking detection; estimate 2D, 3D hand pose; hand activities recognition; Output—quantification sums up the results to show the action ability of the hand. An EV dataset refers to a dataset that is collected from the perspective of a single individual, usually with the use of wearable cameras or other devices that record the individual's view of their surroundings. These datasets typically include video and audio, and they may be used in a variety of applications, such as computer vision, human–computer interaction, and virtual reality. Figure 1 presents hand detection as an important pre-processing step in the application construction process; the detected hand data area is very decisive to the estimation space of the 2D hand pose and 3D hand pose. The problem of hand detection is not a new study; however, the problem persists when it comes to detection in the EV datasets. Since the fingers are obscured by the direction of view or other objects, the visible data area is only the back of the hand. Often, studies using deep learning models for 3D hand pose estimation and hand activity recognition apply third-person viewpoint datasets such as the NYU [1], ICVL [2], and MSRA [3] datasets. These datasets usually have segmented hand data with the environment and hand data not obfuscated and lost data of the fingers, as illustrated in Figure 2.

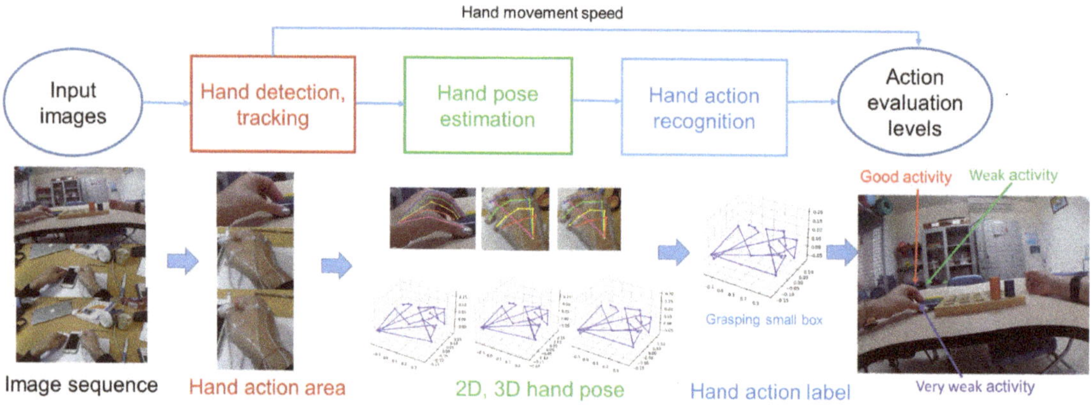

Figure 1. Framework for building an image-based rehabilitation evaluation system of the EV. Hand detection and classification is an important pre-processing step to limit the hand data area for hand pose estimation and activity recognition to assess hand activity levels.

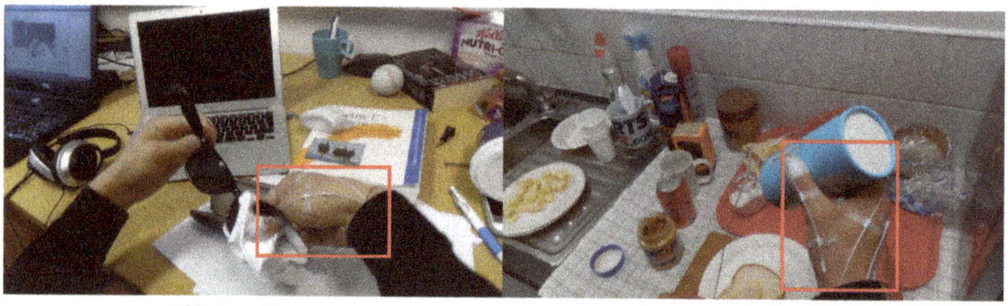

Figure 2. Illustration of obscured fingers in the FPHAB dataset [4].

During the research, we performed a study using Google Mediapipe (GM) [5,6] for hand detection and classification [7] on the HOI4D [8] dataset. The results show that the pre-trained models of models have low results in hand detection and classification (Tables 1 and 2 [7]). Figure 3 shows some cases where the hand is not detected when using the GM on the FPHAB and HOI4D datasets.

Recently, the YOLOv7 model was proposed by Wang et al. [9]. YOLOv7-E6 [9] is more accurate and faster than SWINL Cascade-Mask R-CNN [10] by 2% and 509%, respectively. YOLO v7 is more accurate and faster than other versions of YOLO such as YOLOR [11], YOLOX [12], Scaled-YOLOv4 [13], YOLOv5 [14], DETR [15], and Deformable DETR [16].

FPHAB

HOI4D

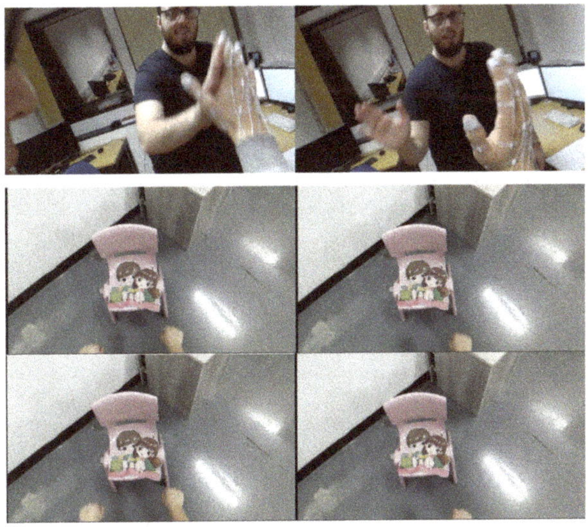

Figure 3. Illustrating some cases where the hand cannot be detected in the image when using the GM on the FPHAB and HOI4D datasets.

In this paper, we are interested in the "hand detection and classification" pre-processing step. We propose using YOLO-family networks with their advantages of accuracy and processing speed for fine-tuning the pre-trained model to detect and classify hand action on many different EV datasets (FPHAB [4], HOI4D [8], RehabHand) with many contexts and different hand movements. FPHAB [4] and HOI4D [8] datasets are two datasets collected from EV and published to evaluate 2D and 3D hand pose estimation models. The RehabHand dataset is also collected from EV mounted on patients who practice grasping rehabilitation at Hanoi Medical University Hospital, Vietnam.

The main contributions of the paper are as follows:

- A framework for building an image-based rehabilitation evaluation system of EV is proposed.
- We systematize the architectures of the YOLO-family networks for object detection.
- We fine-tune hand action detection and classification of the model based on the YOLO-family networks on the first-person viewpoint/EV datasets (FPHAB [4], HOI4D [8], RehabHand [17]).
- We manually mark the hand data area in the datasets for the evaluation of the hand detection results on the FPHAB [4], HOI4D [8], and RehabHand [17] datasets.
- Experiments on hand action detection and classification are presented in detail, and the results of hand action detection and classification are evaluated and compared with YOLO-family networks on the FPHAB [4], HOI4D [8], and RehabHand [17] datasets.

The content of this paper is organized as follows: Section 1 introduces the applications and difficulties of hand action detection and classification on the EV datasets. Section 2 discusses related research in this field. Section 3 presents the process of applying YOLO-family networks to fine-tune the hand action detection and classification models. Section 4 compares the models quantitatively and shows qualitative experiments. Section 5 concludes the contributions and presents the future works.

2. Related Works

The problem of hand detection and classification is not a new research direction. However, the results of hand detection are very important in the process of building applications for human–machine interaction or building support systems. However, when detecting the hand on datasets obtained from EV, there are still some challenges caused by the external conditions, such as fingers being completely obscured due to the viewpoint of the camera and obscured by objects when grasping the object, where the image obtained only has data of the hand palm. There are also now several EV datasets to evaluate computer vision studies. Recently, Marcos et al. [18] published a helpful survey research on activity recognition on EV datasets.

Ren et al. [19] proposed a dataset called Intel EV with 10 video sequences. The total amount of video data is about 120 min, with 100,000 frames; about 70,000 frames contain objects, with about 1600 per object and 42 different hand actions. Fathi et al. [20] published a database under the name GTEA Gaze with more than 30 different types of food and objects. GTEA Gaze includes 94 types of actions and 33 classes of objects. There are also some typical databases collected from EV such as H_2O [21], Meccano [22], etc.

Particularly, Bandini et al. [23] analyzed problems of computer vision based on an EV dataset. The authors focused on three main research directions: localization (hand detection, hand segmentation, hand pose estimation), interpretation (hand gesture recognition, grasping object, hand action recognition, hand activity recognition), and application (hand-based human–computer interaction, healthcare application). The three research directions explored in Bandini et al.'s [23] paper constitute a unified process with the output applications based on hand data obtained from EV.

Today, with the development of computer hardware and the advent of deep learning, researchers have become equipped with novel tools, the most prominent of which are various convolutional networks (CNN). There have been many published CNN-based researches on hand detection such as YOLOv1 [24], YOLOv2 [25], YOLOv3 [26], YOLOv4 [27], YOLOv5 [14,28], YOLOv7 [9], Mask R-CNN [29,30], SSD [31], MobileNetv3 [32], etc. Some of the most prominent results are shown in Figure 1 of Wang et al.'s work [9], where YOLOv7 achieved the best results in terms of accuracy and speed.

More specifically, a study by Gallo et al. [33] used YOLOv7 to evaluate the detection of weeds near plants using images collected from UAVs. The results of the weeds are a mAP@0.5 score of 56.6%, recall of 62.1%, and precision of 61.3%. Huang et al. [34] used YOLOv3 to detect and determine the patient's venous infusion based on flow waveforms. The results were compared with RCNN and Fast-RCNN. Detection results showed a precision of 97.68% and recall of 96.88%. Liu et al. [35] used the YOLOv3 model with four-scale detection layers (FDL) to detect combined B-scan and C-scan GPR images. The proposed method can detect both large particles and small cracks. Recently, Lugaresi et al. [5] and Zhang et al. [6] proposed and evaluated a Mediapipe framework that can perform hand detection on both a CPU and GPU with 95.7% detection accuracy with all types of hand palms in real life.

3. Hand Action Detection and Classification Based on YOLO-Family Networks

3.1. YOLO-Family Networks for Object Detection

Object detection is an important problem in computer vision. YOLO is a convolutional neural network rated with average accuracy; however, the computation speed is very fast and the computation can be performed on a CPU [36]. As studied by Huang et al. [36], when evaluated on the Pascal VOC 2012 dataset, the accuracy results of R-FCN [37], Faster R-CNN [38], SSD [39], and YOLOv3 [26] are 80.5%, 70.4%, 78.5%, and 78.6%, respectively. The processing time results of R-FCN [37], Faster R-CNN [38], SSD [39], and YOLOv3 [26] are 6 fps, 17 fps, 59 fps, and 91 fps, respectively. Before YOLO was born, there were some CNNs such as R-CNN, Fast R-CNN, and Faster R-CNN using a two-stage detector method that obtained very impressive accuracy results but high computation time. To solve the computation time problem, YOLO uses one-stage detectors for object detection.

YOLO version 1 (YOLOv1) [24] uses 24 convolutional layers: 1×1 reduction layers (used to reduce image size) followed by 3×3 convolutional layers, and max pooling layers. The architecture ends with two fully connected layers. The result is a three-dimensional matrix of size $7 \times 7 \times 30$, as illustrated in Figure 4.

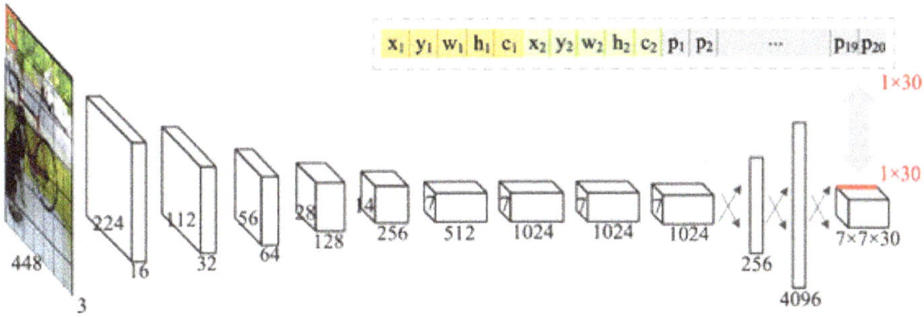

Figure 4. YOLOv1 architecture for object detection [24].

YOLO divides the image into $S \times S$ cells, with each cell being a matrix $\mathbf{A}$. If the center of the object is in the cell (i, j), the corresponding output will be in $A[i, j]$. The prediction process is performed in two steps as follows: the convolutional network performs the feature extraction of the images; extra layers (fully connected layers) analyze and detect the object, then return the output as a matrix $\mathbf{A}$ of the following size:

$$size(\mathbf{A}) = S \times S \times (5 \times B + C) \tag{1}$$

where B is the number of bounding boxes; each bounding box has five components: (x, y, w, h, CS). Confidence Score CS is the probability that the cell contains an object. Finally, the C elements are the representation of the probability distribution of the class object. This C element is a probability distribution p_i and satisfies

$$\sum_{i=0}^{c} p_i = 1 \tag{2}$$

Loss function: YOLO uses the Sum-Squared Error (SSE) function. The values x, y, w, h, C are the values of the ground truth box, and the values $\tilde{x}, \tilde{y}, \tilde{w}, \tilde{h}, \tilde{C}$ are the predicted bounding box.

$$SSE = E_1 + E_2 + E_3 + E_4 + E_5 \tag{3}$$

where

$$E_1 = \lambda_{coord} \sum_{i=0}^{S^2} \sum_{j=0}^{B} \mathbb{LF}_{ij}^{object} [(x_i - \tilde{x}_i)^2 + (y_i - \tilde{y}_i)^2] \tag{4}$$

$$E_2 = \lambda_{coord} \sum_{i=0}^{S^2} \sum_{j=0}^{B} \mathbb{LF}_{ij}^{object} [(\sqrt{w_i} - \sqrt{\tilde{x}_i})^2 + (\sqrt{h_i} - \sqrt{\tilde{h}_i})^2] \tag{5}$$

$$E_3 = \lambda_{coord} \sum_{i=0}^{S^2} \sum_{j=0}^{B} \mathbb{LF}_{ij}^{object} (C_i - \tilde{C}_i)^2 \tag{6}$$

$$E_4 = \lambda_{no\_object} \sum_{i=0}^{S^2} \sum_{j=0}^{B} \mathbb{LF}_{ij}^{no\_object} (C_i - \tilde{C}_i)^2 \tag{7}$$

$$E_5 = \sum_{j=0}^{B} \mathbb{LF}_{i}^{object} \sum_{c \in classes} (p_i(c) - \tilde{p}_i(c))^2 \tag{8}$$

where E_1 is xy_loss when the object exists at box_j in $cell_i$;
E_2 is wh_loss when the object exists at box_j in $cell_i$;
E_3 is $confidence\_loss$ when the object exists at box_j in $cell_i$;
E_4 is $confidence\_loss$ when objects do not exist in the boxes;
E_5 is $class\_probability\_loss$ in the cell where the object exists.

Further, $\mathbb{LF}_{ij}^{object} = 1$ if in the ith cell, there is a jth box containing an object;
$\mathbb{LF}_{ij}^{no\_object}$ is the opposite of $\mathbb{LF}_{ij}^{object}$;
$\mathbb{LF}_{ij}^{object} = 1$ if the ith cell contains an object (otherwise, it is 0);
λ_{coord}, λ_{no_object} is the component weight.

However, even a good model still has a case: predicting multiple bounding boxes for the same object. To solve this problem, YOLO filters out redundant bounding boxes (duplicate and same class) by non-maximum suppression with two steps as follows:

- Boxes with $confidence\_score$ are ranked from high to low $[box\_0, box\_1, \cdots, box\_n]$.
- Traverse from the top of the list, for each box_i, removing box_j that have $IOU(box\_i, box\_j) \geq$ threshold, where $j > i$. The threshold is a pre-selected threshold value. IOU is the formula for calculating the overlap–interference between two bounding boxes, as computed in Equation (10).

YOLOv2 [25] was born to improve on the weaknesses of YOLOv1 [24]. YOLOv2 makes the following improvements:

- Batch Normalization (BN): adding BN to all convolutional layers. This allows weights that would never have been learned without BN to be learned again, and reduces the dependence on the initialization of parameter values.
- High-Resolution Classifier: training the classifier with 224×224 and training with 448×448 at least 10 epochs for object detection.
- Anchor Box: they are pre-generated bounding boxes (not model-predicted bounding boxes). With a grid, it creates some K anchor boxes with different sizes. These anchor boxes will predict whether it contains an object or not, based on the results of the calculation of the IOU between it and the ground truth (if the $IOU > 50\%$, the anchor box is considered to contain the object). Figure 5 shows the process of using anchor boxes for object prediction in an image. YOLOv2 divides the image into 13×13 grid cells; so, the ability to find small objects is higher than that of YOLOv1, which is 7×7. YOLOv2 is trained on images that vary in size from 320×320 up to 640×640. This enables the model to learn more features of the object and have higher accuracy. YOLOv2 uses the Darknet19 with 19 convolution layers along with 5 max-pooling layers (it does not use fully connected layers for prediction but anchor boxes instead). Without using fully connected classes and using anchor boxes instead, the final result of the model will be $13 \times 13 \times 125$. For each tensor of size $1 \times 1 \times 125$, it is calculated as follows: $k\times$ (5 + 20), where $k = 5$ and 20 is the number of pre-trained object classes. Darknet19 is very fast in object recognition; thus, it makes a lot of sense for real-time processing. The architecture is presented in Figure 6.

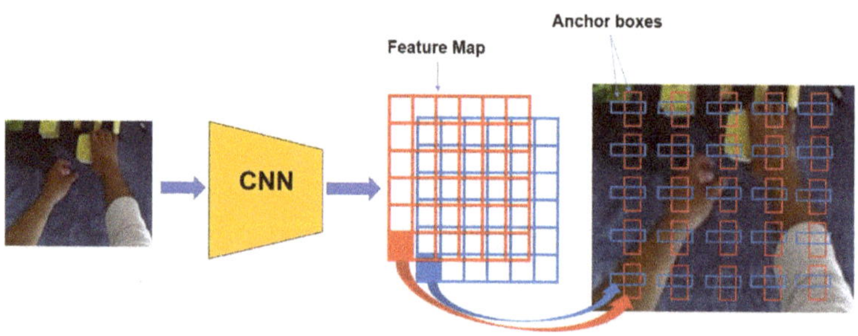

Figure 5. Anchor-based object detector.

Type	Filters	Size/Stride	Output
Convolutional	32	3 × 3	224 × 224
Maxpool		2 × 2/2	112 × 112
Convolutional	64	3 × 3	112 × 112
Maxpool		2 × 2/2	56 × 56
Convolutional	128	3 × 3	56 × 56
Convolutional	64	1 × 1	56 × 56
Convolutional	128	3 × 3	56 × 56
Maxpool		2 × 2/2	28 × 28
Convolutional	256	3 × 3	28 × 28
Convolutional	128	1 × 1	28 × 28
Convolutional	256	3 × 3	28 × 28
Maxpool		2 × 2/2	14 × 14
Convolutional	512	3 × 3	14 × 14
Convolutional	256	1 × 1	14 × 14
Convolutional	512	3 × 3	14 × 14
Convolutional	256	1 × 1	14 × 14
Convolutional	512	3 × 3	14 × 14
Maxpool		2 × 2/2	7 × 7
Convolutional	1024	3 × 3	7 × 7
Convolutional	512	1 × 1	7 × 7
Convolutional	1024	3 × 3	7 × 7
Convolutional	512	1 × 1	7 × 7
Convolutional	1024	3 × 3	7 × 7
Convolutional	1000	1 × 1	7 × 7
Avgpool		Global	1000
Softmax			

Figure 6. Darknet19 architecture.

YOLOv3 [26] was born to improve on the weaknesses of YOLOv1 [24] and YOLOv2 [25]. YOLOv3 uses Darknet53 as the backbone (with 53 convolutional layers), as illustrated in Figure 7. YOLOv3 performs recognition three times on an image with different sizes. YOLOv3 has its output changed to $S \times S \times 255$, with S being the values 13, 26, and 52, respectively. With each grid box, there are nine different anchor boxes with sizes: grid cell 13 × 13: (116 × 90), (156 × 198), (373 × 326); grid cell 26 × 26: (30 × 61), (62 × 45), (59 × 119); grid cell 52 × 52: (10 × 13), (16 × 30), (33 × 23). The training process combines with the k-means clustering algorithm and uses the ground truth to calculate the error between the ground truth and the anchor box by adjusting values (x, y, w, h), thereby learning the features of the object.

	Type	Filters	Size	Output
	Convolutional	32	3 × 3	256 × 256
	Convolutional	64	3 × 3 / 2	128 × 128
	Convolutional	32	1 × 1	
1×	Convolutional	64	3 × 3	
	Residual			128 × 128
	Convolutional	128	3 × 3 / 2	64 × 64
	Convolutional	64	1 × 1	
2×	Convolutional	128	3 × 3	
	Residual			64 × 64
	Convolutional	256	3 × 3 / 2	32 × 32
	Convolutional	128	1 × 1	
8×	Convolutional	256	3 × 3	
	Residual			32 × 32
	Convolutional	512	3 × 3 / 2	16 × 16
	Convolutional	256	1 × 1	
8×	Convolutional	512	3 × 3	
	Residual			16 × 16
	Convolutional	1024	3 × 3 / 2	8 × 8
	Convolutional	512	1 × 1	
4×	Convolutional	1024	3 × 3	
	Residual			8 × 8
	Avgpool		Global	
	Connected		1000	
	Softmax			

Figure 7. Darknet53 architecture.

YOLOv4 [27] was developed to improve the accuracy and processing time of YOLOv3 [26]. YOLOv4 applies the idea of CSPBlock, replaces the usual Residual Block of YOLOv3 to CSPResBlock, and also changes the activation function from LeakyReLU to Mish, creating CSPDarkNet53. The structure of YOLOv4 is divided into four parts:

- Backbone: The backbone can be selected from one of the following three backbones: CSPResNext50, CSPDarknet53, and EfficientNet-B3. CSPDarknet53 is built on a combination of CSP (Cross-Stage-Partial connections) and Darknet53.
  The CSP is derived from the DenseNet architecture that takes the previous input and concatenates it with the current input before moving into the Dense layer. The role of CSP is to remove computational bottlenecks in DenseNet and improve learning by porting an unmodified version of the feature map. DenseNet (Dense-connected convolutional network) is one of the latest networks for visual object recognition. Densenet has a structure of dense blocks and transition layers. With traditional CNN, if we have L layers, there will be L connections; however, in DenseNet, there will be $L(L+1)/2$ connections (i.e., the front layer will be connected with all the layers behind it). Yolov4 uses CSPDarknet53 as the backbone.
  The main idea of CSPBlock of CSPDarknet53 is applied to Residual Block, as presented in Figure 8 [13]. Instead of having only one path from beginning to end, CSPBlock is divided into two paths. By dividing into two such paths, we eliminate the recalculation of the gradient; therefore, the speed of training is increased. Moreover, splitting into two paths, with each path being a part taken from the previous feature map, the number of parameters is also significantly reduced, thereby speeding up the whole inference process.
- Neck: The neck is responsible for mixing and matching the feature maps learned through the feature extraction (backbone) and identification process (YOLOv4, called Dense prediction). YOLOv4 allows customization using the following Neck structures: FPN (Feature Pyramid Networks) [40], PAN (Path Aggregation Networks) [41], NAS-FPN (Neural Architecture Search–Feature Pyramid Networks) [42], Bi-FPN (Bidirectional feature pyramid network) [43], ASFF (Adaptively Spatial Feature Fusion) [44], SFAM (Scale-wise Feature Aggregation Module) [45], SSP (spatial pyramid pooling layer) [46]. In the latter, SSP is a CNN network but is slightly changed; it is no longer about dividing feature maps into bins and then concatenating these bins together to obtain a fixed-dimensional vector.
  SPP, whose input is a feature map, outputs $C \times H \times W$ from the backbone before being fed to the fully-connected layer to perform detection; YOLO applies the spatial pyramid pooling layer to the feature map three times—that is, using the SPP block, as illustrated in Figure 9.
  Yolo-SPP applies a maximum pool with kernels of different sizes. The size of the input feature map is preserved, and the feature maps obtained from applying the max pool (with different kernel sizes) will be concatenated. The architecture of YOLO-SPP is shown in Figure 10. Yolov4 also re-applies this technique.
- Dense prediction: using one-stage detectors; Sparse Prediction: using two-stage detectors such as R-CNN.

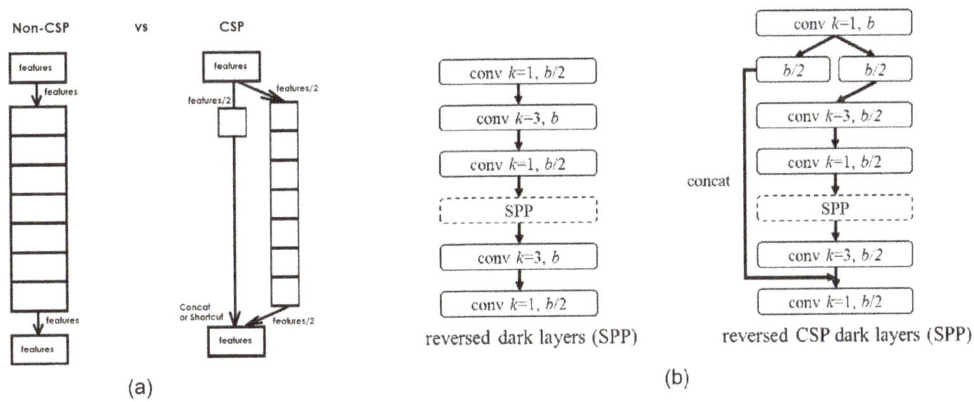

Figure 8. The architecture of CSPBlock in CSPDarknet53 [27]. (**a**) The simple CSP connection. (**b**) A CSP connection in YOLOv4-CSP/P5/P6/P7 [13].

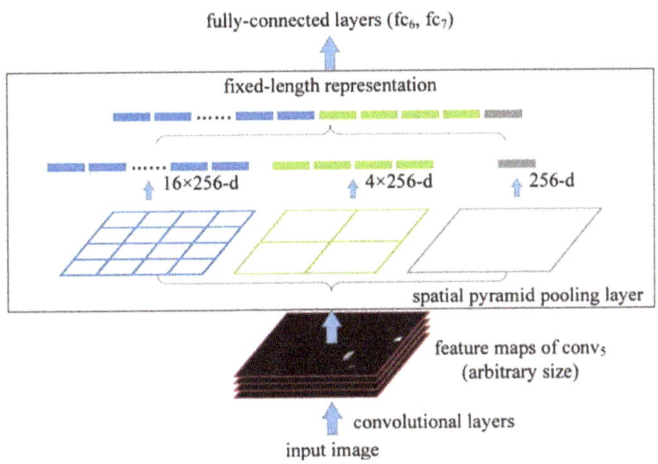

Figure 9. Architecture of SPP [13].

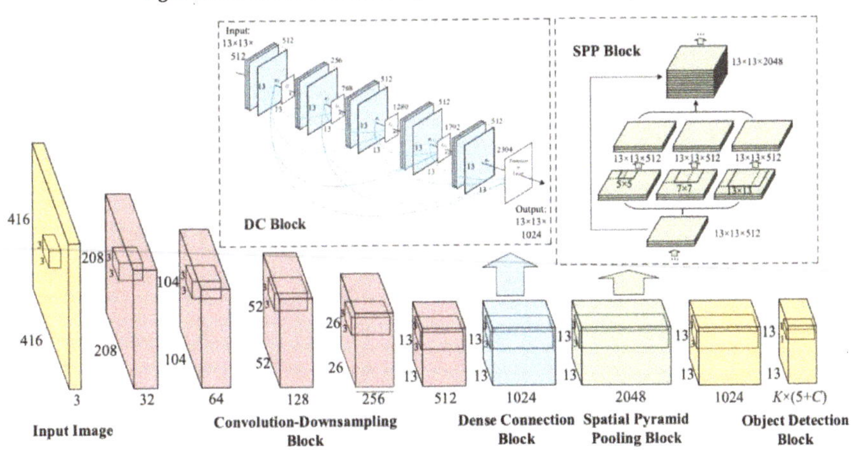

Figure 10. The architecture of YOLO-SPP bypasses the DC Block part [46].

YOLOv5 (v6.0/6.1) [28] has almost the same architecture as YOLOv4 [27] and includes the following components: CSP-Darknet53 as a backbone, SPP and PANet in the model neck, and the head used in YOLOv4. In YOLOv5 (v6.0/6.1), SPPF has been used, which is just another variant of the SPP block, to improve the speed of the network and apply the CSPNet strategy on the PANet model.

- Backbone: YOLOv5 improves YOLOv4's CSPResBlock into a new module, with one less Convolution layer than YOLOv4, called the C3 module. Activation function: YOLOv4 uses the Mish or LeakyReLU for the lightweight version, while in YOLOv5, the activation function used is the SiLU.
- Neck: YOLOv5 adopts a module similar to SPP but twice as fast and calls it SPP-Fast (SPPF). Instead of using parallel max-pooling as in SPP, YOLOv5 SPPF uses sequential max-pooling, as illustrated in Figure 11. The kernel size in SPPF's max-pooling is 5 instead of 5, 9, 13, as in YOLOv4's SPP. Therefore, Neck in YOLOv5 uses SPPF + PAN.

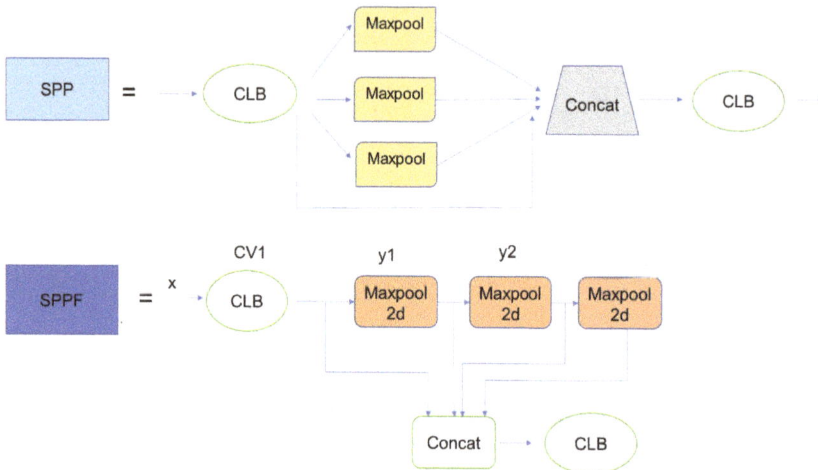

Figure 11. Comparison of the architecture of SPP and SPPF [47].

- Other changes in YOLOv5 include the following:
  - Data Augmentation techniques applied in YOLOv5 include Mosaic Augmentation, Copy–paste Augmentation, and MixUp Augmentation.
  - Loss function: YOLOv5 uses three outputs from PAN Neck to detect objects at three different scales. However, the effect of objects at each scale on Objectness Loss is different; so, the formula for Objectness Loss is changed to Equation (9).

$$\mathbb{LF}_{object} = 4.0 * \mathbb{LF}_{object}^{small} + 1.0 * \mathbb{LF}_{object}^{medium} + 0.4 * \mathbb{LF}_{object}^{large} \quad (9)$$

  - Anchor Box (AB): AB in YOLOv5 received two major changes. The first is to use auto anchor, a technique that applies Genetic Algorithms (GA) to the AB after the k-means step so that the AB works better with custom datasets, and not only works well on the MS COCO dataset. The second is to offset the center of the object to select multiple ABs for an object.

YOLOv7 [9], just like other versions of YOLO, consists of three parts in its architecture, as shown below:

- Backbone: ELAN, E-ELAN;
- Neck: CSP-SPP and (ELAN, E-ELAN)-PAN;
- Head: YOLOR [11] and Auxiliary head.

YOLOv7 includes some major improvements. First, the Efficient Layer Aggregation Networks (ELAN) is proposed to expand to Extended Efficient Layer Aggregation Networks (E-ELAN), where the strategy that learns at more depth with the shortest and longest derivatives along the slope will have a higher probability of convergence. This means not changing the gradient transmission path of the original architecture but increasing the group of convolutional layers of the added features and combining the features of different groups by mixing and merging the cordiality manner, as presented in Figure 12. This way of working can improve the learning efficiency of learned solid maps and improve the use of parameters and calculations. This process increases the accuracy of the learned model without increasing complexity and computational resources.

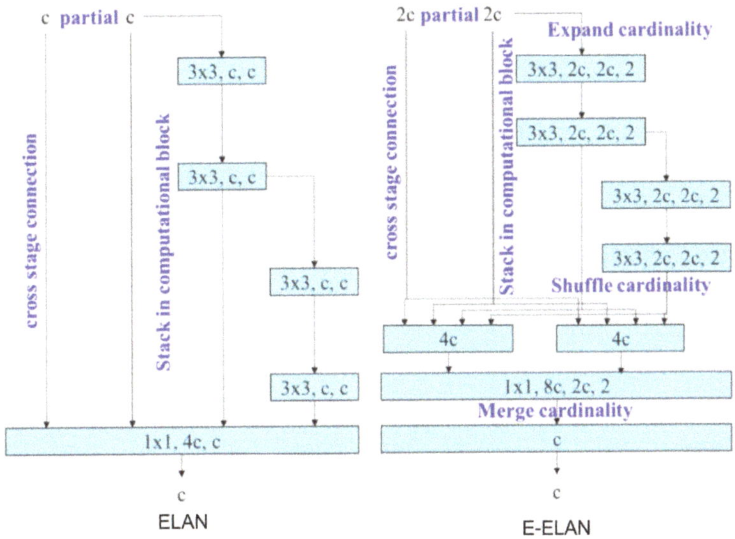

Figure 12. The architecture of ELAN and E-ELAN for efficient learning and faster convergence [9].

Second is the proposed Model Scaling for Concatenation-based Models (MSCM). The main idea of MSCM is based on scaled-YOLOv4 [13] to adjust the number of stages. When increasing the depth of a translation layer, which is immediately after, a concatenation-based computational block will increase, as illustrated in Figure 13a,b. It means the input width of the subsequent transmission layer increases. Therefore, the model scaling on concatenation-based models is proposed. This process only requires the depth in a computational block to be scaled, and the remaining transmission layer is performed with corresponding width scaling, as illustrated in Figure 13c.

The third is to reduce the number of parameters and computation for object detection. YOLOv7 is re-parameterized to combine with a different network. This work can reduce about 40% of the parameters and 50% computation of the object detector, and the detection will be faster and more accurate.

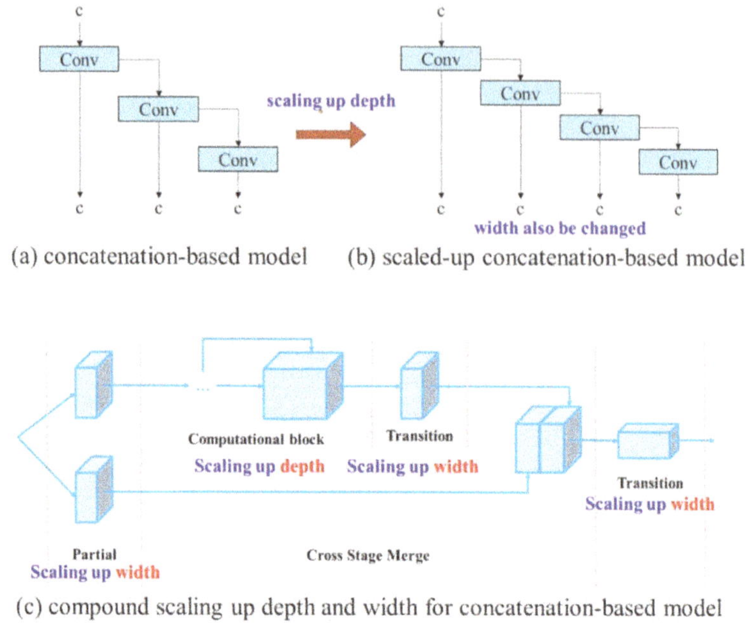

Figure 13. Illustrating of model scaling for concatenation-based models [9].

The fourth is a new label assignment method—as illustrated in Figure 14c,d—that guides both the auxiliary head and lead head by the lead head prediction. This method uses lead head prediction as a guidance to generate coarse-to-fine hierarchical labels, as illustrated in Figure 14e.

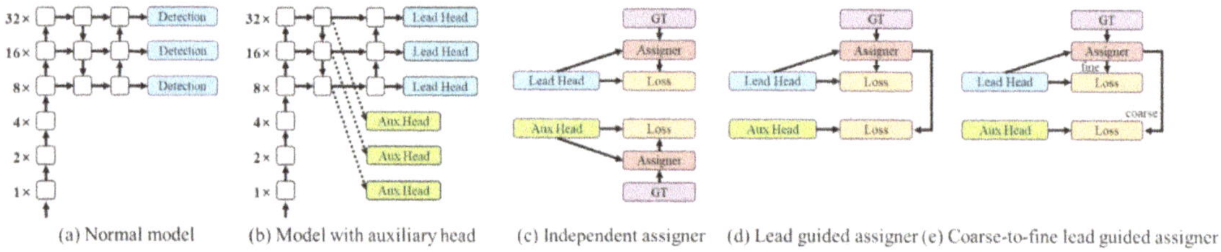

Figure 14. Illustration of coarse for auxiliary and fine for lead head label assigner [9].

3.2. Comparative Study for Hand Detection and Classification

In this paper, we perform a comparative study on YOLO-family networks for hand detection and classification of the EV datasets. The taxonomy of the comparative study is illustrated in Figure 15. In this study, the methods are the YOLO-family networks whose development and improvements are presented in Section 3.1. Two models developed from the YOLO-family networks are hand detection and classification. The hand detection model is the main model tested on all YOLO versions; the hand classification model is only tested from YOLOv3 and later. The datasets used to evaluate the two models are FPHAB, HOI4D, and RehabHand, as presented in Section 4.1. The FPHAB database performs hand detection model evaluation and classifies action hands, background, and other objects. The HOI4D and RehabHand datasets perform the hand detection model assessment and classify left hand, right hand, background, and other objects. In Figure 15, we also present a comparative study of measures and outputs, described in Sections 4.2 and 4.3.

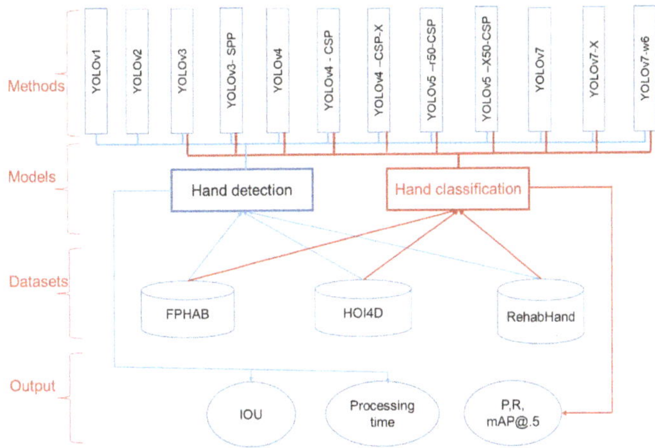

Figure 15. The taxonomy of the comparative study for hand detection and classification is based on the YOLO-family networks.

4. Experimental Results

4.1. Datasets

The FPHAB dataset [4] is the First Person Hand Action Benchmark (FPHAB). This dataset is captured from an Intel RealSense SR300 RGB-D camera attached to the shoulder of a person. The resolutions of the color and depth images are 1920×1080 pixels and 640×480 pixels, respectively. The hand pose is captured using six magnetic sensors; it provides 3D hand pose annotation and intrinsic parameters for converting 2D hand pose annotation. There are several subjects (6 in total) performing multiple activities from 3 to 9 times with 45 hand actions. The number of joints in each 3D hand pose is 21. From attaching the device to mark 3D hand annotation data, the hand data have different characteristics compared to normal hands when obtained from EV. In this paper, we used configurations for training and testing, presented as follows: The configuration ($Conf.$#123) used the first sequence in each subject from $Subj.$#1 to $Subj.$#6 for testing (27,097 samples), the second sequence in each subject for validation (25,475 samples), and the remaining sequence for training (52,887 samples) (the ratio is approximately 1:2.5 for testing and training the model).

The HOI4D dataset [8] is collected and synchronized based on the Kinect v2 RGBD sensor and the Intel RealSense D455 RGB-D sensor. This is a large-scale 4D EV dataset with rich annotation for category-level human–object interaction. HOI4D includes 2.4M RGB-D frames of EV with over 4000 sequences. It is collected from 9 participants interacting with 800 different object instances from 16 categories over 610 different indoor rooms. This dataset provides ground-truth data of the following types: Frame-wise annotations for panoptic segmentation, motion segmentation, 3D hand pose, category-level object pose, and hand action, together with reconstructed object meshes and scene point clouds. The annotation data components are illustrated in Figure 16. To obtain the ground-truth data, which is the bounding boxes of the hand on the image for evaluation, we rely on the hand keypoints named "kps2D" in the 3D hand pose annotation. We take the bounding box of 21 hand keypoint annotations. This process is demonstrated in the source code of the *"get_2D_boundingbox_hand_anntation.py"* file found at the following link (https://drive.google.com/drive/folders/1yzhg5NsalPkOHI6CMkAE07yv5rY63tl7?usp=sharing, accessed on 30 January 2023).

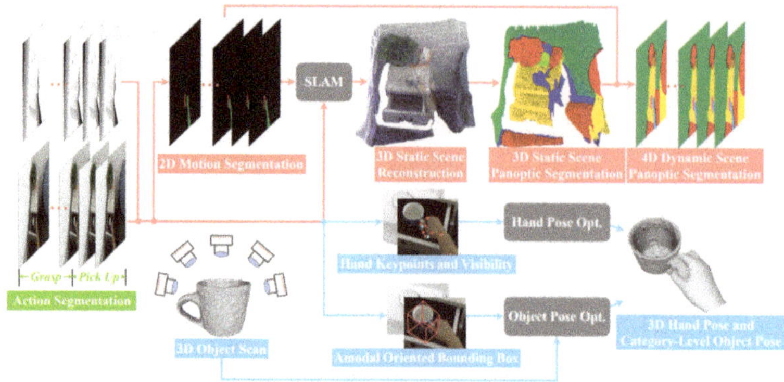

Figure 16. Describing the types of annotation data of the HOI4D dataset [8].

The RehabHand dataset [17] was collected from rehabilitation exercises of patients at Hanoi Medical University Hospital, Vietnam. This dataset consists of frames from the first-person video captured by cameras worn by the patient on the forehead and the chest. The videos are recorded with a 1080p resolution at 30 frames per second. The data were collected using the GoPro Hero4 camera in San Mateo, California, USA. The camera recorded the exercise of 15 patients performing four upper extremity rehabilitation exercises. Each patient performed each exercise five times. The content of exercises related to grasping objects in different positions is presented as follows: exercise 1—practice with the ball, exercise 2—practice with a water bottle, exercise 3—practice with a wooden cube, exercise 4—practice with round cylinders. The collected data include 10 video files in MPEG-4 format with a total duration of 4 h and a total capacity of 53 GB recorded. The data are divided into three subsets for the training set (2220 images), validation set (740 images), and testing set (740 images) with a ratio of 6:2:2, respectively. Figure 17 illustrates the image data of the RehabHand dataset [17].

Figure 17. Illustrating the RGB image data obtained from the EV of the RehabHand dataset [17].

In this paper, we used a server with a NVIDIA GeForce RTX 2080 Ti 12 GB GPU for fine-tuning, training, and testing. The programs were written in the Python language ($\geq$3.7 version) with the support of CUDA 11.2/cuDNN 8.1.0 libraries. In addition, there are a number of other libraries such as OpenCV, Numpy, Scipy, Pillow, Cython, Matplotlib, Scikit-image, Tensorflow $\geq$ 1.3.0, etc.

4.2. Evaluation Metrics

Similar to the evaluation of object detection and classification on images, we perform the calculation of the *IOU* (Intersection over Union) value according to Equation (10).

$$IOU = \frac{B_g \cap B_p}{B_g \cup B_p} \tag{10}$$

where B_g is the ground truth bounding box of hand action and B_p is the predicted bounding box of hand.

To determine whether the bounding box is a true finding, we use a threshold $Thesh_{IOU}$ for the evaluation. If IOU is greater than or equal to $Thesh_{IOU}$, it is a true detection; otherwise, it is false.

In this paper, we also distinguish between the hand action, left hand, right hand, and the background; so, we also use the formulas for precision (P), Recall (R), and F1-Score ($F1$) (Equation (11)) to evaluate the analysis results of hand action classification on the image.

$$P = \frac{TP}{TP + FP}; R = \frac{TP}{TP + FN}; F1 = \frac{2 * (R * P)}{(R + P)} \tag{11}$$

where TP are True Positives, TN are True Negatives, FP are False Positives, and FN are False Negatives. In addition, we also evaluate mAP.5 (mean Average Precision), computed as Equation (12).

$$mAP = \frac{\sum_{i=1}^{c} AP_i}{c} \tag{12}$$

where averaging the average precision (AP) for all classes involved in the trained model yields mAP.

We train YOLO-family networks with 50 epochs and batch size = 4 frames; the size of the image can be $img\_size = 640 \times 640$ or $img\_size = 1280 \times 1280$, $conf\_thres = 0.001$. The hyper-parameter in the feature-extraction phase that the YOLO-family networks uses is the adaptive moment estimation (ADAM) optimizer [48], the learning rate is 0.001, and momentum is 0.937, as illustrated in Figure 18. There are also some other parameters shown in Table 1.

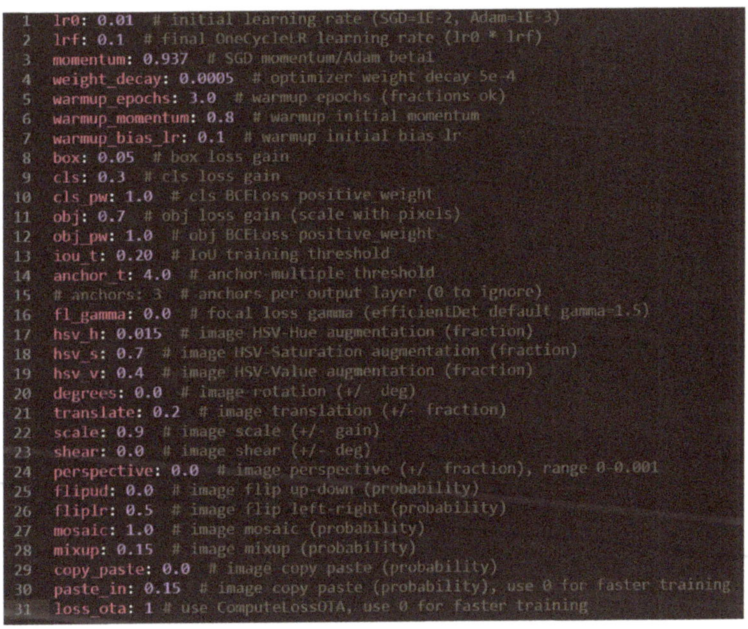

```
lr0: 0.01  # initial learning rate (SGD=1E-2, Adam=1E-3)
lrf: 0.1  # final OneCycleLR learning rate (lr0 * lrf)
momentum: 0.937  # SGD momentum/Adam beta1
weight_decay: 0.0005  # optimizer weight decay 5e-4
warmup_epochs: 3.0  # warmup epochs (fractions ok)
warmup_momentum: 0.8  # warmup initial momentum
warmup_bias_lr: 0.1  # warmup initial bias lr
box: 0.05  # box loss gain
cls: 0.3  # cls loss gain
cls_pw: 1.0  # cls BCELoss positive_weight
obj: 0.7  # obj loss gain (scale with pixels)
obj_pw: 1.0  # obj BCELoss positive_weight
iou_t: 0.20  # IoU training threshold
anchor_t: 4.0  # anchor-multiple threshold
# anchors: 3  # anchors per output layer (0 to ignore)
fl_gamma: 0.0  # focal loss gamma (efficientDet default gamma=1.5)
hsv_h: 0.015  # image HSV-Hue augmentation (fraction)
hsv_s: 0.7  # image HSV-Saturation augmentation (fraction)
hsv_v: 0.4  # image HSV-Value augmentation (fraction)
degrees: 0.0  # image rotation (+/- deg)
translate: 0.2  # image translation (+/- fraction)
scale: 0.9  # image scale (+/- gain)
shear: 0.0  # image shear (+/- deg)
perspective: 0.0  # image perspective (+/- fraction), range 0-0.001
flipud: 0.0  # image flip up-down (probability)
fliplr: 0.5  # image flip left-right (probability)
mosaic: 1.0  # image mosaic (probability)
mixup: 0.15  # image mixup (probability)
copy_paste: 0.0  # image copy paste (probability)
paste_in: 0.15  # image copy paste (probability), use 0 for faster training
loss_ota: 1 # use ComputeLossOTA, use 0 for faster training
```

Figure 18. Illustrating the hyper-parameters of YOLO-family networks.

Table 1. The list of parameters of YOLOv7 and its variants [9], resulting in the processing time of the networks when evaluated on the testing set of the FPHAB dataset.

Methods	Image Size (pixel)	Number of Layers	Number of GFLOPS	Parameters	Number of Epochs	Processing Time for Testing (fps)
YOLOv4-CSP [13]	640 × 640	401	118.9	52,469,023	50	76.9
YOLOv4-CSP-X [13]	640 × 640	493	224.8	96,370,166	50	44
YOLOv3 [26]	640 × 640	261	154.5	61,497,430	50	153
YOLOv3-SPP [49]	640 × 640	269	155.4	62,546,518	50	142
YOLOv4 [13]	640 × 640	401	118.9	52,463,638	50	151
YOLOv5-r50-CSP [28]	640 × 640	314	103.2	36,481,772	50	133
YOLOv5-X50-CSP [28]	640 × 640	560	64.4	33,878,846	50	45
YOLOv7 [9]	640 × 640	314	103.2	36,481,772	50	133
YOLOv7-X [9]	640 × 640	362	188.0	70,782,444	50	98
YOLOv7-w6 [9]	1280 × 1280	370	101.8	80,909,336	50	60

In this paper, we re-trained the YOLO-family networks (YOLOv4-CSP [13], YOLOv4-CSP-X [13], YOLOv3 [26], YOLOv3-SPP [49], YOLOv4 [13], YOLOv5-r50-CSP [28], YOLOv5-X50-CSP [28], YOLOv7 [9], YOLOv7-X [9], YOLOv7-w6 [9]) on the training set of $Conf$.#123 of the FPHAB dataset, the training set of the HOI4D dataset, and the training set of the RehabHand dataset. After that, we evaluated it on the validation set and testing set of configuration $Conf$.#123 of the FPHAB dataset, testing set of the HOI4D dataset, and testing set of the RehabHand dataset. We use the $Thesh_{IOU}$ to evaluate as follows: 0.5, 0.75, 0.95.

4.3. Hand Detection and Classification Results

The result of hand action detection and classification on the $Conf$.#123 of the FPHAB dataset is shown in Table 2. In the FPHAB dataset is the process of detecting the hand action in the image. The action hand detection and classification results on the FPHAB dataset in Table 2 of YOLOv7 and its variants are all greater than 95%. This is a very good result for the following steps on hand activity estimation and recognition.

Table 2. The results of hand detection and classification on the FPHAB dataset when performed on YOLOv7 and YOLO-family networks.

IOU Threshold($Thesh_{IOU}$)/ Precision(P)/ Recall(R)/ Models	0.5			0.75			0.95		
	R (%)	P (%)	mAP@.5 (%)	R (%)	P (%)	mAP@.5 (%)	R (%)	P (%)	mAP@.5 (%)
YOLOv4-CSP [13]	99.8	96.7	98.5	99.7	96.7	98.6	91.4	92.3	94.9
YOLOv4-CSP-X [13]	99.3	96.5	97.2	99.5	96.1	97.2	87.4	90.2	91.8
YOLOv3 [26]	99.9	96.6	98.5	99.9	96.6	98.8	96.4	95.1	97.5
YOLOv3-SPP [49]	99.5	97.6	98.8	99.6	97.3	98.9	95.3	92.2	97
YOLOv4 [13]	99.6	96.4	97.2	99.6	96.2	97.5	84.7	92	92.3
YOLOv5-r50-CSP [28]	99.3	95.2	98.1	99.5	96.5	98.1	92.5	91.4	93.5
YOLOv5-X50-CSP [28]	99.2	94.8	97.6	99.4	96.4	98.3	96.4	92.6	94.1
YOLOv7 [9]	99.7	96.9	98.7	99.4	96.9	99.4	97	93.9	98.2
YOLOv7-X [9]	99.2	98.7	99.1	99.2	98.2	99.5	97.5	94.5	97
YOLOv7-w6 [9]	99.7	96.9	99.8	99.7	96.9	99.7	93.9	98	98.3

In Table 2, it can be seen that the hand action detection and classification results on the FPHAB dataset are very accurate; the results are greater than 95%, even if the $Thesh_{IOU} = 0.95$, which is close to absolute accuracy. Table 2 also shows that P is usually greater than R in most cases. This is because in the image of the FPHAB dataset, there can be two hands and, as a result, there are many background areas that are mistakenly detected as the hand action, so FN increase. Therefore, R is smaller than P in many cases. The processing time of the hand action detection and classification process is shown in Table 1; it is also very fast to ensure the pre-processing step without much impact on the processing time of the construction applications.

Figure 19 shows the results on precision, recall, F1-score, and confusion matrix on the hand action detection on the testing set of FPHAB dataset when $Thesh_{IOU} = 0.5$.

Figure 20 shows the confusion matrix on classifying hand action on the testing set of the FPHAB dataset when $Thesh_{IOU} = 0.5$.

Figure 21 illustrates some results of hand action detection and classification on the testing set of $Conf.$#123 of the FPHAB dataset when $Thesh_{IOU} = 0.5$.

The results of hand detection and classification on the HOI4D dataset [8] are shown in Table 3. Table 3 shows the results of YOLOv7-w6 with the best results ($R = 89.85\%$; $P = 90.55\%$; $mAP@.5 = 88.9\%$) when $Thesh_{IOU} = 0.95$. This is a large dataset with many hand actions, for which the YOLO-family networks still obtain high results even when $Thesh_{IOU} = 0.95$. In this dataset, the YOLO-family networks perform two tasks: detecting and classifying left and right hands. At the same time, the average result (all) of the left and right hands are also computed.

Figure 22 illustrates the results of hand classification on the HOI4D dataset based on YOLOv7. In Figure 22, there are many cases where the subject has the same color as the skin of the hand. However, YOLOv7 still detects and correctly classifies the hand.

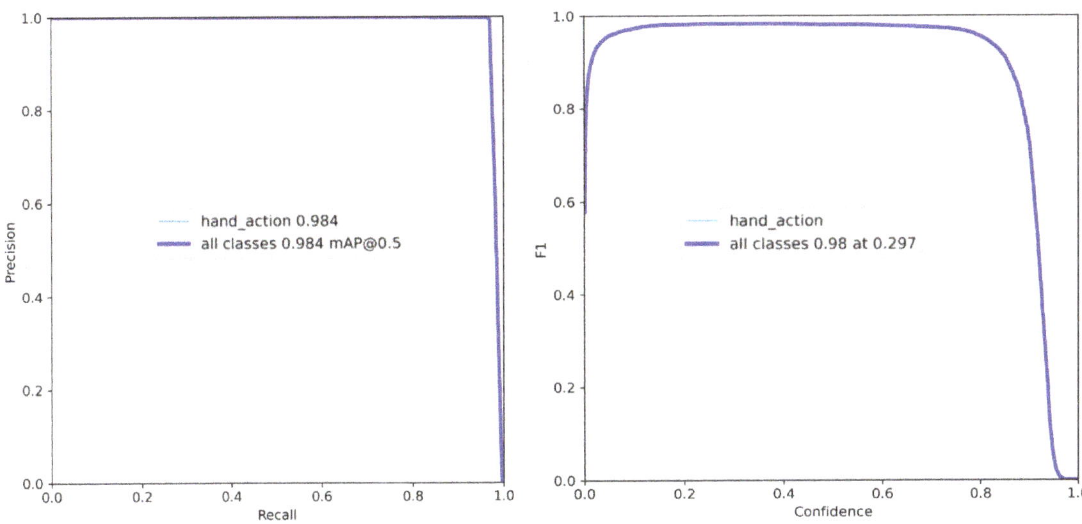

Figure 19. The distribution of precision, recall, and F1-score of hand action detection on the test set of $Conf.$#123 of the FPHAB dataset when $Thesh_{IOU} = 0.5$.

Table 3. The results of hand detection and classification on the HOI4D dataset [8].

IOU Threshold/ ($Thresh_{IOU}$) Precision(P)/ Recall(R)/ Models	Hand	0.5			0.75			0.95		
		R (%)	P (%)	mAP@.5 (%)	R (%)	P (%)	mAP@.5 (%)	R (%)	P (%)	mAP@.5 (%)
YOLOv4-CSP [13]	Right hand	90.7	96.2	95.2	96.4	90.6	95.1	88.9	90.1	91.7
	Left hand	82.5	86	85.5	82.2	85	84.7	76.8	65.4	74.4
	All	88.4	89.4	90.3	89.3	87.8	89.9	82.8	77.7	83.1
YOLOv4-CSP-X [13]	Right hand	96.2	90.7	95	96.2	90.6	94.9	84.2	90.2	90.9
	Left hand	81	86	84.8	82	83.3	84.1	76	62.9	71.9
	All	88.6	88.3	89.9	89.3	87	89.5	80.1	76.6	81.4
YOLOv3 [26]	Right hand	89.9	90.8	94	89.9	90.8	94.2	77.4	89.3	90.7
	Left hand	81.5	81.8	82.4	80.4	81.9	81.6	75.1	61.8	71.2
	All	85.7	86.3	88.2	85.1	86.3	87.9	76.2	75.6	80.9
YOLOv3-SPP [49]	Right hand	88.3	90.8	94.1	89.2	90.7	94.2	69.2	88.8	86
	Left hand	81.6	81.2	82.1	81.4	79.9	81.1	70.3	58.7	65.4
	All	84.9	86	88.1	85.3	85.3	87.6	69.7	73.8	75.7
YOLOv4 [13]	Right hand	89.7	93.4	95.9	90.5	94.6	95.9	71.5	88.8	89.4
	Left hand	82.8	83.5	84.3	84.2	82.4	86.3	72.4	78.3	75.2
	All	86.3	88.5	88.1	87.4	88.5	91.1	72	83.6	82.3
YOLOv5-r50-CSP [28]	Right hand	84.3	87.5	90.9	84.4	87.4	90.9	63	81.2	78.4
	Left hand	79.4	77.6	78.7	78.8	76.9	78.4	61.5	53.6	57.5
	All	81.9	82.5	84.8	81.6	82.1	84.6	62.2	7.4	68
YOLOv5-X50-CSP [28]	Right hand	94.1	90.2	92.7	90.4	89.6	90.8	78.2	88.2	84.4
	Left hand	79.4	77.6	78.7	78.8	76.9	78.4	61.5	73.6	77.5
	All	86.75	83.9	85.7	84.6	83.25	84.6	69.85	80.9	80.95
YOLOv7 [9]	Right hand	87	90.7	93.3	81	90.7	93.4	69.6	89.3	86.4
	Left hand	81.4	78.9	80.7	81.3	78.8	80.8	61.7	56.1	60.8
	All	84.2	84.8	87	84.2	84.8	87.1	65.7	72.7	73.6
YOLOv7-X [9]	Right hand	91.1	90.6	94.1	91.6	90.6	94.2	74.1	89.6	88.4
	Left hand	80.7	81.1	81.2	80.2	80.2	80.6	65.4	59.1	64
	All	85.9	85.9	87.7	85.9	85.4	87.4	69.8	74.3	76.2
YOLOv7-w6 [9]	Right hand	99.3	97.7	97	97.4	94.7	98.7	92.8	94.7	93
	Left hand	86.7	92.3	95.1	85.5	89.3	88.8	86.9	86.4	84.8
	All	93	95	96.05	91.45	92	93.75	89.85	90.55	88.9

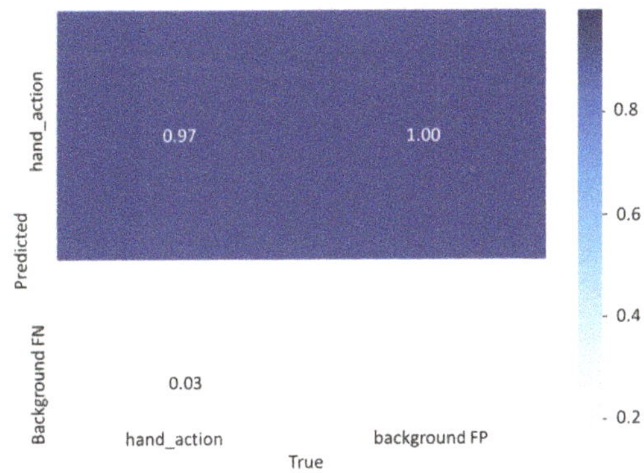

Figure 20. The confusion matrix of hand action classification on the testing set of $Conf.\#123$ of the FPHAB dataset when $Thesh_{IOU} = 0.5$.

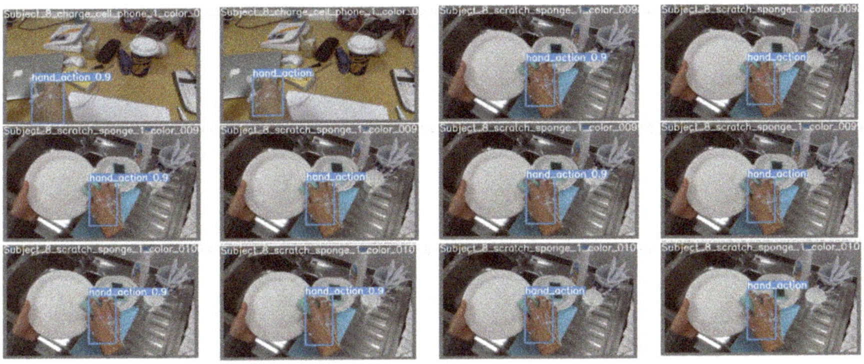

Figure 21. Illustrating some results of hand action detection and classification on the testing set of $Conf.\#123$ of the FPHAB dataset when $Thesh_{IOU} = 0.5$.

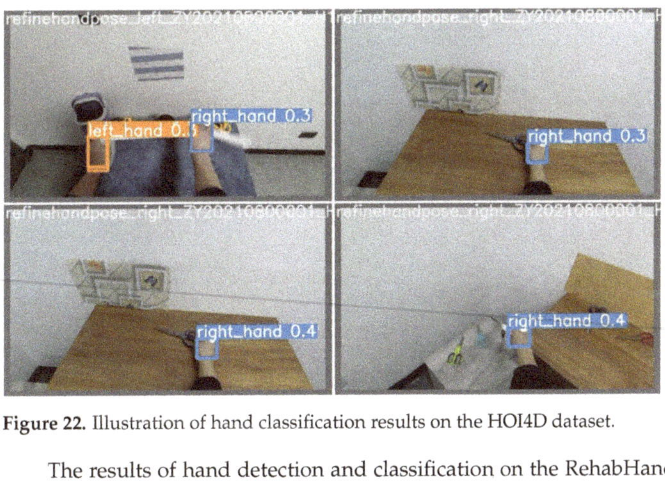

Figure 22. Illustration of hand classification results on the HOI4D dataset.

The results of hand detection and classification on the RehabHand dataset [17] are shown in Table 4. The results in Table 4 show that YOLOv7 has the best results in detecting and classifying with the left hand ($P = 100\%; R = 92.1\%; mAP@.5 = 14\%$ with $Thesh_{IOU} = 0.95$). YOLOv7-X has the best results in detecting and classifying with the right hand ($P = 87.7\%; R = 92.5\%; mAP@.5 = 96.7\%$ with $Thesh_{IOU} = 0.95$), and the average result is also computed. It can be seen that the left hand detection results in some networks are very low because the left hand is as false negative as the right hand, as shown in Table 4.

Table 4. The results of hand detection and classification on the RehabHand dataset [17].

IOU Threshold/ ($Thesh_{IOU}$) Precision(P)/ Recall(R)/ Models	Hand	0.5			0.75			0.95		
		R (%)	P (%)	mAP@.5 (%)	R (%)	P (%)	mAP@.5 (%)	R (%)	P (%)	mAP@.5 (%)
YOLOv4-CSP [13]	Right hand	33.8	55.8	29.6	32.9	54.3	28.4	28.2	36.9	20.8
	Left hand	100	98.9	93.7	100	97.4	92.1	100	92.4	5.1
	All	66.9	77.35	63.45	66.45	75.85	60.25	64.1	66.2	12.95
YOLOv4-CSP-X [13]	Right hand	34.2	56	30.5	33.7	56	30.6	26.6	37.1	21.1
	Left hand	100	97.2	95.4	100	95.7	8.12	100	92.6	5.23
	All	67.1	76.6	62.95	66.8	75.85	19.3	63.3	64.85	13.2
YOLOv3 [26]	Right hand	2.41	94.1	18.9	0.0861	98.8	19.3	0.045	99	11.2
	Left hand	20.8	65.9	17.3	20.5	65.9	17.2	19.4	24.6	12.8
	All	11.6	80	18.1	10.7	82.4	18.3	9.91	61.8	12
YOLOv3-SPP [49]	Right hand	17.5	51.6	17.4	17.6	47	17.5	15.2	45	12.6
	Left hand	100	98.5	90.7	100	96.8	85.2	100	91.7	89.7
	All	58.8	75.5	54.05	58.8	97.9	53.6	58.8	68.35	51.15
YOLOv4 [13]	Right hand	2.26	91.8	21.3	0.91	97.7	21.2	16.4	37	14.5
	Left hand	23.9	55.7	21.9	22.2	55.7	20.8	100	25.8	10.5
	All	13.1	73.8	21.6	11.6	76.7	21	58.2	31.4	12.5
YOLOv5-r50-CSP [28]	Right hand	25.9	68.9	24.1	25.7	67.4	24	0.44	99.1	14.7
	Left hand	100	96.1	92.8	100	94.3	14.6	22.9	40.5	14.1
	All	63	82.5	57.4	62.9	80.85	19.3	11.7	69.8	14.4
YOLOv5-X50-CSP [28]	Right hand	1.69	95.8	19.1	0.62	99.5	17.1	0.33	99.3	6.84
	Left hand	24.2	53.5	23.6	24	53.6	53.6	27.7	23.3	18.9
	All	13	74.7	21.4	12.3	76.6	20.3	14	61.3	12.9
YOLOv7 [9]	Right hand	81.2	96.3	95.2	78.7	91.7	98.8	75.3	96.8	96.7
	Left hand	100	99.2	97.3	100	97.8	96.7	100	92.1	92.4
	All	90.6	97.75	96.25	89.35	94.75	97.75	87.65	94.45	94.55
YOLOv7-X [9]	Right hand	91.3	95.3	93.6	91.1	92.7	93.6	87.7	92.5	96.7
	Left hand	100	96.4	96.2	100	95.5	7.82	100	90.8	90.5
	All	95.65	95.85	95.45	95.55	94.1	50.71	93.85	92.3	93.6
YOLOv7-w6 [9]	Right hand	1.77	96.8	19.9	0.498	99.7	19.5	0.0169	99.5	9.53
	Left hand	25.8	55.5	15.4	24.3	55.6	14.8	15.2	32.9	5.62
	All	13.8	76.1	17.6	12.4	77.6	17.2	7.66	66.2	7.57

Figure 23 illustrates the left hand being negatively classified as the right hand of the RehabHand dataset [17].

The results in Table 4 also show that the RehabHand dataset [17] is very challenging for hand detection and classification. This is a good dataset for evaluating hand detection models, hand pose estimation, and hand activity recognition.

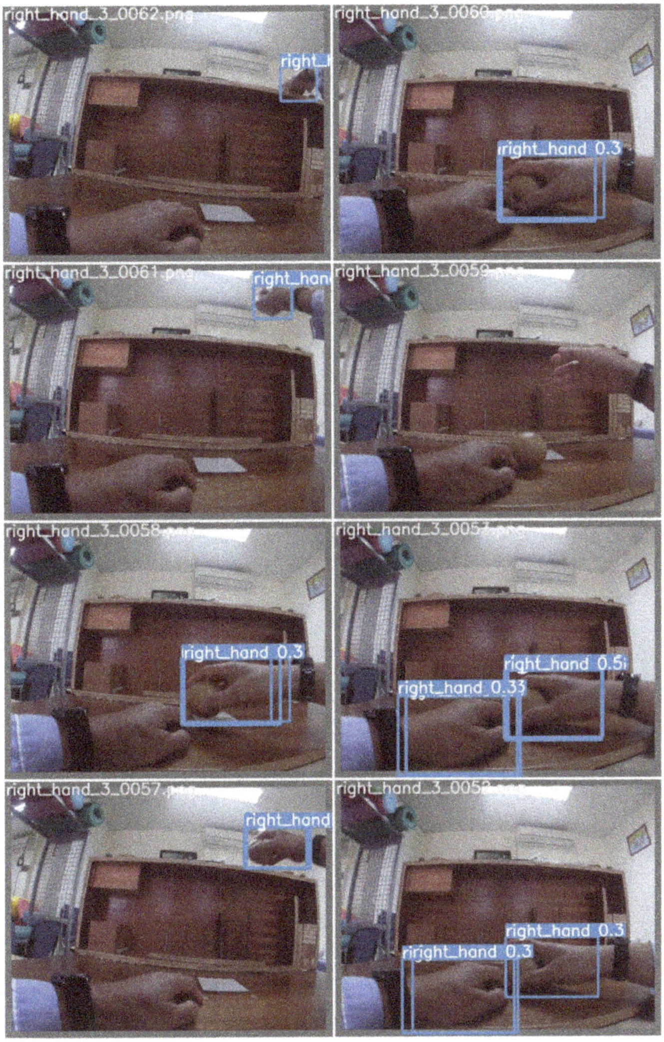

Figure 23. Illustrating the left hand being negatively classified as the right hand of the RehabHand dataset [17].

5. Conclusions and Future Works

Building an application to evaluate the rehabilitation process of the hand using the technology of computer vision and deep learning is a new research area in the medical field. The first step is hand detection, which is a very important pre-processing step. In this paper, we systematize a series of versions of YOLO. We pre-trained hand detection and classification with versions of YOLO on the EV datasets FPHAB, HOI4D, and RehabHand. The results show the performance of the YOLO versions for hand detection and classification. All new versions of YOLO give better results than old versions. The results of YOLOv7 of hand detection and classification on the FPHAB dataset are the best ($P = 96.9\%$ with $Thesh_{IOU} = 0.5$, $P = 96.9\%$ with $Thesh_{IOU} = 0.75$, $P = 93.9\%$ with $Thesh_{IOU} = 0.95$). We apply this model to limit the hand data area, hand pose estimation, and hand activities recognition for evaluation hand function rehabilitation. YOLOv7 and its variations' (YOLOv7-X, YOLOv7-w6) results on the HOI4D and RehabHand datasets are lower (Tables 3 and 4) and unequal (Table 4). We perform pre-training with more epochs and calibrate the model's parameter set to obtain a better model. Further, we compare YOLOv7 with CNN networks such as SSD, Faster R-CNN, and SOTA (State-Of-The-Art) on three datasets: FPHAB, HOI4D, and RehabHand. In the future, we will perform hand detection and tracking, hand pose estimation, and hand activity recognition for assessing the ability of the hand from faculty rehabilitation exercises of patients at Hanoi Medical University Hospital, Huong Sen Rehabilitation Hospital in Tuyen Quang Province in Vietnam [50], as illustrated in Figure 24.

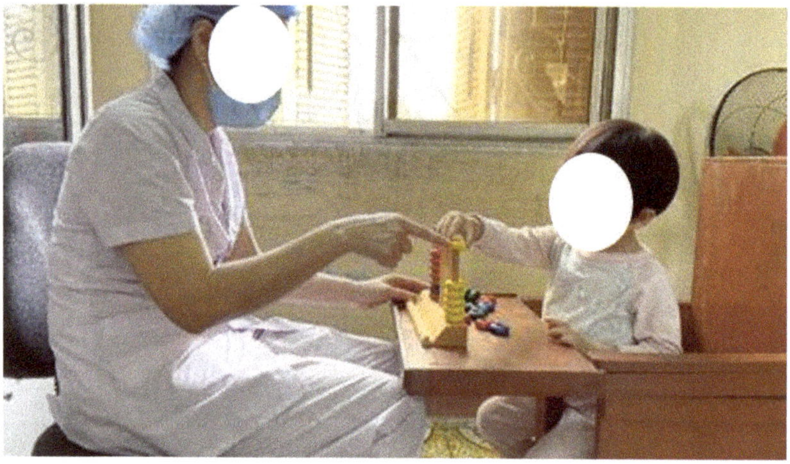

Figure 24. Illustrating the process of the hand rehabilitation exercise [50].

Author Contributions: Methodology, H.-C.N. and V.-H.L.; Data curation, V.-H.L.; Writing—original draft, H.-C.N., T.-H.N., R.S. and V.-H.L.; Writing—review & editing, R.S. All authors have read and agreed to the published version of the manuscript.

Funding: This research is supported by Hung Vuong University under grant number 03/2022/HD-HV03.2022. The project is financed under the program of the Polish Minister of Science and Higher Education under the name "Regional Initiative of Excellence" in the years 2019–2023, project number 020/RID/2018/19; the value of financing is PLN 12,000,000. This research is funded by Tan Trao University in Tuyen Quang province, Vietnam.

Institutional Review Board Statement: Not applicable.

Informed Consent Statement: Not applicable.

Data Availability Statement: Not applicable.

Conflicts of Interest: The paper is our research, not related to any organization or individual.

References

1. Tompson, J.; Stein, M.; Lecun, Y.; Perlin, K. Real-time continuous pose recovery of human hands using convolutional networks. *ACM Trans. Graph.* **2014**, *33*, 169. [CrossRef]
2. Tang, D.; Chang, H.J.; Tejani, A.; Kim, T.K. Latent regression forest: Structured estimation of 3D hand poses. *IEEE Trans. Pattern Anal. Mach. Intell.* **2017**, *39*, 1374–1387. [CrossRef]
3. Sun, X.; Wei, Y.; Liang, S.; Tang, X.; Sun, J. Cascaded hand pose regression. In Proceedings of the IEEE Computer Society Conference on Computer Vision and Pattern Recognition, Boston, MA, USA, 7–12 June 2015; pp. 824–832. [CrossRef]
4. Garcia-Hernando, G.; Yuan, S.; Baek, S.; Kim, T.K. First-Person Hand Action Benchmark with RGB-D Videos and 3D Hand Pose Annotations. In Proceedings of the Proceedings of Computer Vision and Pattern Recognition (CVPR), Salt Lake City, UT, USA, 18–22 June 2018.
5. Lugaresi, C.; Tang, J.; Nash, H.; McClanahan, C.; Uboweja, E.; Hays, M.; Zhang, F.; Chang, C.L.; Yong, M.G.; Lee, J.; et al. MediaPipe: A Framework for Building Perception Pipelines. *arXiv* **2019**, arXiv:1906.08172.

6. Zhang, F.; Bazarevsky, V.; Vakunov, A.; Tkachenka, A.; Sung, G.; Chang, C.L.; Grundmann, M. MediaPipe Hands: On-device Real-time Hand Tracking. In Proceedings of the CVPR Workshop on Computer Vision for Augmented and Virtual Reality, Seattle, WA, USA, 15 June 2020.
7. Le, V.H.; Hoang, D.T.; Do, H.S.; Te, T.H.; Phan, V.N. Real-time hand action detection and classification on the egocentric vision dataset based on Mediapipe. *TNU J. Sci. Technol.* **2022**, *227*, 181–188.
8. Liu, Y.; Liu, Y.; Jiang, C.; Lyu, K.; Wan, W.; Shen, H.; Liang, B.; Fu, Z.; Wang, H.; Yi, L. HOI4D: A 4D Egocentric Dataset for Category-Level Human-Object Interaction. In Proceedings of the Proceedings of the IEEE/CVF Conference on Computer Vision and Pattern Recognition, New Orleans, LA, USA, 18–24 June 2022; pp. 21013–21022.
9. Wang, C.Y.; Bochkovskiy, A.; Liao, H.Y.M. YOLOv7: Trainable bag-of-freebies sets new state-of-the-art for real-time object detectors. *arXiv* **2022**, arXiv:2207.02696.
10. Cai, Z.; Vasconcelos, N. Cascade R-CNN: High quality object detection and instance segmentation. *IEEE Trans. Pattern Anal. Mach. Intell.* **2021**, *43*, 1483–1498. [CrossRef]
11. Wang, C.Y.; Yeh, I.H.; Liao, H.Y.M. You Only Learn One Representation: Unified Network for Multiple Tasks. *arXiv* **2021**, arXiv:2105.04206.
12. Ge, Z.; Liu, S.; Wang, F.; Li, Z.; Sun, J. YOLOX: Exceeding YOLO Series in 2021. *arXiv* **2021**, arXiv:2107.08430.
13. Wang, C.Y.; Bochkovskiy, A.; Liao, H.Y.M. Scaled-yolov4: Scaling cross stage partial network. In Proceedings of the IEEE Computer Society Conference on Computer Vision and Pattern Recognition, Nashville, TN, USA, 19–25 June 2021; pp. 13024–13033. [CrossRef]
14. Jung, H.K.; Choi, G.S. Improved YOLOv5: Efficient Object Detection Using Drone Images under Various Conditions. *Appl. Sci.* **2022**, *12*, 7255. [CrossRef]
15. Carion, N.; Massa, F.; Synnaeve, G.; Usunier, N.; Kirillov, A.; Zagoruyko, S. End-to-End Object Detection with Transformers. *arXiv* **2020**, arXiv:2005.12872.
16. Zhu, X.; Su, W.; Lu, L.; Li, B.; Wang, X.; Dai, J. Deformable DETR: Deformable Transformers for End-to-End Object Detection. In Proceedings of the Published as a Conference Paper at ICLR 2021, Addis Ababa, Ethiopia, 26–30 April 2020; pp. 1–16.
17. Nguyen, S.H.; Vu, H. Hand detection and segmentation in first person images using Mask R-CNN. *J. Inf. Technol. Commun.* **2022**, *2022*, 1–11. [CrossRef]
18. Nunez-Marcos, A.; Azkune, G.; Arganda-Carreras, I. Egocentric Vision-based Action Recognition: A survey. *Neurocomputing* **2022**, *472*, 175–197. [CrossRef]
19. Ren, X.; Philipose, M. Egocentric recognition of handled objects: Benchmark and analysis. In Proceedings of the 2009 IEEE Computer Society Conference on Computer Vision and Pattern Recognition Workshops, CVPR Workshops 2009, Miami Beach, FL, USA, 20–25 June 2009; pp. 49–56. [CrossRef]
20. Fathi, A.; Li, Y.; Rehg, J.M. Learning to recognize daily actions using gaze. In *ECCV 2012: Computer Vision—ECCV 2012, Proceedings of the European Conference on Computer Vision, Florence, Italy, 7–13 October 2012;* Lecture Notes in Computer Science; Springer: Berlin/Heidelberg, Germany, 2020; Volume 7572, pp. 314–327. [CrossRef]
21. Kwon, T.; Tekin, B.; Stühmer, J.; Bogo, F.; Pollefeys, M. H2O: Two Hands Manipulating Objects for First Person Interaction Recognition. In Proceedings of the IEEE International Conference on Computer Vision, Virtual, 11–17 October 2021; pp. 10118–10128. [CrossRef]
22. Ragusa, F.; Furnari, A.; Livatino, S.; Farinella, G.M. The MECCANO Dataset: Understanding human-object interactions from egocentric videos in an industrial-like domain. In Proceedings of the 2021 IEEE Winter Conference on Applications of Computer Vision, WACV 2021, Waikoloa, HI, USA, 3–8 January 2021; pp. 1568–1577. [CrossRef]
23. Bandini, A.; Zariffa, J. Analysis of the hands in egocentric vision: A survey. *IEEE Trans. Pattern Anal. Mach. Intell.* **2020**. [CrossRef]
24. Redmon, J.; Divvala, S.; Girshick, R.; Farhadi, A. You only look once: Unified, real-time object detection. In Proceedings of the IEEE Computer Society Conference on Computer Vision and Pattern Recognition, Las Vegas, NV, USA, 27–30 June 2016; pp. 779–788. [CrossRef]
25. Redmon, J.; Farhadi, A. YOLO9000: Better, Faster, Stronger. In Proceedings of the Cvpr2017, Honolulu, HI, USA, 21–26 July 2016; pp. 187–213.
26. Redmon, J.; Farhadi, A. YOLOv3: An Incremental Improvement. *arXiv* **2018**, arXiv:1804.02767.
27. Bochkovskiy, A.; Wang, C.Y.; Liao, H.Y.M. YOLOv4: Optimal Speed and Accuracy of Object Detection. *arXiv* **2020**, arXiv:2004.10934.
28. Couturier, R.; Noura, H.N.; Salman, O.; Sider, A. A Deep Learning Object Detection Method for an Efficient Clusters Initialization. *arXiv* **2021**, arXiv:2104.13634.
29. He, K.; Gkioxari, G.; Dollar, P.; Girshick, R. Mask R-CNN. In Proceedings of the ICCV, Venice, Italy, 22–29 October 2017.
30. Wu, Y.; Kirillov, A.; Massa, F.; Lo, W.Y.; Girshick, R. Detectron2. 2019. Available online: https://github.com/facebookresearch/detectron2 (accessed on 5 February 2023).
31. Gao, Q.; Liu, J.; Ju, Z.; Zhang, L.; Li, Y.; Liu, Y. Hand Detection and Location Based on Improved SSD for Space Human-Robot Interaction. In *ICIRA 2018: Intelligent Robotics and Applications, Proceedings of the International Conference on Intelligent Robotics and Applications, Newcastle, NSW, Australia, 9–11 August 2018;* Springer: Berlin/Heidelberg, Germany, 2018; Volume 10984, pp. 164–175. [CrossRef]
32. Howard, A.; Sandler, M.; Chen, B.; Wang, W.; Chen, L.C.; Tan, M.; Chu, G.; Vasudevan, V.; Zhu, Y.; Pang, R.; et al. Searching for mobileNetV3. In Proceedings of the IEEE International Conference on Computer Vision, Seoul, Republic of Korea, 27 October–2 November 2019; pp. 1314–1324. [CrossRef]

33. Gallo, I.; Rehman, A.U.; Dehkordi, R.H.; Landro, N.; La Grassa, R.; Boschetti, M. Deep Object Detection of Crop Weeds: Performance of YOLOv7 on a Real Case Dataset from UAV Images. *Remote Sens.* **2023**, *15*, 539. [CrossRef]
34. Huang, L.; Zhang, B.; Guo, Z.; Xiao, Y.; Cao, Z.; Yuan, J. Survey on depth and RGB image-based 3D hand shape and pose estimation. *Virtual Real. Intell. Hardw.* **2021**, *3*, 207–234. [CrossRef]
35. Liu, Z.; Gu, X.; Chen, J.; Wang, D.; Chen, Y.; Wang, L. Automatic recognition of pavement cracks from combined GPR B-scan and C-scan images using multiscale feature fusion deep neural networks. *Autom. Constr.* **2022**, *146*, 104698. [CrossRef]
36. Huang, J.; Rathod, V.; Sun, C.; Zhu, M.; Korattikara, A.; Fathi, A.; Fischer, I.; Wojna, Z.; Song, Y.; Guadarrama, S.; et al. Speed/accuracy trade-offs for modern convolutional object detectors. In Proceedings of the 30th IEEE Conference on Computer Vision and Pattern Recognition, CVPR 2017, Honolulu, HI, USA, 21–26 July 2017; pp. 3296–3305. [CrossRef]
37. Dai, J.; Li, Y.; He, K.; Sun, J. R-FCN: Object Detection via Region-based Fully Convolutional Networks. *arXiv* **2016**, arXiv:1605.06409.
38. Ren, S.; He, K.; Girshick, R.; Sun, J. Faster R-CNN: Towards Real-Time Object Detection with Region Proposal Networks. *IEEE Trans. Pattern Anal. Mach. Intell.* **2017**, *39*, 241–294. [CrossRef] [PubMed]
39. Liu, W.; Anguelov, D.; Erhan, D.; Szegedy, C.; Reed, S.; Fu, C.Y.; Berg, A.C. SSD: Single shot multibox detector. In Proceedings of the European Conference on Computer Vision, Amsterdam, The Netherlands, 11–14 October 2016; Volume 9905, pp. 21–37. [CrossRef]
40. Lin, T.Y.; Dollár, P.; Girshick, R.; He, K.; Hariharan, B.; Belongie, S. Feature Pyramid Networks for Object Detection. *arXiv* **2017**, arXiv:1612.03144
41. Liu, S.; Qi, L.; Qin, H.; Shi, J.; Jia, J. Path Aggregation Network for Instance Segmentation. In Proceedings of the Proceedings of the IEEE Computer Society Conference on Computer Vision and Pattern Recognition, Salt Lake City, UT, USA, 18–22 June 2018; pp. 8759–8768. [CrossRef]
42. Ghiasi, G.; Lin, T.Y.; Le, Q.V. NAS-FPN: Learning scalable feature pyramid architecture for object detection. In Proceedings of the IEEE Computer Society Conference on Computer Vision and Pattern Recognition, Long Beach, CA, USA, 16–17 June 2019; pp. 7029–7038. [CrossRef]
43. Zhu, L.; Deng, Z.; Hu, X.; Fu, C.W.; Xu, X.; Qin, J.; Heng, P.A. Bidirectional feature pyramid network with recurrent attention residual modules for shadow detection. In *ECCV 2018: Computer Vision—ECCV 2018, Proceedings of European Conference on Computer Vision, Munich, Germany, 8–14 September 2018*; Springer: Berlin/Heidelberg, Germany, 2018; Volume 11210, pp. 122–137. [CrossRef]
44. Liu, S.; Huang, D.; Wang, Y. Learning Spatial Fusion for Single-Shot Object Detection. *arXiv* **2019**, arXiv:1911.09516.
45. Zhao, Q.; Sheng, T.; Wang, Y.; Tang, Z.; Chen, Y.; Cai, L.; Ling, H. M2det: A single-shot object detector based on multi-level feature pyramid network. In Proceedings of the 33rd AAAI Conference on Artificial Intelligence, AAAI 2019, 31st Innovative Applications of Artificial Intelligence Conference, IAAI 2019 and the 9th AAAI Symposium on Educational Advances in Artificial Intelligence, EAAI 2019, Honolulu, HI, USA, 29–31 January 2019; pp. 9259–9266. [CrossRef]
46. He, K.; Zhang, X.; Ren, S.; Sun, J. Spatial Pyramid Pooling in Deep Convolutional Networks for Visual Recognition. *IEEE Trans. Pattern Anal. Mach. Intell.* **2015**, *37*, 1904–1916. [CrossRef]
47. YOLOv5. YOLOv5 SPP/SPPF. 2021. Available online: https://blog.csdn.net/weixin_55073640/article/details/122621148 (accessed on 20 November 2022).
48. Kong, S.; Fang, X.; Chen, X.; Wu, Z.; Yu, J. A real-time underwater robotic visual tracking strategy based on image restoration and kernelized correlation filters. In Proceedings of the 30th Chinese Control and Decision Conference, CCDC 2018, Shenyang, China, 9–11 June 2018; pp. 6436–6441. [CrossRef]
49. Zhang, X.; Wang, W.; Zhao, Y.; Xie, H. An improved YOLOv3 model based on skipping connections and spatial pyramid pooling. *Syst. Sci. Control Eng.* **2021**, *9*, 142–149. [CrossRef]
50. Huong Sen Rehabilitation Hospital. Huong Sen Rehabilitation Hospital at Tuyen Quang Province. Available online: http://bv-phcnhuongsentuyenquang.vn/ (accessed on 14 February 2023).

sensors

MDPI

Article

The Successive Next Network as Augmented Regularization for Deformable Brain MR Image Registration

Meng Li, Shunbo Hu *, Guoqiang Li *, Fuchun Zhang, Jitao Li, Yue Yang, Lintao Zhang, Mingtao Liu, Yan Xu, Deqian Fu, Wenyin Zhang and Xing Wang

School of Information Science and Engineering, Linyi University, Linyi 276000, China
* Correspondence: hushunbo@lyu.edu.cn (S.H.); liguoqiang@lyu.edu.cn (G.L.); Tel.: +86-156-5397-6667 (S.H.)

Abstract: Deep-learning-based registration methods can not only save time but also automatically extract deep features from images. In order to obtain better registration performance, many scholars use cascade networks to realize a coarse-to-fine registration progress. However, such cascade networks will increase network parameters by an n-times multiplication factor and entail long training and testing stages. In this paper, we only use a cascade network in the training stage. Unlike others, the role of the second network is to improve the registration performance of the first network and function as an augmented regularization term in the whole process. In the training stage, the mean squared error loss function between the dense deformation field (DDF) with which the second network has been trained and the zero field is added to constrain the learned DDF such that it tends to 0 at each position and to compel the first network to conceive of a better deformation field and improve the network's registration performance. In the testing stage, only the first network is used to estimate a better DDF; the second network is not used again. The advantages of this kind of design are reflected in two aspects: (1) it retains the good registration performance of the cascade network; (2) it retains the time efficiency of the single network in the testing stage. The experimental results show that the proposed method effectively improves the network's registration performance compared to other state-of-the-art methods.

Keywords: brain image registration; generation adversarial network; deep learning

Citation: Li, M.; Hu, S.; Li, G.; Zhang, F.; Li, J.; Yang, Y.; Zhang, L.; Liu, M.; Xu, Y.; Fu, D.; et al. The Successive Next Network as Augmented Regularization for Deformable Brain MR Image Registration. *Sensors* **2023**, *23*, 3208. https://doi.org/10.3390/s23063208

Academic Editors: Euntai Kim, Sangyoun Lee and Kang Ryoung Park

Received: 5 February 2023
Revised: 27 February 2023
Accepted: 8 March 2023
Published: 17 March 2023

1. Introduction

Image registration is one of the basic tasks in medical image processing. It involves the acquisition of a dense deformation field (DDF) when a moving image is matched with a fixed image so that the two to-be-aligned images and their corresponding anatomical structures are aligned accurately in space [1]. The traditional registration method optimizes the cost function through a large number of iterations, a process that usually requires a significant amount of computation and time [2]. With the popularization and application of deep learning in the field of medical image registration, the deep learning registration method is now faster than the traditional image registration method. Therefore, for moving and fixed images, deformation fields can be generated by training a neural network, thus achieving rapid registration for a forward pass in the testing stage. Fan et al. [3] studied the computational costs of seven different deformable registration algorithms. The results showed that the assessed deep-learning network (BIRNet) without any iterative optimization needed the least time. Additionally, the registration accuracy improved after applying the deep learning method. For example, Cao et al. [4] proposed a deep learning method for registering brain MRI images, and it was revealed that the method's Dice coefficient was improved in terms of registering white matter (WM), gray matter (GM), and cerebrospinal fluid (CSF).

The unsupervised learning image registration method has been widely applied because it is not difficult to obtain gold-standard registration [5]. Balakrishnan et al. [6]

optimized the U-Net neural network by defining the loss function as a combination of the mean square error similarity measure and the deformation field's smoothing constraint. de Vos et al. [7] accomplished affine and deformable registration by superimposing several networks through unsupervised training. Kim et al. [8,9] used cyclic consistency to provide implicit regularization for maintaining topology and realizing 2D or 3D image registration. Moreover, a multi-scale strategy was adopted during the experiment to solve the relevant storage problem. Jiang et al. [10] proposed an unsupervised network framework (MJ-CNN) that adopted a multi-scale joint training scheme to achieve end-to-end optimization. Kong et al. [11] designed a cascade-connected channel attention mechanism network. During cascade registration, the attention module is incorporated to learn the features of the input image, thereby improving the expression ability of the image features. Through five iterations of the deformation field, improved bidirectional image registration was realized. Yang et al. [12] used multiple cascaded U-Net models to form a network structure. In their structure, each U-Net is trained with smooth regularization parameters to improve the accuracy of 3D medical image registration. Zhu et al. [13] helped a network develop high-similarity spatial correspondence by introducing a local attention model and integrated multi-scale functionality into the attention mechanism module to achieve the coarse-to-fine registration of local information. Ouyang et al. [14] trained their designed subnetworks synergistically by training the residual recursive cascade network to realize cooperation between the subnetworks. Through the connection of the residual network, the registration speed was accelerated. Guo et al. [15] improved the image registration accuracy and efficiency of CT-MR and used two cyclic consistency methods in a full convolution neural network to generate the spatial deformation field. Sideri-Lampretsa et al. [16] considered that it was easy to obtain edge images, so they used the image's edges to drive the multimodal registration training process and thus help the network learn more effective information. Qian et al. [17] proposed a cascade framework of a registration network, and then registered images in training stages. The authors compared the performance of the cascade network framework with the traditional registration methods, subsequently, it was determined that the registration efficiency of the proposed method was significantly improved. Golkar et al. [18] proposed a hybrid registration framework of vessel extraction and thinning for retinal image segmentation, which improved the registration accuracy of complex retinal vessels.

Inspired by the idea of two-person zero-sum game from game theory, Goodefellow et al. [19] proposed a generation adversarial network (GAN) that used two neural networks for adversarial training and continuously improved the performance of the network in all directions during a game between the two networks. In addition to the in-depth study of the generative adversarial network (GAN), the application of an adversarial network has been integrated with techniques and aims from other fields, for instance, the combination of GAN and image processing. Therefore, GANs are also widely used in image registration. Santarossa et al. [20] used generation adversarial networks combined with ranking loss for multimodal image registration. Fan et al. [21,22] implemented a GAN in the unsupervised deformable registration of 3D brain MR images. In this approach, the discrimination network identifies whether a pair of images are sufficiently similar. The resulting feedback is then used to train the registration network. Simultaneously, GANs have been applied to single- and multi-mode image registration. Zheng et al. [23] used a GAN network to realize symmetric image registration and then transformed the symmetric registration formula of single- and multi-mode images into a conditional GAN. To align a pair of single-mode images, the registration method constitutes a cyclical process of transformation from one image to another and its inverse transformation. To align images with different modes, mode conversion should be performed before registration. In the training process, the method also adopts the semi-supervised method and trains using labeled and unlabeled images. Many registration methods have been produced based on the application of generation adversarial networks [24–28]. Huang et al. [29] fused a difficulty perception model into a cascade neural network composed of three networks. These networks are used to predict the coarse deformation field and the fine deformation field, respectively, so as to achieve accurate registration. GANs showed excellent performance in the aforementioned studies. In the previous study, a GAN based on dual attention mechanisms was proposed, which showed good registration performance in areas with relatively flat edges, but poor registration performance in narrow and long-edge areas. To this end, based on previous research, this paper proposes a method to assist GANs in realizing the registration of long and narrow regions at the peripheries of the brain, which differs from the methods of coarse registration and fine registration. Our main contributions are summarized as follows:

1. During training, the cascade networks are trained simultaneously to save network training time.
2. The second network is used as a loss function. The mean square error loss function added to the second network can constrain the deformation field output by the second network such that it tends to 0. Only the first network is used during testing, which saves testing time.
3. Coupled with the adversarial training of GANs, the registration performance of the first network is further improved.

The rest of this paper is organized as follows. Section 2 introduces the networks proposed in this paper in detail. Section 3 introduces the experimental datasets and evaluation indicators. Section 4 introduces the experimental results obtained from the HBN and ABIDE datasets. In Section 5, we provide a discussion. Finally, the conclusions are given in Section 6.

2. Methodology

This paper proposes a method combining adversarial learning with cascade learning. Joint training of cascaded networks can allow them to predict more accurate deformation fields. The first (registration) network is used to study the deformation field ϕ_1. The second (augmented) network enables the first network to learn more deformations. A discrimination network improves the first network's performance through adversarial training. The structures of each cascading network are similar to those of VoxelMorph [6]. The proposed overall learning framework is illustrated in Figure 1.

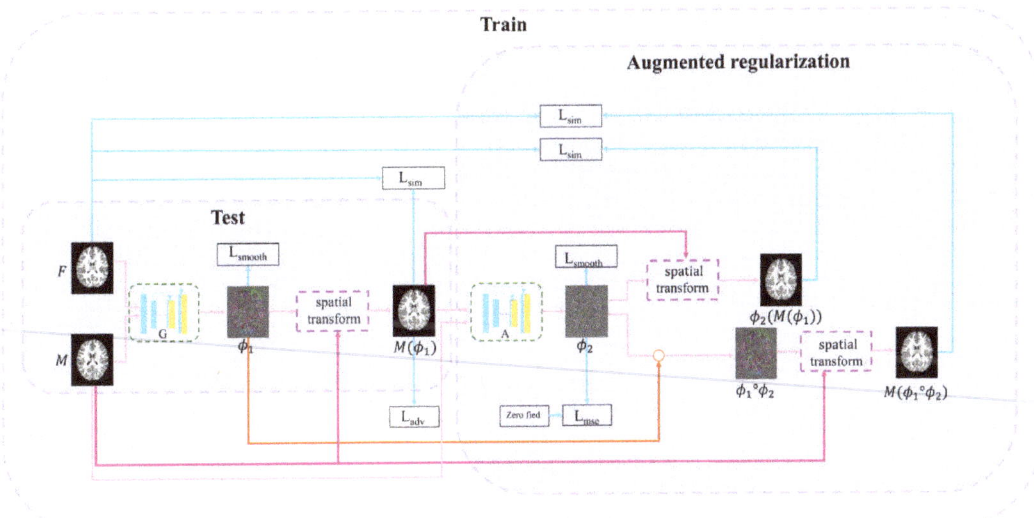

Figure 1. Overall network framework.

2.1. First (Registration) Network

The registration network is the first network in cascading framework. Its inputs are the fixed image F and the moving image M. Its output is the deformation field ϕ_1, i.e., $\phi_1 = G(F, M)$. This network realizes the alignment from M to F, i.e., $F = M(\phi_1)$, where $M(\phi_1)$ is the warped image. Subsequently, the loss function between $M(\phi_1)$ and F is calculated to drive the training process. This loss function includes three parts: intensity similarity loss L_{sim}, adversarial loss L_{adv}, and smooth regularization term L_{smooth}.

The adversarial loss function of the registration network is:

$$L_{adv}(p) = \begin{cases} -\log(1-p), c \in P^+ \\ -\log(p), \quad c \in P^- \end{cases} \tag{1}$$

where p is the output value of the discrimination network and c indicates the registration network input.

Local cross-correlation metric is used to calculate the similarity of the intensity between fixed image F and warped image $M(\phi_1)$. The specific formula of the loss function is:

$$CC(F, M(\phi_1)) = \sum_{p \in \Omega} \frac{\left(\sum_{p_i} (F(p_i) - F(p))(M(\phi(p_i)) - M(\phi(p)))\right)^2}{\left(\sum_{p_i} (F(p_i) - F(p))^2\right)\left(\sum_{p_i} (M(\phi(p_i)) - M(\phi(p)))^2\right)} \tag{2}$$

where p_i denotes the iteration of the n^3 volume center at voxel p, and Ω represents a three-dimensional voxel. In this paper, $n = 9$ $F(p_i)$, and $M(\phi_1(p_i))$ represents the voxel intensities of F and $M(\phi_1)$ at p_i, respectively. $F(p)$ and $M(\phi_1(p))$ are the local mean values of n^3 volume. A higher CC indicates a more accurate alignment. According to the definition of CC, the intensity similarity loss L_{sim} is defined as follows:

$$L_{sim}(F, M(\phi_1)) = -CC(F, M(\phi_1)) \tag{3}$$

Additionally, L2 regularization is implemented to smooth the deformation field ϕ_1:

$$L_{smooth}(\phi_1) = \sum_{p \in \Omega} \|\nabla \phi_1(p)\|^2 \tag{4}$$

2.2. Successive (Augmented) Network

The inputs of the successive network are F and $M(\phi_1)$; the output is DDF ϕ_2. ϕ_2 is used to deform $M(\phi_1)$ to obtain $\phi_2(M(\phi_1))$. Simultaneously, to clarify the warped image, we perform a composed operation on ϕ_1 and ϕ_2, i.e., $\phi_1 \circ \phi_2$. $M(\phi_1 \circ \phi_2)$ is obtained by the moving image M with the composed DDF. Next, two intensity loss functions, namely, $L_{sim}(F, M(\phi_1 \circ \phi_2))$ and $L_{sim}(F, \phi_2(M(\phi_1)))$, are calculated between $M(\phi_1 \circ \phi_2)$ and F and between $\phi_2(M(\phi_1))$ and F, respectively. The DDF ϕ_2 is also constrained as it approaches zero deformation field through the following MSE loss function, allowing the deformation field ϕ_1 to learn more accurate deformations.

The formula of MSE loss function is defined as:

$$L_{mse}(\phi_2) = L_{mse}(\phi_2, 0) = \sum_{p \in \Omega} \|\nabla \phi_2(p)\|^2 \tag{5}$$

Through this function, the output effect of the first network can achieve fine registration after the two networks are connected in series.

The loss function for the registration network is as follows:

$$L_G = \text{Ladv}\ (p) + \alpha L_{sim}(F, M(\phi_1)) + \lambda L_{smooth}(\phi_1) \tag{6}$$

In addition, the loss function used by the second network is:

$$L_A = L_{sim}(F, M(\phi_1)) + L_{sim}(M(\phi_1 \circ \phi_2)) + L_{sim}(F, \phi_2(M(\phi_1))) + L_{mse}(\phi_2) + L_{smooth}(\phi_1) + L_{smooth}(\phi_2) + L_{smooth}(\phi_1 \circ \phi_2) \quad (7)$$

The total loss function is:

$$L_{total} = L_G + L_A \quad (8)$$

2.3. Discrimination Network

The discrimination network consists of four convolutional layers combined with leakyReLU activation layers. Finally, the sigmoid activation function is used to output the probability value. The discrimination network is shown in Figure 2. The discrimination network distinguishes the authenticity of image. The harder it is to distinguish the warped image from the fixed image, the harder it is to judge the authenticity of the image by the discrimination network.

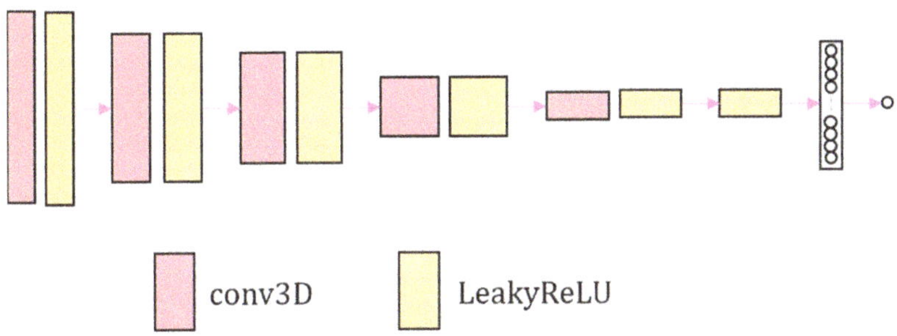

Figure 2. The overall framework of the adversarial network. The adversarial network consists of convolution and LeakyReLU activation layer.

3. Experiment

3.1. Experimental Details

Python and TensorFlow were used to implement the experimental process. The program was trained and tested with GPU NVIDIA GeForce GTX 2080 Ti [30].

In the training process, the patch-based training method is adopted to reduce the occupied memory. Herein, 127 blocks are obtained from each image with a size of 182 × 218 × 182. Each block size is 64 × 64 × 64. The stride is 32. The learning rates for training the registration and discrimination networks are set to 0.00001 and 0.000001, respectively.

The traditional methods of Demons and SyN are used as comparative experiments. The deep learning model VoxelMorph is also trained. VoxelMorph is a model of medical image registration based on unsupervised learning. Therefore, VoxelMorph is selected as the comparative experiment for deep learning. The Dice score, structural similarity, and Pearson's correlation coefficient are used as the evaluation indicators to verify the superiority of the experimental results. Moreover, the influence of the MSE and L_{sim} loss functions on the experimental results is investigated.

3.2. Datasets

To prove the flexibility and superior performance of the proposed method, the HBN [31] and ABIDE datasets [32] are used for training and testing. The HBN dataset consists of brain data obtained from patients with ADHD (aged 5–21 years). Herein, 496 and 31 T1-weighted brain images are selected for training and testing, respectively. ABIDE is a dataset consisting of brain images from patients with autism (aged 5–64 years). Herein, 928 and 60 T1-weighted brain images are used for training and testing, respectively. The fixed image used in training comprises a pair of images randomly selected from the training set such that each image is linearly aligned to the fixed image. The image size of both the

HBN and ABIDE datasets is 182 × 218 × 182 voxels with a resolution of 1 × 1 × 1 mm^3. Both these datasets contain segmentation marker images of CSF, GM, and WM.

3.3. Evaluation Indicators

3.3.1. Dice Score

The Dice coefficient (Dice) index is used to evaluate the degree of overlap between a warped segmentation image and the segmentation image of the fixed image. This index reflects the similarity between the experimental and the standard segmentation images. It is defined as follows:

$$Dice = 2\left|\frac{X_{seg} \cap Y_{seg}}{X_{seg} \cup Y_{seg}}\right| \tag{9}$$

where X_{seg} and Y_{seg} represent the standard and warped segmentation images, respectively. The range of Dice values is 0–1, corresponding to a range in the gap between the warped and the standard segmentation images progressing from large to small values, respectively. Alternatively, the closer the experimental result is to 1, the more similar the warped segmentation image is to the standard segmentation image, and the better is the registration result.

3.3.2. Structural Similarity

The structure similarity index measure [33] can measure the similarity of two images. The SSIM is calculated as:

$$\mathrm{SSIM}(X,Y) = \frac{(2\mu_X\mu_Y + c_1)(2\sigma_{XY} + c_2)}{(\mu_X^2 + \mu_Y^2 + c_1)(\sigma_X^2 + \sigma_Y^2 + c_2)} \tag{10}$$

where X, Y represent the two input 3D images; μ_X and μ_Y represent the average value of X and Y, respectively. σ_X^2 and σ_Y^2 are the variances of X and Y, respectively. σ_X and σ_Y represent the standard deviation of X and Y, respectively. σ_{XY} represents the covariance of X and Y. c_1 and c_2 are constants used to avoid system errors caused by a denominator equal to 0. The SSIM can measure the structural similarity between the real and warped images. A SSIM value close to 1 indicates that the two images have a high degree of similarity.

3.3.3. Pearson's Correlation Coefficient

Pearson's correlation coefficient (PCC) was used to measure the similarity between two 3D images. The calculation formula of PCC is:

$$\rho(X,Y)\frac{\sum_{i=1}^{n}\left(X_i - \bar{X}\right)\left(Y_i - \bar{Y}\right)}{\sqrt{\sum_{i=1}^{n}\left(X_i - \bar{X}\right)^2}\sqrt{\sum_{i=1}^{n}\left(Y_i - \bar{Y}\right)^2}} \tag{11}$$

The closer the value of PCC is to 1, the greater is the correlation. A PCC of 0 indicates no correlation. X, Y refer to the two input 3D images. $\bar{X}$ and $\bar{Y}$ represent the mean value of X and Y, respectively.

4. Results

The proposed methodology is compared with the following approaches: (1) Demons and SyN, two traditional registration methods; (2) Voxelmorph (VM), an unsupervised deep learning registration method; and (3) VM + A, a method consisting of a simultaneously trained registration network and augmented network.

First, the proposed GAN method (VM + A + GAN) is compared with Demons and SyN, which are two traditional methods. Tables 1 and 2 summarize the test results obtained through different datasets, and all indicators show that our experimental results are the best. Figure 3 shows the comparison of the test results of the two datasets. The first row of the experimental image represents the original image obtained from the HBN dataset, and the second row represents the segmentation image corresponding to the original image derived from the HBN dataset. Similarly, the third row represents the original image based on the ABIDE dataset, and the fourth row represents the segmentation image corresponding to the original image derived from the ABIDE dataset. Compared with Demons and SyN, the image obtained by the proposed GAN method is closer in appearance to the fixed image, and the parts with differences are shown in the enlarged image on the right.

Table 1. Dice values obtained with the HBN and ABIDE datasets. Bold numbers indicate the best results.

	HBN			ABIDE		
Methods	**CSF**	**GM**	**WM**	**CSF**	**GM**	**WM**
Demons	0.513 ± 0.158	0.335 ± 0.320	0.345 ± 0.330	0.410 ± 0.228	0.312 ± 0.305	0.332 ± 0.320
SyN	0.585 ± 0.026	0.768 ± 0.022	0.786 ± 0.015	0.593 ± 0.041	0.749 ± 0.019	0.791 ± 0.023
VM + A + GAN	$\mathbf{0.653 \pm 0.041}$	$\mathbf{0.829 \pm 0.029}$	$\mathbf{0.855 \pm 0.015}$	$\mathbf{0.646 \pm 0.046}$	$\mathbf{0.801 \pm 0.024}$	$\mathbf{0.847 \pm 0.019}$

Table 2. SSIM and PCC metrics obtained with the HBN and ABIDE datasets. Bold numbers indicate the best results.

	HBN		ABIDE	
Methods	**SSIM**	**PCC**	**SSIM**	**PCC**
Demons	0.781	0.886	0.763	0.886
SyN	0.904	0.962	0.870	0.958
VM + A + GAN	**0.956**	**0.984**	**0.920**	**0.985**

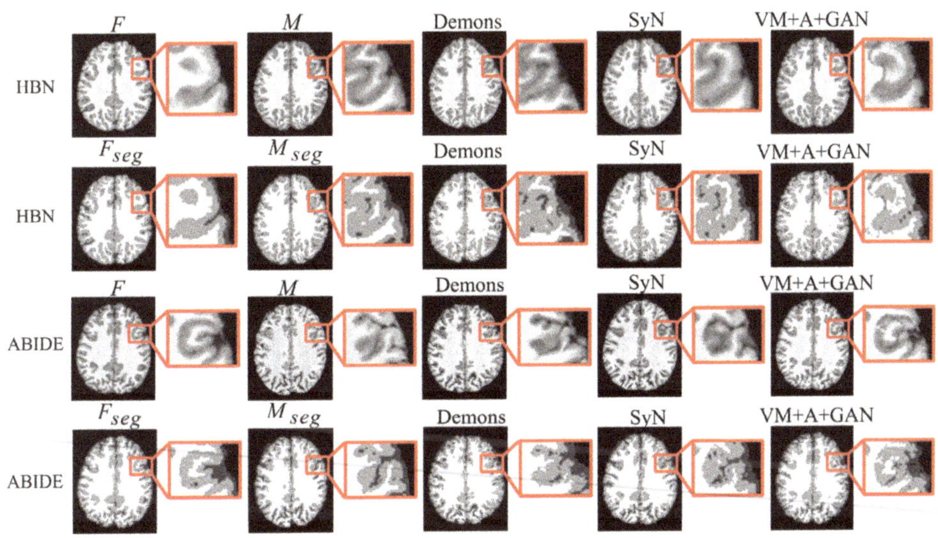

Figure 3. Registration results of Demons, SyN, and our proposed method using HBN and ABIDE datasets.

Second, the proposed GAN method is compared with the VM and VM + A methods. Figure 4 shows the registered moving image and the fixed image. Moreover, the first row represents the original image from the HBN dataset, and the second row represents the segmentation image corresponding to the original image from the HBN dataset. Similarly, the third row represents the original image from the ABIDE dataset, and the fourth row represents the segmentation image corresponding to the original image from the ABIDE dataset. Additionally, the enlarged figure on the right shows that the result for the proposed method regarding the training of the registration, augmented, and discrimination networks together is closer to the fixed image. Through the experimental results, the performance of the registration, augmented, and discrimination networks when trained together is verifiably better than that of the registration network trained individually and of the registration and augmented networks trained simultaneously.

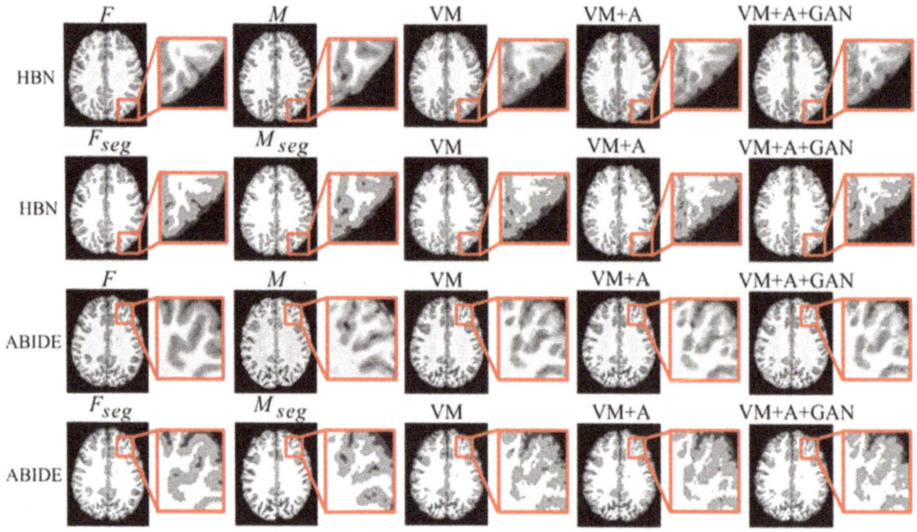

Figure 4. Registration results based on deep learning methods. Among them, VM represents the result obtained by the VoxelMorph method, VM + A represents the result obtained by training the registration network and the enhanced network together, and VM + A + GAN represents the result obtained by our method.

In order to more clearly highlight the effectiveness of the method proposed in this paper, Figure 5 shows the experimental results of the three parts of the brain tissue based on the HBN dataset, and Figure 6 shows the experimental results of the three parts of the brain tissue based on the ABIDE dataset. The dotted circle in the figure is the result obtained by the method proposed in this paper.

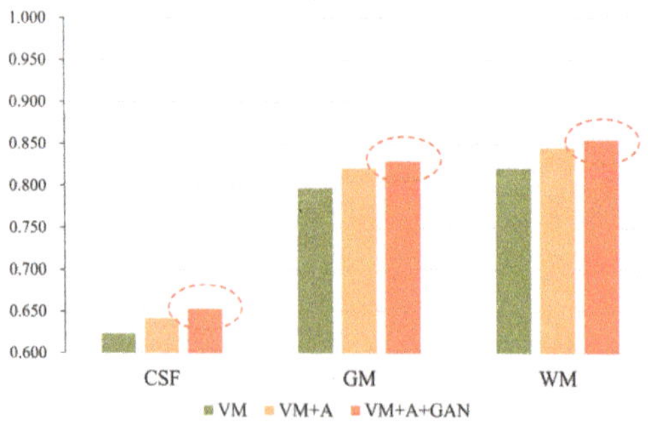

Figure 5. GAN registration performance on the HBN dataset.

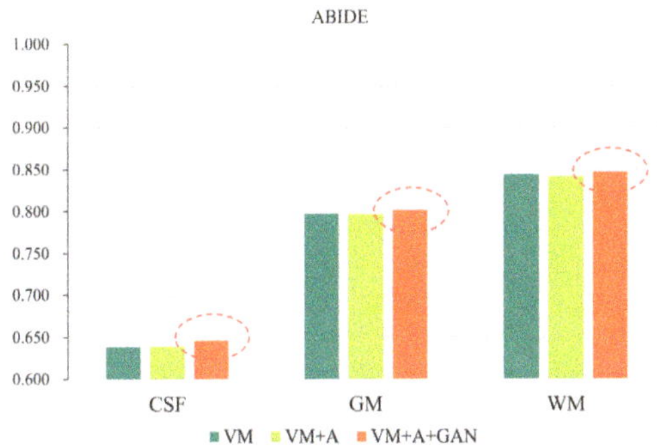

Figure 6. GAN registration performance on the ABIDE dataset.

Tables 3 and 4 summarize the Dice, SSIM, and PCC indices corresponding to the different datasets. Considering Table 3, for the HBN dataset, the proposed method improves the precision values by 0.030, 0.032, and 0.034 compared with the VM method. For the ABIDE dataset, the proposed method improves the accuracies by 0.008, 0.004, and 0.004 compared with the VM method. Considering Table 4, for the HBN dataset, the proposed method increases the SSIM and PCC indices by 0.02 and 0.008, respectively, compared with the VM method. For the ABIDE dataset, the proposed method improves the SSIM and PCC indices by 0.006 and 0.003, respectively, compared with the VM method.

Table 3. Dice indicator based on deep learning. Bold numbers indicate the best results.

	HBN			ABIDE		
Methods	**CSF**	**GM**	**WM**	**CSF**	**GM**	**WM**
VM	0.623 ± 0.037	0.797 ± 0.027	0.821 ± 0.015	0.638 ± 0.048	0.797 ± 0.024	0.843 ± 0.018
VM + A	0.642 ± 0.042	0.821 ± 0.029	0.846 ± 0.014	0.639 ± 0.048	0.796 ± 0.023	0.841 ± 0.018
VM + A + GAN	**0.653 ± 0.041**	**0.829 ± 0.029**	**0.855 ± 0.015**	**0.646 ± 0.046**	**0.801 ± 0.024**	**0.847 ± 0.019**

Table 4. SSIM and PCC metrics for deep-learning-based registration methods. Bold numbers indicate the best results.

	HBN		ABIDE	
Methods	**SSIM**	**PCC**	**SSIM**	**PCC**
VM	0.936	0.976	0.914	0.982
VM + A	0.936	0.976	0.914	0.982
VM + A + GAN	**0.956**	**0.984**	**0.920**	**0.985**

5. Discussion

The usage of a registration and discrimination networks for image registration is a common method. Such a registration method has been investigated experimentally in previous work [34]. However, this adversarial method for training a GAN only limitedly improves a registration network's performance, and the registration capacity in some narrow and long edge areas needs to be further improved. Therefore, this paper proposes a method of training three networks together to allow the registration network to learn more deformations, further improving the registration performance. When the three networks are trained together, the use of different loss functions has a certain impact on the experimental results, which is discussed in the following subsections.

5.1. Importance of MSE

When two networks (VM + A) were trained together, both the L_{smooth} loss function of the deformation field ϕ_2 and the MSE loss function were calculated. An experiment was also performed without the MSE loss function (VM + A − MSE) to verify its effectiveness. Additionally, when the three networks (VM + A + GAN) were trained together, the MSE loss function was removed again (VM + A + GAN − MSE), and experiments were performed to verify the impact of the MSE loss function on the experimental results. Through comparison, the best registration effect was achieved when the three networks were trained together and combined with the MSE loss function. The results are shown in Figure 7.

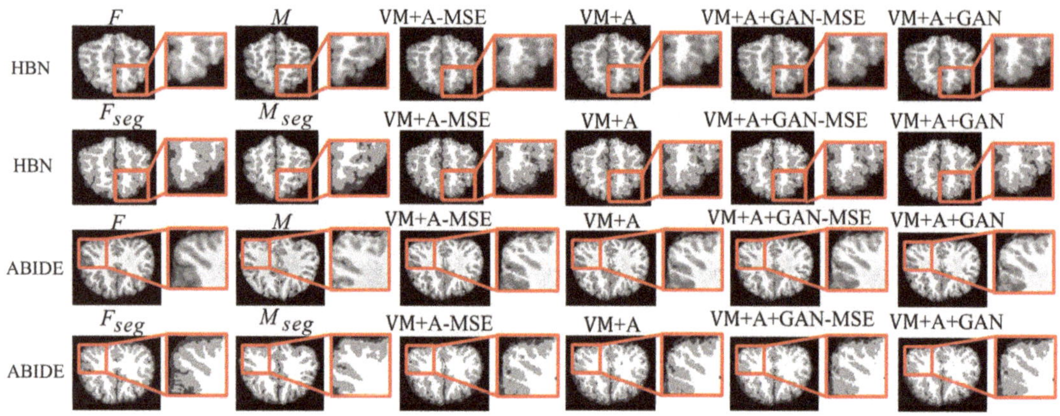

Figure 7. Experimental results regarding the use of the MSE loss function when employing the HBN and ABIDE datasets. Among them, VM + A − MSE indicates that the MSE loss function has been removed when training the registration network and the enhanced network, VM + A indicates the experimental results when the MSE loss function is retained when training the registration network and the enhanced network, VM + A + GAN − MSE indicates our method's experimental results following the removal of the MSE loss function, and VM + A + GAN represents our experimental results with the MSE loss function retained.

Table 5 summarizes the experimental results regarding the removal of the MSE loss function (VM + A − MSE) when two networks were trained together (VM + A) and the removal of the MSE loss function (VM + A + GAN − MSE) when three networks were trained together (VM + A + GAN). When comparing the results, note that the removal of the MSE loss function reduces registration accuracy, thus verifying that registration performance can be improved by adding the MSE loss function when these three networks are trained together. Comparing the SSIM and PCC metrics in Table 6, the loss function used by the proposed method achieves good results. Figure 4 shows the comparison of the experimental results after the MSE loss function was removed (VM + A − MSE) when two networks were trained together and after the MSE loss function was removed (VM + A + GAN − MSE) when three networks were trained together. Evidently, the proposed method obtained a result that is closer to the fixed image, which confirms the effectiveness of training three networks simultaneously; moreover, note that the proposed method intuitively shows a good registration effect in the narrow and long regions of the peripheries of the brain images. The first row of the resulting images represents the original image from the experimental results for the HBN dataset, and the second row represents the segmentation image corresponding to the original image from the experimental results for the HBN dataset. Similarly, the third row represents the original image from the experimental results for the ABIDE dataset, and the second row represents the segmentation image corresponding to the original image from the experimental results for the ABIDE dataset.

Table 5. Dice values when using the MSE loss function and the HBN and ABIDE datasets. Bold numbers indicate the best results.

Methods	HBN			ABIDE		
	CSF	GM	WM	CSF	GM	WM
VM + A − MSE	0.641 ± 0.041	0.821 ± 0.029	0.840 ± 0.014	0.638 ± 0.048	0.797 ± 0.024	0.843 ± 0.018
VM + A	0.642 ± 0.042	0.821 ± 0.029	0.846 ± 0.014	0.639 ± 0.048	0.796 ± 0.023	0.841 ± 0.018
VM + A + GAN − MSE	0.652 ± 0.041	0.829 ± 0.028	0.854 ± 0.014	0.645 ± 0.047	0.800 ± 0.024	0.846 ± 0.019
VM + A + GAN	**0.653 ± 0.041**	**0.829 ± 0.029**	**0.855 ± 0.015**	**0.646 ± 0.046**	**0.801 ± 0.024**	**0.847 ± 0.019**

Table 6. SSIM and PCC values when using the MSE loss function and the HBN and ABIDE datasets. Bold numbers indicate the best results.

Methods	HBN		ABIDE	
	SSIM	PCC	SSIM	PCC
VM + A − MSE	0.950	0.982	0.911	0.982
VM + A	0.952	0.983	0.910	0.981
VM + A + GAN − MSE	**0.957**	**0.985**	0.919	0.985
VM + A + GAN	0.956	0.984	**0.920**	**0.985**

5.2. Importance of L_{sim}

When the three networks (VM + A + GAN) are trained together, the L_{smooth} loss functions between the $\phi_2(M(\phi_1))$ image and the fixed image F as well as the $M(\phi_1 \circ \phi_2)$ image and the fixed image F are removed for experimental comparison. After removing the two L_{sim} loss functions, the registration accuracy decreases significantly. Through this experimental analysis, it is evident that the L_{sim} loss function can restrict the similarity among the images to a certain extent, which proves the effectiveness of adding the L_{sim} loss function. By observing the histogram in Figure 8, it is evident that the proposed method improves the Dice, SSIM, and PCC indices. In Figure 8, note that (a) shows the importance of verifying the L_{sim} loss function for the HBN dataset; (b) shows the difference between verifying the proposed method for the ABIDE dataset and removing the L_{sim} loss function in the Dice index; (c) shows the impact of removing the L_{sim} loss function on the SSIM and PCC indices for the HBN dataset; and (d) shows the impact of removing the L_{sim} loss function on the SSIM and PCC indices for the ABIDE dataset.

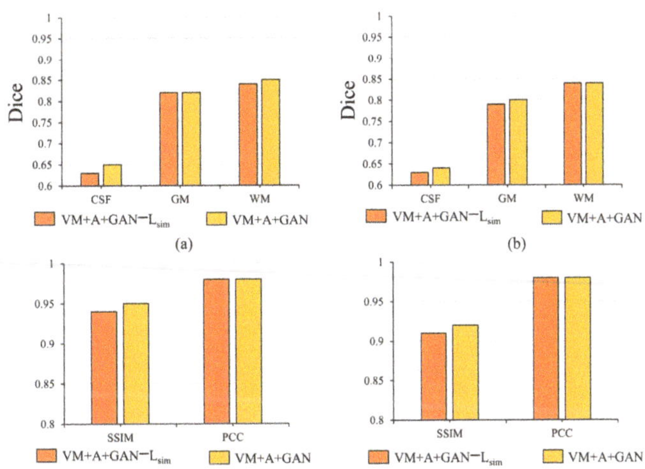

Figure 8. Influence of L_{sim} loss function on registration results.

5.3. *Importance of Different Deformation Fields*

The Dice values for when two networks were trained simultaneously are calculated and discussed next to verify ϕ_1, ϕ_2, and $\phi_1{}^\circ\phi_2$ in the images.

For ϕ_1, the similarity is calculated between the warped moving image segmentation image $M_{seg}(\phi_1)$ and the fixed image segmentation image F_{seg}, expressed as $M_{seg}(\phi_1) - F_{seg}$. For ϕ_2, the similarity is calculated between the warped $(M_{seg}(\phi_1))(\phi_2)$ and the fixed image segmentation image F_{seg}, expressed as $(M_{seg}(\phi_1))(\phi_2) - F_{seg}$. For $\phi_1{}^\circ\phi_2$, the similarity is calculated between the warped moving image segmentation image $M_{seg}(\phi_1{}^\circ\phi_2)$ and the fixed image segmentation image F_{seg}, expressed as $M_{seg}(\phi_1{}^\circ\phi_2) - F_{seg}$.

Considering the Dice values in the Table 7, the deformation field (ϕ_2) still plays a certain role in image registration, but a significantly miniscule role. Therefore, the registration network still allows the deformation field (ϕ_1) to learn more deformations, and the augmented network only plays a secondary role.

Table 7. Test results of output images from registration and augmented network for two datasets. Bold numbers indicate the best results.

Methods	HBN			ABIDE		
	CSF	GM	WM	CSF	GM	WM
$M_{seg}(\phi_1) - F_{seg}$	0.642 ± 0.042	0.821 ± 0.029	0.846 ± 0.014	0.639 ± 0.048	0.796 ± 0.023	0.841 ± 0.018
$(M_{seg}(\phi_1))(\phi_2) - F_{seg}$	**0.644 ± 0.042**	**0.825 ± 0.030**	**0.853 ± 0.015**	**0.642 ± 0.048**	**0.802 ± 0.023**	**0.848 ± 0.018**
$M_{seg}(\phi_1{}^\circ\phi_2) - F_{seg}$	0.643 ± 0.042	0.821 ± 0.028	0.845 ± 0.013	0.640 ± 0.048	0.796 ± 0.023	0.839 ± 0.018

6. Conclusions

In this paper, a method wherein three networks (registration, augmented, and discrimination networks) are trained together is proposed, for which the MSE loss function is introduced into the augmented network to improve the registration network's performance. It was demonstrated that the registration network's performance was further improved when coupled with the adversarial capacity of a GAN. Then, it was proven that the proposed method offers significant advantages over the existing methods. In addition, it was clarified that the proposed training method is easy to implement, and that the implemented loss function is easy to obtain.

In the future, a more novel GAN will be used to further improve image registration performance; moreover, more indicators will be used for comparison. The developed model will then be tested on different datasets to prove its excellent generalizability.

Author Contributions: Conceptualization, writing—original draft preparation, writing—review and editing, M.L. (Meng Li); methodology, S.H.; software, G.L.; validation, M.L. (Mingtao Liu), F.Z. and J.L.; formal analysis, S.H.; investigation, Y.Y.; resources, L.Z.; data curation, Y.X.; visualization, Y.Y.; supervision, D.F.; project administration, W.Z.; funding acquisition, X.W. All authors have read and agreed to the published version of the manuscript.

Funding: This research was funded by National Natural Science Foundation of China under Grant 61771230 and Shandong Provincial Natural Science Foundation: ZR2021MF115.

Informed Consent Statement: Informed consent was obtained from all subjects involved in the study.

Data Availability Statement: All the datasets used to train the model presented in this paper were obtained from the Internet.

Conflicts of Interest: The authors declare no conflict of interest.

References

1. Hu, Y.; Modat, M.; Gibson, E.; Li, W.; Ghavami, N.; Bonmati, E.; Wang, G.; Bandula, S.; Moore, C.M.; Mberton, M.; et al. Weakly-supervised convolutional neural networks for multimodal image registration. *Med. Image Anal.* **2018**, *49*, 1–13. [CrossRef]
2. Shan, S.; Yan, W.; Guo, X.; Chang, E.I.; Fan, Y.; Xu, Y. Unsupervised end-to-end learning for deformable medical image registration. *arXiv* **2017**, arXiv:1711.08608.

3. Fan, J.; Cao, X.; Yap, P.T.; Shen, D. BIRNet: Brain image registration using dual-supervised fully convolutional networks. *Med. Image Anal.* **2019**, *54*, 193–206. [CrossRef]
4. Cao, X.; Yang, J.; Zhang, J.; Wang, Q.; Yap, P.T.; Shen, D. Deformable image registration using a cue-aware deep regression network. *IEEE Trans. Biomed. Eng.* **2018**, *65*, 1900–1911. [CrossRef]
5. Wang, C.; Yang, G.; Papanastasiou, G. Unsupervised image registration towards enhancing performance and explainability in cardiac and brain image analysis. *Sensors* **2022**, *22*, 2125. [CrossRef]
6. Balakrishnan, G.; Zhao, A.; Sabuncu, M.R.; Guttag, J.; Dalca, A.V. VoxelMorph: A learning framework for deformable medical image registration. *IEEE Trans. Med. Imaging* **2019**, *38*, 1788–1800. [CrossRef]
7. de Vos, B.D.; Berendsen, F.F.; Viergever, M.A.; Sokooti, H.; Staring, M.; Išgum, I. A deep learning framework for unsupervised affine and deformable image registration. *Med. Image Anal.* **2019**, *52*, 128–143. [CrossRef]
8. Kim, B.; Kim, D.H.; Park, S.H.; Kim, J.; Lee, J.G.; Ye, J.C. CycleMorph: Cycle consistent unsupervised deformable image registration. *Med. Image Anal.* **2021**, *71*, 102036. [CrossRef]
9. Kim, B.; Kim, J.; Lee, J.G.; Kim, D.H.; Park, S.H.; Ye, J.C. Unsupervised deformable image registration using cycle-consistent cnn. In Proceedings of the International Conference on Medical Image Computing and Computer-Assisted Intervention, Shenzhen, China, 13–17 October 2019; pp. 166–174.
10. Jiang, Z.; Yin, F.F.; Ge, Y.; Ren, L. A multi-scale framework with unsupervised joint training of convolutional neural networks for pulmonary deformable image registration. *Phys. Med. Biol.* **2020**, *65*, 015011. [CrossRef]
11. Kong, L.; Yang, T.; Xie, L.; Xu, D.; He, K. Cascade connection-based channel attention network for bidirectional medical image registration. *Vis. Comput.* **2022**, 1–19. [CrossRef]
12. Yang, J.; Wu, Y.; Zhang, D.; Cui, W.; Yue, X.; Du, S.; Zhang, H. LDVoxelMorph: A precise loss function and cascaded architecture for unsupervised diffeomorphic large displacement registration. *Med. Phys.* **2022**, *49*, 2427–2441. [CrossRef]
13. Zhu, F.; Wang, S.; Li, D.; Li, Q. Similarity attention-based CNN for robust 3D medical image registration. *Biomed. Signal Process. Control.* **2023**, *81*, 104403. [CrossRef]
14. Ouyang, X.; Liang, X.; Xie, Y. Preliminary feasibility study of imaging registration between supine and prone breast CT in breast cancer radiotherapy using residual recursive cascaded networks. *IEEE Access* **2020**, *9*, 3315–3325. [CrossRef]
15. Guo, Y.; Wu, X.; Wang, Z.; Pei, X.; Xu, X.G. End-to-end unsupervised cycle-consistent fully convolutional network for 3D pelvic CT-MR deformable registration. *J. Appl. Clin. Med. Phys.* **2020**, *21*, 193–200. [CrossRef]
16. Sideri-Lampretsa, V.; Kaissis, G.; Rueckert, D. Multi-modal unsupervised brain image registration using edge maps. In Proceedings of the 2022 IEEE 19th International Symposium on Biomedical Imaging (ISBI), Kolkata, India, 28–31 March 2022; pp. 1–5.
17. Qian, L.; Zhou, Q.; Cao, X.; Shen, W.; Suo, S.; Ma, S.; Qu, G.; Gong, X.; Yan, Y.; Jiang, L.; et al. A cascade-network framework for integrated registration of liver DCE-MR images. *Comput. Med. Imaging Graph.* **2021**, *89*, 101887. [CrossRef]
18. Golkar, E.; Rabbani, H.; Dehghani, A. Hybrid Registration of Retinal Fluorescein Angiography and Optical Coherence Tomography Images of Patients with Diabetic Retinopathy. *Biomed. Opt. Express* **2021**, *12*, 1707–1724. [CrossRef]
19. Goodfellow, I.; Pouget-Abadie, J.; Mirza, M.; Xu, B.; Warde-Farley, D.; Ozair, S.; Courville, A.; Bengio, Y. Generative adversarial nets. *Adv. Neural Inf. Process. Syst.* **2014**, *27*, 139–144. [CrossRef]
20. Santarossa, M.; Kilic, A.; von der Burchard, C.; Schmarje, L.; Zelenka, C.; Reinhold, S.; Koch, R.; Roider, J. MedRegNet: Unsupervised multimodal retinal-image registration with GANs and ranking loss. In *Medical Imaging 2022: Image Processing*; SPIE: San Diego, CA, USA, 2022; Volume 12032, pp. 321–333.
21. Fan, J.; Cao, X.; Xue, Z.; Yap, P.T.; Shen, D. Adversarial similarity network for evaluating image alignment in deep learning based registration. In Proceedings of the International Conference on Medical Image Computing and Computer-Assisted Intervention, Granada, Spain, 16–20 September 2018; pp. 739–746.
22. Fan, J.; Cao, X.; Wang, Q.; Yap, P.T.; Shen, D. Adversarial learning for mono-or multi-modal registration. *Med. Image Anal.* **2019**, *58*, 101545. [CrossRef]
23. Zheng, Y.; Sui, X.; Jiang, Y.; Che, T.; Zhang, S.; Yang, J.; Li, H. SymReg-GAN: Symmetric image registration with generative adversarial networks. *IEEE Trans. Pattern Anal. Mach. Intell.* **2021**, *44*, 5631–5646. [CrossRef]
24. Yan, P.P.; Xu, S.; Rastinehad, A.R.; Wood, B.J. Adversarial image registration with application for MR and TRUS image fusion. In Proceedings of the International Workshop on Machine Learning in Medical Imaging, Granada, Spain, 16 September 2018; pp. 197–204.
25. Duan, L.; Yuan, G.; Gong, L.; Fu, T.; Yang, X.; Chen, X.; Zheng, J. Adversarial learning for deformable registration of brain MR image using a multi-scale fully convolutional network. *Biomed. Signal Proces. Control* **2019**, *53*, 101562. [CrossRef]
26. Tanner, C.; Ozdemir, F.; Profanter, R.; Vishnevsky, V.; Konukoglu, E.; Goksel, O. Generative adversarial networks for MR-CT deformable image registration. *arXiv* **2018**, arXiv:1807.07349.
27. Hu, Y.; Gibson, E.; Ghavami, N.; Bonmati, E.; Moore, C.M.; Emberton, M.; Vercauteren, T.; Noble, J.A.; Barrat, D.C. Adversarial deformation regularization for training image registration neural networks. In Proceedings of the International Conference on Medical Image Computing and Computer-Assisted Intervention, Granada, Spain, 16–20 September 2018; pp. 774–782.
28. Lei, Y.; Fu, Y.; Wang, T.; Liu, Y.; Patel, P.; Curran, W.J.; Liu, T.; Yang, X. 4D-CT deformable image registration using multiscale unsupervised deep learning. *Phys. Med. Biol.* **2020**, *65*, 085003. [CrossRef] [PubMed]

29. Huang, Y.; Ahmad, S.; Fan, J.; Shen, D.; Yap, P.T. Difficulty-aware hierarchical convolutional neural networks for deformable registration of brain MR images. *Med. Image Anal.* **2021**, *67*, 101817. [CrossRef] [PubMed]
30. Wu, L.; Hu, S.; Liu, C. Exponential-distance weights for reducing grid-like artifacts in patch-based medical image registration. *Sensors* **2021**, *21*, 7112. [CrossRef] [PubMed]
31. Alexander, L.M.; Escalera, J.; Ai, L.; Andreotti, C.; Febre, K.; Mangone, A.; Vega-Potler, N.; Langer, N.; Alexander, A.; Kovacs, M.; et al. An open resource for transdiagnostic research in pediatric mental health and learning disorders. *Sci. Data* **2017**, *4*, 170181. [CrossRef] [PubMed]
32. Craddock, C.; Benhajali, Y.; Chu, C.; Chouinard, F.; Evans, A.; Jakab, A.; Khundrakpam, B.S.; Lewis, J.D.; Li, Q.; Milham, M.; et al. The Neuro Bureau Preprocessing Initiative: Open sharing of preprocessed neuroimaging data and derivatives. *Front. Neuroinform.* **2013**, *7*, 27.
33. Wang, Z.; Bovik, A.C.; Sheikh, H.R.; Simoncelli, E.P. Image quality assessment: From error visibility to structural similarity. *IEEE Trans. Image Process.* **2004**, *13*, 600–612. [CrossRef]
34. Li, M.; Wang, Y.; Zhang, F.; Li, G.; Hu, S.; Wu, L. Deformable medical image registration based on unsupervised generative adversarial network integrating dual attention mechanisms. In Proceedings of the 14th International Congress on Image and Signal Processing, BioMedical Engineering and Informatics (CISP-BMEI), Shanghai, China, 23–25 October 2021; pp. 1–6.

sensors

MDPI

Article

A Study on the Effectiveness of Deep Learning-Based Anomaly Detection Methods for Breast Ultrasonography

Changhee Yun [1], Bomi Eom [1], Sungjun Park [2], Chanho Kim [2], Dohwan Kim [3], Farah Jabeen [2], Won Hwa Kim [4], Hye Jung Kim [4] and Jaeil Kim [2,*]

1 National Information Society Agency, Daegu 41068, Republic of Korea
2 School of Computer Science and Engineering, Kyungpook National University, Daegu 41566, Republic of Korea
3 Department of Artificial Intelligence, Kyungpook National University, Daegu 41566, Republic of Korea
4 Department of Radiology, Kyungpook National University Chilgok Hospital, Kyungpook National University, Daegu 41404, Republic of Korea
* Correspondence: jaeilkim@knu.ac.kr

Abstract: In the medical field, it is delicate to anticipate good performance in using deep learning due to the lack of large-scale training data and class imbalance. In particular, ultrasound, which is a key breast cancer diagnosis method, is delicate to diagnose accurately as the quality and interpretation of images can vary depending on the operator's experience and proficiency. Therefore, computer-aided diagnosis technology can facilitate diagnosis by visualizing abnormal information such as tumors and masses in ultrasound images. In this study, we implemented deep learning-based anomaly detection methods for breast ultrasound images and validated their effectiveness in detecting abnormal regions. Herein, we specifically compared the sliced-Wasserstein autoencoder with two representative unsupervised learning models autoencoder and variational autoencoder. The anomalous region detection performance is estimated with the normal region labels. Our experimental results showed that the sliced-Wasserstein autoencoder model outperformed the anomaly detection performance of others. However, anomaly detection using the reconstruction-based approach may not be effective because of the occurrence of numerous false-positive values. In the following studies, reducing these false positives becomes an important challenge.

Keywords: breast cancer; ultrasonography; deep learning; anomaly detection; autoencoder

Citation: Yun, C.; Eom, B.; Park, S.; Kim, C.; Kim, D.; Jabeen, F.; Kim, W.H.; Kim, H.J.; Kim, J. A Study on the Effectiveness of Deep Learning-Based Anomaly Detection Methods for Breast Ultrasonography. *Sensors* **2023**, *23*, 2864. https://doi.org/10.3390/s23052864

Academic Editors: Euntai Kim, Sangyoun Lee and Kang Ryoung Park

Received: 3 October 2022
Revised: 4 January 2023
Accepted: 12 January 2023
Published: 6 March 2023

1. Introduction

Recently, deep learning (DL), a branch of machine learning, has attracted considerable attention. This is a technology for hierarchically learning numerous data features through a deep artificial neural network(ANN), extracting from simple features of input data to complex features [1]. In addition, DL performs well in analyzing various data types, such as video, voice, and text. Moreover, it can be applied to various areas, such as image classification, object detection, language translation, sentence classification, voice automatic generation and composition, robotics, medical image analysis, and cybersecurity [2].

In the medical field, various medical imaging techniques, such as magnetic resonance imaging (MRI), X-ray, computed tomography, ultrasound, and endoscopy are used for numerous complicated medical imaging analyses because of their improved diagnosis rates and reduced screening time based on the consistency, scalability, and accuracy of DL. However, it is challenging to apply DL models to numerous medical images using various types of medical equipment without additional information from experts. Consequently, a method for self-learning the inherent features from numerous images without additional expert opinion and maximizing discrimination via a minimal amount of expert judgment has been developed recently [3].

Among the above medical imaging techniques, ultrasound is one of the key diagnostic imaging techniques for the physical examination of various organs, such as abdominal organs, breasts, musculoskeletal systems, heart, and blood vessels [3]. Furthermore, ultrasonic waves can be imaged in real-time and used with existing resources without building a separate environment. However, the quality and interpretation of an image may differ depending on the operator [3,4] and the false-positive rate (FPR), which is the probability of judging a disease-free normal region as an anomaly with a high value [5]. In particular, in breast ultrasonography, it is difficult to detect lesions and accurately diagnose them with a false-negative rate of 50% in dense breasts with a large quantity of mammary tissue and a fairly small quantity of fat[5]. To overcome these limitations, DL technology has been employed to effectively extract biometric information or elaborately visualize anomaly information of organs similar to masses and tumors to aid diagnosis.

Therefore, in this study, DL models were applied to breast ultrasound images to learn the image features. Using anomalous data, the results of applying deep learning-based anomaly detection methods for ultrasound images were verified. Thus, DL-based anomalous region detection technology can automatically detect anomalous regions with tumors or masses in ultrasound images. Moreover, we aim to study the effectiveness of this technology in practical applications, e.g., whether it can be used as a computer-aided diagnostic tool to detect anomalous regions more quickly in ultrasound diagnosis and more accurately by visually presenting the anomalous region to the user than those of the other tools.

2. Related Work

2.1. Deep Learning-Based Anomaly Detection

An anomaly is generally defined as the contrary conception of the normal defined in a field or problem. Anomalies can be largely categorized into point, contextual, and collective anomalies [5]. Point anomalies represent irregularities or diversions; individual data can be linked from given data without a particular interpretation and are considered anomalies. Contextual anomalies are also called conditional anomalies; data are judged to be anomalous in certain situations and are identified in consideration of contextual, behavioral, and operational attributes. Collective anomalies may not be anomalies for individual data; however, data related to each other show anomalous characteristics within an entire group and are judged as anomalous.

Anomaly detection means finding an unusual pattern unless the expected behavior in the data is followed, defining a region representing normal behavior, and considering data that do not belong to the specific region as anomalous and finding them [6]. These detection methods have long been applied in various fields, e.g., medicine, transportation, cyber intrusion, telephone or insurance fraud, and industrial control system detection, playing a crucial role as the demand increases and applications become widespread [7].

DL is a type of ANN that resembles human cognitive function as a machine learning technique [8]. This is to achieve flexibility by learning how to express data in an overspread hierarchical structure and ensure excellent performance in learning complex data characteristics such as high-dimensional, temporal, spatial, and graphic data on its purpose of analysis [7]. DL-based anomaly detection applies DL technology to the anomaly detection method. A deep ANN algorithm comprising artificial neurons stacked between the input and output layers is applied to determine whether there is an anomaly.

This method is further classified into supervised, semisupervised, and unsupervised learning according to the learning approach. Besides, this method is utilized to supervise outlier detection according to the presence or absence of label data, which is used for learning data [6].

Unsupervised Deep Anomaly Detection

Supervised and semisupervised deep anomaly detection approaches require securing labels for learning data. Because obtaining labeled data is complex, research is actively being conducted to enable learning without obtaining separate label data, assuming that most data are normal [9]. The objective of unsupervised anomaly detection is to detect previously unseen rare objects or events without prior knowledge about them, meaning it only requires a single labeling process to train a model. Consequently, high accuracy is not achieved because the restoring performance of the original data depends on the degree of compression of input data.

The reconstruction methodology for deep anomaly detection has been implemented for unsupervised-based deep anomaly detection. The authors of [10] assumed that learned traditional structures are well-remodeled and reconstructed; however, abnormal structures were difficult to reconstruct. Specifically, in images, a significant difference was visible between the input data and the anomalous region reconstructed using the data that can be determined using an object. The core model of unsupervised-based deep anomaly detection is an AE[11]. As shown in Figure 1, an AE is a generative unsupervised DL algorithm for reconstructing high-dimensional input data. An AE uses an NN with a narrow bottleneck layer in the middle that contains the latent that compresses features and then decodes data to reconstruct the original input. The encoder maps the input data features to a low-dimensional latent space, and the decoder is trained to restore the low-dimensional features most similar to the input data through reverse processing.

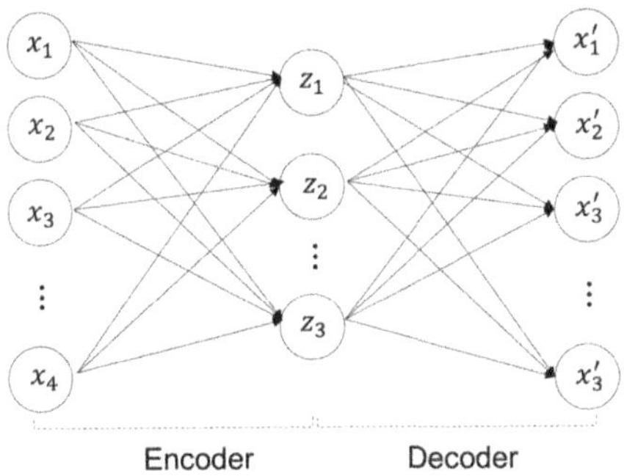

Figure 1. Autoencoder (AE) Architecture.

The encoder maps high-dimensional data into a low-dimensional latent space as shown in Equation (1), and the decoder reconstructs and restores the compressed low-dimensional data as shown in Equation (2) into high-dimensional data [1]. In Equations (1) and (2), the encoder parameters are $\{\mathbf{W}, b\}$ and the decoder parameters are $\{\mathbf{W}', b'\}$. The activation function is α [1].

$$z = encoder(x) = \alpha(\mathbf{W}x + b) \tag{1}$$

$$x' = decoder(z) = \alpha(\mathbf{W}'z + b') \tag{2}$$

As shown in Equation (3), the purpose of the AE model is to minimize the reconstruction errors using the difference between the restored images and the input image that mainly uses mean square error (MSE) and cross-entropy error.

$$L(x, x') = argmin\frac{1}{n}\sum_{i=1}^{n}\|x - x'\|^2 \tag{3}$$

We consider three typical AE models applied to unsupervised-based deep anomaly detection: variational AE (VAE), general adversarial network (GAN), and sliced-Wasserstein AE (SWAE).

The VAE model was proposed by D. Kingma and M. Welling[12] in 2014; it is a generative model that learns the probability distribution of data and generates new data from the learned probability distribution. The structure is shown in Figure 2 and comprises a network structure of an encoder and a decoder, as shown in the AE model. The encoder extracts potential features by abstracting input data, and the decoder restores these potential features to the original data. At this time, the decoder generates data on the premise of a normal distribution with the average (μ) and variance (σ) of the latent features created by the encoder as parameters.

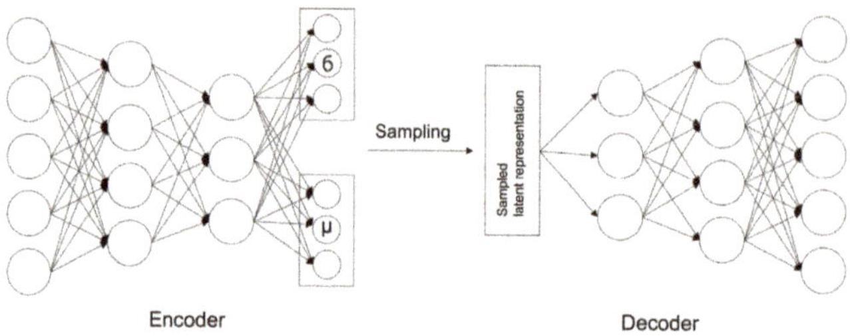

Figure 2. VAE Architecture.

The loss function of the VAE model is shown in Equation (4), which computes the errors in the two optimization tasks. It comprises a sum of reconstruction errors, indicating how well the input image has been restored, and Kullback–Leibler divergence (KLD) errors, indicating how closely the latent variable matched the Gaussian distribution, i.e., the latent space probability distribution.

$$L_i(\theta,\phi) = -E_z \sim q_\theta(z|x_i)[\log_{p_\phi}(x_i|z)] + KL(q_\theta(z|x_i)|p(z)), \tag{4}$$

where x is an input value, and z represents a sampled latent variable. θ is the encoder parameter, ϕ is the decoder parameter; the encoder and decoder can be expressed as $q_\theta(z|x)$ and $p_\phi(x|z)$, respectively.

The SWAE model enables the shaping of the latent space distribution into a samplable probability distribution without the need to train an adversarial network [12]. Similar to the VAE model, the sample data distribution is enforced. However, in the normalization process, there is a difference between the usage of the Wasserstein distance (WD) and not the KLD. Both the KLD and WD measure the distance between probability distributions. However, the KLD is θ when the two probability distributions overlap, as shown in Equation (5), and $+\infty$ when they do not overlap. Thus, learning becomes problematic if the probability distribution is not continuous. However, the WD (EM distance) maintains a constant $|\theta|$ regardless of whether the two probability distributions overlap, as shown in Equation (6). Hence, it is easy to use it in learning because probability distributions that do not converge with other distances or divergences can converge with it.

$$KL(P_\theta||P_0) = KL(P_0||P_\theta) = \begin{cases} +\infty & \text{if } \theta \neq 0, \\ 0 & \text{if } \theta = 0 \end{cases} \tag{5}$$

$$W(P_0, P_\theta) = |\theta| \tag{6}$$

To minimize the sliced-WD (SWD) between the distribution of encoded learning data and the prior distribution, the distance used in sliced-Wasserstein is the same as that in Equation (7). It refers to the lower limit when the expected value of the distance is the smallest in the combined probability distributions $\gamma(x,y)$ combining the two probability densities P_r and P_g.

$$W(P_r, P_g) = \inf_{\gamma \in \Pi(P_r,P_g)} (E_{(x,y)\gamma}[||x-y||^p]^{\frac{1}{p}}) \tag{7}$$

However, because it is impossible to find the minimum in all combinations of probability distributions, we calculate the value for the 1-Lipschitz function $\|f\|_L \leq 1$, which is the upper limit where the average rate of change between any two points does not exceed 1, using the Kantorovich–Rubinstein equation:

$$W(P_r, P_g) = \sup_{\|f\|_L \leq 1} E_{x \sim P_r}[f(x)] - E_{x \sim P_g}[f(x)] \tag{8}$$

The SWD projects high-dimensional probability densities such as $P_r and P_g$ in the distribution of Equation (8) from the WD into one-dimensional (1D) peripheral distributions and compares these peripheral distributions through the WD.

For the two probability distributions R and G, the Wasserstein-2 distance is calculated as Equation (9), and the SWD is approximated to W_2^2 as shown in Equation (10) and optimized as Equation (11).

$$W_2^2(R, G) = \frac{1}{|G|} \min_M \sum_{i=1}^{|G|} \sum_{j=1}^{|R|} M_{i,j} \|R_j - G_i\|_2^2, M : \int, doubly\ stochastic \tag{9}$$

$$\widetilde{W_2^2}(R, G) = \int_{w \in \Omega} W_2^2(R^w, G^w) dw, R^w = w^T R_{i_{i=1}}^{|R|}, G^w = w^T G_{i_{i=1}}^{|G|}, \Omega : unit\ sphere \tag{10}$$

$$\min_\theta \frac{1}{\widehat{|\Omega|}} W_2^2(R^w, G^w(\theta)) dw \tag{11}$$

The 1D peripheral distribution of the high-dimensional probability densities may be defined as follows:

$$R_{PX}(t; \theta) = \int_X PX(x) \delta(t - \theta \bullet x) dx, \forall \theta \in S^{d-1}, \forall t \in R, \tag{12}$$

where S^{d-1} means a unit sphere of d-dimensional, and for fixed $\theta \in S^{d-1}$, $R_{PX}(\bullet; \theta)$ is a 1D slice of PX distribution. That is, $R_{PX}(\bullet; \theta)$ is obtained by integrating a hyperplane PX orthogonal to θ. The following Equation (13) is the sliced-WD defined from the peripheral distribution of Equation (12).

$$SW_c(PX, PY) = \int_{S^{d-1}} W_c(R_{PX}(\bullet; \theta), R_{PY}(\bullet; \theta)) d\theta \tag{13}$$

According to Soheil Kolouri[13], the SWAE is calculated as follows to optimize the model to the minimum SWD value:

$$argmin_{\phi, \psi} W_c(PX, PY) + \lambda SW_c(pz, qz), \tag{14}$$

where ϕ represents an encoder, ψ represents a decoder, PX represents a data distribution, PY represents a distribution of data through an encoder and a decoder; pz is the encoded data distribution, and qz represents a predefined sampling distribution; λ represents the relative importance of the loss function. The model structure is shown in Figure 3.

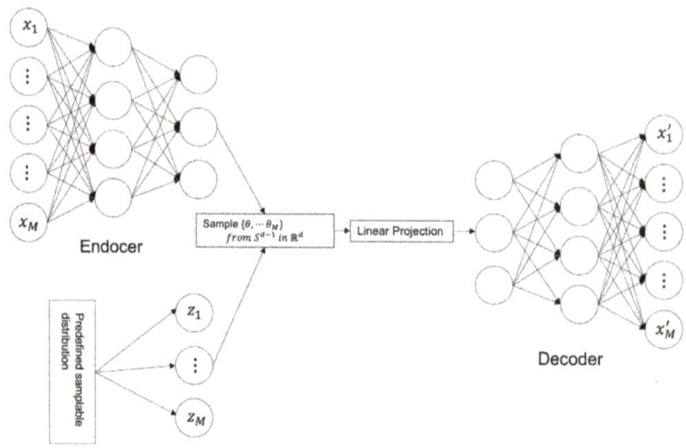

Figure 3. SWAE Architecture.

Most DL-based anomaly detection models learn using one of the aforementioned three learning approaches and determine whether it is abnormal through output values. According to the result, an abnormal score that can be determined based on a specific reference value is defined for a given problem to determine its abnormality.

2.2. Deep Learning-Based Anomaly Detection for Medical Images

In the medical field, DL-based anomaly detection methods have been applied to improve classification performance by learning the characteristics of complex and abstract medical images and spatially transforming lesions to contribute to the characteristics, which is helpful for prevention treatments [14].

Data imbalance due to the variety of data is a common issue in the medical field. It is challenging to collect disease data compared with normal data due to practical limitations in detecting and classifying lesions. Recently, DL methods have been implemented for anomaly detection for various medical images modalities, such as brain MRI, retinal optical coherence tomography (OCT), hand X-ray, chest X-ray, skin disease, and muscle ultrasound [15–26].

Unsupervised anomaly detection based on implicit field learning was recently proposed for high-resolution three-dimensional volume images [27]. The implicit field learning was implemented to learn a mapping of latent features and coordinates to a data point intensity class so that the encoding module preserves as much information as possible in the original image. The implicit field learning approach with AE achieved state-of-the-art performance in anomaly detection for brain cancer MRI. GAN-based architectures have also been employed in various anomaly detection studies. In [28], the GANomaly architecture was applied to detect chronic brain infarcts. In [29], a unified GAN and VAE architecture was proposed to identify chest radiographs with abnormal lesions.

DL methods, especially AE and GAN architectures, learn normal image patterns of human organs in medical images without lesions. In the process of reconstructing a given image, they have the advantage of using the difference between the input image and the reconstructed image to determine the abnormality of the input. However, although various AEs have been proposed, the FPR is still high in pixel-wise anomaly detection. In this study, the effectiveness of the SWAE in anomaly detection, which is known to have better reconstruction quality than other AE variants, is validated through comparative studies with the VAE and conventional AE models.

3. Materials & Methods

3.1. Materials

In this study, we retrospectively collected 1147 breast ultrasound images comprising 947 normal breast ultrasound images and 200 abnormal ultrasound images from Kyungpook National University Hospital in the Republic of Korea. The images consist of 113 benign tumors and 87 malignant tumors. The size of all data is $224 \times 224 \times 3$; 853 normal breast ultrasound data and 94 normal data for model training and verification. Data with anomalous region (region of interest: ROI) label values were used for model evaluation.

The ultrasound images used in the experiment were cut into specific areas. Some normal ultrasound images were used for learning via applying Gaussian filters for noise removal, and gamma correction with 0.5 and 1.5 gamma values, which decide to express the dark areas of ultrasound in more detail. The input data were used by dividing the values of 0–255 pixels into 255 values and converting them into values between 0 and 1.

3.2. Reconstruction-Based Anomaly Detection

The method of detecting an anomalous region applied in this study is to detect an unrestored region by considering it as abnormal using an error image between an input image and a reconstructed image (Figure 4). The learning process uses a modified SWAE model based on AE, a representative generation model of ANNs, and the conventional AE, which obtains latent features for the summit through input. In the evaluation process, anomalous data are input to the learned model, and an anomalous region is detected through the restored results. The difference between the input image and the restored image is calculated to derive an anomaly map, which is an error image. The anomaly map is binary divided based on a specific threshold to detect the anomalous region. This process was applied to the three models to compare and analyze their detection performances and investigate the factors influencing anomalous region detection in breast ultrasound images.

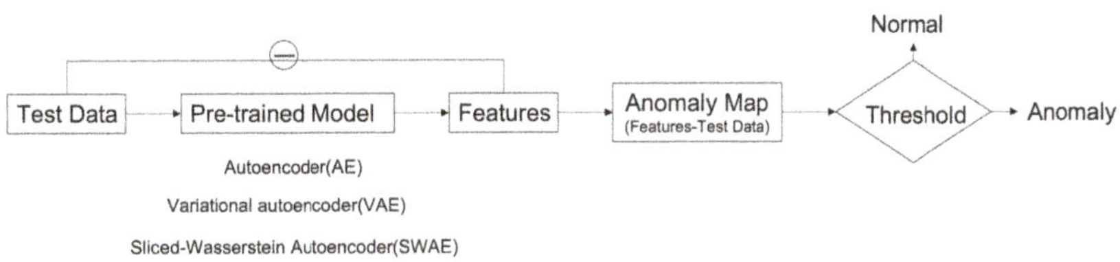

Figure 4. Deep Learning-based Anomalous Region Detection Process.

3.2.1. Hyperparameter Tuning

In this study, the hyperparameters of the implemented models are as shown in Tables 1–3. We tuned the hyperparameters by the grid search method.

Table 1. Hyperparameter setting of the AE model.

Hyper Parameter	Value
Activation Function	LeakyReLU
Output Function	Sigmoid
Loss Function	L1 distance
Optimizer	Adam
Batch Size	16
Epochs	150
Learning Rate	0.0002

Table 2. Hyperparameter setting of the VAE model.

Hyper Parameter	Value
Activation Function	LeakyReLU
Output Function	Sigmoid
Loss Function	Reconstruction Error + KLD
Optimizer	Adam
Batch Size	16
Epochs	150
Learning Rate	0.0002

Table 3. Hyperparameter setting of the SWAE model.

Hyper Parameter	Value
Activation Function	LeakyReLU
Output Function	Sigmoid
Loss Function	Reconstruction Error + SWD
Optimizer	Adam
Batch Size	16
Epochs	150
Learning Rate	0.0002

3.2.2. Model Architecture of Anomaly Detection Model for Breast Ultrasound

The implemented models comprise encoders and decoders with multiple hidden layers. In the learning process, the encoders map normal ultrasound images into low-dimensional spaces to represent them as key features of the latent space; meanwhile, the decoders update and restore weight to some extent according to input. The process for detecting the anomalous region calculates a pixel unit error over the reconstructed, restored image and the input image (Figure 5). The anomaly map detects an anomalous region by binary division based on a specific threshold. It considers the region abnormal if it is larger than the threshold value and normal otherwise.

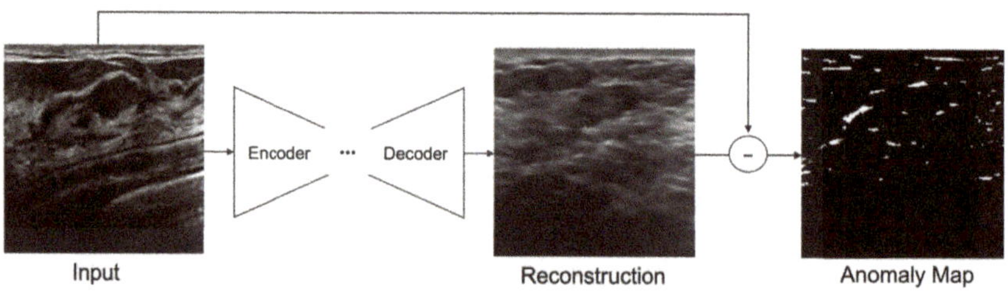

Figure 5. Anomaly detection by pixel difference between an original image and reconstructed image on ultrasonography.

Autoencoder (AE) Model

Figure 6 describes the AE model, comprising different filter sizes and convolutional layers that are added to the encoder and decoder to extract features. Therefore, the batch normalization layer is used to normalize the power value. The LeakyRelu activation function is used with a slight slope to convert the calculated input value into the power value. In this model, input data are converted to values between 0 and 1 through normalization, and a sigmoid function is used as the output layer.

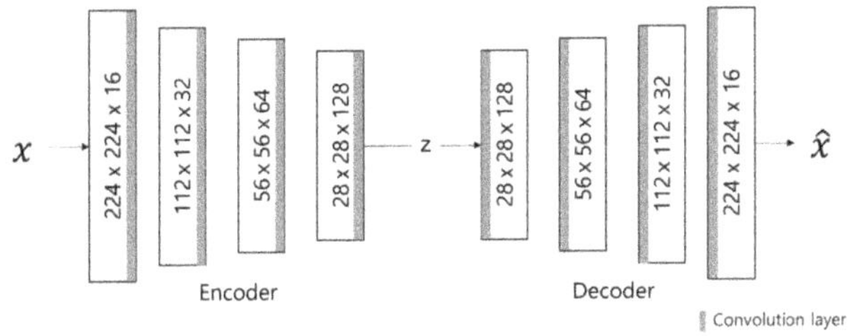

Figure 6. AE model architecture.

The loss value L of the AE model is calculated using the L1 distance loss function to indicate the abnormal score by the difference in pixel values. This is calculated as the sum of the absolute values of the difference between the restored image $\hat{x}$ and the input image x (Equation (15); the smaller the loss value, the better the model performance. The Adam optimizer is used for model optimization. The learning rate is set to the maximum initial value of 0.0002. The cosine annealing method, which can improve accuracy by adjusting the learning rate in a cosine function, is applied.

$$L(x, \hat{x}) = \sum_{i=1}^{n} |x_i - \hat{x}_i| \tag{15}$$

Variational Autoencoder (VAE) Model

The VAE model comprises an encoder and a decoder similar to the AE model. The only difference is the AE model is used to map Gaussian distribution and noise for normalization to the latent space (Figure 7). It is to generate similar data using the latent variable z by allowing the encoder to return the distribution of the latent space instead of a single point. Changing the parameter can be ideal for the probability distribution. In this case, the distribution returned from the encoder is close enough to the standard normal distribution. In this study, we assumed a Gaussian distribution. Because the immediate differential calculation is impossible in the latent variable sampling stage. Thus, the latent variable is converted into $z = \mu + \epsilon\sigma(sample\ \epsilon \sim N(0,1))$ using the reparameterization trick for optimization to enable backpropagation.

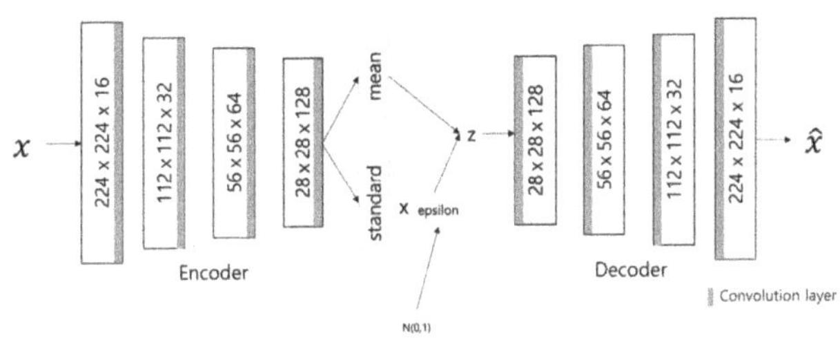

Figure 7. VAE model architecture.

The input data are converted into values between 0 and 1 through normalization, and the output layer of the model uses a sigmoid function. The loss value L for model optimization comprises the sum of reconstruction errors using L1 distances as shown in Equation (16) and the KLD terms for normalization. As in the AE model, the learning rate is set to 0.0002 and adjusted by applying cosine annealing for accurate learning. The parameters are updated using the Adam optimizer for model optimization.

$$
\begin{aligned}
L &= Regularization\ Parameter + Reconstruction\ Error \\
&= D_{KL}(q_{\varnothing}(z|x) \| p_{\theta}(z|x)) + L(\theta, \varnothing, x) \\
&= D_{KL}(N(\mu, \textstyle\sum) \| N(0,1)) + E_{q_{\varnothing}}[\log_{p_{\theta}}(x|z)] \\
&= -\frac{1}{2}\sum_{j=1}^{J}(1 + \log(\sigma_j^2) - \mu_j^2 + \sigma_j^2) + E[\sum_{i=1}^{D}(x_i \log_{y_i} + (1 - x_i) \bullet \log(1 - y_i))]
\end{aligned} \tag{16}
$$

SWAE Model

Similar to the VAE model, the SWAE model is a generative model comprising an encoder and a decoder, which allows the latent space to be formed into a sampling probability distribution. However, the only difference is normalizing reconstruction losses using the SWD between the encoded learning sample distribution and the predefined sampling distribution. Figure 8 shows the SWAE architecture.

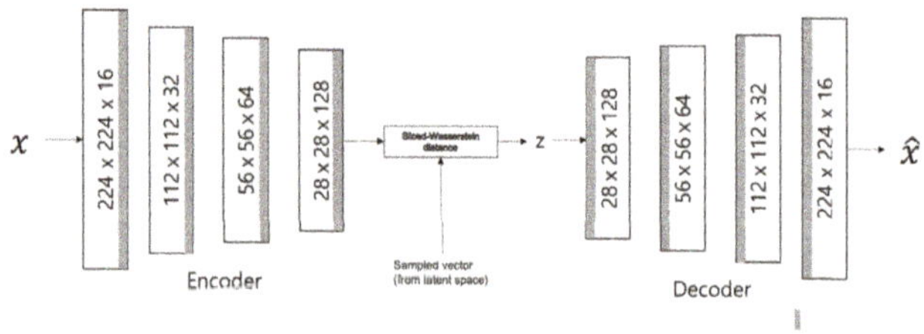

Figure 8. SWAE model architecture.

Ultrasonography data converted to values between 0 and 1 are used as input, and the configuration and output of each model layer are configured the same as those of the AE and VAE models. The loss value L is calculated as the sum of the reconstruction error and the SWD of the 1D projection for normalization (Equation (17)). The maximum value of the learning rate is set to 0.0002, and cosine annealing is applied and adjusted to increase accuracy.

$$
\begin{aligned}
L &= L_{rec} + Sliced - Wasserstein\ distance \\
&= \frac{1}{n}\sum_{i=1}^{n}(x - \hat{x})^2 + SW(P_x, P_{\hat{x}}) \\
&= argmin_{Enc,Dec} W(P_x, P_{\hat{x}}) + \lambda SW(pz, qz)
\end{aligned} \tag{17}
$$

In the loss function calculation, L_{rec} evaluates the error between the input and reconstructed images as a pixel-by-pixel MSE, and the SWE is applied by projecting the difference between the encoded data distribution pz and predefined sampling distribution qzin dimensions.

$$
MSE = \frac{1}{n}\sum_{i=1}^{n}(x_i - \hat{x}_i)^2 \tag{18}
$$

3.2.3. Validation of Anomaly Detection Method for Breast Ultrasonography

Anomalous data are input to the learned model to detect the anomalous region of an ultrasound image, and the output is a difference image between the restored and input images. The anomalous region is detected by a binary division based on a specific threshold. For performance verification, the ROI label data, extracted from a tumor region of the breast ultrasound image, is used. Indicators such as similarity (Dice), sensitivity (true-positive rate (TPR)), and FPR are calculated using overlapping pixel value information in the anomalous region of the label data and the binary-split image obtained from the models. Further, these indicators are employed to compare and analyze the detection results of each model. In addition, factors influencing the anomalous region detection results in an ultrasound image are identified.

Performance Evaluation of Anomaly Detection

In this study, three models were used to detect anomalous regions using the error value between the input and reconstructed images. This should restore the normal ultrasound image input for learning, and the abnormal ultrasound image input for testing should restore the anomalous region close to normal. The role of restoration is essential for successful anomalous region detection by applying a reconstruction-based approach to ultrasound images. Accordingly, the restoration results for each model for normal and abnormal ultrasound images are compared and analyzed using the root MSE (RMSE) values that minimize the error between the input and reconstructed images (Equation (19)).

$$RMSE = \sqrt{\frac{1}{n}\sum_{i=1}^{n}(Reconstruction - Input)^2} \tag{19}$$

Restoration performance by RMSE value-based model can be considered as a model with improved learning when learning with normal data, a high RMSE value when evaluated with anomalous data, and failure to restore results and can be attributed to a well-trained model for anomalous region detection.

In addition, three indicators, Dice, TPR, and FPR, belonging to the overlap-based evaluation index group, were used to evaluate anomaly detection performance. Dice is calculated from Equation (20) using true positive (TP), false positive (FP), false negative (FN), and true negative (TN), which are components of the diffusion matrix. It is an indicator that checks the similarity with the correct answer by directly comparing the division results of the two images. TPR is an indicator of sensitivity, and by predicting the actual anomalous region abnormal, the anomalous region detection results can be confirmed. Moreover, FPR is an indicator of the normal region classified above [30]. Performance is measured based on the indicator values for each model derived by inputting anomalous data into the model, which are evaluation data. Indicator values are also compared and analyzed to verify whether the reconstruction-based approach of unsupervised learning is suitable for anomaly detection in ultrasound images.

$$Dice = \frac{2TP}{2TP + FN + FP},\ TPR = \frac{TP}{TP + FN},\ FPR = \frac{FP}{FP + TN} \tag{20}$$

Analysis of Factor Influencing Anomalous Region Detection

To measure the anomaly detection performance of the reconstruction-based approach, we analyzed the effects of threshold setting and model-specific latent variables on reconstruction [17] and tumor and mass size of ultrasound images on anomaly detection.

As for the threshold for determining the anomalous region, the difference between the mean values of the individual anomaly maps and the overall anomaly map of the validation data is calculated using 94 normal data points for validation, as shown in Algorithm 1, and the maximum value calculated by applying the Relu function is set as the reference threshold [31]. However, in this study, the Relu function applied to obtain the threshold value treats the negative value of the vector as 0. Hence, the threshold value becomes relatively large, resulting in a region that treats the abnormality as normal. Therefore, by supplementing this, three additional thresholds, 0.1, 0.2, and 0.3, which can more accurately detect anomalous regions in ultrasound images, were applied and compared.

Algorithm 1 Find threshold for anomaly detection

Input: anomaly map of validation dataset
Output: threshold

$Max\_relu \leftarrow 0$
calculate an average of anomaly map
for v **in** validation set **do**
  $relu\_th \leftarrow ReLU(v - \text{average})$
  if $Max\_relu < max(relu\_th)$ **then**
    $Max\_relu \leftarrow max(relu\_th)$
  end if
end for
  return Max_relu

Other influencing factors include the latent variable dimension of the latent space. The results are analyzed by limiting the structure of latent features through whether the encoder that generates latent variables for each model reduces dimensions. A reconstructed image is derived by varying the latent space dimensions of the three models. Anomalous region detection was performed by setting the latent space to a low dimension. In addition, the encoder and anomalous region detection results were confirmed by setting the latent space to a high dimension. Furthermore, changes in indicators according to the ROI sizes, such as masses and tumors of abnormal images used in the evaluation process, were examined. We also confirmed that ROI affects anomalous region detection.

4. Experimental Results and Analysis

4.1. Experimental Overview and Environment

In our experiment, AE, VAE, and SWAE models were implemented by applying the reconstruction-based approach of unsupervised learning. The detection performance of each model was measured. In addition, the effect of anomaly detection application in ultrasound was confirmed by comparison based on the performance evaluation values for each model.

The experimental environment used is the programming language Python 3.6.9 version, DL framework Pytorch 1.6 version, CUDA 10.0 version for GPU operation, and cuDNN 7.6.5 version library. A model's learning, evaluation, and outcome analysis are performed in an environment using Intel(R) Core(TM) i7-1065G7 CPU @ 1.30 GHz 1.50 GHz and GeForce GTX Titan Xp 440.100 versions.

4.2. Evaluation of Anomalous Region Detection in Ultrasonography

4.2.1. Reconstruction Performance by Model

The reconstruction performances of the models are presented in Table 4 by comparing the average RMSE of the verification process using normal ultrasound images and the average RMSE of abnormal ultrasound images. In the image reconstruction process by an AE, the smaller the RMSE value, the better the reconstruction performance. However, in a test process for abnormal ultrasonic images, a larger RMSE value indicates that the input image is not well-reconstructed. This means that the input image contains abnormal features that are difficult to reconstruct by the model. The pixel-wise differences between the input and reconstructed images would be suitable for identifying an anomalous region. In the comparison experiment for the three models, the RMSE value increases in the order of SWAE, VAE, and AE, and the anomalous region detection performance is found to be the best in the SWAE model. Examples of the image reconstruction results for each model are shown in Figure 9 below.

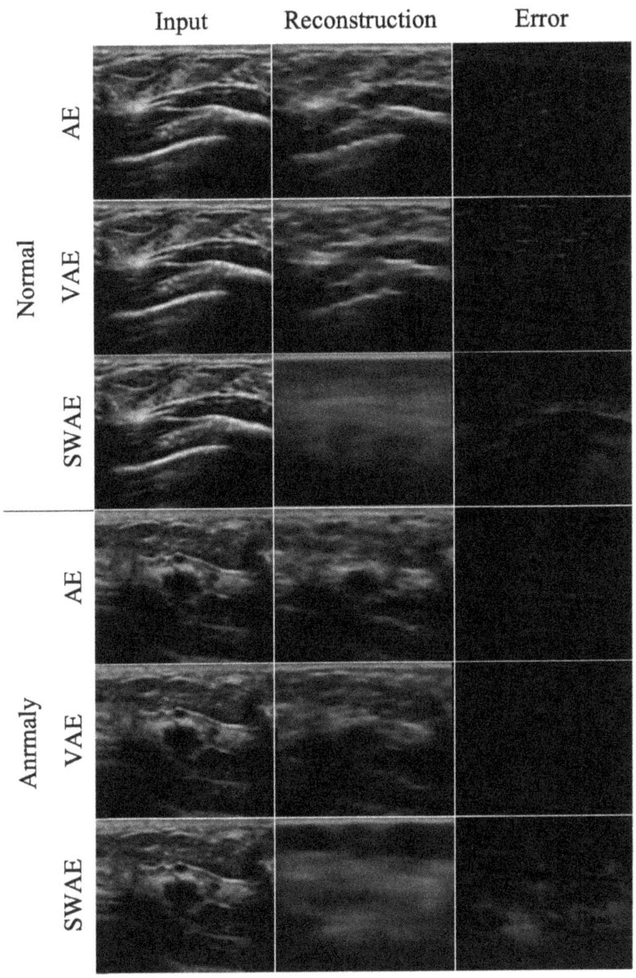

Figure 9. Reconstructed images by model.

Table 4. Reconstruction performances of models.

Model	Normal Ultrasound RMSE	Abnormal Ultrasound RMSE
AE	0.077	0.072
VAE	0.089	0.084
SWAE	0.139	0.139

We confirmed that the AE model with the smallest RMSE value yielded restoration as the input. For the VAE model, although the normalization value was considered in learning, the results were similar to those of the AE model. This shows that it is difficult to find an anomalous region in an error image by restoring the anomalous region similar to the input as a result of the test by inputting an abnormal image. Conversely, the reconstructed images of the SWAE model, which showed the highest RMSE value in the evaluation process, did not restore abnormal features. The anomalous region could be verified in the different maps more accurately.

4.2.2. Anomalous Region Detection

To evaluate the anomaly detection performance of the three models, we used three indicators, Dice, TPR, and FPR, as described in Section N. The results of detecting anomalous regions by the three models based on an arbitrary threshold of 0.2 are shown in Table 5.

Table 5. Indicators of anomalous region detection results of models.

Model	Similarity (Dice)	True Positive Rate (TPR)	False Positive Rate (FPR)
AE	0.000017	0.001995	0.001494
VAE	0.00005	0.005804	0.001616
SWAE	0.001252	0.312863	0.043162

Similarity generally showed low values in the three models. However, they were the lowest in the AE model, and all indicator values showed the highest results in the SWAE model. The SWAE model showed relatively high sensitivity and good performance, but the FPR value was relatively low. Figure 10 shows each model's anomalous region detection performance.

The AE model, which has the smallest similarity, sensitivity, and performance values, restored an input very similarly. It can be seen that there is almost no region indicating an abnormality in the case of binary division based on a specific threshold of 0.2. The VAE model restored the input image similar to the AE model, and both the error and binary-split images, and the indicator values, showed similar results to the AE model. The SWAE model shows the most significant result in all three indicator values. The anomalous region is most clearly detected and displayed in the error and binary-split images.

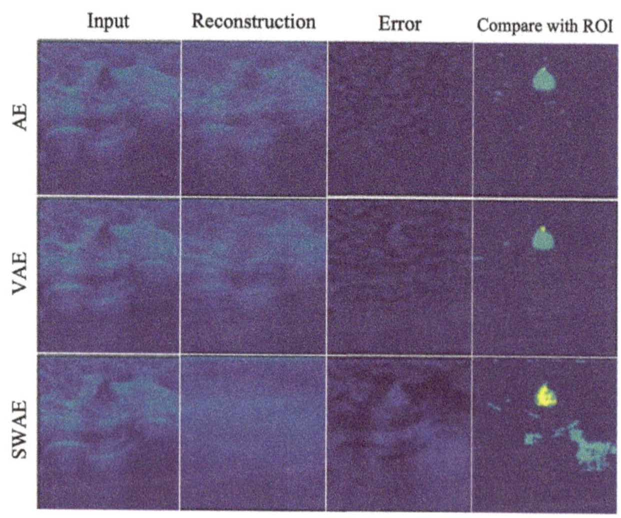

Figure 10. Reconstructed result images by models.

4.3. Analysis of Factor Influencing Anomalous Region Detection in Ultrasonography

4.3.1. Threshold

As a result of detecting anomalous regions of the models, the reconstruction-based approach is considerably affected by the threshold value. Figure 11 shows the change in indicators for each arbitrary threshold.

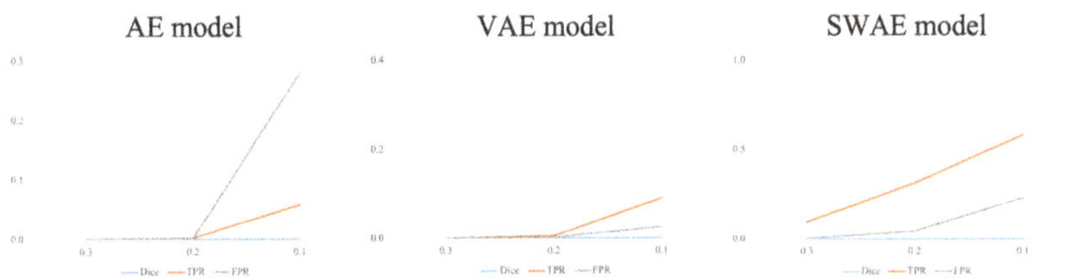

Figure 11. Changes in indicators according to the threshold for each model.

In all three models, the smaller the threshold, the larger the region, which is considered abnormal, indicating an increase in the TPR and FPR values. In the AE model, the FPR value increases significantly more than the TPR value because the FPR value, which considers typical abnormalities as normal, is larger than the TPR value, which considers abnormalities as abnormalities. It is difficult to say that the anomalous region was well-detected. The VAE and SWAE models show that the TPR value increases more than the FPR value as the threshold value decreases. In particular, for the SWAE model, the TPR value increases the most, indicating that the anomalous region was well-detected by considering the actual abnormality as abnormal. As shown in Figure 11, thresholds play an important role in anomalous region detection, thus, we did not use arbitrary thresholds. We applied the method using the validation data mentioned in Algorithm 1 of the Research Methodology to derive thresholds. The derived thresholds are shown in Table 6.

The method applied in Figure 6 uses the Relu function. The application method shows a relatively significant threshold value because the negative number is treated as 0 in the vector value of the error image. A significant threshold may occur in a region where the abnormality is treated as normal during the binary division of an error image. Figure 12 demonstrates the anomalous region detection results. Figure 12 shows that most results compared with the ROI are considered normal in the error image, resulting in the anomalous region not occurring and no overlapping area with the ROI occurring, which further indicates that it is difficult to detect the anomalous region.

Table 6. Comparison of thresholds by models.

Threshold	AE Model	VAE Model	SWAE Model
Applying Relu	0.52675	0.559735	0.497874

When the average value of the verified data error image was used without applying the Relu function to calculate the threshold value for detecting the anomalous region of the breast ultrasonography, a threshold value, somewhat lower than that of applying the Relu function, was derived, indicating relatively good results for anomalous region detection. However, for small thresholds, the FPR value increases as the increase of FPs, indicating the limitation of anomalous detection.

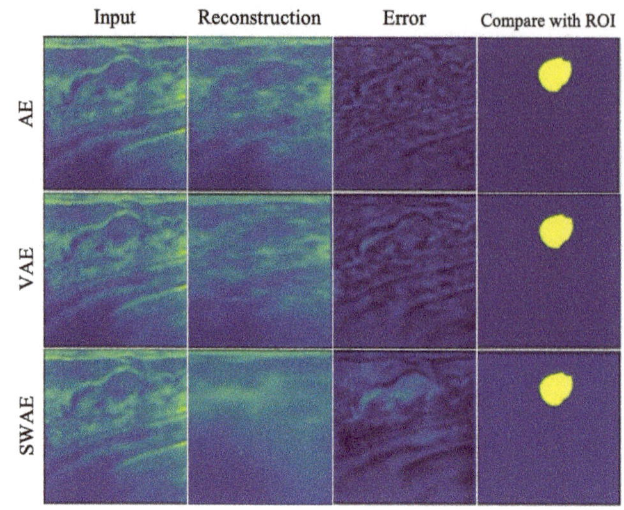

Figure 12. Anomalous region detection results with respect to threshold with applying Relu function.

4.3.2. Size of Tumor

The number of pixels in the ROI image representing the tumor was calculated to confirm the effect of tumor size on anomalous region detection. The tumor size was divided into ranges according to the number of pixels, and the averages of the Dice scores and TPR values in the corresponding range were calculated to compare the performance of each model. Figure 13 shows the change in indicators according to tumor size at a corresponding threshold for each model.

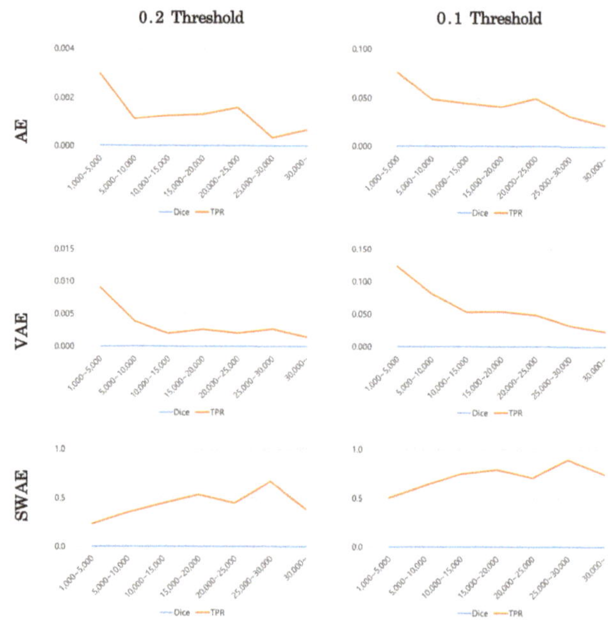

Figure 13. Changes in indicators according to tumor size by model.

Dice scores were small in all models, making it difficult to compare, but TPR values showed similar patterns for each model. The error image is binary divided based on a specific threshold, hence, the TPR value can be calculated somewhat larger at a smaller threshold. However, the TPR value according to tumor size showed a similar pattern depending on the model's threshold value. In the AE and VAE models, the TPR value decreased as the tumor size increased. Meanwhile, in the SWAE model, the TPR value increased as the tumor size increased to a specific range; in general, the larger the tumor size, the larger the TPR value.

5. Conclusions

In this study, we have used the reconstruction-based approach of unsupervised learning to confirm the effect of using deep learning-based technology to detect anomalies in breast ultrasound images. Three models–AE, VAE, and SWAE–were used to compare the results of anomalous region detection based on calculated specific threshold similarity (Dice), sensitivity (TPR), and FPR indicators. The performance results of restoring ultrasound images were good in the order of AE, VAE, and SWAE; however, abnormal images could not be restored in the anomalous region detection.

In addition, we confirmed that the SWAE model, which represents a more significant TPR value than the FPR value, exhibited relatively good performance in anomalous region detection. Meanwhile, the VAE model, which performed similar learning as the SWAE model by adding normalization values, failed to enforce the distribution of sample data, a characteristic of the model, resulting in similar results to the AE model.

The anomalous region detection technology applied in this study has a threshold-dependent limitation because based on a specific threshold, it determines whether an error image is abnormal by dividing it. This resulted in a higher TPR value with a decreasing threshold value. However, the FPR value that could detect non-tumor regions as tumors also increased and that was not a good result.

Changes in the Dice and TPR indicators according to the tumor size were confirmed to check the effect of tumor size on detecting anomalous regions. Although the indicator values might differ due to the difference in anomalous regions according to the threshold value, similar patterns were observed for each model. In the AE and VAE models, the larger the tumor size, the fewer the detected anomalous regions. This is observed as a result of a restoration similar to the anomalous region, resulting in a smaller region considered abnormal. Furthermore, because the reconstruction in the SWAE model was restored to map the anomalous region to normal, the overall anomalous region was detected. The larger the tumor size, the more overlapping parts occurred, and the higher the TPR value was.

In this study, we detected anomalous regions such as tumors and masses in ultrasound images and checked whether they could be visually presented. The results of anomalous region detection using the SWAE model showed the best performance in ultrasound images among the three AE-based models.

Further research is required to reduce learning through securing various samples, FPR values, and increasing TPR values to detect anomalous regions with improved performance on breast ultrasound images with high variance characteristics. Moreover, because the threshold setting considerably influences the anomalous region detection results, visual presentation of anomalous regions for ultrasound images will be possible if additional methods are applied to determine anomalies without a separate threshold setting.

Author Contributions: Conceptualization, B.E., S.P., C.Y., and J.K.; data curation, B.E., S.P., and J.K.; formal analysis, B.E., S.P., and J.K.; investigation, B.E., S.P., and C.Y.; methodology, B.E. and S.P.; project administration, B.E., S.P., and J.K.; supervision, J.K.; validation, B.E., S.P., and J.K.; visualization, B.E., S.P., and J.K.; writing—original draft, B.E.; writing—review and editing, C.Y., D.K., C.K., J.K., and F.J.; Discussion and Editing, W.H.K. and H.J.K.; and CREDENCE and ICONIC investigators. All authors have read and agreed to the published version of the manuscript.

Funding: This research was supported by Kyungpook National University Research Fund, 2018.

Institutional Review Board Statement: This manuscript contains data from an IRB approved study (Kyungpook National University Chilgok Hospital). The study received ethical approval from the local ethics committee. All data reported here were anonymized and stored in line with data privacy regulations in South Korea.

Informed Consent Statement: Not applicable.

Data Availability Statement: Not applicable.

Acknowledgments: The authors would like to acknowledge the participants in the study.

Conflicts of Interest: The authors declare no conflict of interest. The funders had no role in the design of the study; in the collection, analyses, or interpretation of data; in the writing of the manuscript; or in the decision to publish the results.

References

1. Edwards, B.I.; Khougali, N.H.O.; Cheok, A.D. Trends in Computer-Aided Diagnosis Using Deep Learning Techniques: A Review of Recent Studies on Algorithm Development. *Preprints* **2017**, 2017100117 . [CrossRef]
2. Alom, M.Z.; Taha, T.M.; Yakopcic, C.; Westberg, S.; Sidike, P.; Nasrin, M.S.; Hasan, M.; Van Essen, B.C.; Awwal, A.A.; Asari, V.K. A state-of-the-art survey on deep learning theory and architectures. *Electronics* **2019**, *8*, 292. [CrossRef]
3. Shin, S.J.; Jeong, B.J. Principle and comprehension of ultrasound imaging. *J. Korean Orthop. Assoc.* **2013**, *48*, 325–333. [CrossRef]
4. Berg, W.A.; Blume, J.D.; Cormack, J.B.; Mendelson, E.B. Operator dependence of physician-performed whole-breast US: Lesion detection and characterization. *Radiology* **2006**, *241*, 355–365. [CrossRef] [PubMed]
5. Boyd, N.F.; Martin, L.J.; Yaffe, M.J.; Minkin, S. Mammographic density and breast cancer risk: Current understanding and future prospects. *Breast Cancer Res.* **2011**, *13*, 223. [CrossRef] [PubMed]
6. Chandola, V.; Banerjee, A.; Kumar, V. Anomaly detection: A survey. *ACM Comput. Surv. CSUR* **2009**, *41*, 1–58. [CrossRef]
7. Pang, G.; Shen, C.; Cao, L.; Hengel, A.V.D. Deep learning for anomaly detection: A review. *ACM Comput. Surv. CSUR* **2021**, *54*, 1–38. [CrossRef]
8. Lee, J.G.; Jun, S.; Cho, Y.W.; Lee, H.; Kim, G.B.; Seo, J.B.; Kim, N. Deep learning in medical imaging: General overview. *Korean J. Radiol.* **2017**, *18*, 570–584. [CrossRef] [PubMed]
9. Ruff, L.; Vandermeulen, R.; Goernitz, N.; Deecke, L.; Siddiqui, S.A.; Binder, A.; Müller, E.; Kloft, M. Deep one-class classification. In Proceedings of the International Conference on Machine Learning. PMLR, Stockholm, Sweden, 10–15 July 2018; pp. 4393–4402.
10. Zong, B.; Song, Q.; Min, M.R.; Cheng, W.; Lumezanu, C.; Cho, D.; Chen, H. Deep autoencoding Gaussian mixture model for unsupervised anomaly detection. In Proceedings of the International Conference on Learning Representations, Vancouver, BC, Canada, 30 April–3 May 2018.
11. Chalapathy, R.; Chawla, S. Deep learning for anomaly detection: A survey. *arXiv* **2019**, arXiv:1901.03407.
12. Kingma, D.P.; Welling, M. Auto-encoding variational bayes. *arXiv* **2013**, arXiv:1312.6114.
13. Kolouri, S.; Pope, P.E.; Martin, C.E.; Rohde, G.K. Sliced Wasserstein auto-encoders. In Proceedings of the International Conference on Learning Representations, Vancouver, BC, Canada, 30 April–3 May 2018.
14. Liao, S.; Gao, Y.; Oto, A.; Shen, D. Representation learning: A unified deep learning framework for automatic prostate MR segmentation. In Proceedings of the International Conference on Medical Image Computing and Computer-Assisted Intervention, Nagoya, Japan, 22–26 September 2013; Springer: Berlin/Heidelberg, Germany, 2013; pp. 254–261.
15. Vasilev, A.; Golkov, V.; Meissner, M.; Lipp, I.; Sgarlata, E.; Tomassini, V.; Jones, D.K.; Cremers, D. q-Space novelty detection with variational autoencoders. In *Computational Diffusion MRI*; Springer: Berlin/Heidelberg, Germany, 2020; pp. 113–124.
16. Chen, X.; Konukoglu, E. Unsupervised detection of lesions in brain MRI using constrained adversarial auto-encoders. *arXiv* **2018**, arXiv:1806.04972.
17. Baur, C.; Wiestler, B.; Albarqouni, S.; Navab, N. Deep autoencoding models for unsupervised anomaly segmentation in brain MR images. In Proceedings of the International MICCAI Brainlesion Workshop, Granada, Spain, 16 September 2018; Springer: Berlin/Heidelberg, Germany, 2018; pp. 161–169.
18. Vu, H.S.; Ueta, D.; Hashimoto, K.; Maeno, K.; Pranata, S.; Shen, S.M. Anomaly detection with adversarial dual autoencoders. *arXiv* **2019**, arXiv:1902.06924.
19. Schlegl, T.; Seeböck, P.; Waldstein, S.M.; Schmidt-Erfurth, U.; Langs, G. Unsupervised anomaly detection with generative adversarial networks to guide marker discovery. In Proceedings of the International Conference on Information Processing in Medical Imaging, Boone, NC, USA, 25–30 June 2017; Springer: Berlin/Heidelberg, Germany, 2017; pp. 146–157.
20. Seeböck, P.; Waldstein, S.M.; Klimscha, S.; Bogunovic, H.; Schlegl, T.; Gerendas, B.S.; Donner, R.; Schmidt-Erfurth, U.; Langs, G. Unsupervised identification of disease marker candidates in retinal OCT imaging data. *IEEE Trans. Med Imaging* **2018**, *38*, 1037–1047. [CrossRef] [PubMed]
21. Seeböck, P.; Orlando, J.I.; Schlegl, T.; Waldstein, S.M.; Bogunović, H.; Klimscha, S.; Langs, G.; Schmidt-Erfurth, U. Exploiting epistemic uncertainty of anatomy segmentation for anomaly detection in retinal OCT. *IEEE Trans. Med. Imaging* **2019**, *39*, 87–98. [CrossRef] [PubMed]

22. Zhou, K.; Gao, S.; Cheng, J.; Gu, Z.; Fu, H.; Tu, Z.; Yang, J.; Zhao, Y.; Liu, J. Sparse-gan: Sparsity-constrained generative adversarial network for anomaly detection in retinal oct image. In Proceedings of the 2020 IEEE 17th International Symposium on Biomedical Imaging (ISBI), Iowa City, IA, USA, 3–7 April 2020; pp. 1227–1231.
23. Davletshina, D.; Melnychuk, V.; Tran, V.; Singla, H.; Berrendorf, M.; Faerman, E.; Fromm, M.; Schubert, M. Unsupervised anomaly detection for X-ray images. *arXiv* **2020**, arXiv:2001.10883.
24. Tataru, C.; Yi, D.; Shenoyas, A.; Ma, A. Deep Learning for abnormality detection in Chest X-Ray images. In Proceedings of the IEEE Conference on Deep Learning, Cancun, Mexico, 18–21 December 2017.
25. Lu, Y.; Xu, P. Anomaly detection for skin disease images using variational autoencoder. *arXiv* **2018**, arXiv:1807.01349.
26. Burlina, P.; Joshi, N.; Billings, S.; Wang, I.J.; Albayda, J. Unsupervised deep novelty detection: Application to muscle ultrasound and myositis screening. In Proceedings of the 2019 IEEE 16th International Symposium on Biomedical Imaging (ISBI 2019), Venice, Italy, 8–11 April 2019; pp. 1910–1914.
27. Naval Marimont, S.; Tarroni, G. Implicit field learning for unsupervised anomaly detection in medical images. In Proceedings of the International Conference on Medical Image Computing and Computer-Assisted Intervention, Strasbourg, France, 27 September–1 October 2021; Springer: Berlin/Heidelberg, Germany, 2021; pp. 189–198.
28. van Hespen, K.M.; Zwanenburg, J.J.; Dankbaar, J.W.; Geerlings, M.I.; Hendrikse, J.; Kuijf, H.J. An anomaly detection approach to identify chronic brain infarcts on MRI. *Sci. Rep.* **2021**, *11*, 7714. [CrossRef] [PubMed]
29. Nakao, T.; Hanaoka, S.; Nomura, Y.; Murata, M.; Takenaga, T.; Miki, S.; Watadani, T.; Yoshikawa, T.; Hayashi, N.; Abe, O. Unsupervised deep anomaly detection in chest radiographs. *J. Digit. Imaging* **2021**, *34*, 418–427. [CrossRef] [PubMed]
30. Kim, J.; Kim, J. Review of evaluation metrics for 3D medical image segmentation. *J. Korean Soc. Imaging Infor. Med.* **2017**, *23*, 14–20.
31. Jang, J. Deep Learning Algorithms for Visual Inspection. Ph.D. Thesis, Seoul National University Graduate School, Seoul, Republic of Korea, 2020.

sensors

MDPI

Article

Facial Expression Recognition Robust to Occlusion and to Intra-Similarity Problem using Relevant Subsampling

Jieun Kim and Deokwoo Lee *

Department of Computer Engineering, Keimyung University, Daegu 42601, Republic of Korea
* Correspondence: dwoolee@kmu.ac.kr; Tel.: +82-53-580-5268

Abstract: This paper proposes facial expression recognition (FER) with the wild data set. In particular, this paper chiefly deals with two issues, occlusion and intra-similarity problems. The attention mechanism enables one to use the most relevant areas of facial images for specific expressions, and the triplet loss function solves the intra-similarity problem that sometimes fails to aggregate the same expression from different faces and vice versa. The proposed approach for the FER is robust to occlusion, and it uses a spatial transformer network (STN) with an attention mechanism to utilize specific facial region that dominantly contributes (or that is the most relevant) to particular facial expressions, e.g., anger, contempt, disgust, fear, joy, sadness, and surprise. In addition, the STN model is connected to the triplet loss function to improve the recognition rate which outperforms the existing approaches that employ cross-entropy or other approaches using only deep neural networks or classical methods. The triplet loss module alleviates limitations of the intra-similarity problem, leading to further improvement of the classification. Experimental results are provided to substantiate the proposed approach for FER, and the result outperforms the recognition rate in more practical cases, e.g., occlusion. The quantitative result provides FER results with more than 2.09% higher accuracy compared to the existing FER results in CK+ data sets and 0.48% higher than the accuracy of the results with the modified ResNet model in the FER2013 data set.

Keywords: facial expression recognition; spatial transformation network; attention mechanism; triplet loss function; intra-similarity problem

Citation: Kim, J.; Lee, D. Facial Expression Recognition Robust to Occlusion and to Intra-Similarity Problem using Relevant Subsampling. *Sensors* **2023**, *23*, 2619. https://doi.org/10.3390/s23052619

Academic Editors: Euntai Kim, Kang Ryoung Park and Sangyoun Lee

Received: 4 February 2023
Revised: 22 February 2023
Accepted: 25 February 2023
Published: 27 February 2023

1. Introduction

Recognition problems have always been issues in computer vision and pattern recognition. In particular, face or facial expression recognition is considered the most widely explored topic in research and industrial fields. Computer vision, pattern recognition, and imaging-related technologies have achieved impressive performance from quantitative and qualitative perspectives in recent years with the appearance of end-to-end learning frameworks, such as deep neural network models. Among numerous practical applications of computer vision and pattern recognition, image-based, automatic, and intelligent facial expressions are considered one of the most popular topics because facial expression conveys emotional states and can play a key role in detecting, analyzing, and predicting emotional or behavioral states. In facial expression recognition (FER), researchers have usually dealt with discrete facial expressions, such as happiness, surprise, neutral, sadness, fear, disgust, and anger [1]. Thus, FER aims to achieve accurate classification among different facial expressions, i.e., maximize inter-class distance and minimize intra-class distance.

Over a few decades, numerous approaches have been proposed, and they are categorized into two groups, conventional ones, and deep-learning-based ones. Similar to a field of object recognition, conventional FER is usually composed of three major steps: (1) preprocessing of an image containing a facial image followed by detection of the face region, (2) extracting features of a face, and (3) classification and recognition of expressions. From a technological perspective, FER is similar to face recognition (FR) [2], but FER is different from FR in that FER chiefly deals with the seven target expressions mentioned above. Moreover, facial expressions play a more important role in human communication or other interactions between human–machine and human–human. FR usually plays a key role in human identification or authentication rather than interaction activities.

1.1. Traditional Methods

Pre-processing in FER requires reliable quality of image data so that feature extraction and face detection can be accurately achieved. Noise reduction (or removal) is carried out before detecting the region of interest, e.g., face detection. Various types of filters usually categorize in low pass filters, such as Gaussian filter, Laplacian of Gaussian (LOG) filter or bilateral filter. Histogram data are sometimes utilized to enhance image quality, e.g., histogram stretching, histogram equalization, etc. If a facial image contains illumination, varying pose, or occlusion, more complex preprocessing techniques are required [3]. Furthermore, a face image can be acquired using various types of sensors or using a combination of multiple sensors, e.g., fusion of RGB and IR (infrared red) sensors, leading to increasing complexity of algorithms. In the course of recognition, it can successfully begin with accurate detection of the region of interest (ROI). In FER, accurate detection of the face region needs to be carried out before performing expression recognition. Human face detection has also been one of the most important processes in face recognition, expression recognition, gesture recognition, etc.

In conventional approaches of FER or FR (without using deep learning), tremendous work for face detection was proposed [4], and it can be categorized into feature-based and image-based approaches. The former includes active shape model (ASM), low-level analysis (color, motion, or edge-based analysis), and feature analysis (Viola Jones detector, AdaBoost, local binary pattern, Gabor feature-based method, constellation method, etc.). The latter includes more recent approaches that use training and test data sets to perform the matching procedure for the detection (neural network, principal component analysis (PCA), and support vector machine (SVM)).

The image-based approach also contains a sub-space-based method and statistical approach (PCA and SBVM are also included in this category). In feature-based face detection, accurate feature extraction (invariant feature points) is desired, while the image-based approach achieves accuracy and computational efficiency by performing dimensionality reduction. Once the region of a face is detected, feature extraction is carried out. Accurate feature extraction is crucial for diverse applications of image processing and computer vision. Almost all imaging and vision technologies require highly accurate feature extraction results. Feature extraction has been one of the most significant contributions to FER, and there have been extensive research activities to propose accurate feature extraction algorithms. In FER or FR, feature extraction and face detection are very closely related and highly correlated, and some of the algorithms are overlapped. In addition, extracted landmarks are important in many facial tasks [5,6]. In feature extraction of a facial image, applying proper spatial filters to a facial image is a very basic and simple approach. The Gabor filter, local binary pattern (LBP), scale-invariant feature transform (SIFT), speed-up robust features (SURF), and histograms of oriented gradients (HOG) are the most popularly used ones. Encoding based on a code-book is another approach for feature extraction, composed of a training phase and an encoding phase. K-means algorithm, Gaussian mixture model (GMM), and Fisher Vector (FV) are encoding-based approaches. Spatial pooling and holistic encoding also play roles in feature extraction.

Classification is the final stage for recognition. In the recognition, inter-class distance is to be maximized, while intra-class distance is to be minimized. Numerous conventional approaches have been used for recognition recently, e.g., Hausdorff distance (HD), Euclidean distance (ED), SVM, PCA, hidden Markov model (HMM), hidden conditional random fields (HCRF), etc. [7–11].

1.2. Deep-Learning-Based Methods

Although tremendous efforts have been made to improve performance of traditional FER from qualitative and quantitative perspectives, it still lacks recognition accuracy when used in an uncontrolled experimental environment, with images that belong to a wild setting, or with unrefined input images. Similar to other image processing, computer vision, and pattern recognition problems, recent FER has shown remarkable improvements by employing deep learning models [12]. Deep-learning-based FER uses a deep neural network that has various types of structures each of which has its strengths. Proper selection of the model can significantly improve the performance of face detection, feature extraction, and classification for the recognition. A deep neural network (DNN), with a sufficiently large amount of data, provides an end-to-end framework for FER tasks. The recent state-of-the-art approaches have verified the advantages in the fields of visual object recognition, pose estimation, depth estimation and others [13]. Deep-learning-based FER aims at classifying facial expressions using a single image or sequence of images, and the neural network structures learn characteristics or information contained in image data sets. Even if not under controlled experimental environments, deep-learning-based FER provides accurate and reliable recognition results. In other words, in contrast to the traditional methods, deep-learning-based FER shows less dependence on data sets. Moreover, deep-learning-based FER, in contrast to the traditional methods, does not have to consider three major steps (face detection or localization, feature extraction, and classification) separately because the DNN model has the capability to learn a sufficient amount of information to classify seven facial expressions in an end-to-end manner. Among several DNN models, the convolutional neural network (CNN) model is the most popularly employed, especially in the case of static input images. Convolution is a very well-known arithmetic operation in signal processing and image processing when spatial and time-domain data are directly utilized. Before the CNN model was popularly used, frequency domain analysis, e.g., Fourier Transform, was one of the most popular approaches. The CNN model enables direct use of spatial numeric data, i.e., pixel values, for detection, feature extraction, recognition, and classification work. In addition to these, almost all of the fields related to computer vision, image, and signal processing have significant benefits from the CNN model. Usual CNN-based FER takes static images or a set of static images as an input to the network model that is composed of more than three layers, called hidden layers, each of which provides convolution results with the output data of the previous steps. Various structures of filters are convolved with the input data or the output data of the previous layer, leading to an increase in computational complexity which has been resolved with the improvement of hardware infrastructures and the algorithms for developing a light structure of DNN models. Each layer contains the result of the convolution operation providing feature maps followed by generating fully connected layers to proceed to conduct classification. In the recognition work that uses static images as input data, CNN-based approaches have been considered a main method [14]. In practice, recognition tasks in a wild environment may require detection followed by classification in a real-time manner because the input image data varies over time. If DNN models are required to train input face (or facial expression) images with the variation of the expressions over time, i.e., input data has spatiotemporal features, the recurrent neural network (RNN) model is considered more appropriate for the recognition work [15]. In this case, sequences of facial expression data have a temporal dependency in addition to a spatial one, so the additional dependency is taken into account during the classification and recognition process.

In this paper, we present a novel approach to automatic FER using a spatial transformer network with a triplet loss function. To this end, the proposed method aims at accurate and efficient FER by focusing on the relevant region for each facial expression while robust to occlusion. A flow diagram of the proposed approach is shown in Figure 1.

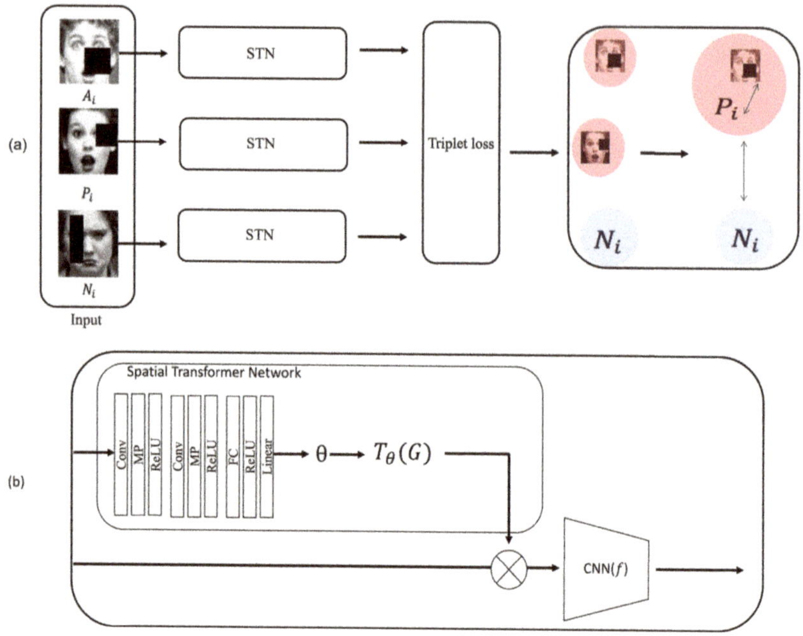

Figure 1. An overview of the proposed model for FER that solves the intra-similarity problem and is robust to occlusion. (**a**) Three classes of facial images with occlusion (anchor, positive and neutral) are classified using the proposed model, STN-TL. (**b**) Architecture of spatial transformer network is basically used in the proposed FER.The images in this figure are from public data set (CK+) [16].

The rest of this paper is organized as follows. Section 2 briefly introduces related work to the field of FER using deep neural networks and STN. This section also introduces the loss functions that have been applied to the recognition work. In Section 3, we introduce our proposed model TL-STN in detail. Then in Section 4, we introduce the data set used in our experiments. In addition in this section, we describe a comparison between the cross-entropy loss function and triplet loss function and also between occlusion data and non-occlusion data. Then, we compare the state-of-the-art model and TL-STN.

2. Related Work

Transformer architecture that has been widely used in the field of natural language processing (NLP) shows exceptionally well-performed results in the recognition task, especially in the case of using sequences of images as input data [17]. Recently, a new learnable module, spatial transformer network (STN) was proposed to provide robust performance by allowing spatial manipulation of input image data. STN is inserted into existing network models, e.g., CNN, and it enables one to achieve robust training results from invariance to the spatial transformation of input data, e.g., translation, rotation, warping, etc. [18]. The STN model has been applied to many practical problems among numerous cases of the recognition problems with encouraging results [19].

Another transformer model, vision transformer (ViT) [20], has gained attention in the recognition field and has been proposed as an alternative to the existing DNN models.

Although the existing FER work has achieved significantly improved results from quantitative and qualitative perspectives, it is still a challenging task due to the existence of uncontrolled external environments, pose variations, or occlusion that degrade the performance of FER results. More complex scenarios of FER need to be dealt with for high-quality FER from practical perspectives. Thus, it is worth investigating FER methods using STN, which has gained attention in the area of deep learning. In this paper, inspired by a spatial transformer network module, FER is performed efficiently by selecting the most relevant part of a facial image followed by applying the triplet loss function. The result is compared to the results using cross-entropy and to the other FER results using state-of-the-art algorithms. The proposed method shows superior recognition results in FER, particularly in the case of facial images having occlusion areas. A spatial transformer network that is included in standard neural network structure has advantages in case of rotation, cropping, scaling, and non-rigid deformation of images that sometimes happen to face images in practice.

As very well-known, traditional methods for the FER employ pixel-level, geometric model, or object-level-based approaches. Recent approaches are usually categorized as deep-learning-based approaches. In the past few years, learning-based approaches have witnessed a significant improvement in recognition tasks, especially in the areas of face recognition, facial expression recognition, activity recognition, etc. In the deep-learning-based approaches, convolutional neural networks and recurrent neural network models are the most widely used. More recently, a transformer network has been considered one of the alternatives to the CNN and RNN-based approaches. In this paper, we are interested in FER using deep-learning-based approaches and the spatial transformer network (STN) with a triplet loss function which improves the success rate of the recognition. The deep neural network (DNN) model enables one to perform automated FER that has long been an interesting and challenging task in the field of recognition problems. Instead of extracting feature points from facial images using a specific mathematical or statistical model, the DNN-based recognition approach extracts diverse and numerous feature points using large numbers of hidden layers that contribute to feature extraction with a brain-like mechanism.

Contrary to the traditional approach that always tries to achieve minimum intra-class distance and maximum inter-class distance in the recognition problem using an analytical model (mathematics, statistics, etc), the recent deep-learning-based approach is able to find abstraction and complex patterns that are inherent in real facial images.

Traditional approaches for the recognition of a face, facial expression, activity, or object lack generalizability due to the variations of a pose or scale and randomly additive noise. In addition, due to the non-existence of data sets, almost all of the recognition task was based on the manual or analytic model-based extraction of feature points. Inspired by the advent of deep learning, CNN-based models have shown robustness to the abovementioned variations, so FER has employed a CNN model to achieve higher accuracy of the recognition rate. CNN-based analysis of facial data has appeared in the work by Lawrence [21], LeCun [22] and Fasel [23] whose work has utilized less than five hidden layers in their network models. Almost all of the FER algorithms also use those works as a baseline to propose novelty or further improvement in the accuracy of the recognition. Since the work of FER in the early stages, significant progress has been achieved in more practical and wild-setting circumstances by utilizing the DNN model. The CNN model is one of the earliest ones that deeply learns and extracts facial feature points that have subtle expression changes that are difficult to extract using traditional recognition methods. Since the introduction of CNN for the recognition work, FER has also employed CNN structure by adding more layers, leading to deep CNN architectures that improved FER results [24]. In the beginning stage of FER using CNN, a limited number of image data sets were used and a specific expression was a target to be recognized. Subject independence and translation, rotation, and scale-invariant FER using CNN has been proposed to discriminate smiling from talking based on the saliency of visual cues [25]. Inspired by the expressions of real emotion, FER has been extended to micro-expression (ME) recognition using deep learning methods [26]. A single deep learning network structure that consists of two convolution layers followed by max pooling and four inception layers was introduced in the early stages of FER using DNN, but this work uses a registered face image data-set and the landmark is extracted a priori [27]. Much research has been conducted to solve wild data set FER problems. The work in [28], proposes a multi-task learning (MTL) framework that exploits the dependencies between these two models using a graph convolutional network

(GCN) to recognize facial expressions in the wild. The work in [29], proposes a visual-based end-to-end emotion recognition framework, which consists of the robust pre-trained backbone model and temporal sub-system to model temporal dependencies across many video frames. In addition, facial expressions can be applied in many applications. The work in [30] used facial expression recognition to analyze students' behavior in the e-learning environment. They used EfficientNet-B2 to extract emotional features in each frame. The sequence of facial images (video sequence) can be used as inputs to the DNN model, and in this case, the temporal relations between frames need to be taken into account, leading to the necessity of a long short-term-memory (LSTM) unit being additionally employed [31]. Another popular deep learning model is the recurrent neural network (RNN) model that is more suitable to temporal, sequential data, such as video, voice, text, etc., leading to the superior performance of the prediction task [32]. Transformer architecture was first proposed as a sequence transduction model based only on attention, and the spatial transformer network (STN) model chiefly deals with images with spatial transformation, so it shows a geometric invariant generalization of differentiable attention that is robust to any spatial transformation. Basic CNN models have the inevitable drawbacks of precise localization of important parts, particularly in the case of small objects. To the best of the authors' knowledge, there have been fewer research activities on STN-based FER with adaptive loss function. Thus, this section introduces STN-based recognition work (not limited to FER) focusing on the recent literature. The beginning state of deep-learning-based FER could be enhanced by adding an attention mechanism because it can focus on the most important sub-region of a facial image. In FER, the need for an attention mechanism has consequently brought the proposal of the STN model [18]. In [33], the attention mechanism uses a semi-supervised localizer that precisely detects salient regions, and the STN model is inserted into the existing DNN model (e.g., CNN model) to solve the recognition and detection problem in case of spatially transformed input images. The work shows STN incorporates into the basic CNN model, and the whole architecture is composed of three parts, localization, sampling grid, and image sampling. However, it can localize the rough position of the target (e.g., the jersey number of a soccer player), and there is no work on selecting an adaptive loss function for optimization. An attentional convolutional network has been introduced to classify facial expressions where the number of classes is smaller than the usual cases of classification problems [34]. In the work of [34], the authors used less than 10 hidden layers and added an attention mechanism for efficient FER, and the proposed approach reported better accuracy than state-of-the-art results. However, the accuracy shows oscillations, and the work reports that there is a trade-off between the recognition rate and the speed of convergence. Occlusion or pose variations are two major factors that degrade recognition accuracy, so region attention networks (RAN) have been employed for robust FER by adaptively capturing the important facial regions [35]. As FER in the wild is a challenging task, an attention mechanism with a basic CNN model (ACNN) has been proposed to perceive occlusion regions while focusing on the most unoccluded facial regions [36]. In [37], an extension to the basic STN model is proposed by adding procedures for capturing effective attentional regions using facial landmarks or facial visual saliency maps. In [38], STN-based FER was added to the CNN model with spatial and channel attention, and further improvement could be possibly achieved using the proposed GELU (Gaussian error linear unit) activation function. Multimodal emotion recognition that uses speech and facial images has been proposed. In this approach, pre-trained STN for saliency maps and bi-LSTM for the attention mechanism is proposed for emotion recognition [39]. However, in this work, the input image is transformed into mel-spectograms, leading to an increase in computational complexity.

Despite efforts in FER using DNN with an attention mechanism, there is room for further improvement, and our proposed method yields FER that is robust to occlusion and efficient by focusing on the most relevant facial region for specific expression by adding STN. In addition to the methods using STN, our approach employs an adaptive loss function and a triplet loss function that improves recognition accuracy in case of occlusion.

3. Proposed Method

Deep-learning-based facial expression recognition research shows high accuracy and performance. Nevertheless, there is still a problem in that it is hard to accurately recognize wild data sets due to external factors such as occlusion, pose, and illumination. Our proposed method is robust to occlusion using STN with an attention mechanism and triplet loss function that achieves optimized recognition accuracy. In practical cases, in addition to the occlusion problem, FER struggles with minimizing intra-class distance, i.e, existing FER algorithms sometimes fail to recognize the same expression. Some examples are depicted in Figure 2a,b.

Different from previous FER images, Figure 2a contains non-formalized FER images. In this case, although the facial images belong to the same class (e.g., fear, happiness, and neutral) the existing FER methods do not successfully classify or recognize the expression. Figure 2b shows another difficult classification problem between "sad" and "angry". Our proposed method solves this classification problem as well as the occlusion problem using a model called TL-STN which combines the spatial transformer network (STN) and a triplet loss function. In this section, the proposed model TL-STN is briefly explained, and the spatial transformer network and triplet loss are described in detail in Section 3.1 (Figure 1).

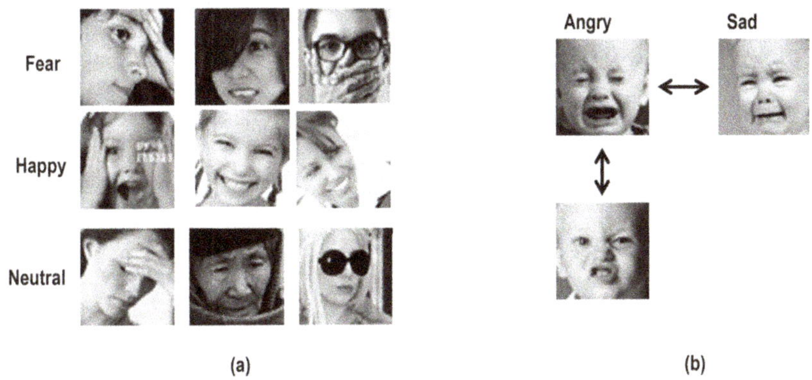

Figure 2. In practice, existing FER algorithms sometimes struggle with intra-similarity problems: (**a**) same expressions from different peoples' faces, and the existing algorithms sometimes consider them as different expressions; (**b**): different expressions from the same person's face. Existing work sometimes considers them as the same expressions. The images in this figure are from public data set (FER2013) [40].

3.1. Overview of TL-STN

In this section, we introduce an overview of a model that combines a spatial transformer network and triplet loss to solve problems affected by external environments, such as occlusion, pose, and illumination among facial recognition problems. In addition, the proposed method alleviates the limitation of recognizing the same expressions of wild data sets by combining STN and the triplet loss function. In particular, the triplet loss function contributes to aggregation of similar expressions.

As shown in Figure 1a, anchor (A_i), positive (P_i), and negative (N_i) images are used as input data for training the triplet loss function. Anchor data (A_i) stands for the original data which we want to classify. Positive data stands for the data belonging to the same class as the anchor data. Negative data stands for different class data from the anchor data. In this study, we used three input data for training. Positive and negative data were sequentially picked randomly from the same class and different classes.

Each facial data image is fed to the spatial transformer network followed by a triplet loss function so that the distance between P_i and A_i is minimized and the distance between

P_i and N_i is maximized. Furthermore, input image (A_i, P_i and N_i) with occlusion is fed to the STN that is combined with ResNet [41]. To this end, the classification of facial expression with occlusion and under the wild circumstance can be achieved with high accuracy in place of using only deep neural networks (e.g., CNN-based, RNN-based, STN only, etc) with the facial data acquired under the controlled circumstances.

3.2. *Spatial Transformer Network*

When carrying out the classification of facial images based on deep learning technologies, it is important that accurate classification be performed in realistic conditions, such as pose change, occlusion, and missing some parts of the facial image. CNN-based models use a pooling layer to solve these spatial variance problems. The spatial transformer network is a recent classification method [18], which can be utilized in the fields of image classification, co-localization, and spatial attention, and it can solve the spatial invariance problem by transforming specific parts that are required in the learning tasks. Our proposed approach to facial expression recognition combines STN and triplet loss instead of only using a single deep neural network model so that robust recognition can be performed in case of occlusion. In addition to robustness to occlusion, the triplet loss function alleviates the limitation of the existing FER methods that sometimes fail to categorize the same expression as shown in Figure 2. The combination of STN and triplet successfully aggregates the same expression that is captured in wild-set environments.

Figure 1b shows the STN in detail. It consists of a localization network, a grid generator, and a sampler. Conv, MP, ReLU, Linear, and CNN stand for convolution layer, max pooling layer, ReLU activation function, fully connected layer, and CNN model, respectively. ResNet has been used in the STN because it shows the best accuracy of classification in case of occlusion and wild-set environments. The localization network returns parameters required for spatial transformation, and the grid generator returns the grid required for transformation. In the course of transformation, we used affine transformation. The sampler samples the grid and input image generated through the grid generator. The localization network is constructed by adding max pooling and ReLU activation functions with convolution layers and one fully connected layer (Figure 1b).

Through the localization network, six parameters required for the affine grid (A_θ), written as Equation (1), are returned as outputs.

$$A_\theta = \begin{bmatrix} \theta_{11} & \theta_{12} & \theta_{13} \\ \theta_{21} & \theta_{22} & \theta_{23} \end{bmatrix} \tag{1}$$

The affine grid (A_θ) is generated through the six parameters returned from the localization network. The generated grid and input image are sampled through grid sampling to generate a final conversion grid necessary for learning, as written by Equation (2). In Equation (2), x_i and y_i stand for the coordinates of a horizontal and vertical axis of the generated grid. $T_\theta(G_i)$ stands for the grid generator, and A_θ stands for the affine grid.

$$\begin{pmatrix} x_i \\ y_i \end{pmatrix} = T_\theta(G_i) = A_\theta \begin{pmatrix} x_i \\ y_i \\ 1 \end{pmatrix} \tag{2}$$

3.3. *Triplet Loss*

FER frequently shows limitations in that different classes (e.g., expressions) of the same person do not show maximized distance, i.e, the existing methods fail to classify different expressions if those are from the same person. FER also shows a limitation in that the existing method fails to aggregate the same expressions of different people. The triplet loss function alleviates this limitation.

Triplet loss is a loss function for metric learning. Based on the anchor data, the triplet loss function enables one to minimize the distance between the same expressions of different people and to maximize the distance between the different expressions of the same person. By using this triplet loss, the Euclidean distance is decreased for facial expression data belonging to the same class, and the Euclidean distance is increased for facial expression data belonging to different classes. Data are organized as follows. The data belonging to the same class as the anchor data are composed of positive data, the data belonging to the other class are composed of negative data, the three sets of data are learned by each STN model, and the output result value is calculated as the Euclidean distance through the triplet loss function, written as

$$\sum_{i}^{N}[||f(x_i^a) - f(x_i^p)||_2 - ||f(x_i^a) - f(x_i^n)||_2 + \alpha] \quad (3)$$

where N is the number of data; f is the STN model; a, p, and n are anchor, positive, and negative, respectively,$||.||_2$ means L2 normalization. α is a hyperparameter representing the margin, and in this experiment, it is set to 1.0. In the experiment, back-propagation learning is performed through the above equation for positive data, negative data and anchor data that have passed each STN model.

4. Experimental Results

This section details the results with the used data set for the experiments, experimental setup, and environments. To validate the proposed approach in this work, we compare various image data, e.g., occlusion and non-occlusion data. The comparison is carried out through ablation studies, followed by a comparison with the state-of-the-art (SOTA) model.

4.1. Data Set

CK+: The Extended CohnKanada (CK+) database [16] is the most widely used laboratory control database in the field of FER. CK+ contains 593 video sequences of 123 topics. The sequences vary in duration from 10 to 60 frames and show transitions from neutral to peak facial expressions. In this video, seven basic facial expression labels (anger, contempt, disgust, fear, happiness, sadness, surprise) are classified based on the FACS (Facial Action Coding System). In this paper, a total of 981 frames were extracted and used in the experiment. Here, 800 training images and 181 test images were randomly divided into experiments.

FER2013: The FER2013 database [40] is the data used in ICML2013 Challenges in Presentation Learning. FER2013 is a large database that is automatically collected by the Google Image Search API. All images are scaled to a size of 48 × 48 pixels and consist of 7 expression labels (anger, disgust, fear, happiness, sadness, surprise, neutrality). It consists of 28,709 training images, 3589 verification images, and 3589 test images.

4.1.1. Experimental Environment

In this paper, the image size of the data set was adjusted to 224 × 224. Anchor, positive, and negative data were used with batch size eight. Based on the anchor data, images belonging to the same or different classes were randomly extracted to form positive and negative data, respectively. Each of the three data (anchor, positive, negative) is trained through the STN model combined with modified ResNet-18. The triplet loss was calculated through the three output values obtained by the model. We initialized the learning rate to 0.001, and the Adam optimizer was applied. The modified ResNet-18 layers are shown in Table 1. The existing ResNet model is modified by removing the number of layers in the model of ResNet-18, which includes the smallest number of layers among ResNet models. Then, the number of layers becomes smaller.

Table 1. Modified size of each layer of ResNet structure.

Layer Type	Output Size	Patch Size, Channel
Convolution layer 1	24 × 24	7 × 7, 64, stride 2
Convolution layer 2	12 × 12	3 × 3, 64, 3 × 3, 64
Convolution layer 3	6 × 6	3 × 3, 128, 3 × 3, 128
Convolution layer 4	3 × 3	3 × 3, 256, 3 × 3, 256
Convolution layer 5	2 × 2	3 × 3, 512, 3 × 3, 512
Average Pool	1 × 1	-

4.1.2. Ablation Studies

Figure 3 visualizes the distribution of image data used for the comparison experiments. The comparison is performed from two perspectives. One is a comparison between occluded facial images and non-occluded ones. The other is a comparison between using cross-entropy and using the triplet loss function. Figure 3a,b represent the case of using the cross-entropy loss function, and (c) and (d) represent the case of using the triplet loss function. Here, (a) and (c) show the visualization of experimental results using occlusion data, and (b) and (d) show the experimental results using non-occlusion data. It seems that it is difficult to clearly distinguish between different classes when a cross-entropy loss is used (Figure 3a,b). On the other hand, when triplet loss is used (Figure 3c,d), it can be seen that classification is more successful. We can see Figure 3a,c show a more clearly distinguished result than (b) and (d) show, which are trained by occlusion data. Nevertheless, when using triplet loss, classes are more clearly distinguished in occlusion data than when using the cross-entropy loss function, and we can see that they are more clearly aggregated between the same classes (indices of vertical and horizontal axes just present relative locations of points of each expression).

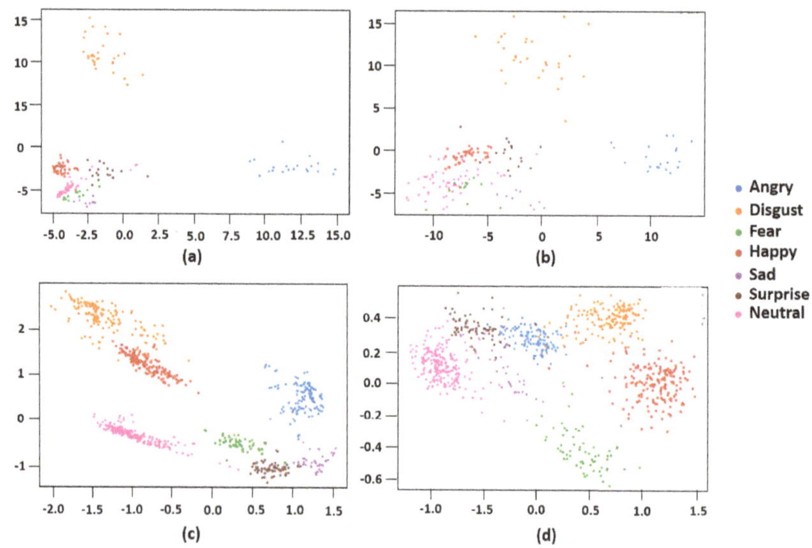

Figure 3. Visualization of data distribution under loss function and occlusion: (**a**,**b**) show classification using the cross-entropy loss function; (**c**,**d**) show the result using triplet loss; (**a**,**c**) visualize the result with non-occlusion data; (**b**,**d**) visualize the result with occlusion data.

Table 2 compares the accuracy of the original ResNet-18 and the modified ResNet-18 using randomly erased CK+ data and compares the accuracy of using cross-entropy loss and triplet loss. Through the experiment, it is confirmed that when the modified ResNet-18 is combined with STN, the accuracy achieved is 2.09% higher than the model combined with STN and the original ResNet-18. Through this experiment, it can be confirmed that when modified ResNet-18 is combined with STN, it shows better performance. In other words, the results imply that the model with fewer layers in the network model shows higher accuracy when combined with STN. In addition, through the experiment, it was confirmed that the use of the triplet loss function showed a classification accuracy of 99.44%, which is 0.48% higher than with the cross-entropy loss. Although our approach uses the ResNet-18 structure, the result is comparable to the result using the ResNet-34 structure (LHC-Net). Unfortunately, our approach struggles with the optimization of the recognition result with an increase in the number of layers. These results verify that in the case of facial expression recognition, using the triplet loss function further improves the accuracy compared to using the cross-entropy loss function.

Table 2. Ablation study of ResNet model and using different loss function on CK+ data set.

ResNet	Loss Function	Accuracy (%)
Orignal ResNet-18	CrossEntropy	96.87
Modify ResNet-18	CrossEntropy	98.96
Modify ResNet-18	**Triplet**	**99.41**

4.1.3. Comparison Result

Experimental results using CK+ and FER2013 data sets are shown in Figure 3 and Tables 2 and 3. Table 2 chiefly compares the accuracy between using the cross-entropy and triplet loss function in the ablation study. Table 3 comprehensively shows the comparison results between the SOTA models and our proposed approach. In the case of using the CK+ data set, the proposed model, TL-STN, is compared with FER-IK [42], IPA2LT [43] , lp-norm MKL multiclass-SVM [44], twofold random forest classier [45], and the self-supervised learning (SLL) puzzling model [46].

In the CK+ data set, ViT+SE [47] and FAN [48] show high accuracy, but we excluded it from the comparison table because ViT+SE uses 10-fold cross-validation and FAN uses video sequences as input data which is a different setup from our proposed approach. FER-IK is known as a knowledge-augmented image-based FER model, and IPA2LT is known as inconsistent pseudo annotations to the latent truth model. The lp-norm MKL multiclass-SVM is known as multiple kernel learning (MKL) in multiclass support vector machines (SVM). Twofold random forest classier is known as a model which recognizes AUs from image sequences using a twofold random forest classifier. SSL puzzling is known as a nonlinear evaluation in supervised learning (SL) and the self-supervised learning (SSL) puzzling model.

In the case of using the FER 2013 data set, our proposed model, STN with modified ResNet-18 and cross-entropy, is compared to LHC-Net [49], CNN [50], GoogleNet [51] , ResNet [41], VGGNet [52], and STN with a cross-entropy loss function.

In the FER2013 data set, Ensemble ResMaskingNet with six other CNNs [40] and Local Learning Deep+BOW [53] showed high accuracy, but we excluded it from the comparison table because these models use machine learning, unlike our proposed model. In addition, simple comparisons are impossible because we propose a loss function using attention-focused mechanism-based models and metric learning.

In this experiment, our approach does not show an improved result, but the proposed model shows superior accuracy compared to the result of using the original ResNet-18 model. TL-STN with the CK+ data set achieves the best recognition accuracy despite using randomly erased facial images.

Table 3. Performances comparison with state-of-the-art methods.

Model	Datasets	Accuracy (%)
FER-IK [42]	CK+	97.59
IPA2LT [43]		91.67
lp-norm MKL multiclass-SVM [44]		93.6
Twofold random forest classier [45]		96.38
Nonlinear eval on SL + SSL Puzzling [46]		98.23
TL-STN (ours)		**99.41**
LHC-Net [49]	FER2013	74.42
CNN [50]		62.44
GoogleNet [51]		65.20
ResNet [41]		72.4
VGGNet [52]		73.28
STN (w/orignal ResNet-18) + TL (ours)		**72.30**
STN (w/modified ResNet-18) + TL (ours)		**73.31**

5. Conclusions

This paper presents facial expression recognition based on deep learning technology. Since the advent of deep neural network models, diverse applications using image data have shown significant improvement from theoretical and practical perspectives. However, a lot of challenges remain due to the unexpected factors that degrade the performance of recognition. Furthermore, facial expression directly reflects human emotion which is a very qualitative component. Facial expression and human emotion are very delicate, leading to the technical difficulty in analysis and quantification. In this paper, the proposed approach chiefly contributes to two problems, one is occlusion, and the other one is a classification of expression (intra-class similarity problem) in practical cases (Figure 2). Exsiting FER methods usually employ cross-entropy loss function which helps reduce the difference between ground truth values and the estimated (or predicted) ones that are similar to other image recognition fields. The cross-entropy loss function shows a high recognition accuracy for objects that do not change the appearance of objects in the image, but it is difficult to classify when there are various features in the same class, such as facial expressions. In this paper, to solve these problems, the experiments were conducted by using the triplet loss function which was the first suggested in the field of facial expression recognition, and the proposed one can be applied to diverse practical fields. The triplet loss function with a STN (w/modified ResNet-18) alleviates the abovementioned limitations. The proposed model solves occlusion and illumination, poses change issues, and shows superior results to the existing work. To verify the benefit of the modified ResNet-18 model, a comparison was performed that showed a 1.01% improvement on the FER2013 data set. When the triplet loss function and the modified ResNet-18 were combined, they yielded 99.41% accuracy. The experiment with a randomly erased pre-processed CK+ data set showed the highest accuracy compared to SOTA models which were performed with the original CK+ data set. Through these experiments, it was confirmed that even for data with occlusion, our model shows high performance in FER. In addition, the proposed model shows the availability of a metric-learning-based loss function.

In future work, we will more deeply focus on enhancement of the recognition accuracy with the more delicate differences of facial expression, as well as more practical issues in recognition problems. Specifically, we will analyze the practical limitations existing in the proposed approach and will try to solve them through contrastive loss functions such as triplet loss. Through this, we plan to see if we can solve other recognition problems by applying our methods to fine-grain recognition problems that use small data sets, such as medical diagnosis, gender classification, etc. [54–56].

Author Contributions: Software, J.K.; Validation, J.K.; Investigation, J.K.; Writing—original draft, J.K.; Writing—review & editing, J.K. and D.L.; Project administration, D.L.; Funding acquisition, D.L. All authors have read and agreed to the published version of the manuscript.

Funding: This research was supported by the Bisa Research Grant of Keimyung University in 2023 (No. 20220208).

Data Availability Statement: The datasets generated during and/or analysed during the current study are publicly available(FER2013 and CK+).

Conflicts of Interest: The authors declare no conflict of interest.

References

1. Ekman, P. An argument for basic emotions. *Cogn. Emot.* **1992**, *6*, 169–200. [CrossRef]
2. Kyeong Tae, K.; Jae Young, C. Development of Semi-Supervised Deep Domain Adaptation Based Face Recognition Using Only a Single Training Sample. *J. Korea Multimed. Soc.* **2022**, *25*, 1375–1385.
3. Yoon, A.K.-Y.; Park, K.-C.; Lee, B.-C.; Jang, J.-H. A Study on Overcoming Disturbance Light using Polarization Filter and Performance Improvement of Face Recognition System. *J. Multimed. Inf. Syst.* **2020**, *7*, 239–248. [CrossRef]
4. Ruyang, Z.; Eung-Joo, L. Face Recognition Research Based on Multi-Layers Residual Unit CNN Model. *J. Korea Multimed. Soc.* **2022**, *25*, 1582–1590.
5. Arunkumar, P.M.; Sangeetha, Y.; Raja, P.V.; Sangeetha, S.N. Deep Learning for Forgery Face Detection Using Fuzzy Fisher Capsule Dual Graph. *Inf. Technol. Control* **2022**, *51*, 563–574. [CrossRef]
6. Wei, W.; Ho, E.S.L.; McCay, K.D.; Damaševičius, R.; Maskeliūnas, R.; Esposito, A. FAssessing Facial Symmetry and Attractiveness using Augmented Reality. *Pattern Anal. Appl.* **2022**, *25*, 635–651. [CrossRef]
7. Henrikson, J. FER-net: Completeness and total boundedness of the Hausdorff metric. *MIT Undergrad. J. Math.* **1999**, *1*, 10.
8. Cortes, C.; Vapnik, V. Support-vector networks. *Mach. Learn.* **1995**, *20*, 273–297. [CrossRef]
9. Abdi, H.; Williams, L.J. Principal component analysis. *Wiley Interdiscip. Rev. Comput. Stat.* **2010**, *2*, 433–459. [CrossRef]
10. Eddy, S.R. Hidden markov models. *Curr. Opin. Struct. Biol.* **1996**, *6*, 361–365. [CrossRef]
11. Wang, S.B.; Quattoni, A.; Morency, L.P.; Demirdjian, D.; Darrell, T. Hidden conditional random fields for gesture recognition. *IEEE Comput. Soc. Conf. Comput. Vis. Pattern Recognit.* **2006**, *2*, 1521–1527.
12. Li, S.; Deng, W. Deep facial expression recognition: A survey. *IEEE Trans. Affect. Comput.* **2020**, *13*, 1195–1215. [CrossRef]
13. Jiao, L.; Zhang, F.; Liu, F.; Yang, S.; Li, L.; Feng, Z.; Qu, R. A survey of deep-learning-based object detection. *IEEE Access* **2019**, *7*, 128837–128868. [CrossRef]
14. LeCun, Y.; Boser, B.; Denker, J.S.; Henderson, D.; Howard, R.E.; Hubbard, W.; Jackel, L.D. Backpropagation applied to handwritten zip code recognition. *Neural Comput.* **1989**, *1*, 541–551. [CrossRef]
15. Hochreiter, S.; Schmidhuber, J. Long short-term memory. *Neural Comput.* **1997**, *9*, 1735–1780. [CrossRef] [PubMed]
16. Lucey, P.; Cohn, J.F.; Kanade, T.; Saragih, J.; Ambadar, Z.; Matthews, I. The extended cohn-kanade data set (ck+): A complete data set for action unit and emotion-specified expression. In Proceedings of the 2010 IEEE Computer Society Conference on Computer Vision and Pattern Recognition-Workshops, San Francisco, CA, USA, 13–18 June 2010; pp. 94–101.
17. Vaswani, A.; Shazeer, N.; Parmar, N.; Uszkoreit, J.; Jones, L.; Gomez, A.N.; Kaiser, Ł.; Polosukhin, I. Attention is all you need. In Proceedings of the 31st International Conference on Neural Information Processing Systems, Long Beach, CA, USA, 4–9 December 2017.
18. Jaderberg, M.; Simonyan, K.; Zisserman, A.; Kavukcuoglu, K. Spatial Transformer Networks. In Proceedings of the 28th International Conference on Neural Information Processing Systems, Montreal, QC, Canada, 7–12 December 2015.
19. Niu, Z.; Zhong, G.; Yu, H. A review on the attention mechanism of deep learning. *Neurocomputing* **2021**, *452*, 48–62. [CrossRef]
20. Dosovitskiy, A.; Beyer, L.; Kolesnikov, A.; Weissenborn, D.; Zhai, X.; Unterthiner, T.; Dehghani, M.; Minderer, M.; Heigold, G.; Gelly, S.; et al. An image is worth 16 × 16 words: Transformers for image recognition at scale. *arXiv* **2020**. [CrossRef]
21. Lawrence, S.; Giles, C.L.; Tsoi, A.C; Back, A.D. Face rccognition: A convolutional neural network approach. *IEEE Trans. Neural Netw.* **1997**, *8*, 98–113. [CrossRef]
22. LeCun, Y.; Bengio, Y. Convolutional Networks for Iamges, Speech and Time-Series. In *The Handbook of Brain Theory and Neural Networks*; MIT Press: Cambridge, MA, USA, 1995.
23. Fasel, B. Robust Face Analysis using Convolutional Neural Networks. In Proceedings of the 2002 International Conference on Pattern Recognition, Quebec City, QC, Canada, 11–15 August 2002.
24. Lecun, Y. Generalization and Network Design Strategies. *Connect. Perspect.* **1989**, *19*, 18.
25. Matsugu, M.; Mori, K.; Mitari, Y.; Kaneda, Y. Subject independent facial expression recognition with robust face detection using a convolutional neural network. *Neural Netw.* **2003**, *16*, 555–559. [CrossRef]

26. Zhao, Y.; Xu, J. A Convolutional Neural Network for Compound Micro-Expression Recognition. *Sensors* **2019**, *19*, 5553. [CrossRef] [PubMed]
27. Mollahosseini, A.; Chan, D.; Mahoor, M.H. Going deeper in facial expression recognition using deep neural networks. In Proceedings of the 2016 IEEE Winter Conference on Applications of Computer Vision (WACV), Lake Placid, NY, USA, 7–10 March 2016.
28. Panagiotis, A.; Panagiotis, F.; Petros, M. Exploiting Emotional Dependencies with Graph Convolutional Networks for Facial Expression Recognition. *arXiv* **2021**, arXiv:2106.03487.
29. Ryumina, E.; Dresvyanskiy, D.; Karpov, A. In search of a robust facial expressions recognition model: A large-scale visual cross-corpus study. *Neurocomputing* **2022**, *514*, 435–450. [CrossRef]
30. Savchenko, A.V.; Savchenko, L.V.; Makarov, I. Classifying emotions and engagement in online learning based on a single facial expression recognition neural network. *IEEE Trans. Affect. Comput.* **2022**, *13*, 2132–2143. [CrossRef]
31. Hasani, B.; Mahoor, M.H. Facial Expression Recognition Using Enhanced Deep 3D Convolutional Neural Networks. In Proceedings of the IEEE Conference on Computer Vision and Pattern Recognition (CVPR) Workshops, Honolulu, HI, USA, 21–16 26 2017; pp. 30–40.
32. Zaremba, W.; Sutskever, I.; Vinyals, O. Recurrent neural network regularization. *arXiv* **2014**, arXiv:1409.2329.
33. Li, G.; Xu, S.; Liu, X.; Li, L.; Wang, C. Jersey Number Recognition with Semi-Supervised Spatial Transformer Network. In Proceedings of the IEEE Conference on Computer Vision and Pattern Recognition Workshops, Salt Lake City, UT, USA, 18–22 June 2018; pp. 690–696.
34. Minaee, S.; Minaei, M.; Abdolrashidi, A. Deep-Emotion: Facial Expression Recognition Using Attentional Convolutional Network. *Sensors* **2021**, *21*, 3046. [CrossRef]
35. Wang, K.; Peng, X.; Yang, J.; Meng, D.; Qiao, Y. Region Attention Networks for Pose and Occlusion Robust Facial Expression Recognition. *IEEE Trans. Image Process.* **2020**, *29*, 4057–4069. [CrossRef]
36. Li, Y.; Zeng, J.; Shan, S.; Chen, X. Occlusion Aware Facial Expression Recognition Using CNN With Attention Mechanism. *IEEE Trans. Image Process.* **2018**, *28*, 2439–2450. [CrossRef]
37. Luna-Jiménez, C.; Cristóbal-Martín, J.; Kleinlein, R.; Gil-Martín, M.; Moya, J.M.; Fernández-Martínez, F. Guided Spatial Transformers for Facial Expression Recognition. *Appl. Sci.* **2021**, *11*, 7217. [CrossRef]
38. Wang, C.; Wang, Z.; Cui, D. Facial Expression Recognition with Attention Mechanism. In Proceedings of the 2021 14th International Congress on Image and Signal Processing, BioMedical Engineering and Informatics (CISP-BMEI), Shanghai, China, 23–25 October 2021; pp. 1–6.
39. Luna-Jiménez, C.; Griol, D.; Callejas, Z.; Kleinlein, R.; Montero, JFernández-Martínez, F. Multimodal Emotion Recognition on RAVDESS Dataset Using Transfer Learning. *Sensors* **2021**, *21*, 7665. [CrossRef]
40. Goodfellow, I.J.; Erhan, D.; Carrier, P.L.; Courville, A.; Mirza, M.; Hamner, B.; Cukierski, W.; Tang, Y.; Thaler, D.; Lee, D.-H.; et al. Challenges in representation learning: A report on three machine learning contests. In Proceedings of the Neural Information Processing: 20th International Conference, ICONIP 2013, Daegu, Republic of Korea, 3–7 November 2013; Springer: Berlin/Heidelberg, Germany, 2013; pp. 117–124.
41. He, K.; Zhang, X.; Ren, S.; Sun, J. Deep residual learning for image recognition. In Proceedings of the IEEE Conference on Computer Vision and Pattern Recognition, Las Vegas, NV, USA, 27–30 June 2016; pp. 770–778.
42. Cui, Z.; Song, T.; Wang, Y.; Ji, Q. Knowledge augmented deep neural networks for joint facial expression and action unit recognition. *Adv. Neural Inf. Process. Syst.* **2020**, *33*, 14338–14349.
43. Zeng, J.; Shan, S.; Chen, X. Facial expression recognition with inconsistently annotated data sets. In Proceedings of the European Conference on Computer Vision, Munich, Germany, 8–14 September 2018; pp. 222–237.
44. Zhang, X.; Mahoor, M.H.; Mavadati, S.M. Facial expression recognition using lp-norm MKL multiclass-SVM. *Mach. Vis. Appl.* **2015**, *26*, 467–483. [CrossRef]
45. Pu, X.; Fan, K.; Chen, X.; Ji, L.; Zhou, Z. Facial expression recognition from image sequences using twofold random forest classifier. *Neurocomputing* **2015**, *168*, 1173–1180. [CrossRef]
46. Pourmirzaei, M.; Montazer, G.A.; Esmaili, F. Using Self-Supervised Auxiliary Tasks to Improve Fine-Grained Facial Representation. *arXiv* **2021**, arXiv:2105.06421.
47. Aouayeb, M.; Hamidouche, W.; Soladie, C.; Kpalma, K.; Seguier, R. Learning vision transformer with squeeze and excitation for facial expression recognition. *arXiv* **2017**, arXiv:2107.03107.
48. Meng, D.; Peng, X.; Wang, K.; Qiao, Y. Frame attention networks for facial expression recognition in videos. In Proceedings of the 2019 IEEE International Conference on Image Processing (ICIP), Taipei, Taiwan, 22–25 September 2019; pp. 3866–3870.
49. Pecoraro, R.; Basile, V.; Bono, V. Local multi-head channel self-attention for facial expression recognition. *Information* **2022**, *13*, 419. [CrossRef]
50. Liu, K.; Zhang, M.; Pan, Z. Facial expression recognition with CNN ensemble. In Proceedings of the 2016 International Conference on Cyberworlds (CW), Chongqing, China, 28–30 September 2016; pp. 163–166.
51. Giannopoulos, P.; Perikos, I.; Hatzilygeroudis, I. Deep learning approaches for facial emotion recognition: A case study on FER-2013. In *Advances in Hybridization of Intelligent Methods: Models, Systems and Applications*; Springer: Cham, Switzerland, 2018; pp. 1–16.
52. Simonyan, K.; Zisserman, A. Very deep convolutional networks for large-scale image recognition. *arXiv* **2014**, arXiv:1409.1556.

53. Georgescu, M.I.; Ionescu, R.T.; Popescu, M. Local learning with deep and handcrafted features for facial expression recognition. *IEEE Access* **2019**, *7*, 64827–64836. [CrossRef]
54. Yazdani, A.; Fekri-Ershad, S.; Jelvay, S. Diagnosis of COVID-19 Disease in Chest CT-Scan Images Based on Combination of Low-Level Texture Analysis and MobileNetV2 Features. *Comput. Intell. Neurosci.* **2022**, *2022*, 1658615. [CrossRef]
55. Mohan, K.; Seal, A.; Krejcar, O.; Yazidi, A. FER-net: Facial expression recognition using deep neural net. *Neural Comput. Appl.* **2021**, *33*, 9125–9136. [CrossRef]
56. Van der Maaten, L.; Hinton, G. Visualizing data using t-SNE. *J. Mach. Learn. Res.* **2008**, 9, 2579–2605.

sensors

MDPI

Article

A Feature-Trajectory-Smoothed High-Speed Model for Video Anomaly Detection

Li Sun [1,†], Zhiguo Wang [1,†], Yujin Zhang [1] and Guijin Wang [1,2,*]

[1] Department of Electronic Engineering, Tsinghua University, Beijing 100084, China
[2] Shanghai AI Laboratory, Shanghai 200232, China
* Correspondence: wangguijin@tsinghua.edu.cn
† These authors contributed equally to this work.

Abstract: High-speed detection of abnormal frames in surveillance videos is essential for security. This paper proposes a new video anomaly–detection model, namely, feature trajectory–smoothed long short-term memory (FTS-LSTM). This model trains an LSTM autoencoder network to generate future frames on normal video streams, and uses the FTS detector and generation error (GE) detector to detect anomalies on testing video streams. FTS loss is a new indicator in the anomaly–detection area. In the training stage, the model applies a feature trajectory smoothness (FTS) loss to constrain the LSTM layer. This loss enables the LSTM layer to learn the temporal regularity of video streams more precisely. In the detection stage, the model utilizes the FTS loss and the GE loss as two detectors to detect anomalies. By cascading the FTS detector and the GE detector to detect anomalies, the model achieves a high speed and competitive anomaly-detection performance on multiple datasets.

Keywords: anomaly detection; generation error; feature trajectory smoothness; surveillance video

Citation: Sun, L.; Wang, Z.; Zhang, Y.; Wang, G. A Feature-Trajectory-Smoothed High-Speed Model for Video Anomaly Detection. *Sensors* **2023**, *23*, 1612. https://doi.org/10.3390/s23031612

Academic Editors: Euntai Kim, Sangyoun Lee and Kang Ryoung Park

Received: 18 November 2022
Revised: 20 January 2023
Accepted: 27 January 2023
Published: 2 February 2023

1. Introduction

Surveillance cameras are widely used in people's daily lives. Detecting anomalies in surveillance videos is important for safe-protection and crime prevention. Anomalies in videos generally refer to events that have low probabilities of occurrence [1], or patterns that do not conform to expected behaviors [2].

Abnormal event detection is of great significance in many scenarios. For example, in office areas, illegal intrusion, theft, and fire are anomalies; in transportation scenes, traffic violations and traffic accidents are anomalies [3–5]; in public areas, terrorist attacks, robbery, and fare evasion are anomalies. Thus, improving the detection ability of surveillance video in public areas garners attention in research [6,7]. Detecting anomalies in surveillance videos is a challenging task because (1) surveillance videos are private property and (2) anomalous events have rarity, diversity, and scene-dependent properties. It is almost infeasible to gather all kinds of abnormal events and tackle the problem of anomaly detection with a simple classification method [8].

Video anomaly-detection methods can be classified into three categories, i.e., supervised methods, unsupervised methods and semisupervised methods. Supervised methods transform the anomaly-detection task into a binary or multiclassification task, by collecting and annotating a large number of normal and abnormal video samples. Ullah et al. proposed a lightweight model for anomaly detection [9], which works for a real-world surveillance network and employs the residual attention-based long short-term memory (LSTM) which can effectively learn temporal context information and precisely recognize anomalous events. Dubey et al. proposed an innovative framework called DMRMs, which was tested on the UCF–crime and ShanghaiTech datasets [10]. The results and ablation study demonstrated their effectiveness when compared with other methods. The disadvantages of this kind of method include the facts that the workload of sample collection and annotation is huge, and the generalization of detecting unknown abnormal events is poor. The unsupervised method analyzes the distribution of sample space and judges a small number of samples far away from the majority of samples as anomalies. Ionescu et al. proposed a novel framework for abnormal event detection in the video that requires no training sequences [11]. The disadvantages of this kind of method include a large amount of computation, poor real-time performance, and poor anomaly-detection. The semisupervised method transforms anomaly detection into a classification task by only collecting a large number of normal samples. They study the patterns of normal samples and identify those that do not follow normal patterns as abnormal. This kind of method has a small sample collection and sample labeling workload, has good generalization for unknown anomalies, and good real-time anomaly-detection speed. This has gained the most attention among the three kinds of methods.

The semisupervised surveillance video-anomaly detection algorithm has been developed for a long time. Recently, with the excellent performances of deep learning in many computer vision tasks, deep-learning-based semisupervised surveillance video anomaly detection (DSAD) algorithms have gained much attention. These methods use neural networks to learn the manifold distribution of normal samples, and then judge the samples that deviate from the normal manifold distribution as anomalies. Based on the types of indicators in anomaly detection, the semi-supervised methods can be classified into four categories: the deep distance-based method [12–14], the deep probability-based method [15,16], the deep generation error-based (GE-based) method [17–20], and and the aggregation method [21–23]. The deep distance-based method clusters samples to multiple groups by the deep neural network (DNN), and judges the samples that are outliers of all normal clusters as anomalies. The deep probability-based method learns the probability distribution of normal video samples, and take samples with low distribution probabilities (DPs) as anomalies. The deep GE-based method trains generative models to generate normal video frames and judge testing frames with large GE errors as anomalies. The aggregation methods train no less than two detectors that belong to the above three methods to detect the video anomaly events.

In the DSAD method, the GE indicator is a very important indicator because of its good anomaly detection and location performances. It usually plays a major role in aggregation methods. In order to improve the anomaly-detection effect of GE, many improvement strategies have been proposed. One important and fundamental improvement strategy is to capture videos' temporal regularity. In the surveillance video anomaly-detection field, many previous works such as [24–26] have proven that LSTM has a solid ability to capture video temporal regularity. These LSTM methods [24–26] utilized autoencoder models to generate normal video frames, adopted GE loss to constrain models' generation performances, and asserted LSTM layers between the encoder and decoder modules to capture videos' temporal regularity. However, the GE loss does not constrain videos' features directly, and is not powerful enough to force the maintainance of videos' temporal regularity in the feature space. Thus, these LSTM methods would not capture videos' temporal regularity precisely. As a result, the LSTM layer could not effectively improve the anomaly-detection performance of the model. In addition, deep neural networks usually face the problem of large amounts of computation. The way to further reduce the amount of computation and improve the abnormal detection speed of neural networks is a problem that requires constant attention.

In order to solve the aforementioned problems, this paper proposes a new detection model, namely, the feature trajectory-smoothed long short-term memory (FTS-LSTM). In the training stage, the model imposes a temporal smoothing loss on the feature space of the LSTM layer, which enables features to maintain the videos' temporal regularity better and thus enables the LSTM layer to learn videos' temporal regularity more precisely. In the detecting stage, the model utilizes the feature-trajectory smoothness (FTS) loss as a new anomaly-detection indicator. The FTS indicator judges frames with high FTS losses as anomalies. It can detect anomalies quickly because of its low computation cost. The generation error (GE) indicator can detect anomalies precisely [19,27]. By cascading the FTS and the GE indicators, the proposed model achieves fast and accurate anomaly-detection performances.

The contributions of the paper are summarized as follows.

- A a video anomaly-detection model, namely, FTS-LSTM, is proposed. In this model, an FTS loss is designed to enable the LSTM layer to learn videos' temporal regularity better.
- A new indicator to detect anomalies, namely, the FTS indicator, is proposed. It can detect anomalies precisely with a high speed.
- This work has good generalization capability and can easily transfer to other models with LSTM layers.

The overall structure of the article is summarized below. In Section 2, we discuss the development of existing techniques concerning anomaly detection in surveillance videos. Section 3 describes the detail of the novel FTS-LSTM method. In Section 4, the model implementation and experimental results, along with the evaluation of the proposed model are discussed. Finally, the conclusion and future work are given in Section 5.

2. Related Work

The development of semisupervised anomaly-detection algorithms can be classified into two stages, namely, the stage of traditional machine learning methods and the stage of deep learning methods. Furthermore, the traditional machine learning methods can be classified into three broad research areas, and the deep learning methods can be classified into four broad research areas.

2.1. Traditional Machine Learning Stage

In the traditional machine learning stage, many studies extract features manually and use traditional machine learning models to detect anomalies. Anomaly-detection indicators in this stage can be roughly classified into distance-based (DB) methods, probability-based (PB) methods, and reconstruction error (RE) methods.

The distance-based method [28,29] detects anomalies by using distances from test samples to normal samples or clusters of normal samples. This type of methods usually includes a step of clustering. Before model training, the normal samples are divided into multiple clusters, and then the samples far away from all normal clusters are judged as abnormal. Ionescu et al. [28] used k-means to cluster samples and one-class support vector machines (OC-SVM) to detect outliers. Hinami et al. [29] trained a multitask fast recurrent convolutionary neural network (RCNN) model to extract features. They grouped features into different clusters by k-means and used kernel density estimation (KDE) to detect anomalies on all clusters.

The probability-based method [30,31] learns the distribution probability density of the sample feature space or the inferred relationship between normal features through the model, and then takes the samples with low distribution probability density or those which do not obey the normal inferred relationship as abnormal. Hu X. et al. [32] modeled the distribution of normal sample feature spaces with models in question. They first proposed a local binary pattern feature with a squirrel cage structure, and then modeled the feature space of normal samples with a model in question. Weixin Li et al. [33] used the mixture dynamic texture (MDT) model to construct transition rules for normal sample feature sequences. MDT consists of k-linear dynamic systems, which are used to capture k-state transition laws of normal sample features. When the test sample does not meet any of the normal transition rules, the algorithm judges it as an abnormal event.

The reconstruction error method [34] used the common factors shared by the normal samples to reconstruct normal samples, but abnormal samples cannot be reconstructed because they do not share any common factors. Cong et al. [35] proposed a sparse coding method that weighs word anomalies so that different words have different anomaly weights. Chu et al. [36] proposed a recurrent framework that combines deep feature extraction with sparse coding. They put the module for training 3D convolutional neural networks to extract deep features and the module for learning sparse coding dictionaries with deep features under the same loop framework to be iteratively optimized, so that the features extracted by the network are the features most suitable for the sparse coding method, in order to achieve better performance in terms of good anomaly detection.

2.2. Deep Learning Stage

In the deep learning stage, many studies train DNNs to detect anomalies in the end-to-end manner. The indicators can be classified into four categories based on their characters, i.e., the deep distance-based (DDB) method, the deep probability-based (DPB) method, the deep generation error-based (DGE) method, and the aggregation method.

The deep distance-based method [12–14] in the deep learning stage clusters samples to multiple groups by DNN in an end-to-end manner. It judges the samples that are outliers of all normal clusters as anomalies. Fan et al. [37] trained a Gaussian mixture fully convolutional variational autoencoder (GMFC-VAE) to map samples to multiple clusters in the latent space and judged samples that have low condition probabilities with any existing clusters as anomalies. Wu et al. [14] trained a deep one-class neural network (DeepOC) to map normal samples into a single hypersphere and judged the samples mapped out of the hypersphere as anomalies.

The deep probability-based method [20,38,39] learns the probability distribution of normal videos and judges samples with low distribution probabilities as anomalies. It uses the discriminator to output the DPs of the video frames to detect anomalies. Ravanbakhsh et al. [39] trained two GANs to generate motion images from appearance images which were generated from motion images. They combined two DP score maps generated by two discriminators to detect anomalies.

The deep generation error-based method [17–20,22,24–26,40–43] trains generative models to generate normal video frames and judges testing frames with large GE errors as anomalies. Hasan et al. [26] first introduced the autoencoder(AE) to video anomaly detection. Gong et al. [40] proposed a memory-augmented autoencoder (MemAE) to limit the AE's generalization ability. Zhou et al. [41] proposed an attention-driven training loss to alleviate the imbalance problem between the foreground and stationary background. In order to capture videos' spatiotemporal regularity, many methods [18,21,22,24,25,42,43] have utilized the LSTM-AE to detect anomalies. There are some works which train no less than two detectors to disclose the video anomaly events which belongs to deep generation error-based method.

The aggregation method [21–23] trains no less than two detectors to disclose the video anomaly events. Lee et al. proposed a spatiotemporal adversarial network to detect anomalies [21]. The algorithm extracts two anomaly detectors which are a generative error detector and a generative adversarial network (GAN) probabilistic detector. The two detectors disclose anomalies with a weighted sum of the anomaly scores of the two detectors. Wang et al. proposed an integrated approach called primary–auxiliary fusion [23]. The core detector is a video anomaly detector based on the pixel generation error, and the auxiliary detector is a detector with high accuracy in detecting strong normality and strong anomaly. The algorithm extracts this decision ability from the auxiliary detector and weighs it with the outlier score in the main detector to obtain an integrated detector.

3. Method

The pipeline of the proposed work is illustrated in Figure 1. It uses normal videos to train the model and detect anomalies in the testing videos. This section introduces the proposed work in three aspects, i.e, the network structure, the training process, and the detecting process.

3.1. Network Structure

As shown in Figure 1, the proposed method consists of three network modules, which are the encoder module, the ConvLSTM module, and the decoder module, repectively.

There is a skip connection from the encoder to the decoder, which can improve the model ability to transmit more information from the encoder to the decoder.

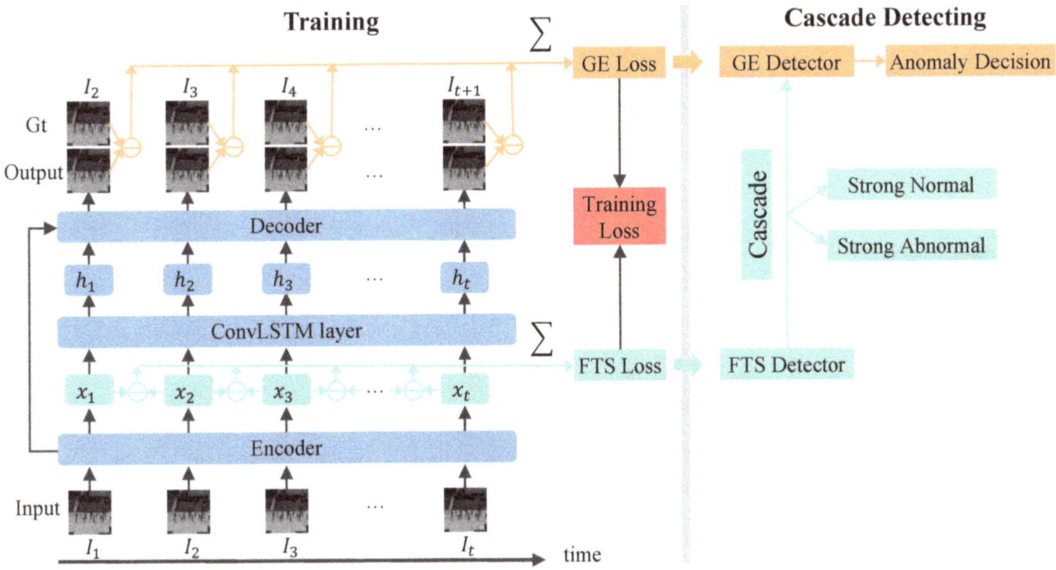

Figure 1. Pipeline of the proposed method. FTS-LSTM trains an LSTM-AE to predict future frames for input frames. FTS-LSTM uses two losses to constrain the model: a GE loss and a FTS loss. The GE loss enables the model to predict future frames precisely. The FTS loss enables features to maintain videos' temporal regularity. In the testing period, the FTS loss and the GE loss as indicators are utilized to detect anomalies. FTS-LSTM cascades the FTS indicator and the GE indicator to achieve fast and accurate performances.

3.1.1. Encoder Module

The encoder module extracts spatial features for input frames. It consists of several 2D spatial convolution layers. Let $\mathcal{E}$ express the encoder, and $\{I_1, \ldots, I_t, \ldots, I_T\}$ be T consecutive input video frames. The feature of the frame I_t can be represented as

$$x_t = \mathcal{E}(I_t), \tag{1}$$

where x_t is the extracted feature for frame I_t. Therefore, we can get T consecutive features $\{x_1, \ldots, x_t, \ldots, x_T\}$ for $\{I_1, \ldots, I_t, \ldots, I_T\}$.

3.1.2. ConvLSTM Module

The ConvLSTM module aims to capture videos' temporal regularities in the feature space. The ConvLSTM is widely used in many video processing tasks. The process of the ConvLSTM module can be expressed as

$$\hat{C}_t = relu(W_C \odot [h_{t-1}, x_t] + b_C) \tag{2}$$
$$i_t = \sigma(W_i \odot [h_{t-1}, x_t] + b_i) \tag{3}$$
$$f_t = \sigma(W_f \odot [h_{t-1}, x_t] + b_f) \tag{4}$$
$$C_t = f_t * C_{t-1} + i_t * \hat{C}_t \tag{5}$$
$$o_t = \sigma(W_o \odot [h_{t-1}, x_t] + b_o) \tag{6}$$
$$h_t = o_t * relu(C_t), \tag{7}$$

where i_t, f_t and o_t are the input gate, forget gate, and output gate at time t; $\hat{C}_t$ is the input information of the LSTM at time t; C_t is the cell state at time t (it stores the information of history frames $[I_{T-4}, I_{T-1}]$); h_t is the output of the LSTM layer at time t; W_C, W_i, W_f, W_o are the weights metrics; b_C, b_i, b_f, b_o are the biases of ConvLSTM; $\odot$ and $*$ represent the convolution operation and pointwise multiplication, respectively; and σ and $relu$ represent the sigmoid and ReLU [44] activation function. The LSTM network is shown in Figure 2. We use $\mathcal{H}$ to represent the ConvLSTM module. At time t, the ConvLSTM's processing function can be simply expressed as

$$h_t = \mathcal{H}(x_t, h_{t-1}), \tag{8}$$

where x_t is the input at time t; h_{t-1} is the hidden state at time $t-1$; and h_t is the hidden state at time t. Based on (8), we get T consecutive hidden states $\{h_1, \ldots, h_t, \ldots, h_T\}$ for consecutive features $\{x_1, \ldots, x_t, \ldots, x_T\}$.

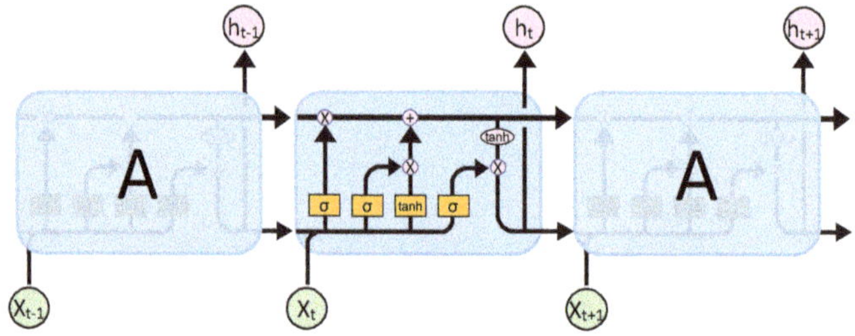

Figure 2. LSTM structure.

3.1.3. Decoder Module

The decoder module plays the role of a generator. It predicts future frames for input frames given $\{h_1, \ldots, h_t, \ldots, h_T\}$. It consists of several 2D convolution layers and 2D deconvolution layers. We utilize $\mathcal{D}$ to express the decoder, and use $\hat{I}_{t+1}$ to represent the prediction result for frame I_t. We have

$$\hat{I}_{t+1} = \mathcal{D}(h_t), \tag{9}$$

where $\mathcal{D}$ is the decoder and $\hat{I}_{t+1}$ is the output of $\mathcal{D}$, whose ground truth is I_{t+1}.

3.2. *The Training Process*

In the training process, we use a GE loss and an FTS loss to constrain the model to learn videos' normal regularity.

3.2.1. The GE Loss

The GE loss consists of two sub-GE losses, l_{int} and l_{gdl}, whose functions are represented as follows,

$$L_{GE} = l_{int} + l_{gdl}, \tag{10}$$

$$l_{int} = \sum_{t=1}^{T} \|\hat{I}_{t+1} - I_{t+1}\|_2, \tag{11}$$

$$l_{gdl} = \sum_{t=1}^{T} (\|\nabla_x(\hat{I}_{t+1}) - \nabla_x(I_{t+1})\|_1 + \|\nabla_y(\hat{I}_{t+1}) - \nabla_y(I_{t+1})\|_1), \tag{12}$$

where l_{int} is the intensity loss, which is applied to penalize the losses on pixels' intensities; l_{gdl} is the gradient loss which is applied to penalize errors around edges; and ∇_x and ∇_y represent the spatial derivatives along the x-axis and y-axis, respectively.

The purpose of GE loss is to enable the model to accurately generate normal samples. It does not constrain videos' features directly, because there is a decoder module between the feature space and the GE loss. As a result, the GE loss is not powerful enough to force features maintaining videos' temporal regularity, and the LSTM layers would not capture videos' temporal regularity precisely.

3.2.2. FTS Loss

In order to capture the videos' temporal regularity precisely, we present an FTS loss to constrain the feature space directly. The content of the video frames changes smoothly over time. Therefore, the features of video frames should also change smoothly in the feature space.

Based on this point, we design the FTS loss to force temporal-consecutive features to be similar. We use the Euclidean distance to measure the similarity between features and accumulate the distances between all temporal-neighbored features to formulate the FTS loss. The FTS loss is expressed as

$$L_{FTS} = \sum_{t=1}^{T-1} \|x_{t+1} - x_t\|_2. \tag{13}$$

3.2.3. Global Training Loss

We combine the GE loss and FTS loss to train the model. The global training loss has a coefficient that is called λ, and it can be represented as

$$L_{train} = L_{GE} + \lambda * L_{FTS}. \tag{14}$$

3.3. *Detecting Process*

In the detecting period, we design a GE detector and FTS detector based on the GE loss and the FTS loss, respectively. We cascade these two detectors to achieve faster and better anomaly detections.

This section first introduces the GE detector's and the FTS detector's working mechanisms, then analyses why the FTS loss is helpful to improve GE detector's anomaly-detection performance.

3.3.1. The GE Detector

The model is trained to predict normal samples. It cannot predict anomalous samples well. We use the l_{int} of the last frame to detect anomalies. Considering that anomalies usually occur in local areas, the maximum of block-level GEs in a frame is used to detect anomalies [45], which is defined as

$$GE_{map}(t) = \sum_{c} \|\hat{I}_{t+1} - I_{t+1}\|_2, \tag{15}$$

$$S_{GE}(t) = max(mean_{bl\_size}(GE_{map}(t))), \tag{16}$$

where $GE_{map}(t)$ is the GE map of the predicted frame $\hat{I}_{t+1}$; $S_{GE}(t)$ is the anomaly score for frame I_{t+1} in the GE detector; $mean_{bl\_size}$ indicates a mean filter with kernel size *bl_size*; and c indicates the number of channels of a frame.

3.3.2. The FTS Detector

The DNN learns the mapping function between two manifold distributions, which is only applicable to samples that obey the manifold distributions. When a sample does not obey the input manifold distribution, its mapping position will deviate from its target position on the output distribution. We call the difference between the mapping position of the sample and the target mapping position as a mapping error. In FTS-LSTM, the encoder learns a mapping function from the manifold of normal frames to a feature space. When an abnormal sample (outliers of the normal manifold) is input to the encoder, there will be a large number of mapping errors in the feature space, and anomalous videos FTS loss will increase. Therefore, the FTS loss can be used to detect abnormalities. Based on this point, we use the FTS loss as an indicator to detect anomalies and judge the samples with large FTS losses as anomalies. Considering that anomalies occur in local areas, we use the maximum value of the FTS loss map to detect anomalies. The FTS detector is defined as

$$FTS_{map}(t) = \sum_{c} \|x_t - x_{t-1}\|_2, \tag{17}$$

$$S_{FTS}(t) = max(FTS_{map}(t)), \tag{18}$$

where $FTS_{map}(t)$ is the FTS-loss-map of I_t; $S_{FTS}(t)$ is the anomaly score for I_t in the FTS detector; and c indicates the number of channels of the feature map.

As shown in (17), the FTS detector detects anomalies by detecting the difference between the apparent characteristics of the target over time. Therefore, the detector is suitable to detecting dynamic anomalies (the abnormal targets having motion in the scene).

3.3.3. Cascade

The FTS detector detects anomalies in the feature space. It is faster than the GE detector. The FTS detector can be cascaded with the GE detector to detect anomalies. When a sample is input into the model, its features are extracted and then the SN and SA samples are detected with the FTS detector. Afterward, the remaining features are fed to the following network modules and the GE detector is used to make the final decision. In the cascading process, it is essential to set suitable thresholds for FTS detector In this paper, we set the SA threshold thr^a and the SN threshold thr^n based on the FTS anomaly scores of the training data. We have

$$thr^a = max(S^{train}_{FTS}) + (max(S^{train}_{FTS}) - min(S^{train}_{FTS})) * \gamma_a, \tag{19}$$

$$thr^n = min(S^{train}_{FTS}) + (max(S^{train}_{FTS}) - min(S^{train}_{FTS})) * (1 - \gamma_n), \tag{20}$$

where $max(scores)$ and $min(scores)$ indicates the maximum value and the minimum value of the $scores$, respectively; S^{train}_{FTS} indicates the FTS anomaly scores of the training data; γ_a and γ_n indicate the strict coefficients for thr^a and thr^n, respectively.The higher the γ_a and γ_n, the more credible the extracted SA and SN samples. Generally, γ_a and γ_n are in the range of $[0,1]$.

As shown in (19), we set the maximum value of normal training samples' FTS loss, $max(S^{train}_{FTS})$, as the base value of the SA threshold. We added the second term, $(max(S^{train}_{FTS}) - min(S^{train}_{FTS})) * \gamma_a$, as the strengthen value. The strengthen value is calculated by the max–min difference value multiplying a ratio. As shown in (20), we set the minimum value of normal training samples' FTS loss, $min(S^{train}_{FTS})$, as the base value of the SN threshold. It is too strict to detect SN samples. Therefore, we added the second term, $(max(S^{train}_{FTS}) - min(S^{train}_{FTS})) * (1 - \gamma_n)$, as the relaxing value. The relaxing value is calculated by the $(max(S^{train}_{FTS}) - min(S^{train}_{FTS}))$ difference value multiplying a ratio.

3.3.4. Discussion

The GE detector can detect both temporal and spatial anomalies in videos. Its anomaly-detection mechanism is analyzed as follows. Let us substitute Equations (8) and (9) into Equation (16). Then the GE detector can be expressed as

$$S_{GE} = max(mean(\sum_{c} |\mathcal{D}(\mathcal{H}(h_{t-1}, x_t)) - I_{t+1}|^2)). \tag{21}$$

As shown in (21), the GE is generated by $\hat{I}_{t+1}$ and $\hat{I}_{t+1}$ is generated from h_t. The h_t has two information sources: the x_t and the h_{t-1}.

The x_t supplies the spatial information of the current input frame I_t. It is generated by the encoder module. The encoder module is trained to extract spatial features for normal frames; it cannot extract features correctly for abnormal frames. Therefore, there will be information differences between the extracted features and the aiming features for abnormal frames. The information differences in x_t will lead to the large GEs in $\hat{I}_{t+1}$. Therefore, the GE loss can be used to detect spatial anomalies.

The h_{t-1} supplies history information including $I_{t-4}, I_{t-3}, I_{t-2}, I_{t-1}$, respectively . The h_{t-1} captures history information by the memory cell C_t and three gates i_t, f_t, o_t in the LSTM module. In the training process, the memory cell and three gates are trained to capture information from sequences of historical features that obey normal temporal regularities. When features do not obey normal temporal regularities, the three gates will capture incorrect information from historical features. Thus, there will be errors of information in h_{t-1}. The error of information in h_{t-1} will lead to the larger GE losses in $\hat{I}_{t+1}$. Therefore, the GE loss can be used to capture temporal anomalies.

As analyzed above, the better the LSTM layer learns normal videos' temporal regularity, the better the performance the GE detector can capture videos' temporal anomalies. The better FTS loss enables feature space to maintain normal videos temporal regularity, the better the LSTM layer can learn videos' temporal regularity. Therefore, the FTS loss can help the GE detector to achieve better anomaly-detection performances.

4. Results

In this section, we carry out experiments to demonstrate the effectiveness of the proposed method.

4.1. Datasets

We evaluate our method on three popular public datasets.

UCSD dataset [46] has two subdatasets: The UCSD Pedestrian 1 (Ped1) dataset and the UCSD Pedestrian 2 (Ped2) dataset. The Ped1 dataset contains 34 training videos and 36 testing videos. The Ped2 dataset contains 16 training videos and 12 testing videos. The two datasets are captured from different scenarios. Their abnormal events include cycling, skateboarding, crossing lawns, cars, etc. These two subdatasets are usually used separately.

The CUHK Avenue dataset [34] contains 16 training videos and 21 testing videos. The abnormal events include running, throwing schoolbag, throwing papers, etc. The size of people may change with the positions and angles of the camera.

The ShanghaiTech (SH) dataset [19] contains 330 training videos and 107 testing videos. The videos are captured from 13 different scenes. The abnormal events include running, cars, throwing schoolbag, etc.

4.2. Implementation Details

In all experiments, video frames are resized to 256 × 256 pixels, the pixel values of video frames are normalized to $[-1, 1]$, the LSTM layer's length $T = 5, minibatch = 2$, and $\lambda = 100$. In the training process, the Adam algorithm [47] is utilized as the optimizer. Each dataset trains for 200,000 iterations with minibatch=2 on a single GTX 1080 GPU. The learning rate is set 1×10^{-4} when the iteration is low than 40,000, which is set to 1×10^{-5} when the iteration is high than 40,000. In the testing stage, set $bl\_size = 30, \gamma_a = 0.2$. In Ped1 and Ped2 datasets, $\gamma_n = 0.8$. In Avenue and SH datasets, $\gamma_n = 0.4$ to achieve better performances.

The detail of FTS-LSTM network is shown in Figure 3. All the kernel sizes and strides of the convolution layers are $(3, 3)$ and $(1, 1)$, respectively. All the kernel sizes and strides of the transpose convolution layers are $(2, 2)$ and $(2, 2)$, respectively. The pool size and strides of the polling layers are $(2, 2)$ and $(2, 2)$, respectively. We adopt the Relu activation function in all convolution layers. The green rectangles indicate the tensor obtained by the convolution operation, and the orange rectangles indicate the tensor obtained by deconvolution. In the deconvolution process, the number of tensor channels is halved, and the height and width of tensors are doubled. The function of concatenate is to transmit more information from the encoder to the decoder so that the decoder can obtain a better generation effect and better anomaly-detection effect [8].

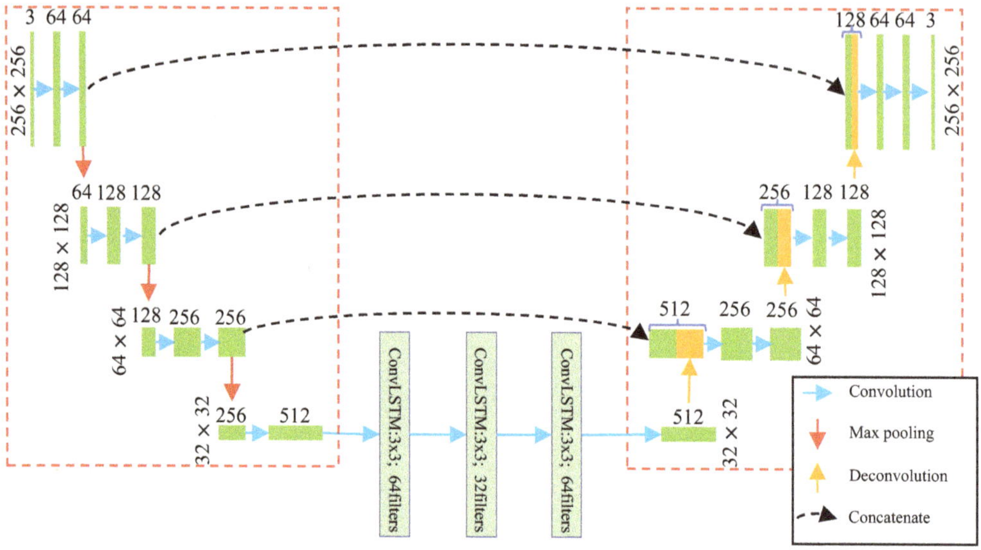

Figure 3. The detail of the network structure of our work. There are three zones in the network, in which the left zone is called the encoder, the right zone is called the decoder, and the rest of the structure in the middle is the LSTM network.

As shown in Figure 3, The entire network contains 21 layers of convolution or deconvolution operations: seven layers of 3 × 3 convolution operations in the encoder module, three layers of 3 × 3 convolution operations in the LSTM module, three deconvolution operations in the Decoder network, and eight convolution operations in the decoder network.

4.3. Evaluation Metric

In video anomaly detection, the most commonly used evaluation metric is the receiver operation characteristic (ROC) curve and the area under this curve (AUC). A higher AUC value indicates better anomaly-detection performance. This paper adopts the frame-level AUC to evaluate anomaly-detection performances.

4.4. Anomaly-Detection Performances

Table 1 shows anomaly detection ROC/AUC performances of the proposed model, comparing with some state-of-the-art (SOTA) and classic methods, including DDB [14], DPB [20], DGE [8,19,40,41,48], and the aggregation methods [21–23]. In the Table, the optimal performance in each dataset is marked with bold font, and the suboptimal performance is marked with bold italic font. The proposed model achieves optimal and suboptimal performances on Ped2, Avenue, and SH datasets. Meanwhile, its detection speed is 117 FPS on average, which is far faster than other algorithms. These performances demonstrate the superiority of the proposed method.

Frame-level anomaly-detection scores (between 0 and 1) provided by our FST-LSTM framework are shown in Figure 4. The cyan zone represents the ground-truth abnormal events and our scores are illustrated in red. The pictures in the figure are the frames of the Avenue dataset captured from test video 4 to test video 6, which illustrate the effect of our framework. Anomaly-detection heatmaps of videos are shown in Figure 5. As shown in Figure 5b,c, the FTS loss in anomalous areas are higher than that in normal areas. They demonstrate that the FTS loss can detect and localize anomalies. Figure 5d,e show intensity maps and heatmaps of the GE indicator. They demonstrate the anomaly-detection performances of the GE indicator.

4.5. Ablation Study

This section carries out experiments to demonstrate the problems proposed in the introduction and prove the effectiveness of the proposed model in solving these problems.

4.5.1. Feature Space TSNE Visualization

Figure 6 visualizes two video features in the model's feature space. As shown in Figure 6a, when the model is trained without utilizing the FTS loss, video features are randomly distributed in the feature space. It indicates that the feature space does not maintain videos' temporal regularity precisely. As shown in Figure 6b, when the model is trained with utilizing the FTS loss, video features are distributed in the feature space in an orderly manner. The features of different videos are separable from each other. It indicates that the model's feature space maintained videos' temporal regularity. The visualization verified the effectiveness of the FTS loss on maintaining videos' temporal regularity. The result demonstrates the proposed model can solve the question when utilizing LSTM layer to detect anomalies.

Table 1. Frame-level ROC/AUCs of different methods. The bold font represent the optimal performance, and the bold italic font represent the suboptimal performance.

Method	–	Ped1	Ped2	Avenue	SH	Speed
Deep Distance-based	DeepOC [14]	83.5	96.9	86.6	–	***40 FPS***
Deep Probability-based	Tang et al. [20]	84.7	96.3	85.1	71.5	30 FPS
Aggregation methods	STAN [21]	82.1	96.5	87.2	–	–
	TAM-Net [22]	83.5	98.1	78.3	–	–
	MAAS [23]	**85.8**	**99.0**	**92.1**	69.7	4 FPS
Deep Generation-error-based	Unet [8]	83.1	95.4	85.1	72.8	12 FPS
	Ts-Unet [48]	–	97.8	88.4	–	12 FPS
	sRNN [19]	–	92.2	83.5	69.6	10 FPS
	MemAE [40]	–	94.1	83.3	71.2	38 FPS
	Zhou et al. [41]	83.9	96.0	86.0	–	–
	FTS-LSTM (ours)	83.5	***98.3***	***91.1***	***72.9***	**117 FPS**

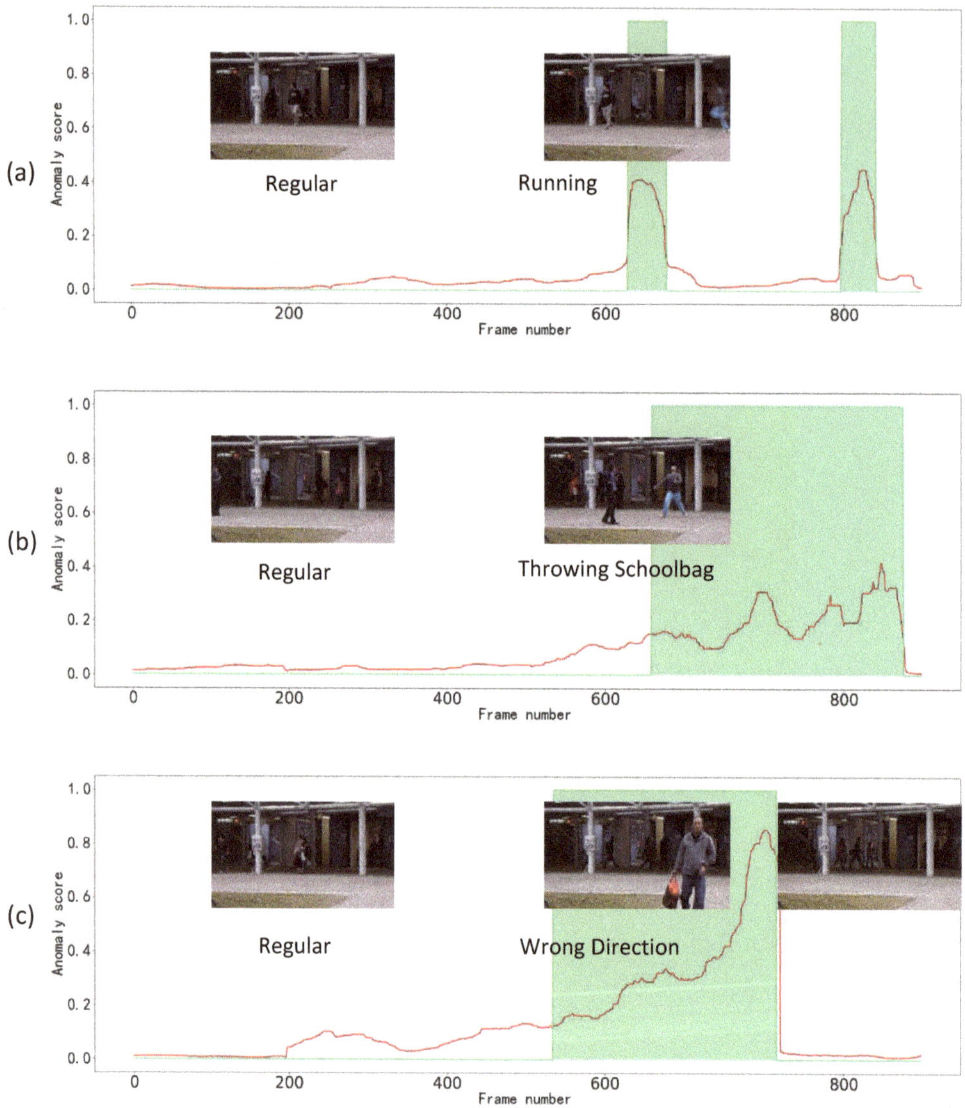

Figure 4. Frame-level anomaly-detection scores (between 0 and 1) provided by our FST-LSTM framework based on the late fusion strategy, for test in the Avenue dataset. The green lines and green zone represent the ground truth abnormal events. The red lines represent our scores. (**a**) Test video 4 in the Avenue dataset. (**b**) Test video 5 in the Avenue dataset. (**c**) Test video 6 in the Avenue dataset.

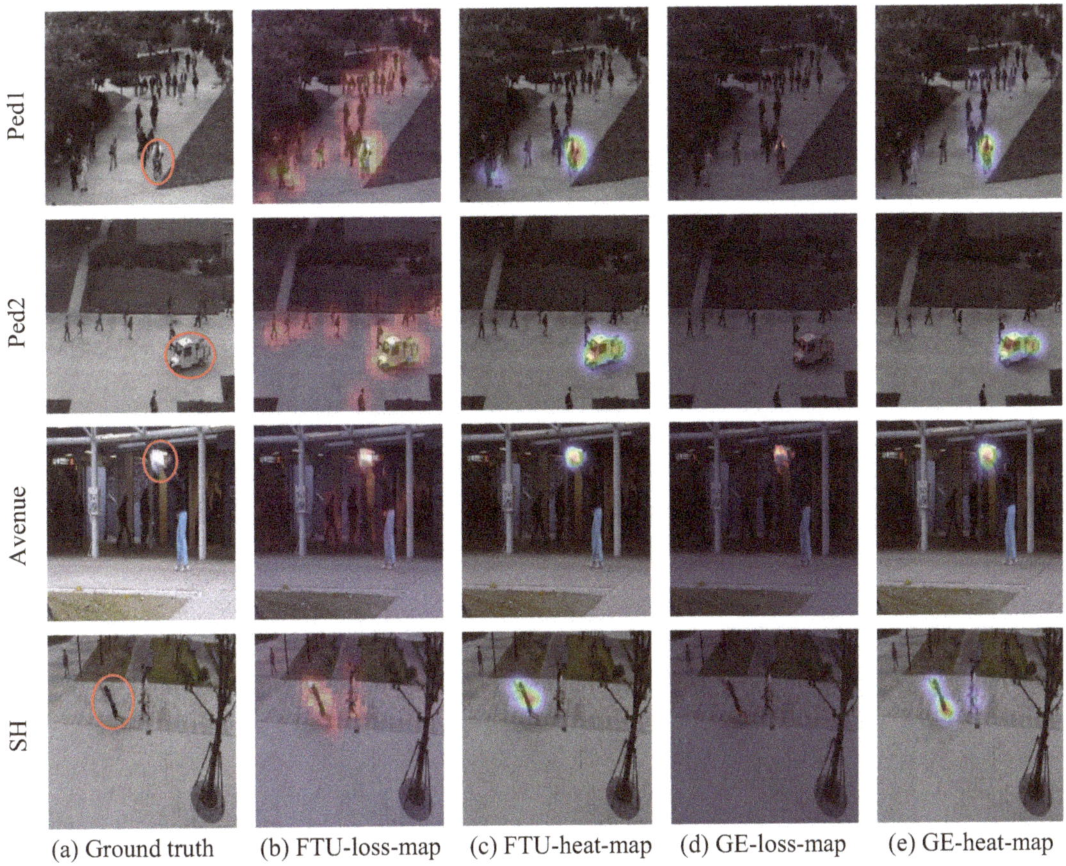

Figure 5. Anomaly-detection visualization. (**a**) Anomalous frames in different datasets. The contents in red circles are anomalous events. (**b**) FTS loss's intensity map. (**c**) FTS loss's heatmap. (**d**) GE loss's intensity map. (**e**) GE loss's heatmap.

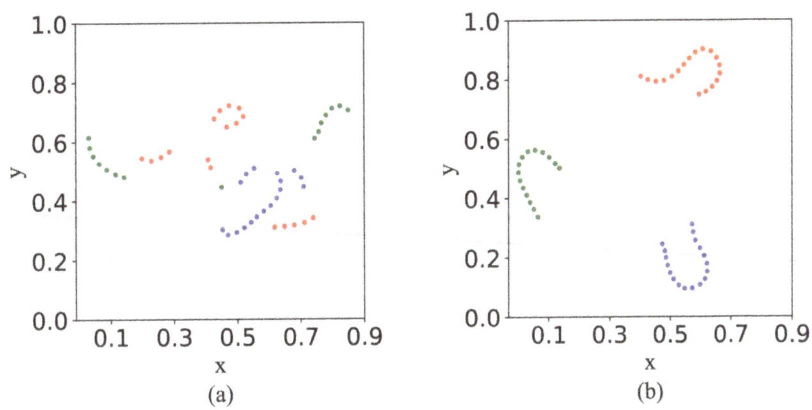

Figure 6. Dots with different colors indicates features belonging to different videos. (**a**) Without FTS loss. (**b**) With FTS loss.

4.5.2. Impact of FTS Loss on the GE Detector

The FTS loss enables LSTM layer to learn videos' temporal regularity more precisely. It increases GE detector's anomaly-detection performance. Table 2 and Figure 7 show the anomalous frames' GE saliencies in models trained with and without utilizing the FTS loss and shows the ROC/AUCs of corresponding models. The table demonstrates that the FTS loss improves anomalous frames' GE saliencies and improves GE detector's anomaly-detection performances.

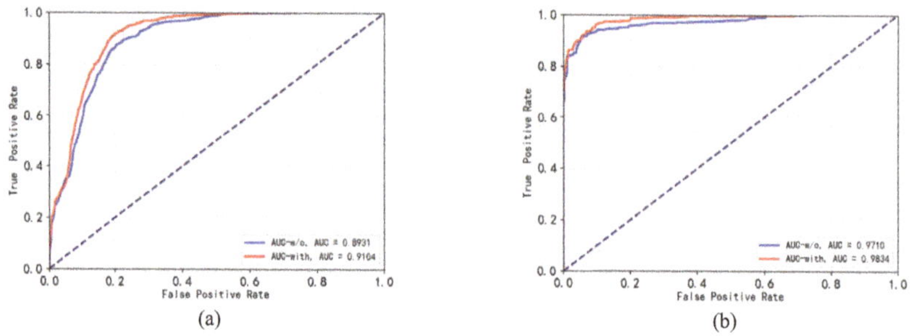

Figure 7. The ROC/AUC curves of the GE detectors trained with and without utilizing the FTS loss on multiple datasets. The red curve represents the detector trained with FTS loss. The blue curve represents the detector trained without FTS loss. (**a**) The ROC/AUC curves on Avenue dataset. (**b**) The ROC/AUC curves on Ped2 dataset. The dashed blue line represent the ROC curve of a completely random classifier.

Table 2. Frame-level GE saliency and ROC/AUCs of the GE detectors on multiple datasets. The bold font represent GE saliency of anomalous frames and ROC/AUC performances utilizing the FTS loss.

	FTS Loss	Ped1	Ped2	Avenue	SH
GE saliency of Anomalous frames	w/o	1.930	3.657	2.645	1.184
	with	**2.205**	**3.985**	**2.656**	**1.366**
ROC/AUC	w/o	82.73	97.10	89.31	71.20
	with	**83.51**	**98.34**	**91.04**	**72.92**

4.5.3. Impact of the FTS Loss on FTS Detector

The DNN trained on normal samples cannot maintain relationships among abnormal samples. Table 3 calculates the FTS loss saliencies of anomalous frames compared with normal frames. As shown in the table, all the FTS loss anomaly saliencies are positive, which indicates that the FTS losses of the anomalous frames are higher than that of the normal frames. It indicates that the FTS loss can be used to detect anomalies, which proves our analysis.

Table 3 and Figure 8 show anomaly-detection performances of the FTS detectors. The FTS loss strengthened the encoder to maintain more relationships among normal frames. It increased the anomaly saliencies of the anomalous frames in FTS.

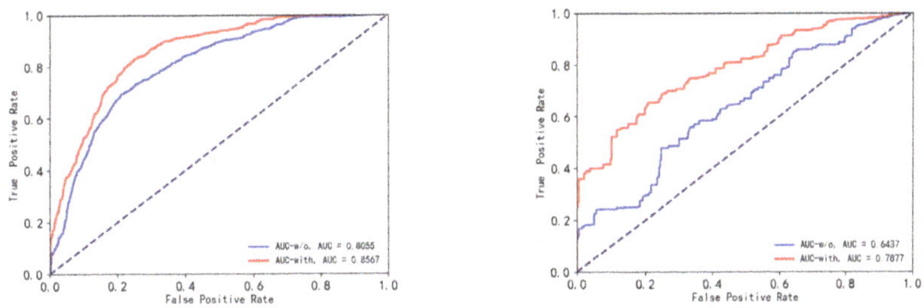

(a) (b)

Figure 8. The ROC/AUC curves of the FTS detectors trained with and without utilizing the FTS loss on multiple datasets. The red curve is represents the detector trained with FTS loss. The blue curve represents the detector trained without FTS loss. (**a**) The ROC/AUC curves on Avenue dataset. (**b**) The ROC/AUC curves on Ped2 dataset. The dashed blue line represent the ROC curve of a completely random classifier.

Table 3. Frame-level FTS saliency and ROC/AUCs of the FTS detectors on multiple datasets. The bold font represent FTS saliency of anomalous frames and ROC/AUC performances utilizing the FTS loss.

	FTS Loss	**Ped1**	**Ped2**	**Avenue**	**SH**
FTS saliency of Anomalous frames	w/o	0.086	0.055	0.342	0.342
	with	**0.162**	**0.122**	**0.639**	**0.374**
ROC/AUC	w/o	64.02	64.37	80.55	67.22
	with	**70.22**	**78.77**	**85.67**	**68.71**

4.5.4. Detection Speed Analysis

By cascading the FTS and GE detectors, the proposed model achieves fast and precise performances. Table 4 shows anomaly-detection ROC/AUCs and speeds of different detectors. It demonstrates that, by cascading the FTS and the GE detectors, the model maintains GE detector's ROC/AUC and achieves a faster speed than the GE detector.

As shown in Table 4, this work can achieve a speed of 117 FPS, and this high detection speed mainly benefits from the low computational complexity of the FTS detector. The FTS detector only calls the encoder module of the network (7 layers 3×3 convolution operations) to detect anomalies and can filter out most video frames in anomaly detection. Only a small number of video frames are transmitted to the subsequent network module, which greatly reduces the amount of calculation in the anomaly-detection process.

Table 4. Frame-level ROC/AUCs of the cascaded detector on multiple datasets

	ROC/AUC				**Speed**
	Ped1	**Ped2**	**Avenue**	**SH**	
FTS Detector	70.22	78.77	85.67	68.71	186 FPS
GE Detector	83.51	98.34	91.04	72.92	50 FPS
Cascade	83.51	98.34	91.14	72.92	117 FPS

4.5.5. Impact of Weight λ

Figure 9 shows the anomaly-detection ROC/AUC of GE metrics and FTS metrics under different λ. This figure proves that the FTS loss can robustly improve the anomaly-detection performance of the model.

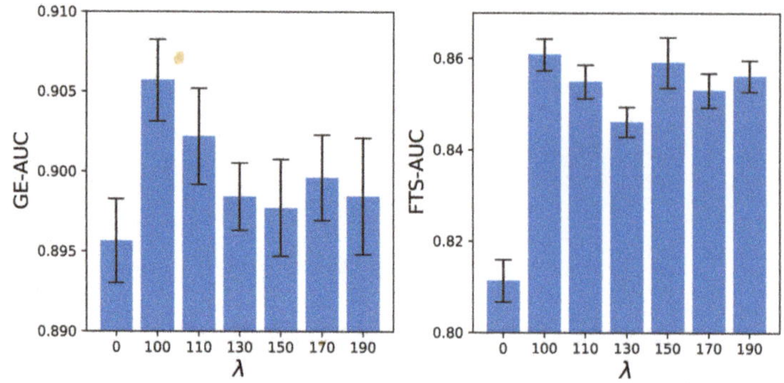

Figure 9. Frame-level ROC/AUCs of the GE and FTS detectors under different FTS loss weights.

4.5.6. Generality

Table 5 shows anomaly-detection saliency and ROC/AUC with or without applying FTS loss in the LSTM model [24]. The anomaly-detection performance and anomaly saliency of the the LSTM model have been significantly improved with FTS loss. This result proves that the temporal smoothing loss in the feature space is general for improving the anomaly-detection performance of the generative model by restraining generated errors.

Table 5. Saliency and ROC/AUC of the LSTM model with or without applying FTS loss. The bold font represent saliency of anomalous frames and ROC/AUC performances utilizing the FTS loss.

	FTS Loss	Ped2	Avenue	Average
Saliency of Anomalous frames	w/o	0.9278	1.086	1.0007
	with	**1.104**	**1.192**	**1.148**
ROC/AUC	w/o	76.51	79.18	77.85
	with	**82.25**	**81.62**	**81.94**

4.6. Limitation

As described above, our proposed method achieves relatively better performance on the UCSD dataset and ShanghaiTech dataset. However, this method might not be good at detecting static anomaly time. For example, the car parked on the sidewalk, the FTS can detect the object in to scene but cannot respond to the static car out because the target brings no changes to the frame's apparent feature. Generally, abnormal events occur along with a dynamic process. Therefore, this limitation is acceptable to surveillance video anomaly detection.

5. Conclusions

This paper proposes a FTS-LSTM method for video anomaly detection. It trains a LSTM-AE to generate normal videos and to detect anomalies. In the training process, it uses the FTS loss and the GE loss to constrain the model. In the detecting process, it cascades the FTS and the GE indicators to detect anomalies. Experiments on multiple datasets reveal the proposed method's effectiveness and efficiency. The shortcoming of the FTS indicator is that it cannot detect static anomalies. In general monitoring scenarios, the occurrence of abnormal events generally have a dynamic process. Therefore, this shortcoming can be ignored. In the future, we will combine the FTS loss with Transformer and the GRU method to explore the proposed method's generalization, and we will study the solution of combining the FTS detector with a static anomaly-detection method to improve the algorithm's ability.

Author Contributions: Conceptualization, L.S., Z.W. and G.W.; funding acquisition, G.W.; methodology, L.S. and Z.W.; project administration, G.W.; resources, G.W.; software, L.S. and Z.W.; supervision, Y.Z.; writing—original draft, L.S.; writing—review & editing, L.S. and Z.W. All authors have read and agreed to the published version of the manuscript.

Funding: This research received no external funding.

Institutional Review Board Statement: Not applicable.

Informed Consent Statement: Not applicable.

Data Availability Statement: Not applicable.

Conflicts of Interest: The authors declare no conflict of interest.

References

1. Xiao, T.; Zhang, C.; Zha, H. Learning to detect anomalies in surveillance video. *IEEE Signal Process. Lett.* **2015**, *22*, 1477–1481. [CrossRef]
2. Prasad, N.R.; Almanza-Garcia, S.; Lu, T.T. Anomaly detection. *Comput. Mater. Contin.* **2009**, *14*, 1–22. [CrossRef]
3. Kim, I.; Jeon, Y.; Kang, J.W.; Gwak, J. RAG-PaDiM: Residual Attention Guided PaDiM for Defects Segmentation in Railway Tracks. *J. Electr. Eng. Technol.* **2022**. [CrossRef]
4. Kang, J.; Kim, C.S.; Kang, J.W.; Gwak, J. Recurrent Autoencoder Ensembles for Brake Operating Unit Anomaly Detection on Metro Vehicles. *Comput. Mater. Contin.* **2022**, *73*, 1–4. [CrossRef]
5. Kang, J.; Kim, C.S.; Kang, J.W.; Gwak, J. Anomaly detection of the brake operating unit on metro vehicles using a one-class lstm autoencoder. *Appl. Sci.* **2021**, *11*, 9290. [CrossRef]
6. Zhang, T.; Aftab, W.; Mihaylova, L.; Langran-Wheeler, C.; Rigby, S.; Fletcher, D.; Maddock, S.; Bosworth, G. Recent Advances in Video Analytics for Rail Network Surveillance for Security, Trespass and Suicide Prevention—A Survey. *Sensors* **2022**, *22*, 4324. [CrossRef]
7. Khan, S.W.; Hafeez, Q.; Khalid, M.I.; Alroobaea, R.; Hussain, S.; Iqbal, J.; Almotiri, J.; Ullah, S.S. Anomaly Detection in Traffic Surveillance Videos Using Deep Learning. *Sensors* **2022**, *22*, 6563. [CrossRef]
8. Liu, W.; Luo, W.; Lian, D.; Gao, S. Future Frame Prediction for Anomaly Detection—A New Baseline. In Proceedings of the 2018 IEEE/CVF Conference on Computer Vision and Pattern Recognition, Salt Lake City, UT, USA, 18–22 June 2018; pp. 6536–6545.
9. Ullah, W.; Ullah, A.; Hussain, T.; Khan, Z.A.; Baik, S.W. An efficient anomaly recognition framework using an attention residual lstm in surveillance videos. *Sensors* **2021**, *21*, 2811. [CrossRef]
10. Dubey, S.; Boragule, A.; Gwak, J.; Jeon, M. Anomalous event recognition in videos based on joint learning of motion and appearance with multiple ranking measures. *Appl. Sci.* **2021**, *11*, 1344. [CrossRef]
11. Ionescu, R.T.; Smeureanu, S.; Alexe, B.; Popescu, M. Unmasking the Abnormal Events in Video. In Proceedings of the 2017 IEEE International Conference on Computer Vision (ICCV), Venice, Italy, 22–29 October 2017; pp. 2914–2922. [CrossRef]
12. Oza, P.; Patel, V.M. One-Class Convolutional Neural Network. *IEEE Signal Process. Lett.* **2019**, *26*, 277–281.
13. Weixiang, J.; Gong, L. One-class neural network for video anomaly detection and localization. *Electron. Meas. Instrum.* **2021**, *35*, 60–65.
14. Wu, P.; Liu, J.; Shen, F. A Deep One-Class Neural Network for Anomalous Event Detection in Complex Scenes. *IEEE Trans. Neural Netw. Learn. Syst.* **2019**, *31*, 2609–2622. [CrossRef] [PubMed]
15. Abati, D.; Porrello, A.; Calderara, S.; Cucchiara, R. Latent space autoregression for novelty detection. In Proceedings of the IEEE Computer Society Conference on Computer Vision and Pattern Recognition, Long Beach, CA, USA, 16–17 June 2019; IEEE Computer Society: Washington, DC, USA, 2019; Volume 2019, pp. 481–490.
16. Wang, T.; Xu, X.; Shen, F.; Yang, Y. A Cognitive Memory-Augmented Network for Visual Anomaly Detection. *IEEE/CAA J. Autom. Sin.* **2021**, *8*, 1296–1307. [CrossRef]
17. Sabokrou, M.; Pourreza, M.; Fayyaz, M.; Entezari, R.; Fathy, M.; Gall, J.; Adeli, E. AVID: Adversarial Visual Irregularity Detection. In *Computer Vision—ACCV 2018, Proceedings of the 14th Asian Conference on Computer Vision, Perth, Australia, 2–6 December 2018*; Lecture Notes in Computer Science (Including Subseries Lecture Notes in Artificial Intelligence and Lecture Notes in Bioinformatics); Springer: Cham, Switzerland, 2018; Volume 11366 LNCS, pp. 488–505.
18. Song, H.; Sun, C.; Wu, X.; Chen, M.; Jia, Y. Learning Normal Patterns via Adversarial Attention-Based Autoencoder for Abnormal Event Detection in Videos. *IEEE Trans. Multimed.* **2020**, *22*, 2138–2148. [CrossRef]
19. Luo, W.; Liu, W.; Lian, D.; Tang, J.; Duan, L.; Peng, X.; Gao, S. Video Anomaly Detection with Sparse Coding Inspired Deep Neural Networks. *IEEE Trans. Pattern Anal. Mach. Intell.* **2021**, *43*, 1070–1084. [CrossRef]
20. Tang, Y.; Zhao, L.; Zhang, S.; Gong, C.; Li, G.; Yang, J. Integrating prediction and reconstruction for anomaly detection. *Pattern Recognit. Lett.* **2020**, *129*, 123–130. [CrossRef]
21. Lee, S.; Kim, H.G.; Ro, Y.M. STAN: Spatio-Temporal Adversarial Networks for Abnormal Event Detection. In Proceedings of the 2018 IEEE International Conference on Acoustics, Speech and Signal Processing (ICASSP), Calgary, AB, Canada, 15–20 April 2018; pp. 1323–1327.

22. Ji, X.; Li, B.; Zhu, Y. TAM-Net: Temporal Enhanced Appearance-to-Motion Generative Network for Video Anomaly Detection. In Proceedings of the 2020 International Joint Conference on Neural Networks (IJCNN), Glasgow, UK, 19–24 July 2020; pp. 1–8. [CrossRef]
23. Wang, Z.; Zhang, Y.; Wang, G.; Xie, P. Main-Auxiliary Aggregation Strategy for Video Anomaly Detection. *IEEE Signal Process. Lett.* **2021**, *28*, 1794–1798. [CrossRef]
24. Chong, Y.S.; Tay, Y.H. Abnormal Event Detection in Videos Using Spatiotemporal Autoencoder. In *Advances in Neural Networks—ISNN 2017*; Lecture Notes in Computer Science; Springer: Cham, Switzerland, 2017; Volume 10262, pp. 189–196. [CrossRef]
25. Luo, W.; Liu, W.; Gao, S. Remembering history with convolutional LSTM for anomaly detection. In Proceedings of the 2017 IEEE International Conference on Multimedia and Expo (ICME), Hong Kong, China, 10–14 July 2017; pp. 439–444. [CrossRef]
26. Hasan, M.; Choi, J.; Neumann, J.; Roy-Chowdhury, A.K.; Davis, L.S. Learning Temporal Regularity in Video Sequences. In Proceedings of the 2016 IEEE Conference on Computer Vision and Pattern Recognition (CVPR), Seattle, WA, USA, 27–30 June 2016; pp. 733–742.
27. Huang, C.; Wen, J.; Xu, Y.; Jiang, Q.; Yang, J.; Wang, Y.; Zhang, D. Self-Supervised Attentive Generative Adversarial Networks for Video Anomaly Detection. *IEEE Trans. Neural Networks Learn. Syst.* **2022**, 1–15. [CrossRef]
28. Ionescu, R.T.; Smeureanu, S.; Popescu, M.; Alexe, B. Detecting Abnormal Events in Video Using Narrowed Normality Clusters. In Proceedings of the 2019 IEEE Winter Conference on Applications of Computer Vision (WACV), Waikoloa Village, HI, USA, 7–11 January 2019; pp. 1951–1960.
29. Hinami, R.; Mei, T.; Satoh, S. Joint Detection and Recounting of Abnormal Events by Learning Deep Generic Knowledge. In Proceedings of the 2017 IEEE International Conference on Computer Vision (ICCV), Venice, Italy, 22–29 October 2017; pp. 3639–3647.
30. Pruteanu-Malinici, I.; Carin, L. Infinite Hidden Markov Models for Unusual-Event Detection in Video. *IEEE Trans. Image Process.* **2008**, *17*, 811–822. [CrossRef]
31. Xiang, T.; Gong, S. Incremental and adaptive abnormal behaviour detection. *Comput. Vis. Image Underst.* **2008**, *111*, 59–73. [CrossRef]
32. Hu, X.; Huang, Y.; Gao, X.; Luo, L.; Duan, Q. Squirrel-cage local binary pattern and its application in video anomaly detection. *IEEE Trans. Inf. Forensics Secur.* **2019**, *14*, 1007–1022. [CrossRef]
33. Gnouma, M.; Ejbali, R.; Zaied, M. Video Anomaly Detection and Localization in Crowded Scenes. *Adv. Intell. Syst. Comput.* **2020**, *951*, 87–96. [CrossRef]
34. Lu, C.; Shi, J.; Jia, J. Abnormal Event Detection at 150 FPS in MATLAB. In Proceedings of the 2013 IEEE International Conference on Computer Vision, Sydney, NSW, Australia, 1–8 December 2013; pp. 2720–2727.
35. Cong, Y.; Yuan, J.; Liu, J. Sparse reconstruction cost for abnormal event detection. In Proceedings of the IEEE Computer Society Conference on Computer Vision and Pattern Recognition, Colorado Springs, CO, USA, 20–25 June 2011; pp. 3449–3456. [CrossRef]
36. Chu, W.; Xue, H.; Yao, C.; Cai, D. Sparse Coding Guided Spatiotemporal Feature Learning for Abnormal Event Detection in Large Videos. *IEEE Trans. Multimed.* **2019**, *21*, 246–255. [CrossRef]
37. Fan, Y.; Wen, G.; Li, D.; Qiu, S.; Levine, M.D.; Xiao, F. Video anomaly detection and localization via Gaussian Mixture Fully Convolutional Variational Autoencoder. *Comput. Vis. Image Underst.* **2020**, *195*, 102920.
38. Sabokrou, M.; Khalooei, M.; Fathy, M.; Adeli, E. Adversarially Learned One-Class Classifier for Novelty Detection. In Proceedings of the 2018 IEEE/CVF Conference on Computer Vision and Pattern Recognition, Salt Lake City, UT, USA, 18–22 June 2018; pp. 3379–3388.
39. Ravanbakhsh, M.; Nabi, M.; Sangineto, E.; Marcenaro, L.; Regazzoni, C.; Sebe, N. Abnormal event detection in videos using generative adversarial nets. In Proceedings of the 2017 IEEE International Conference on Image Processing (ICIP), Beijing, China, 17–20 September 2017; pp. 1577–1581. [CrossRef]
40. Gong, D.; Liu, L.; Le, V.; Saha, B.; Mansour, M.R.; Venkatesh, S.; Van Den Hengel, A. Memorizing Normality to Detect Anomaly: Memory-Augmented Deep Autoencoder for Unsupervised Anomaly Detection. In Proceedings of the 2019 IEEE/CVF International Conference on Computer Vision (ICCV), Seoul, Republic of Korea, 27 October–2 November 2019; Volume 2019, pp. 1705–1714.
41. Zhou, J.T.; Zhang, L.; Fang, Z.; Du, J.; Peng, X.; Xiao, Y. Attention-Driven Loss for Anomaly Detection in Video Surveillance. *IEEE Trans. Circuits Syst. Video Technol.* **2020**, *30*, 4639–4647. [CrossRef]
42. Medel, J.R.; Savakis, A. Anomaly Detection in Video Using Predictive Convolutional Long Short-Term Memory Networks. *arXiv* **2016**, arXiv:1612.00390.
43. Lu, Y.; Kumar, K.M.; Nabavi, S.S.; Wang, Y. Future Frame Prediction Using Convolutional VRNN for Anomaly Detection. In Proceedings of the 2019 16th IEEE International Conference on Advanced Video and Signal Based Surveillance (AVSS), Taipei, Taiwan, 18–21 September 2019; pp. 1–8.
44. Nair, V.; Hinton, G.E. Rectified Linear Units Improve Restricted Boltzmann Machines. In Proceedings of the 27th International Conference on International Conference on Machine Learning (ICML'10), Madison, WI, USA, 21–24 June 2010; pp. 807–814.
45. Wang, Z.; Yang, Z.; Zhang, Y.J. A promotion method for generation error-based video anomaly detection. *Pattern Recognit. Lett.* **2020**, *140*, 88–94.

46. Mahadevan, V.; Li, W.; Bhalodia, V.; Vasconcelos, N. Anomaly detection in crowded scenes. In Proceedings of the 2010 IEEE Computer Society Conference on Computer Vision and Pattern Recognition, San Francisco, CA, USA, 13–18 June 2010; pp. 1975–1981.
47. Kingma, D.P.; Ba, J. Adam: A Method for Stochastic Optimization. *arXiv* **2014**, arXiv:1412.6980.
48. Wang, Z.; Yang, Z.; Zhang, Y.; Su, N.; Wang, G. Image and Graphics. In *Ts-Unet: A Temporal Smoothed Unet for Video Anomaly Detection, Proceedings of the 11th International Conference on Image and Graphics, Shanghai, China, 13–15 September 2017*; Springer: Berlin/Heidelberg, Germany, 2017; Volume 10666, pp. 447–461. [CrossRef]

sensors

MDPI

Article

VERD: Emergence of Product-Based Video E-Commerce Retrieval Dataset from User's Perspective

Gwangjin Lee [1], Won Jo [2] and Yukyung Choi [1,2,*]

1 Department of Intelligent Mechatronics Engineering, Sejong University, Seoul 05006, Republic of Korea
2 Department of Artificial Intelligence, Sejong University, Seoul 05006, Republic of Korea
* Correspondence: ykchoi@sejong.ac.kr

Abstract: Customer demands for product search are growing as a result of the recent growth of the e-commerce market. According to this trend, studies on object-centric retrieval using product images have emerged, but it is difficult to respond to complex user-environment scenarios and a search requires a vast amount of data. In this paper, we propose the Video E-commerce Retrieval Dataset (VERD), which utilizes user-perspective videos. In addition, a benchmark and additional experiments are presented to demonstrate the need for independent research on product-centered video-based retrieval. VERD is publicly accessible for academic research and can be downloaded by contacting the author by email.

Keywords: computer vision; information retrieval; content-based video retrieval

Citation: Lee, G.; Jo, W.; Choi, Y. VERD: Emergence of Product-Based Video E-Commerce Retrieval Dataset from User's Perspective. *Sensors* **2023**, *23*, 513. https://doi.org/10.3390/s23010513

Academic Editor: Giorgio Fumera

Received: 25 November 2022
Revised: 28 December 2022
Accepted: 30 December 2022
Published: 3 January 2023

1. Introduction

Image retrieval aims to search the database for images that are similar to a given query image. This technology has been used for automatic checkouts, which scan for products at supermarket cash registers. However, as a result of the recent expansion of the e-commerce market induced by the development of communication technology, research has been conducted to find similar items in online shopping malls.

The goal of the offline datasets [1–5] used for image retrieval in conventional stores is to identify the products at the checkout counter in order to complete automatic payment. These datasets comprise images of products placed on shelves in supermarkets in order to find which items are placed on the checkout desk. Due to the characteristics of this product arrangement, the images in datasets have a uniform background due to the fact that they were filmed in a confined space, although there could be subtle variations in illumination and background depending on the display condition. In addition, they consist of photos taken from the product's front that clearly depict the brand and its characteristics in order to facilitate automatic checkout.

As previously stated, the growth of the e-commerce market has resulted in the emergence of online datasets [6–9] used to find similar products in images. These datasets have a more complex background than prior datasets used for automatic checkout. This is because the datasets consist of both images captured by users and uploaded by the sellers to promote the item. The data obtained from the user are realistic, but the product image processed by the seller may include marketing text or other effects. There is a difference in the angle of view, illumination, color, and background between the objects photographed by the actual user and the objects photographed by the seller. These differences make it difficult to identify real objects.

Multimodal online datasets [10–12] have appeared to tackle these problems and enable a more sophisticated product search. In contrast to previous studies that relied solely on images, most of the multimodal online datasets contain text and image information that can be used for retrieval, and audio and video information are also being utilized in research [13].

By using extra information from multimodal online datasets, retrieval can be achieved even when images have insufficient information to distinguish products. However, because the datasets were derived from data processed by the seller, they lack the very same data as the user's search environment. In addition, it is inconvenient that users must provide additional data in addition to image data in a real-world retrieval environment.

To handle the limitations that existing datasets are not comparable to actual search environments and that multimodal datasets do not reduce search complexity, we propose a dataset named Video E-commerce Retrieval Dataset (VERD). Figure 1 shows the difference between the VERD and existing datasets. Traditional datasets collect data based on images, while VERD collects data based on video reviews, which are increasingly popular on the e-commerce platform. This was performed to leverage the idea that video reviews are filmed from a variety of viewpoints and contain a wealth of product-related information. These video reviews were not filmed by sellers, but, rather, by users with the devices that were used to conduct actual searches. Therefore, unlike the data of sellers, filmed in uniform environments, video reviews include a wide variety of backgrounds and camera angles. Based on these attributes, VERD is comparable to the data used in a real-world search, and it aims to conduct retrieval using only the visual information provided by the video, without providing any additional information. Lastly, we present benchmark performance through the existing video retrieval methods [14–16] on VERD. We believe that VERD and benchmarks will encourage research on video-based product retrieval.

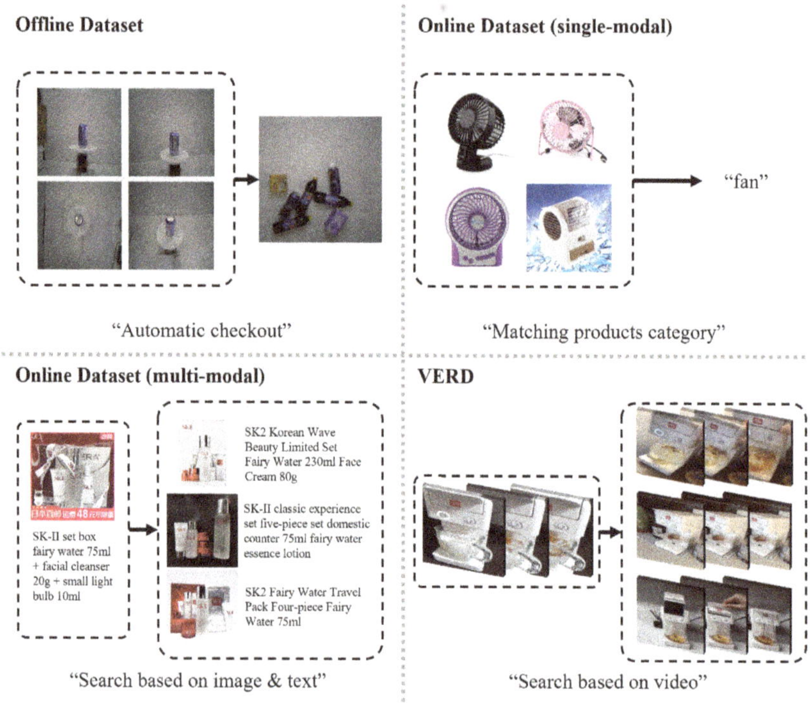

Figure 1. Comparison of datasets related to object-centric retrieval.

2. Related Work

2.1. Datasets

2.1.1. Offline Dataset

Offline datasets are configured to perform tasks such as automatic payment or shop management in conventional grocery stores. Merler et al. [1] proposed the Grozi-120 dataset, which contains images of all products taken in real marketplaces and ideal studios. Jund et al. [2] suggested the Freiburg Grocery dataset, which gathered images from the real-world environments of various shops and apartments to identify various common items, including groceries. Klasson et al. [3] proposed the Grocery Store dataset with hierarchical label information that can combine visual and semantic information on supermarket groceries. Georgiadis et al. [4] suggested the Products-6k dataset, which was created by capturing photos containing product brand names or product descriptions for large-scale product recognition in a supermarket environment. Wei et al. [5] proposed the Retail Product Checkout (RPC) dataset for automatic checkout, which consists of images of objects taken from multiple angles.

2.1.2. Online Dataset

Online datasets, as opposed to offline datasets, comprise data acquired in a varied environment because they collect data uploaded to the e-commerce market. These online datasets can be broadly categorized into two types. The first type of online datasets is a single modality dataset comprising images. Song et al. [6] suggested the Stanford Online Product (SOP) dataset, which has five photos per class but a vast number of classes collected from an e-commerce website. Liu et al. [7] offered the Deepfashion dataset, which contains a variety of images of fashion items, ranging from posed store images to unsupervised consumer photographs. Ge et al. [8] proposed the Deepfashion2 dataset, which includes numerous landmarks and skeletons extracted from fashion-related images. Bai et al. [9] proposed the Product-10k dataset, which consists of photographs of frequently purchased e-commerce product classes across multiple categories, such as food, fashion, and household products.

The second type of online dataset is a multimodal dataset, mainly consisting of text and image data. Corbiere et al. [10] proposed the Dress Retrieval dataset, a noisy image–text multimodal dataset for e-commerce website catalog product descriptions. Chen et al. [11] offered the MEP-3M dataset, which applied hierarchical labels to image–text pair data acquired from Chinese online shopping websites. Zhan et al. [12] proposed the Product 1M dataset containing extensive cosmetic data by gathering textual descriptions of cosmetics and product displays. Dong et al. [13] suggested the M5 product dataset with several modalities, including audio, video, and text, utilizing data uploaded by online retailers.

2.2. *Methods*

2.2.1. Image-Based Retrieval

Traditionally, image-based product retrieval studies [17,18] were conducted in the offline market for applications such as automatic checkout or store management. George et al. [17] proposed a genetic algorithm optimized by multilabel image classification to identify products on shelves. Li et al. [18] proposed the Data Priming Network (DPNet) for automatic checkout to pick reliable samples utilizing the detection and counting collaborative learning strategy during the training process.

In addition, research is extending to include online shopping malls due to the expansion of the e-commerce market. These methods [19–21] are typically employed to recommend similar products to users, as well as to locate and recommend similar products, by combining various models that can extract varied product attributes. Shankar et al. [19] introduced VisNet, an end-to-end DCNN architecture comprising deep and shallow networks. Yang et al. [20] and Hu et al. [21] developed a visual search system that uses a reranking mechanism that can be can be applied to large search engines.

2.2.2. Video-Based Retrieval

The majority of research on video-based retrieval focuses on video copy detection for video copy protection and verification, and also content-based video retrieval for video recommendation. These studies can be classified into two categories based on the similarity-calculating method.

The first methods [15,22,23] extract frame-level features, conduct interframe similarity calculations, and then aggregate the results into video-level similarities. Tan et al. [22] proposed a temporal network (TN) using graphs generated by keyframe matching. Chou et al. [23] proposed dynamic programming (DP), which extracts the diagonal pattern from a frame-level similarity map to detect a spatiotemporal pattern. Kordopatis et al. [15] proposed video similarity learning (ViSiL), which employs metric learning combining chamfer similarity to calculate pairwise similarities on an interframe similarity map.

The second methods [14,16] encode video-level features by aggregating frame-level features derived from images and calculating video-level similarity by comparing the obtained features. Kordopatis et al. [14] proposed deep metric learning (DML) utilizing L_N-iMAC [24]. Shao et al. [16] proposed temporal context aggregation (TCA), which utilizes the self-attention mechanism to integrate long-range temporal information between frame-level features.

2.2.3. Multimodal-Based Retrieval

Recently, with the emergence of datasets that support various modalities, studies using various modality information have emerged. Shin et al. [25] proposed e-CLIP, which can be deployed on multiple e-commerce downstream tasks, based on an approach [26] that utilizes both visual and language information. Dong et al. [13] proposed the Self-harmonized Contrastive Learning (SCALE) framework, which unifies the several modalities into a unified model through an adaptive mechanism for fusing features.

3. Proposed Dataset

3.1. Video Collection

This section discusses the data collection procedure in the Video E-commerce Retrieval Dataset (VERD). We aimed to create a dataset with scenarios resembling those in which consumers look for objects in video. To accomplish this objective, VERD was collected using recently introduced video-based product reviews from online shopping malls (https://shopping.naver.com (accessed on 31 May 2022)).

These product reviews were freely filmed to describe the things that consumers purchased. Due to the various viewpoints, it has a complex background as well as differences in illumination and color. In addition, despite being a review of the same product, the captured area varies according to what the buyer wants to show. As shown in Figure 2, these characteristics allowed us to collect realistic data from the same environment as the user's search devices.

3.2. Annotation Process

This section explains the processing of the dataset. Due to the flexibility of user-uploaded video reviews, we find that reviews are sometimes irrelevant to the product or inadequately depict the product during the data collection section. To address these issues, we conducted a four-step preprocessing procedure to obtain a clean dataset.

The first step is to remove duplicate videos. Occasionally, the same video was reused for many reviews on the e-commerce platform. To eliminate these duplicate videos, Video Duplicate Finder (https://github.com/0x90d/videoduplicatefinder (accessed on 7 July 2022)) was employed. Additionally, visually similar but nonidentical videos were deemed irrelevant and removed.

In a second step, the face-containing video was excluded. We found that in some video product reviews, the user's face was captured with the product. These reviews contain products, but they are not filmed around the items themselves, making it difficult to identify objects. To filter these videos for object-centric video retrieval, RetinaFace [27] was used to recognize video frames containing faces. If a video had an identifiable face in even a single frame, it was excluded from the dataset.

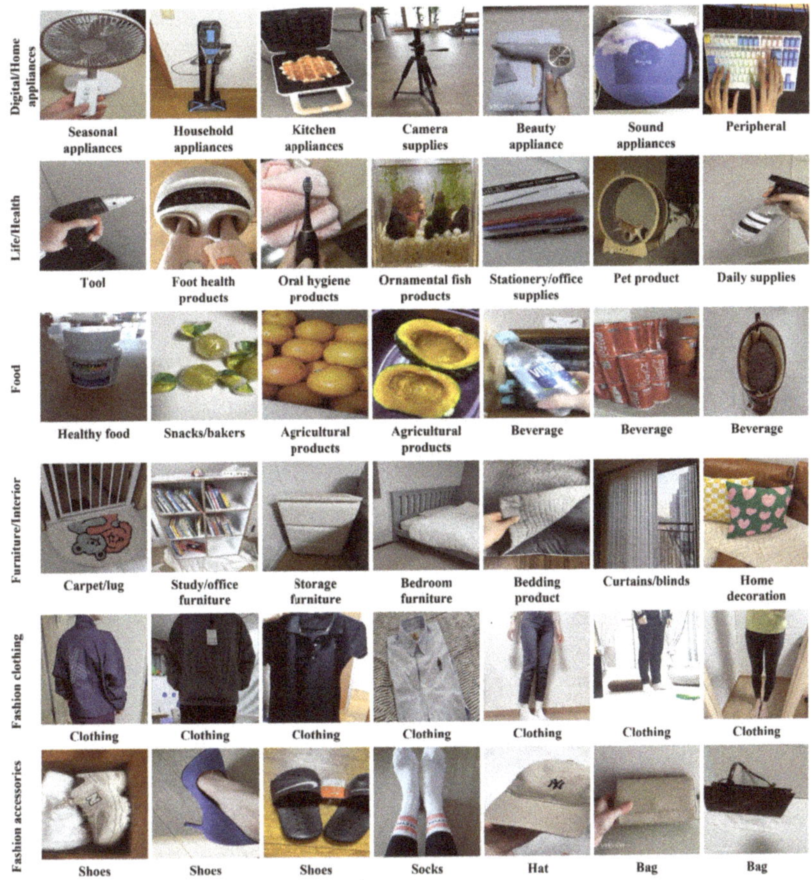

Figure 2. Videos of the hierarchical category of VERD. The category to the left of images represents the Level-0 category, and the category below images represents the Level-1 category.

In the third step, videos captured away from the object's center were discarded. Typically, this is the case for a long-form review. Long-form reviews provide a comprehensive explanation of the product from the perspective of a product review. However, these reviews contain numerous frames that are irrelevant to the item from the perspective of product search. Therefore, these videos were omitted from the dataset because they did not align with the goal of the dataset collection.

In the final phase, labels were adjusted based on their visual similarity with hierarchical category labels. In fact, videos in the category "coffee" can be divided into a subcategory "capsule coffee" and "cold brew coffee". These two items were labeled as the same product up to the level of subdivision, although their physical properties were different. Therefore, some labels were reclassified as distinct goods to allow a more detailed search.

Through this annotation process, it was possible to construct a precise dataset with less noise by excluding videos that did not correspond to the data collection goal. In conclusion, VERD includes a total of 41,570 videos and 187 categories.

3.3. Hierarchical Category Labeling

Following the annotation process, this section describes the category configuration of the VERD. In the majority of datasets, a label associated with a product relates to a fixed value. This fixed label is inappropriate from the perspective of the retrieval task, which needs to search for related objects. Therefore, we adopted hierarchical category labeling to understand the relationship between products, taking into account the nature of the e-commerce market that sells a wide range of goods.

The hierarchical category labeling that we established is a new definition of product taxonomy. Generally, e-commerce markets employ product taxonomy to facilitate the sale of goods. However, the existing product taxonomy has separate categories for products with similar visual qualities or is unable to distinguish between products within the same category. To overcome these difficulties, we created a new product taxonomy based on whether a product can be visually classified.

The hierarchical category is separated into four levels, whereby the higher the level, the more specific the product classification. From Level-0 to Level-3, there are 6 categories, 44 categories, 119 categories, and 91 categories, respectively. Every video has a hierarchical category with a minimum Level-1 and a maximum Level-3. Figure 3 provides a detailed illustration of hierarchical category. Even though "fan" and "air circulator" have the same Level-2 category, "fan" is subcategorized further for "air circulator", which works similarly to a fan but differs visually. However, there were occasions in which products in the same class could be visually distinguished from one another. Figure 3 provides another example of this scenario. The "humidifier", which is designated as a Level-2 category, could be further defined based on how the product performs. In this case, it was modified to add subcategories so that it could be classified into other categories.

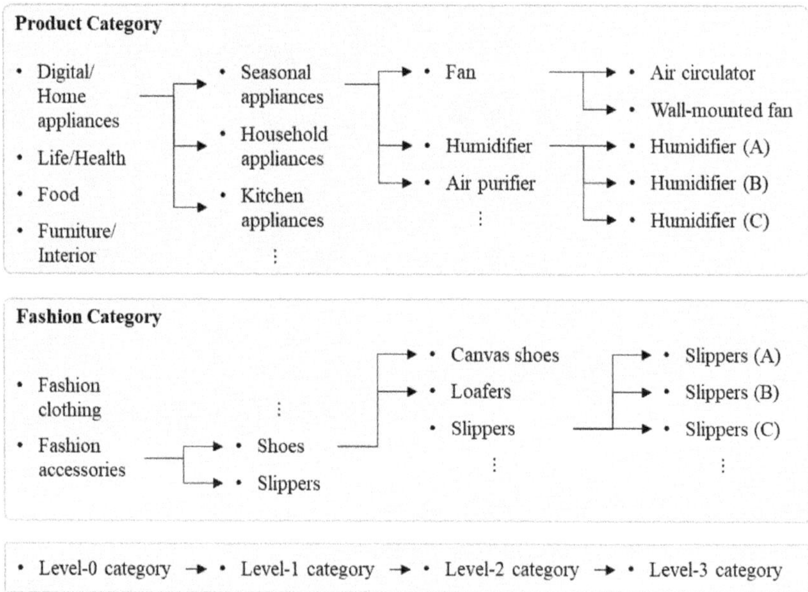

Figure 3. An example of the hierarchical categories of VERD.

3.4. Dataset Statistics

This section explains the video statistics of VERD. The dataset contains 41,570 videos. Videos consist of short clips that average 9.8 s. The large majority of videos are under 10 s, and videos under 30 s comprise 94% of the dataset. This demonstrates that most of the videos were filmed around the product rapidly to introduce it.

The dataset can be separated mainly into product-related and fashion-related categories. The product-related category covers the Level-0 categories "digital/home appliances" (10,135), "life/health" (6327), "food" (5754), and "furniture/interior" (1240). Following that, the fashion-related category contains "fashion accessories" (10,283) and "fashion clothing" (7831), for a total of 18,114 videos. This demonstrates that the dataset is dispersed rather equally.

3.5. Dataset Characteristics

VERD attempted to construct a dataset that simulates the scenario in which a user conducts an object search through a video. From this perspective, the data can be broadly separated into seller-centric data and user-centric data. In this part, we discuss in detail how user-centric data differ from seller-centric data in terms of the information they may provide.

Differences in illumination and color: Lighting variance can be the most significant difference between the environment presented by the seller and the user. Figure 4A illustrates these attributes. There are instances in which it is difficult to understand the properties of a product due to the surrounding lighting, which is not simply a matter of dark or bright illumination. Even when it was the same product, it occasionally offered various colors. VERD has invested a significant amount of time in collecting these videos so that related products can be identified based on their visual characteristics.

Complex backgrounds: Figure 4B shows examples of various backgrounds within the sample videos. In general, seller-centric data exclude a background to emphasize the product. Due to the fact that consumers take shots in a variety of locations, such as their homes and workplaces, multiple items are captured alongside the product. In the videos shown in Figure 4B, it can be verified that the backgrounds are distinctive and do not match. VERD has obtained videos in these varied contexts.

Variety of viewpoints: The majority of the information on the page for product sales is taken from the front in order to make the product seem more attractive. However, users do not consider these factors when capturing the product. In order to address this issue, Figure 4C provides examples of videos collected from a variety of perspectives within the dataset. In the example, filming began on the front of the product but was finished by moving the camera upward so that the mechanical part of the product could be seen clearly. In this real-world scenario, including the search for various product parts, the video-based VERD can work effectively.

We illustrate numerous examples of the user-filmed environment by describing Figure 4 and the characteristics of the dataset. They may have a complicated history with irrelevant items and diverse viewpoints. These characteristics suggest that VERD is suitable for real-world scenarios.

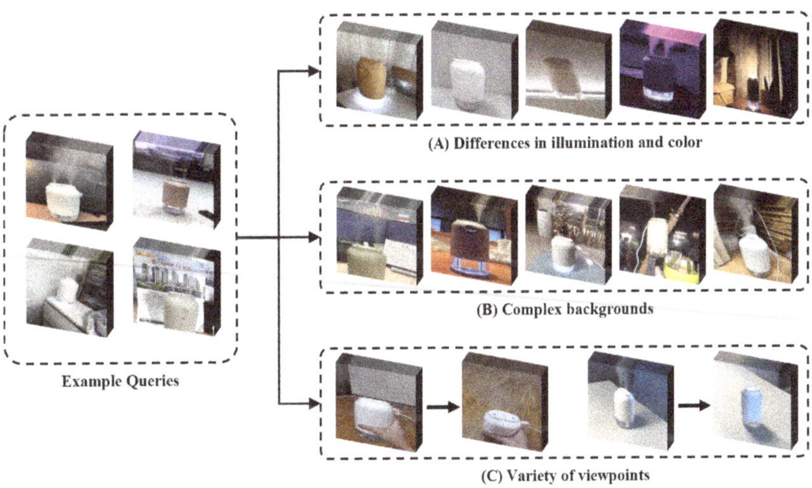

Figure 4. An example of VERD in "Humidifier" category.

4. Experiments

4.1. Setup

In this section, we propose a benchmark performance with several video retrieval systems. Among these methods, we conducted experiments on DML [14], ViSiL [15], and TCA [16] that published codes. Following the previous approach, the performance was also reported as mean average precision (mAP).

Due to the absence of available training datasets for object-centric video studies, K-fold cross-validation was applied as the evaluation approach. We fixed query videos in the dataset and set K to 5 to split the database. In order to ensure that a sufficient quantity of data is used in the search, the experiment was constructed so that while one fold was used for learning, the remaining fold was used for evaluation.

4.2. Benchmark

Table 1 shows benchmark results for existing video retrieval models on VERD. Benchmark experiments were conducted using the authors' provided code, with only a few hyperparameters modified. All performances were evaluated by choosing the methodology for which the highest performance was reported for each method (ViSiL$_v$, TCA$_f$, DML$_{late}$).

Table 1. Benchmark results of applying VERD to existing video retrieval methods.

Method	Category	Fold					Mean
		1	2	3	4	5	
DML [14]	Product	0.081	0.080	0.087	0.077	0.083	0.082
	Fashion	0.090	0.077	0.097	0.093	0.092	0.090
ViSiL [15]	Product	0.309	0.310	0.311	0.311	0.309	0.310
	Fashion	0.159	0.159	0.158	0.159	0.161	0.159
TCA [16]	Product	0.290	0.292	0.293	0.293	0.294	0.292
	Fashion	0.175	0.182	0.183	0.184	0.184	0.181

Benchmark performance was obtained by separating the product category and the fashion category. This is due to the fact that the two categories have different visual qualities. As a result, items in the fashion category have varied shapes based on whether or not they are worn by humans, whereas the visual aspects of products change based on location but the shape of the item does not change. Therefore, the overall performance of the fashion category was deemed to be inferior to that of the product category.

On the other hand, it is noticeable that the performance, in general, is insufficient. This demonstrates that the existing video-to-video retrieval model did not acquire the properties required by object-centric video datasets such as VERD, as it was mainly researched using incident-centric videos. Consequently, the experimental result shows the need for future independent object-centric video retrieval study.

4.3. Analysis

4.3.1. Feature Comparison

Most video retrieval methods use frame-level features or video-level features to calculate video similarities. The frame-level feature calculates the similarity between each frame to determine the similarity of the video, while the video-level feature compresses the feature representation of the video to determine the similarity.

Table 2 presents the performance based on the feature difference in TCA [16] to determine the difference between frame-level and video-level feature presentation in object-centric video retrieval. Experiments indicate that the type of feature has a negligible impact on the feature's performance. This indicates that VERD was taken around an object, allowing the model to understand the expression of the object in the majority of video frames.

Moreover, despite the fact that frame-level features perform better in incident-centric video retrieval studies, video-level features are appropriate for e-commerce platforms that need speedy search when performance gaps among feature types are considered.

Table 2. Performance comparison between frame-level feature (TCA_f) and video-level feature (TCA_c).

Descriptor	Category	Fold					Mean
		1	2	3	4	5	
Frame-level	Product	0.290	0.292	0.293	0.293	0.294	0.292
	Fashion	0.175	0.182	0.183	0.184	0.184	0.181
Video-level	Product	0.288	0.290	0.290	0.290	0.292	0.290
	Fashion	0.173	0.181	0.182	0.184	0.183	0.181

4.3.2. Modality Comparison

To demonstrate that video clips have a higher volume for visual representation than images, analysis was conducted to compare the performance of image-based and video-based retrieval in Table 3.

The experiment employed the same K-fold cross-validation as Section 4.2; however, in the analysis experiment, only evaluation sets were used identically since there was no training engaged. Moreover, since there was no corresponding dataset for images and videos, a pseudo image dataset was created in VERD for the experiment. This dataset was processed by extracting images from the video's intermediate frame.

Table 3. Performance evaluation of the VERD using video-based and image-based retrieval.

Method	Category	Fold					Mean
		1	2	3	4	5	
Image-based	Product	0.191	0.192	0.192	0.192	0.192	0.192
	Fashion	0.100	0.100	0.099	0.100	0.100	0.100
Video-based	Product	0.291	0.291	0.292	0.293	0.291	0.292
	Fashion	0.158	0.158	0.157	0.158	0.159	0.158

Using the method of [15], a simple video search model was built in order to evaluate the performance of the dataset created in this approach. Similarity was calculated using chamfer similarity with L4-iMAC as a feature.

Table 3 demonstrates that video-based methods consistently outperform image-based methods. This difference in performance is because the video was taken from multiple perspectives, allowing it to be responded to even if the front and side visual characteristics of the product are varied. This means that the video contains more information than the image, as this paper suggests. In the case of existing video search models, where the focus is on incident-centric video retrieval, Table 1 indicates that the performance does not improve significantly, even after training. This demonstrates the necessity for independent research on object-centric video retrieval.

5. Conclusions

Object-centric retrieval in the user environment is a major task that can be handled in the expanding e-commerce industry. According to this trend, research on single and multimodal search based on product images emerged, but the challenge was that it was difficult to respond to complex scenarios or that the quantity of data required for a search was massive. Therefore, we propose the Video E-commerce Retrieval Dataset (VERD), comprising videos that have not been utilized in previous studies. We present benchmark performance experiments applying the proposed dataset to existing video search methodologies, and additional experiments indicate the better performance of videos relative to images, demonstrating the need for video-based research.

Author Contributions: Conceptualization, G.L. and W.J.; methodology, G.L.; software, G.L.; validation, G.L.; formal analysis, G.L. and W.J.; investigation, Y.C.; resources, Y.C.; data curation, G.L.; writing—original draft preparation, G.L.; writing—review and editing, G.L. and W.J.; visualization, G.L.; supervision, Y.C.; project administration, Y.C.; funding acquisition, Y.C. All authors have read and agreed to the published version of the manuscript.

Funding: This work was partly supported by the Institute of Information and Communications Technology Planning and Evaluation (IITP) grant funded by the Korea government (MSIT) (No.2021-0-02067, Next Generation AI for Multi-purpose Video Search, 50%) and (RS-2022-00156345, ICT Challenge and Advanced Network of HRD, 50%).

Institutional Review Board Statement: Not applicable.

Informed Consent Statement: Not applicable.

Data Availability Statement: The data presented in this study are available on request from the corresponding author. The provided data can be only used for nonprofit purposes.

Conflicts of Interest: The authors declare no conflict of interest.

References

1. Merler, M.; Galleguillos, C.; Belongie, S. Recognizing groceries in situ using in vitro training data. In Proceedings of the 2007 IEEE Conference on Computer Vision and Pattern Recognition, Minneapolis, MN, USA, 17–22 June 2007; pp. 1–8.
2. Jund, P.; Abdo, N.; Eitel, A.; Burgard, W. The freiburg groceries dataset. *arXiv* **2016**, arXiv:1611.05799.
3. Klasson, M.; Zhang, C.; Kjellström, H. A hierarchical grocery store image dataset with visual and semantic labels. In Proceedings of the 2019 IEEE Winter Conference on Applications of Computer Vision (WACV), Waikoloa Village, HI, USA, 7–11 January 2019; pp. 491–500.
4. Georgiadis, K.; Kordopatis-Zilos, G.; Kalaganis, F.; Migkotzidis, P.; Chatzilari, E.; Panakidou, V.; Pantouvakis, K.; Tortopidis, S.; Papadopoulos, S.; Nikolopoulos, S.; et al. Products-6K: A Large-Scale Groceries Product Recognition Dataset. In Proceedings of the The 14th PErvasive Technologies Related to Assistive Environments Conference, Virtual Event, 29 June–2 July 2021; pp. 1–7.
5. Wei, X.S.; Cui, Q.; Yang, L.; Wang, P.; Liu, L.; Yang, J. RPC: A Large-Scale and Fine-Grained Retail Product Checkout Dataset. *arXiv* **2022**, arXiv:1901.07249.
6. Oh Song, H.; Xiang, Y.; Jegelka, S.; Savarese, S. Deep metric learning via lifted structured feature embedding. In Proceedings of the IEEE Conference on Computer Vision and Pattern Recognition, Las Vegas, Nevada, USA, 27–30 June 2016; pp. 4004–4012.
7. Liu, Z.; Luo, P.; Qiu, S.; Wang, X.; Tang, X. Deepfashion: Powering robust clothes recognition and retrieval with rich annotations. In Proceedings of the IEEE Conference on Computer Vision and Pattern Recognition, Las Vegas, Nevada, USA, 27–30 June 2016; pp. 1096–1104.
8. Ge, Y.; Zhang, R.; Wang, X.; Tang, X.; Luo, P. Deepfashion2: A versatile benchmark for detection, pose estimation, segmentation and re-identification of clothing images. In Proceedings of the IEEE/CVF Conference on Computer Vision and Pattern Recognition, CA, USA, 16–20 June 2019; pp. 5337–5345.
9. Bai, Y.; Chen, Y.; Yu, W.; Wang, L.; Zhang, W. Products-10k: A large-scale product recognition dataset. *arXiv* **2020**, arXiv:2008.10545.
10. Corbiere, C.; Ben-Younes, H.; Rame, A.; Ollion, C. Leveraging Weakly Annotated Data for Fashion Image Retrieval and Label Prediction. In Proceedings of the IEEE International Conference on Computer Vision (ICCV) Workshops, Venice, Italy, 22–29 October 2017.
11. Chen, D.; Liu, F.; Du, X.; Gao, R.; Xu, F. MEP-3M: A Large-scale Multi-modal E-Commerce Products Dataset. In Proceedings of the IJCAI 2021 Workshop on Long-Tailed Distribution Learning, Virtual Event, 21 August 2021.
12. Zhan, X.; Wu, Y.; Dong, X.; Wei, Y.; Lu, M.; Zhang, Y.; Xu, H.; Liang, X. Product1M: Towards Weakly Supervised Instance-Level Product Retrieval via Cross-Modal Pretraining. In Proceedings of the IEEE/CVF International Conference on Computer Vision (ICCV), Seoul, South Korea, 27 October–2 November 2021; pp. 11782–11791.
13. Dong, X.; Zhan, X.; Wu, Y.; Wei, Y.; Kampffmeyer, M.C.; Wei, X.; Lu, M.; Wang, Y.; Liang, X. M5Product: Self-Harmonized Contrastive Learning for E-Commercial Multi-Modal Pretraining. In Proceedings of the IEEE/CVF Conference on Computer Vision and Pattern Recognition (CVPR), New Orleans, LA, USA, 18–24 June 2022; pp. 21252–21262.
14. Kordopatis-Zilos, G.; Papadopoulos, S.; Patras, I.; Kompatsiaris, Y. Near-duplicate video retrieval with deep metric learning. In Proceedings of the IEEE International Conference on Computer Vision Workshops, Venice, Italy, 22–29 October 2017; pp. 347–356.
15. Kordopatis-Zilos, G.; Papadopoulos, S.; Patras, I.; Kompatsiaris, I. Visil: Fine-grained spatio-temporal video similarity learning. In Proceedings of the IEEE/CVF International Conference on Computer Vision, Seoul, Korea, 27–28 October 2019; pp. 6351–6360.
16. Shao, J.; Wen, X.; Zhao, B.; Xue, X. Temporal context aggregation for video retrieval with contrastive learning. In Proceedings of the IEEE/CVF Winter Conference on Applications of Computer Vision, Waikoloa, HI, USA, 3–8 January 2021; pp. 3268–3278.

17. George, M.; Floerkemeier, C. Recognizing products: A per-exemplar multi-label image classification approach. In Proceedings of the European Conference on Computer Vision, Zurich, Switzerland, 6–12 September 2014; pp. 440–455.
18. Li, C.; Du, D.; Zhang, L.; Luo, T.; Wu, Y.; Tian, Q.; Wen, L.; Lyu, S. Data priming network for automatic check-out. In Proceedings of the 27th ACM International Conference on Multimedia, Nice, France, 21–25 October 2019; pp. 2152–2160.
19. Shankar, D.; Narumanchi, S.; Ananya, H.; Kompalli, P.; Chaudhury, K. Deep learning based large scale visual recommendation and search for e-commerce. *arXiv* **2017**, arXiv:1703.02344.
20. Yang, F.; Kale, A.; Bubnov, Y.; Stein, L.; Wang, Q.; Kiapour, H.; Piramuthu, R. Visual search at ebay. In Proceedings of the 23rd ACM SIGKDD International Conference on Knowledge Discovery and Data Mining, Halifax, NS, USA, 13–17 August 2017; pp. 2101–2110.
21. Hu, H.; Wang, Y.; Yang, L.; Komlev, P.; Huang, L.; Chen, X.; Huang, J.; Wu, Y.; Merchant, M.; Sacheti, A. Web-scale responsive visual search at bing. In Proceedings of the 24th ACM SIGKDD International Conference on Knowledge Discovery & Data Mining, London, UK, 19–23 August 2018; pp. 359–367.
22. Tan, H.K.; Ngo, C.W.; Hong, R.; Chua, T.S. Scalable detection of partial near-duplicate videos by visual-temporal consistency. In Proceedings of the 17th ACM international conference on Multimedia, Columbia, BC, Canada, 19–24 October 2009; pp. 145–154.
23. Chou, C.L.; Chen, H.T.; Lee, S.Y. Pattern-based near-duplicate video retrieval and localization on web-scale videos. *IEEE Trans. Multimed.* **2015**, *17*, 382–395. [CrossRef]
24. Kordopatis-Zilos, G.; Papadopoulos, S.; Patras, I.; Kompatsiaris, Y. Near-duplicate video retrieval by aggregating intermediate cnn layers. In Proceedings of the International Conference on Multimedia Modeling, Reykjavik, Iceland, 4–6 January 2017; pp. 251–263.
25. Shin, W.; Park, J.; Woo, T.; Cho, Y.; Oh, K.; Song, H. e-CLIP: Large-Scale Vision-Language Representation Learning in E-commerce. In Proceedings of the 31st ACM International Conference on Information & Knowledge Management, Atlanta, GA, USA, 17–21 October 2022; pp. 3484–3494.
26. Radford, A.; Kim, J.W.; Hallacy, C.; Ramesh, A.; Goh, G.; Agarwal, S.; Sastry, G.; Askell, A.; Mishkin, P.; Clark, J.; et al. Learning transferable visual models from natural language supervision. In Proceedings of the International Conference on Machine Learning, Virtual Event, 18–24 July 2021; pp. 8748–8763.
27. Deng, J.; Guo, J.; Ververas, E.; Kotsia, I.; Zafeiriou, S. Retinaface: Single-shot multi-level face localisation in the wild. In Proceedings of the IEEE/CVF Conference on Computer Vision and Pattern Recognition, Seattle, WA, USA, 13–19 June 2020; pp. 5203–5212.

sensors

Article

PRAGAN: Progressive Recurrent Attention GAN with Pretrained ViT Discriminator for Single-Image Deraining

Bingcai Wei, Di Wang, Zhuang Wang and Liye Zhang *

College of Computer Science and Technology, Shandong University of Technology, Zibo 255000, China
* Correspondence: zhangliye@sdut.edu.cn; Tel.: +86-15653380753

Abstract: Images captured in bad weather are not conducive to visual tasks. Rain streaks in rainy images will significantly affect the regular operation of imaging equipment; to solve this problem, using multiple neural networks is a trend. The ingenious integration of network structures allows for full use of the powerful representation and fitting abilities of deep learning to complete low-level visual tasks. In this study, we propose a generative adversarial network (GAN) with multiple attention mechanisms for image rain removal tasks. Firstly, to the best of our knowledge, we propose a pretrained vision transformer (ViT) as the discriminator in GAN for single-image rain removal for the first time. Secondly, we propose a neural network training method that can use a small amount of data for training while maintaining promising results and reliable visual quality. A large number of experiments prove the correctness and effectiveness of our method. Our proposed method achieves better results on synthetic and real image datasets than multiple state-of-the-art methods, even when using less training data.

Keywords: deep learning; image deraining; neural network; vision transformer; generative adversarial network

Citation: Wei, B.; Wang, D.; Wang, Z.; Zhang, L. PRAGAN: Progressive Recurrent Attention GAN with Pretrained ViT Discriminator for Single-Image Deraining. *Sensors* **2022**, *22*, 9587. https://doi.org/10.3390/s22249587

Academic Editors: Christoph M. Friedrich, Euntai Kim, Sangyoun Lee and Kang Ryoung Park

Received: 11 October 2022
Accepted: 1 December 2022
Published: 7 December 2022

1. Introduction

Rain patterns in an image will affect the visibility of the image and cause considerable trouble to imaging instruments. Degradation phenomena, such as rain streaks and fog, will greatly decrease the accuracy of visual tasks, especially for high-level tasks. Therefore, removing rain from rainy images has become classical in down-stream visual tasks, while, single-image deraining is a challenging task in low-level visual research fields.

Deep learning, relying on its strong representation and mapping fitting ability, has made great achievements in the field of computer vision in recent years. Not only in high-level visual tasks, such as image classification [1], object detection [2], semantic segmentation [3], and person reidentification [4], has deep learning occupied a dominant achievement, but also in the low-level visual tasks. For visual representations, the depth of network is very important [5], but simply deepening the neural network will make it difficult to train. Since ResNet [6] solved this problem, the application of convolutional neural network (CNN) in computer vision has shown a spurt of development [7,8]. Later researchers mimicked human visual attention by adding attention mechanisms [9,10] to CNN, allowing it to allocate more computing resources to parts that contain significant information based on dynamic weight scores [11]. Recently, with the excellent performance of self-attention [12], ViT [13] has re-examined the choices of network backbone. Meanwhile, CNN can also be combined with GAN and recurrent neural network (RNN), respectively. Using the powerful generation ability of GAN and the outstanding temporal modeling capability of RNN, attractive achievements have been made in image generation [14] and deblurring [15], video super-resolution [16], and denoising [17] tasks.

Single-image deraining is a hot issue because the images captured in rainy days will be significantly degraded by rain patterns, so computer vision tasks are difficult to perform.

In contrast to the model-based or prior-based methods in traditional algorithms, learning-based methods are applied to image rain removal, and can achieve more promising results with better generalization ability, while requiring no prior knowledge. In detail, combined with image processing domain knowledge, Fu et al. [18] proposed a modestly sized CNN to modify the objective function for image deraining. Yang et al. [19] created a recurrent rain detection and removal network that could jointly detect and remove rain from single images. Zhang et al. [20] proposed a density-aware, multi-stream, densely connected network for joint rain density estimation and deraining that can automatically determine the rain-density information. As for single-image deraining, Zhang et al. [21] proposed ID-CGAN (image deraining conditional generative adversarial network) by leveraging the powerful generative modeling capabilities of conditional GAN. Ren et al. [22] proposed a progressive recurrent network that can take the advantage of recursive computation while exploiting the dependencies of deep features across stages. Attention mechanisms, such as CNN, GAN, RNN, and ViT, are all excellent components of deep learning, which can be used as components to design a network that combines the advantages and characteristics of a variety of structures. The use of these network structures alone cannot obtain a satisfactory effect, therefore, our motivation was to give full play to the advantages of various network structures by integrating and collocating multiple network structures. On the other hand, training of complex neural networks requires a lot of data, which means it takes a lot of time simultaneously. Therefore, the efficient use of data will make training easier. Given that generators are the more important member, there have been few studies on discriminators and the stability of their training. In order to solve the above problems, in this study, we propose a progressive recurrent attention generation adversarial network, the generator for which includes a convolutional block attention module [10] (CBAM) and convolutional LSTM [23] (ConvLSTM). At the same time, a pretrained ViT is proposed as a discriminator to organize the adversarial training with the generator. Finally, we introduce a training method that can use only a portion of training image pairs while obtaining results beyond the amount of data. Detailed ablation experiments and comparative experiments have proven the rationality and effectiveness of our proposed method.

The main contributions of this work are as follows:

1. We propose an adversarial model using a pretrained ViT discriminator. We utilize ViT's powerful fitting ability in computer vision while minimizing its drawback of requiring large amounts of data for pretraining. To our best knowledge, there has been little work to improve the performance of discriminators in image deraining, and we are the first to propose a pretrained ViT discriminator to improve the overall performance of GAN.
2. We propose a data reselection algorithm, called DRA. To be specific, the training data are reselected at a specific time in the process of network training. Compared with the fixed part of training data, the rain removal effect of our model can be significantly improved by using this algorithm.
3. A large number of comparative experiments and ablation experiments on synthetic and real datasets prove the effectiveness and rationality of our proposed method.

1.1. Single-Image Deraining

Compared with video deraining tasks, which that can use inter-frame temporal information, significantly less information can be fully utilized in individual images for single-image deraining. Therefore, it is obviously more difficult and challenging to remove rain streaks in single images. In early studies, the rain model is usually simply expressed as Formula (1):

$$O = B + \widetilde{S} \tag{1}$$

where O is the input image with rain streaks, B is the background image, and $\widetilde{S}$ is the rain streak layer. Yang et al. [6] proposed a new model in order to realistically simulate the rain streak phenomena in the real world. By accommodating streak accumulation and overlapping rain streaks with different directions, this model can both comprise of multiple layers of rain streaks and represent diversity of rain streaks. The new rain model is expressed as Formula (2):

$$O = \alpha(B + \sum_{t=1}^{s} \widetilde{S}_t R) + (1 - \alpha)A \tag{2}$$

where $\widetilde{S}_t$ is the rain streak layer in the same direction, which has effects of atmospheric shading; S is the maximum number of rain streak layers; and t is the index of these layers. R represents binary values of 0 or 1, 0 representing areas without rain and 1 representing areas with rain. α represents the atmospheric propagation transmittance that is common in image dehazing and A represents the global atmospheric light value.

1.2. ConvLSTM and GAN

To solve the problem that storing information over extended time intervals is time-consuming, Sepp et al. [24] proposed long short-term memory (LSTM). As a recurrent version of the cascade correlation learning architecture, recurrent cascade correlation can learn from examples to map an input sequence to the desired output sequence while preserving the benefits of cascade correlation, such as fast learning. LSTM can lead to more successful runs than recurrent cascade correlation while learning much faster. However, the fully connected LSTM (FC-LSTM) cannot encode spatial information in handling spatiotemporal data. To overcome this major drawback of LSTM, Shi et al. [23] proposed ConvLSTM, which is more suitable for spatiotemporal data than FC-LSTM while preserving the advantages of it. ConvLSTM consists of an input gate i_t, an output gate o_t, a forget gate f_t, and a memory cell C_t [25]. The key equations of ConvLSTM are shown in Formula (3):

$$\begin{aligned} i_t &= \sigma(W_{xi} * \mathcal{X}_t + W_{hi} * \mathcal{H}_{t-1} + W_{ci} \circ \mathcal{C}_{t-1} + b_i) \\ f_t &= \sigma(W_{xf} * \mathcal{X}_t + W_{hf} * \mathcal{H}_{t-1} + W_{cf} \circ \mathcal{C}_{t-1} + b_f) \\ \mathcal{C}_t &= f_t \circ \mathcal{C}_{t-1} + i_t \circ tanh(W_{xc} * X_t + W_{hc} * \mathcal{H}_{t-1} + b_c) \\ o_t &= \sigma(W_{xo} * \mathcal{X}_t + W_{ho} * \mathcal{H}_{t-1} + W_{co} \circ \mathcal{C}_t + b_o) \\ \mathcal{H}_t &= o_t \circ tanh(\mathcal{C}_t) \end{aligned} \tag{3}$$

where $\circ$ and $*$ denote Hadamard product and convolution operator. X_t, H_t, W_*, b_* are input tensor, hidden state tensor, network weights, and bias terms, respectively.

By simultaneously training a generative model G and a discriminative model D via an adversarial process, GAN can represent even degenerate distributions with no approximate inference better than methods based on Markov chains [26]. The training objective of D is to distinguish between data generated by G and real data as much as possible. The training goal of G is to make D unable to distinguish between them. The adversarial process is shown as a two-player minimax game in Formula (4):

$$\min_G \max_D V(D,G) = \mathbb{E}_{x \sim P_{data(x)}}[\log D(x)] + \mathbb{E}_{z \sim P_{z(x)}}[\log(1 - D(G(z)))] \tag{4}$$

where $P_{data(x)}$ and $P_{z(x)}$ are the distributions of real data and generated data, meanwhile, $D(x)$ and $D(G(z))$ are the probabilities of the discriminator judging real or generated data as true, respectively. GAN has a disadvantage that D must be synchronized well with G during training [26] while suffering from training instability [27]. Therefore, the structures of G and D must be well-designed, and the components used in the proposed network will be described in the following section.

1.3. CBAM and ViT

Hu et al. [12] proposed the SE module, which uses global average-pooled features to compute channel-wise attention for exploiting the inter-channel relationship. However, the SE module is suboptimal because it only focuses on the channel dimension. CBAM [10] can sequentially infer attention maps along not only the channel but also the spatial dimension to get better inter-dependencies than [9]. The overall attention process of CBAM [10] is shown in Formulas (5) and (6):

$$\begin{aligned} F' &= M_c(F) \otimes F \\ F'' &= M_s(F') \otimes F' \end{aligned} \tag{5}$$

where F' and F'' are the intermediate feature map and the final refined output, while $\otimes$ denotes element-wise multiplication, in which:

$$\begin{aligned} M_c(F) &= \sigma(W_1(W_0(F^c_{avg})) + W_1(W_0(F^c_{\max}))) \\ M_s(F) &= \sigma(f^{7\times7}([F^s_{avg}; F^s_{\max}])) \end{aligned} \tag{6}$$

where F^*_{avg}, $F^*_{\max}$, W_*, and $f^{7\times7}$ denote average-pooled features, max-pooled features, CNN's weights, and convolution operations with a 7×7 filter, respectively. Further, the structure of the CBAM is shown in Figure 1.

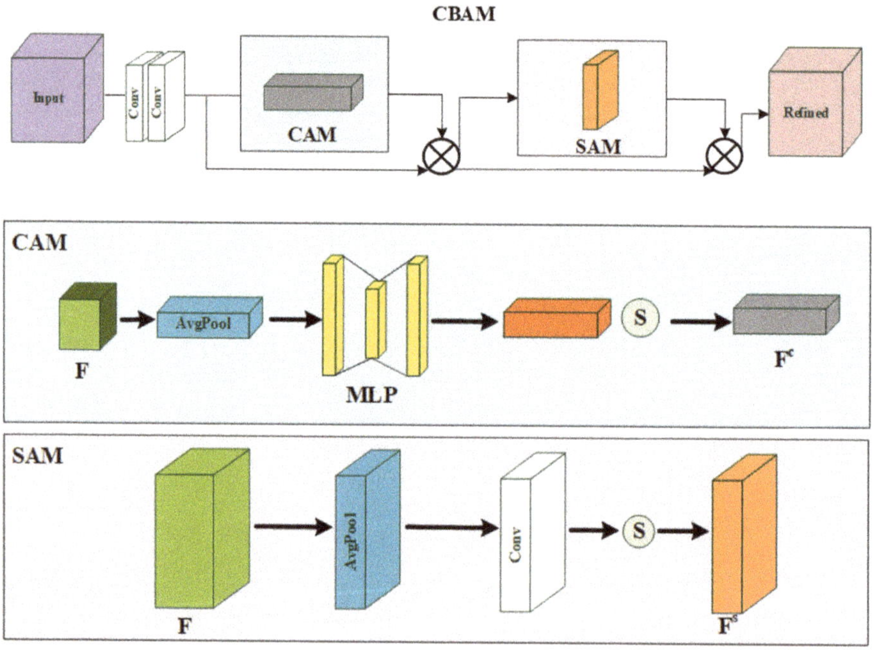

Figure 1. Detailed structure of CBAM.

Based solely on self-attention mechanisms, transformer [12] is the de facto standard for natural language processing tasks. Applications of pure transformer [13] or its variants [28,29] to computer vision tasks prove the superiority of transformer over CNN and RNN. By flattening the loss landscapes [30], multi-head self-attentions (MSAs) in transformer improve not only accuracy but also generalization, which gives transformer excellent fitting and representation abilities. As a discriminator, we only used the transformer encoder, which includes a MSA module and a feed-forward network (FFN). The size of input f_{pi} is the same as that of output patch in encoder $f_{E_i} \in \mathbb{R}^{P^2 \times C}$, and the whole calculation of transformer can be formulated in Formula (7):

$$\begin{aligned} y_0 &= [E_{p1} + f_{p1}, E_{p2} + f_{p2}, \ldots, E_{pn} + f_{pn}], \\ q_i &= k_i = v_i = LN(y_{i-1}), \\ y_i' &= MSA(q_i, k_i, v_i) + y_{i-1}, \\ y_i &= FFN(LN(y_i')) + y_i', i = 1, \ldots, l \\ [f_{E1}, f_{E2}, \ldots, f_{E_n}] &= y_1 \end{aligned} \tag{7}$$

in which the self-attention in MSA can be unified as (8):

$$Attention(Q, K, V) = soft\max(\frac{Q \cdot K^T}{\sqrt{d_k}}) \tag{8}$$

where q, k, v are query, key, and value used in MSA; Q, K, V are vectors packed together into three different matrices which are derived from different inputs [31], respectively. In addition, l denotes the number of layers in the encoder and LN is the layer normalization [32] applied before every block.

2. Proposed Method

In the second chapter, we mainly introduce three parts. Firstly, we mention the overall network structure and progressive recurrent loss function. The second part introduces a confrontation model using a pretrained ViT discriminator. Finally, we introduce an effective training method: reselecting data progressively.

2.1. Network and Loss Function

We promote the guiding role of loss function [33] for end-to-end single-image rain removal. The whole structure of our generator is shown in Figure 2, the generator was inspired by the manner of progressively coarse-to-fine restoration from degraded to sharp images in [34–36], methods of sharing network parameters in [37], and RNN for deraining [22,36].

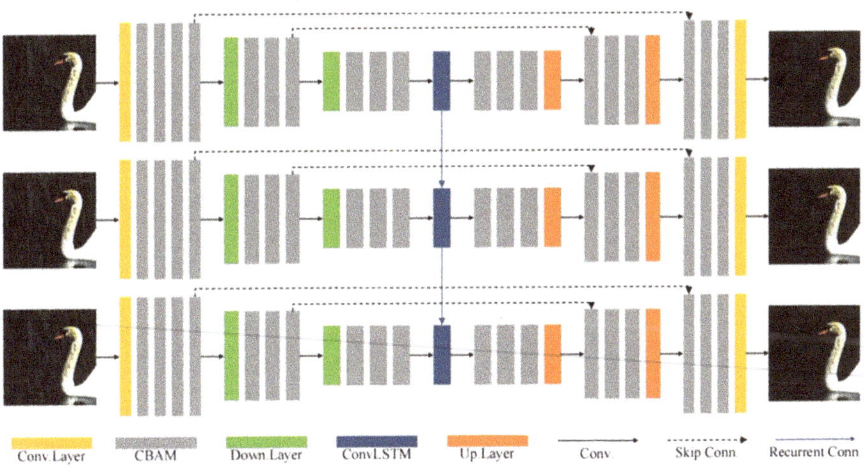

Figure 2. The overall structure of our proposed generator. The CBAM part is shown in Figure 1. Our generator is a variant of the recurrent neural network, which offers three cycles in the figure above. The parameters of the three cycles are shared, that is, only one-third of the parameters of the overall network. The generator does not need to be pretrained. In the adversarial training, the pretrained discriminator is used to conduct adversarial training with the generator proposed above.

We applied one loss to each loop in the training process of the generator to achieve a progressive recurrent loss. Specifically, in the first loop, we used MSE loss, which is expressed in Equation (9):

$$\mathcal{L}_{mse} = \frac{1}{N}\|B_1 - B\|_2^2 \tag{9}$$

where B_1 is the output of first loop and N is the number of elements in B_1 to normalize. In the second loop, we employed the EDGE loss, which is expressed in Equation (10):

$$\mathcal{L}_{edge} = \sqrt{(Lap(B) - Lap(B_2))^2 + \varepsilon^2} \tag{10}$$

where $Lap(*)$ denotes the edge maps extracted from images via Laplacian operator [38] and ε is set to 0.001. In the last loop, as the final result, we chose structural similarity (SSIM) [39] loss, which can take into account the overall coordination between predicted deraining images and labels. The SSIM between image X and image Y can be expressed as Formula (11):

$$SSIM(X,Y) = l(X,Y)c(X,Y)s(X,Y) \tag{11}$$

where $l(X, Y)$, $c(X, Y)$, and $s(X, Y)$ are luminance component, contrast component, and structure component of SSIM, respectively. The SSIM loss between final output B_3 and label can be defined as Formula (12):

$$\mathcal{L}_{SSIM} = 1 - SSIM(B_3, B) \tag{12}$$

In order to avoid the burden of fine-tuning parameters [22], we conducted a prior analysis of the above three loss values, as shown in Figure 3. Therefore, we simply arranged the order by numerical value from small to large. Therefore, the final loss used for our model is defined as Formula (13):

$$Loss_{all} = \mathcal{L}_{mse} + \mathcal{L}_{edge} + \mathcal{L}_{SSIM} \tag{13}$$

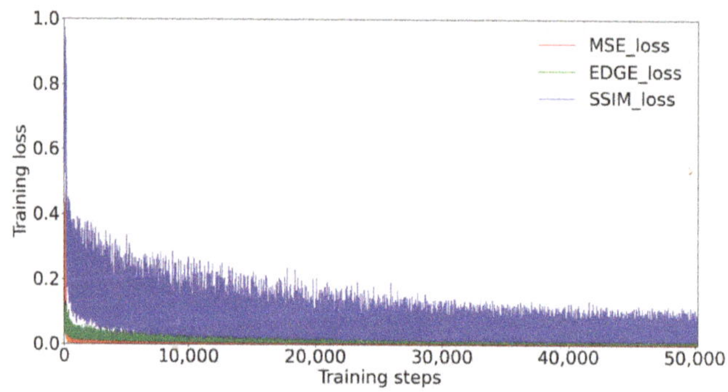

Figure 3. Numerical comparison of three losses during model training.

2.2. Discriminator: Pretrained ViT

Due to their capacity for long-range representation [40] and faculty for flattening loss landscapes [30], transformer-based models show high performance for visual tasks with less need for vision-specific induction [31]. Multiple tasks [41–43] have revealed that transformer-based models heavily rely on massive datasets for large-scale training, which may be the key to achieving its inductive bias [13]. However, pretraining [44] on large-scale datasets (e.g., ImageNet [45]) is both very demanding on hardware and does not necessarily improve the final target task accuracy [46].

In this section, we give a detailed description of a proposed strategy that uses our pretrained ViT as the discriminator of GAN. Compared with the large-scale dataset that includes over tens of millions of images, we used less than 3×10^4 images for training.

Given that this pretraining process can be regarded as a binary classification task, the number of training iterations is small while the effect is good. To demonstrate the superiority of ViT over CNN, we also trained a classical CNN, called PatchGAN [47], which is often used as a discriminator in image restoration tasks [48,49]. The PatchGAN [47] network mainly includes: $C64 - C128 - C256 - C512$. C_k presents a 4×4 Convolution + BatchNorm + LeakyReLU block with stride two and k filters. The parameters of these LeakyReLU activation functions were set to 0.2 and the last two layers of this network are made up of a 4×4 convolution layer, for which stride and filter number were set to one, and an average pooling layer. Meanwhile, the ViT used as a discriminator has 16 patch sizes, 768 embedding dimensions, 6 MSAs Blocks, and 12 attention heads. The detailed structures of ViT [13] and PatchGAN [47] are shown in Figures 4 and 5. By recording the loss function, as shown in Figure 6, ViT [13] converges faster and is more stable than CNN during training. Further, as shown in Figure 7, by testing the trained network on whole data, we found that, as a discriminator, pretraining ViT can better distinguish images with rain from clear images. After pretraining, this ViT has been fully equipped with the ability to distinguish whether the training data contain rain.

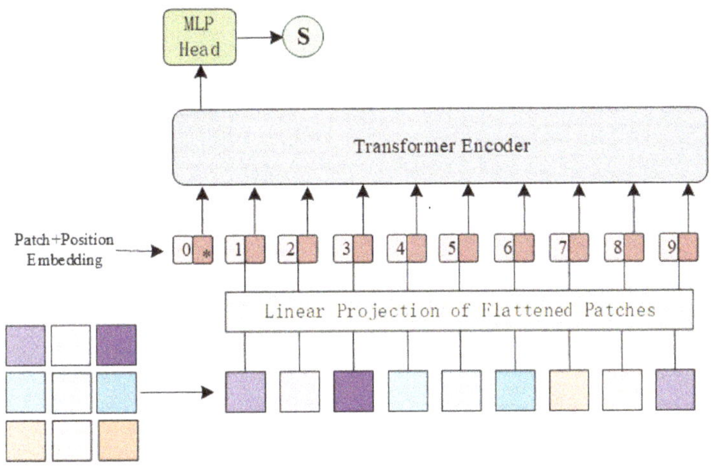

Figure 4. The detailed structure of the transformer discriminator used in this article. In addition, the '*' symbol represents class token.

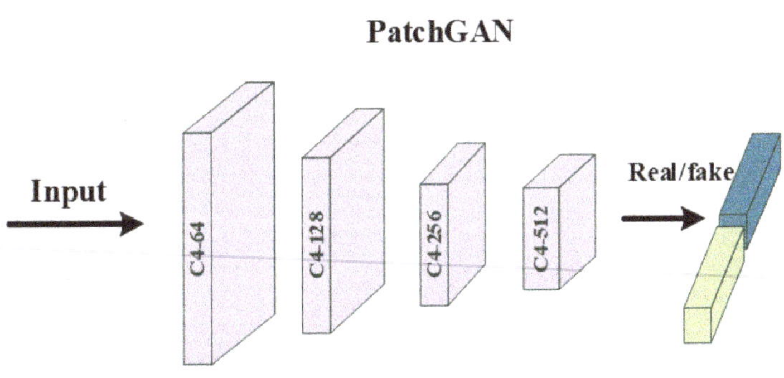

Figure 5. The detailed structure of the classical PatchGAN.

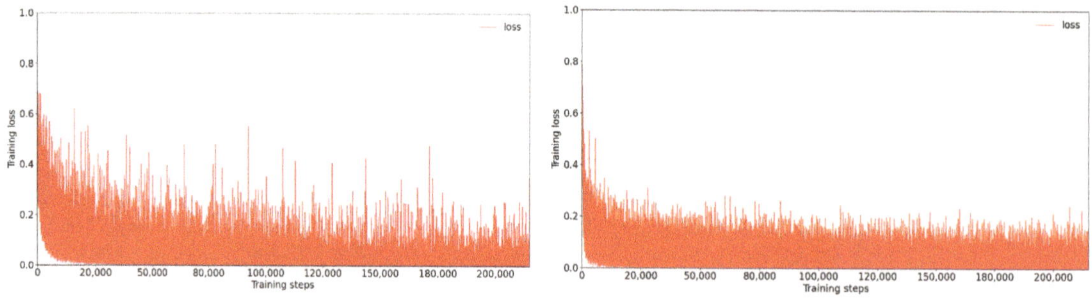

(**a**) (**b**)

Figure 6. Comparison of losses convergence in pretraining. Note, our aim is not to make a performance comparison between these two, but rather to explore the wider use of ViT [13] and pretraining for generative tasks from the perspective of a GAN's discriminator. (**a**) PatchGAN [47] on 128 × 128 patches. (**b**) ViT [13] on 128 × 128 patches.

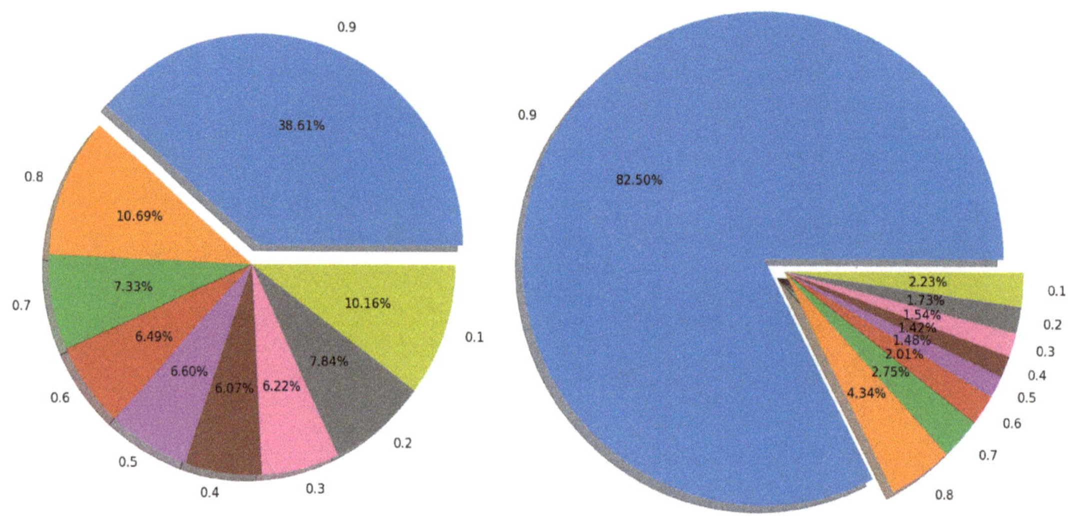

(**a**) (**b**)

Figure 7. The difference between the predicted values of all image pairs after training. The goal of both networks is to return 1 to the image without rain and 0 to the image with rain. The number in the figure is the return value of the image without rain minus the return value of the image with rain. That is, the larger the difference is, the stronger the network's discrimination ability is. As shown in the figure, on the image patches of 128 × 128, ViT [13] performs better than PatchGAN [47]. (**a**) PatchGAN [47] on 128 × 128 patches. (**b**) ViT [13] on 128 × 128 patches.

2.3. Reselecting Data Progressively: Train More Effectively

Nowadays, deep neural networks often require a large amount of data for training to converge. As described in the previous section, pretraining on large-scale datasets requires fairly good hardware conditions and very long time, but does not necessarily improve final target task accuracy [46]. Not only that, in order to comprehensively explore the competence, models for single image deraining also require massive data for training [36], which also increases the difficulty for this task to a certain extent.

To solve these problems, we propose an algorithm for progressively random reselection of data, which is inspired by the coarse-to-fine principle that has been proved to be effective [50] by other image restoration tasks [35,51]. Specifically, randomly select a portion from the entire training set at the beginning and then reselect it several times. By reselecting training data at the end of a specific training epoch, we can achieve better results than using the same amount of training data without reselecting. In addition, in accordance with the principle of coarse-to-fine, we interval different training epochs to reselect the data, which makes the intervals change from large to small. At the end stage of network training, the data are reselected every two epochs, while in the initial stage of network training, data are reselected every twenty-five epochs. From the perspective of network generalization performance, using different data for training every once in a while can simply inhibit over-fitting. At the same time, in contrast to the discriminator, our generator does not need pretraining, although pretraining will not automatically help reduce overfitting [46]. Each process is carried out before one training epoch; compared with the time required for training, time consumption of reselecting data can be ignored, but it can perform better results. The process of reselecting data is summarized in Algorithm 1:

Algorithm 1: Reselecting Data Progressively.

Parameters:
M = 251: total epoch number for training,
E = 50: number of epochs included in one stage of reselecting data progressively,
D: all the training data,
R = 4: ratio of overall data selection,
List = [25, 10, 5, 2]: a list of epoch values for reselecting data,
S = [0, *List*[0]]: a list for saving the number of rounds for which data should be reselected,
L = [0, ... , len(D)]: a list of integers from 0 to the length of D,
Loader: dataloader in Pytorch

```
for i = 0 to len(List) do
    while S[−1] < (E*(i + 2)) do
        S.append(List[i] + S[−1])
    end while
end for
for i = 0 to M do
    if i in S then
        Shuffle(L)
        Part = [D[j] for j in L[0:(len(D)//R)]]
        Loader(Part)
    end if
    Train one epoch
end for
```

3. Experimental Results

3.1. Implementation Details

We implemented our model with the pytorch library. The generator was able to be divided into three stages based on the size of feature map. After each down-sampling, the number of channels in the convolution layer was twice that before. The number of channels in the convolution layer at the beginning of the network was 32, and the convolution kernel size of all convolution layers was 3. The image patches used in all experiments were 256 $\times$ 256. Due to hardware limitations, specific ablation experiments may use different batch sizes. All the generators in different ablation experiments used Adam [52] optimizer for training, and the initial learning rate was 0.0002, which steadily decreased to 1×10^{-6} using the cosine annealing strategy [53]. In contrast to the generator, the initial learning rate of the discriminator during pretraining was 2×10^{-5}, and AdamW [54] optimizer was used for optimization. Horizontal and vertical flips were randomly applied for data augmentation. In addition to pretraining the discriminator, our experiments were conducted on an NVIDIA RTX 3060 GPU. Further details may be found in [55].

On several synthetic datasets, our proposed PRAGAN was compared with seven state-of-the-art models, i.e., DerainNet [5], RESCAN [55], DIDMDN [7], UMRL [56], SEMI [57], PreNet [22] and MSPFN [36]. All other methods were configured as in [36], and we used the results provided by [36] to establish a re-evaluation of image quality by employing the peak signal to noise ratio (PSNR) and SSIM in scikit-image. All datasets used for training included Rain14000 [58], Rain1800 [6], Rain800 [21], and Rain12 [59], with a total of 13,712 image pairs, which we call MIX.

On real-world rainy image datasets, according to the configuration in [60], we only trained the proposed model on the Rain100L [6] training set, we call it Train200. Train200 has a total of 200 image pairs, and PRAGAN was tested on Internet-Data [57] and SPA-Data [61]. These two datasets contain 147 rainy images and 1000 image pairs, respectively. Given that Internet-Data [57] has no ground truth, we only provide visual comparison with several state-of-the-art models in Figure 7.

3.2. *Ablation Studies*

In this section, we provide the contributions of different designs quantitatively.

3.2.1. Network Structure and LOSS

By removing CBAM and ConvLSTM, we verified the necessity of using them. For progressive recurrent loss, experiments have shown that this loss can achieve better results than adding three losses to one loop or using MSE loss to measure the prediction value of each loop directly. Finally, it should be noted that our network only inputs the original rain image each loop, rather than the predicted value of the previous loop. We found through experiments that for PRAGAN, doing so will bring performance losses. The training set and testing set used in all experiments in this section were Rain800 [21] and Test100 [21]; mini-batch size and training epoch were 1 and 101. All the results are shown in Table 1.

Table 1. Ablation studies on network structure and loss function. A1 represents the results of removing the CBAM model and A2 shows the predicted value of the last loop as the next input of the network. A3 is the case where MSEloss is used to measure the training effects of three loops. A4 represents the results of removing ConvLSTM. A5 and A6 are the results of adding three losses to the same loop for training and then performing one and three inferences. A7 is the overall network structure with progressive recurrent loss.

	A1	A2	A3	A4	A5	A6	A7
PSNR	24.21	24.31	24.58	24.75	24.77	24.77	24.79
SSIM	0.861	0.869	0.846	0.872	0.871	0.872	0.874

3.2.2. Pretraining ViT as Discriminator

In this section, we compare the ViT pretrained on 128×128 and 256×256 image patches, as shown in Tables 2 and 3. For the smaller image patches, we set batch size to 64, while for the larger image patches, due to the hardware limitation, we set batch size to 16. The number of training epochs and the initial learning rate were 502 and 2×10^{-5}, respectively. When the discriminator was used for adversarial training, the initial learning rate was 1×10^{-5}. Our discriminator used AdamW [54] as optimizer in pretraining and adversarial learning. During pretraining and adversarial training, the loss of both patch sizes for discriminator was BCEloss. The training dataset for pretraining was MIX. In order to better display the superiority of pretraining for ViT, as for the smaller patch, we trained the network on a quarter of the MIX training set. Meanwhile, for the larger patch, we used a quarter of Rain1800 [19] for training. Pretraining ViT can effectively help the generator to improve the performance of image deraining.

Table 2. Ablation studies on pretraining the discriminator or not. ViT discriminator can make the generator perform better in image deraining tasks through pretraining on 128 × 128 image patches.

	Test100 PSNR/SSIM	Rain100H PSNR/SSIM	Rain100L PSNR/SSIM	Test1200 PSNR/SSIM
No pretraining	21.89/0.837	22.99/0.799	24.47/0.861	25.71/0.873
Pretraining	22.55/0.855	23.54/0.813	25.94/0.890	25.95/0.877

Table 3. Performance comparison of pretrained ViT discriminator on 256 × 256 image patches. The model was trained on Rain800 [21] and tested on Test100 [21].

	PSNR	SSIM
No pretraining	24.78	0.870
Pretraining	24.87	0.871

3.2.3. Reselecting Data Algorithm

In this part, we studied the reselecting data algorithm with small and large amounts of training data to better demonstrate its effectiveness. Specifically, the smaller one was Rain800 [21] and the larger one was a quarter of MIX, including 700 and 3426 image pairs, respectively. Batch size of the former was 1, the latter was 2. The number of training epochs and the size of image patches were 251 and 256, respectively, and relevant results are shown in Tables 4 and 5. With the increase of the amount of training data, the corresponding image evaluation index will also increase. Meanwhile, using same amount of data, by employing a reselecting data algorithm, the deraining task can obtain better results.

Table 4. Studies of reselecting data on small-scale training set. The model was trained on Rain800 [21] and tested on Test100 [21]. r represents the proportion of reselected data to the total and 1/4 means fixed quarter of total data.

	r = 20	r = 10	1/4	r = 4
PSNR	23.87	24.83	25.79	26.08
SSIM	0.844	0.866	0.888	0.889

Table 5. Studies of reselecting data on a large-scale training set. r represents the proportion of reselected data to the total and 1/4 means fixed quarter of total data.

	Test100 PSNR/SSIM	Rain100H PSNR/SSIM	Rain100L PSNR/SSIM	Test2800 PSNR/SSIM
r = 20	23.97/0.868	25.52/0.838	28.80/0.898	27.76/0.899
r = 10	24.94/0.885	26.69/0.861	30.32/0.929	27.85/0.902
1/4	27.25/0.911	27.09/0.877	31.22/0.933	27.90/0.905
r = 4	27.54/0.912	27.51/0.884	32.77/0.955	27.97/0.906

3.3. *Comparison with Other Methods*

3.3.1. Synthetic Images

Through training on one quarter of the MIX training set, combined with DRA and pretraining of the ViT discriminator, we obtained the best results with the proposed method. We compared it with eight state-of-the-art methods. Due to the relatively long time, we remeasured the image quality, which may be different from the previous study. We used the results provided by [36] to perform a re-evaluation of all methods, as shown in Table 6. Meanwhile, visualized images shown in Figures 8 and 9 match well with the quantitative results, which shows PRAGAN's superior deraining ability and favorable image restoration capability. Note that most other methods used all MIX training sets, while PRAGAN never used all 13,712 images for training, and only 1/4 of the data can achieve the best results.

Table 6. Comparative results on synthetic deraining datasets, all models were directly tested on Test1200 [20]. For MPRNet [59], we retrained it with the same number of iterations using the same experimental configuration as our proposed method. Specifically, MPRNet [59] was trained for 63 epochs with all training data.

	Test100 PSNR/SSIM	Rain100H PSNR/SSIM	Rain100L PSNR/SSIM	Test1200 PSNR/SSIM
DerainNet [18]	21.90/0.837	13.67/0.573	26.36/0.873	22.24/0.848
DDC [58]	22.63/0.825	14.51/0.499	26.75/0.858	27.59/0.882
DIDMDN [20]	21.56/0.811	16.31/0.556	23.71/0.804	27.00/0.883
SEMI [57]	21.39/0.781	15.50/0.519	24.05/0.820	24.95/0.841
RESCAN [55]	23.09/0.830	24.86/0.783	27.46/0.864	27.14/0.869
UMRL [56]	23.92/0.883	24.85/0.835	27.73/0.929	29.59/0.922
PreNet [22]	24.03/0.872	25.75/0.861	31.64/0.949	30.86/0.926
MSPFN [36]	26.97/0.898	27.42/0.864	31.66/0.921	31.59/0.928
MPRNet [59]	26.24/0.894	27.73/0.867	32.65/0.951	31.84/0.929
PRAGAN	27.71/0.916	27.41/0.883	32.54/0.957	32.20/0.934

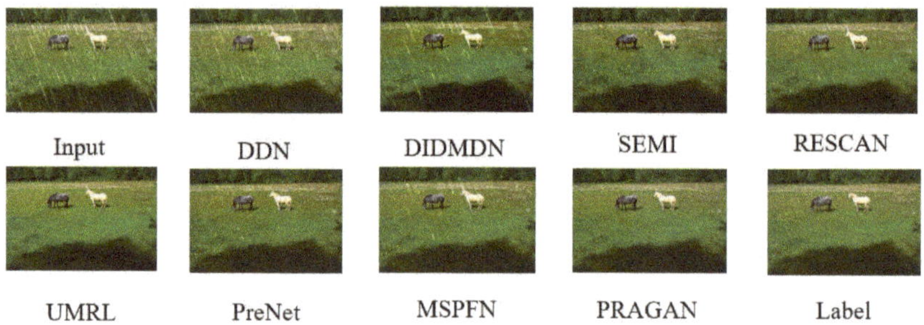

Figure 8. Deraining results from the Rain100L [19] testing set. Rain100L [19] consists of 100 image pairs for testing with one type of rain streak. It can be seen from the figure that most of the methods can remove rain streaks to a certain extent, but our PRAGAN can almost remove all the rain streaks compared with other methods, and restore images closer to ground truth.

Figure 9. Deraining results from the Rain100H [19] testing set. In contrast to the relatively simple Rain100L [19], Rain100H [19] contains five types of streak directions, so part of the rain removal method was not effective. Our method needed only a quarter of the 13,712 image pairs for training.

3.3.2. Real Images

Due to the inevitable difference between synthetic rain streaks and real data, this section lists the comparison results of our proposed PRAGAN with other methods on real deraining datasets. According to the results provided by [61], we conducted experiments on two datasets, namely Internet-Data [57] and SPA-Data [62]. For Internet-Data [57], we only provide visual comparison, given that it has no ground truth to allow a quantitative comparison. We pretrained the ViT discriminator for this section with a new dataset that contained Train200 and Internet-Data, for which the mini-batch size was 32, while other configurations were the same as the previous pretraining. In the adversarial training, given that the overall dataset Train200 has only 200 image pairs, we did not use the reselecting data algorithm. PSNR and SSIM comparisons on SPA-Data [62] are shown in Table 7 and a visual comparison on Internet-Data [57] is displayed in Figure 10.

Table 7. Comparisons on real-world dataset SPA-Data [61].

Methods	PSNR	SSIM
Input	34.15	0.927
DSC [63]	34.95	0.942
GMM [60]	34.30	0.943
JCAS [64]	34.95	0.945
Clear [5]	32.66	0.942
DDN [58]	34.70	0.934
RESCAN [55]	34.70	0.938
JORDER_E [65]	34.34	0.936
SIRR [57]	34.85	0.936
PRAGAN	34.96	0.951

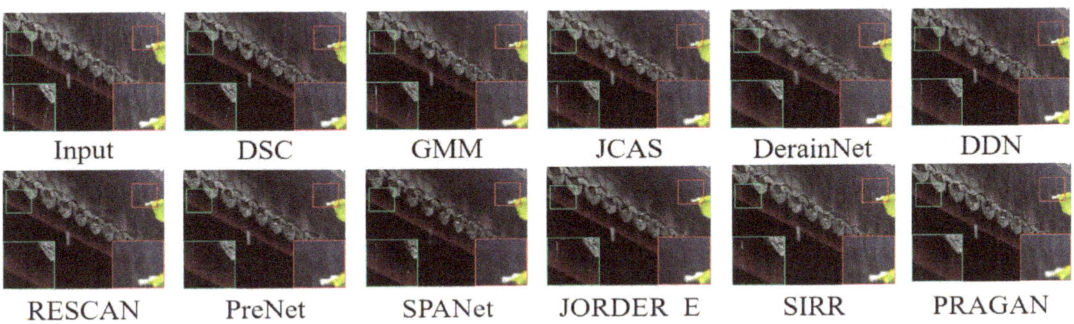

Figure 10. Deraining results on Internet-Data [57] testing set. Best viewed when zoomed in and in color.

4. Conclusions

In this study, we propose a novel generative adversarial network consisting of a pretrained ViT discriminator and a progressive recurrent attention generator for single-image deraining tasks. First of all, we propose a parameter sharing recurrent neural network for image deraining. Secondly, we propose a new pretrained ViT discriminator for image deraining in a GAN. Compared with PatchGAN, ViT in the pretrained stage shows more stable convergence. Finally, we propose a data reselecting algorithm DRA, which can not only make efficient use of training data on small datasets, but also promote the deraining performance of our model on large datasets. We have shown extensive ablation studies and comparative experiments to fully validate the effectiveness of our proposed PRAGAN on both synthetized and real datasets. A more in-depth investigation on image deraining and GAN will be carried out in the future.

Author Contributions: Conceptualization, B.W.; methodology, B.W.; software, B.W.; validation, B.W., D.W. and Z.W.; formal analysis, L.Z.; investigation, B.W.; resources, B.W.; data curation, B.W.; writing—original draft preparation, B.W.; writing—review and editing, B.W.; visualization, B.W.; supervision, D.W.; project administration, D.W.; funding acquisition, L.Z. All authors have read and agreed to the published version of the manuscript.

Funding: This research was funded by Shandong Provincial Natural Science Foundation, China (Grant Number ZR2019BF022) and National Natural Science Foundation of China (Grant Number 62001272).

Data Availability Statement: All the datasets used for training the model of this paper are from Internet.

Conflicts of Interest: The authors declare no conflict of interest.

References

1. Howard, A.G.; Zhu, M.; Chen, B.; Kalenichenko, D.; Wang, W.; Weyand, T.; Adam, H. Mobilenets: Efficient Convolutional Neural Networks for Mobile Vision Applications. *arXiv* **2017**, arXiv:1704.04861.
2. Girshick, R. Fast R-CNN. In Proceedings of the IEEE International Conference on Computer Vision, Santiago, Chile, 7–13 December 2015; pp. 1440–1448. [CrossRef]
3. Ronneberger, O.; Fischer, P.; Brox, T. U-Net: Convolutional Networks for Biomedical Image Segmentation. In Proceedings of the International Conference on Medical Image Computing and Computer-Assisted Intervention, Munich, Germany, 5–9 October 2015; Springer: Cham, Switzerland, 2015; pp. 234–241. [CrossRef]
4. Zhang, X.; Luo, H.; Fan, X.; Xiang, W.; Sun, Y.; Xiao, Q.; Sun, J. Alignedreid: Surpassing Human-Level Performance in Person Re-Identification. *arXiv* **2017**, arXiv:1711.08184.
5. Simonyan, K.; Zisserman, A. Very Deep Convolutional Networks for Large-Scale Image Recognition. *arXiv* **2014**, arXiv:1409.1556.
6. He, K.; Zhang, X.; Ren, S.; Sun, J. Deep Residual Learning for Image Recognition. In Proceedings of the IEEE Conference on Computer Vision and Pattern Recognition, Las Vegas, NV, USA, 26 June–1 July 2016; pp. 770–778. [CrossRef]
7. Huang, G.; Liu, Z.; van der Maaten, L.; Weinberger, K.Q. Densely Connected Convolutional Networks. In Proceedings of the IEEE Conference on Computer Vision and Pattern Recognition, Honolulu, HI, USA, 21–26 July 2017; pp. 4700–4708. [CrossRef]
8. Xie, S.; Girshick, R.; Dollár, P.; Tu, Z.; He, K. Aggregated Residual Transformations for Deep Neural Networks. In Proceedings of the IEEE Conference on Computer Vision and Pattern Recognition, Honolulu, HI, USA, 21–26 July 2017; pp. 1492–1500. [CrossRef]
9. Hu, J.; Shen, L.; Sun, G. Squeeze-and-Excitation Networks. In Proceedings of the IEEE Conference on Computer Vision and Pattern Recognition, Salt Lake City, UT, USA, 18–22 June 2018; pp. 7132–7141. [CrossRef]
10. Woo, S.; Park, J.; Lee, J.Y.; Kweon, I.S. CBAM: Convolutional Block Attention Module. In Proceedings of the European Conference on Computer Vision (ECCV), Munich, Germany, 8–14 September 2018; pp. 3–19. [CrossRef]
11. Guo, M.H.; Xu, T.X.; Liu, J.J.; Liu, Z.N.; Jiang, P.T.; Mu, T.J.; Hu, S.M. Attention mechanisms in computer vision: A survey. *Comput. Vis. Media* **2022**, *8*, 331–368. [CrossRef]
12. Vaswani, A.; Shazeer, N.; Parmar, N.; Uszkoreit, J.; Jones, L.; Gomez, A.N.; Polosukhin, I. Attention is all you need. *Adv. Neural Inf. Process. Syst.* **2017**, *30*. [CrossRef]
13. Dosovitskiy, A.; Beyer, L.; Kolesnikov, A.; Weissenborn, D.; Zhai, X.; Unterthiner, T.; Houlsby, N. An Image is Worth 16 × 16 Words: Transformers for Image Recognition at Scale. *arXiv* **2020**, arXiv:2010.11929.
14. Zhu, J.Y.; Park, T.; Isola, P.; Efros, A.A. Unpaired Image-to-Image Translation Using Cycle-Consistent Adversarial Networks. In Proceedings of the IEEE International Conference on Computer Vision, Venice, Italy, 22–29 October 2017; pp. 2223–2232. [CrossRef]
15. Zhang, K.; Luo, W.; Zhong, Y.; Ma, L.; Stenger, B.; Liu, W.; Li, H. Deblurring by Realistic Blurring. In Proceedings of the IEEE/CVF Conference on Computer Vision and Pattern Recognition, Seattle, WA, USA, 14–19 June 2020; pp. 2737–2746. [CrossRef]
16. Haris, M.; Shakhnarovich, G.; Ukita, N. Recurrent Back-Projection Network for Video Super-Resolution. In Proceedings of the IEEE/CVF Conference on Computer Vision and Pattern Recognition, Long Beach, CA, USA, 16–20 June 2019; pp. 3897–3906. [CrossRef]
17. Tassano, M.; Delon, J.; Veit, T. DVDNET: A Fast Network for Deep Video Denoising. In Proceedings of the 2019 IEEE International Conference on Image Processing (ICIP), Taipei, Taiwan, 22–25 September 2019; pp. 1805–1809. [CrossRef]
18. Fu, X.; Huang, J.; Ding, X.; Liao, Y.; Paisley, J. Clearing the skies: A Deep Network Architecture for Single-Image Rain Removal. *IEEE Trans. Image Process.* **2017**, *26*, 2944–2956. [CrossRef] [PubMed]
19. Yang, W.; Tan, R.T.; Feng, J.; Liu, J.; Guo, Z.; Yan, S. Deep Joint Rain Detection and Removal from a Single Image. In Proceedings of the IEEE Conference on Computer Vision and Pattern Recognition, Honolulu, HI, USA, 21–26 July 2017; pp. 1357–1366. [CrossRef]
20. Zhang, H.; Patel, V.M. Density-Aware Single Image De-Raining Using a Multi-Stream Dense Network. In Proceedings of the IEEE Conference on Computer Vision and Pattern Recognition, Salt Lake City, CA, USA, 18–22 June 2018; pp. 695–704. [CrossRef]

21. Zhang, H.; Sindagi, V.; Patel, V.M. Image De-Raining Using a Conditional Generative Adversarial Network. *IEEE Trans. Circuits Syst. Video Technol.* **2019**, *30*, 3943–3956. [CrossRef]
22. Ren, D.; Zuo, W.; Hu, Q.; Zhu, P.; Meng, D. Progressive Image Deraining Networks: A Better and Simpler Baseline. In Proceedings of the IEEE/CVF Conference on Computer Vision and Pattern Recognition, Long Beach, CA, USA, 16–20 June 2019; pp. 3937–3946. [CrossRef]
23. Shi, X.; Chen, Z.; Wang, H.; Yeung, D.Y.; Wong, W.K.; Woo, W.C. Convolutional LSTM network: A machine learning approach for precipitation nowcasting. *Adv. Neural Inf. Process. Syst.* **2015**, *28*. [CrossRef]
24. Graves, A. Long Short-Term memory. In *Supervised Sequence Labelling with Recurrent Neural Networks;* Springer: Berlin/Heidelberg, Germany, 2012; pp. 37–45. [CrossRef]
25. Azad, R.; Asadi-Aghbolaghi, M.; Fathy, M.; Escalera, S. Bi-Directional ConvLSTM U-Net with Densley Connected Convolutions. In Proceedings of the IEEE/CVF International Conference on Computer Vision Workshops, Seoul, Republic of Korea, 27 October–2 November 2019. [CrossRef]
26. Goodfellow, I.; Pouget-Abadie, J.; Mirza, M.; Xu, B.; Warde-Farley, D.; Ozair, S.; Bengio, Y. Generative adversarial networks. *Commun. ACM* **2020**, *63*, 139–144. [CrossRef]
27. Gulrajani, I.; Ahmed, F.; Arjovsky, M.; Dumoulin, V.; Courville, A.C. Improved training of wasserstein gans. *Adv. Neural Inf. Processing Syst.* **2017**, *30*. [CrossRef]
28. Liu, Z.; Lin, Y.; Cao, Y.; Hu, H.; Wei, Y.; Zhang, Z.; Guo, B. Swin Transformer: Hierarchical Vision Transformer Using Shifted Windows. In Proceedings of the IEEE/CVF International Conference on Computer Vision, Montreal, QC, Canada, 10–17 October 2021; pp. 10012–10022. [CrossRef]
29. Liang, J.; Cao, J.; Sun, G.; Zhang, K.; van Gool, L.; Timofte, R. Swinir: Image Restoration Using Swin Transformer. In Proceedings of the IEEE/CVF International Conference on Computer Vision, Montreal, QC, Canada, 10–17 October 2021; pp. 1833–1844. [CrossRef]
30. Park, N.; Kim, S. How Do Vision Transformers Work? *arXiv* **2022**, arXiv:2202.06709.
31. Han, K.; Wang, Y.; Chen, H.; Chen, X.; Guo, J.; Liu, Z.; Tao, D. A Survey on Vision Transformer. *IEEE Trans. Pattern Anal. Mach. Intell.* **2022**, *1*, 87–110. [CrossRef] [PubMed]
32. Ba, J.L.; Kiros, J.R.; Hinton, G.E. Layer Normalization. *arXiv* **2016**, arXiv:1607.06450.
33. Zhao, H.; Gallo, O.; Frosio, I.; Kautz, J. Loss Functions for Image Restoration with Neural Networks. *IEEE Trans. Comput. Imaging* **2016**, *3*, 47–57. [CrossRef]
34. Tao, X.; Gao, H.; Shen, X.; Wang, J.; Jia, J. Scale-Recurrent Network for Deep Image Deblurring. In Proceedings of the IEEE Conference on Computer Vision and Pattern Recognition, Salt Lake City, CA, USA, 18–22 June 2018; pp. 8174–8182. [CrossRef]
35. Nah, S.; Kim, T.H.; Lee, K.M. Deep Multi-Scale Convolutional Neural Network for Dynamic Scene Deblurring. In Proceedings of the IEEE Conference on Computer Vision and Pattern Recognition, Honolulu, HI, USA, USA 21–26 July 2017; pp. 3883–3891. [CrossRef]
36. Jiang, K.; Wang, Z.; Yi, P.; Chen, C.; Huang, B.; Luo, Y.; Jiang, J. Multi-Scale Progressive Fusion Network for Single Image deraining. In Proceedings of the IEEE/CVF Conference on Computer Vision and Pattern Recognition, Seattle, WA, USA, 14–19 June 2020; pp. 8346–8355. [CrossRef]
37. Gao, H.; Tao, X.; Shen, X.; Jia, J. Dynamic Scene Deblurring with Parameter Selective Sharing and Nested Skip Connections. In Proceedings of the IEEE/CVF Conference on Computer Vision and Pattern Recognition, Long Beach, CA, USA, 16–20 June 2019; pp. 3848–3856. [CrossRef]
38. Kamgar-Parsi, B.; Rosenfeld, A. Optimally Isotropic Laplacian Operator. *IEEE Trans. Image Process.* **1999**, *8*, 1467–1472. [CrossRef]
39. Wang, Z.; Bovik, A.C.; Sheikh, H.R.; Simoncelli, E.P. Image Quality Assessment: From Error Visibility to Structural Similarity. *IEEE Trans. Image Process.* **2004**, *13*, 600–612. [CrossRef]
40. Xu, Y.; Wei, H.; Lin, M.; Deng, Y.; Sheng, K.; Zhang, M.; Xu, C. Transformers in computational visual media: A survey. *Comput. Vis. Media* **2022**, *8*, 33–62. [CrossRef]
41. Devlin, J.; Chang, M.W.; Lee, K.; Toutanova, K. Bert: Pre-Training of Deep Bidirectional Transformers for Language Understanding. *arXiv* **2018**, arXiv:1810.04805.
42. Floridi, L.; Chiriatti, M. GPT-3: Its nature, scope, limits, and consequences. *Minds Mach.* **2020**, *30*, 681–694. [CrossRef]
43. He, K.; Chen, X.; Xie, S.; Li, Y.; Dollár, P.; Girshick, R. Masked Autoencoders are Scalable Vision Learners. In Proceedings of the IEEE/CVF Conference on Computer Vision and Pattern Recognition, New Orleans, LA, USA, 19–24 June 2022; pp. 16000–16009.
44. Long, J.; Shelhamer, E.; Darrell, T. Fully Convolutional Networks for Semantic Segmentation. In Proceedings of the IEEE Conference on Computer Vision and Pattern Recognition, Boston, MA, USA, 7–12 June 2015; pp. 3431–3440. [CrossRef]
45. Russakovsky, O.; Deng, J.; Su, H.; Krause, J.; Satheesh, S.; Ma, S.; Fei-Fei, L. Imagenet large scale visual recognition challenge. *Int. J. Comput. Vis.* **2015**, *115*, 211–252. [CrossRef]
46. He, K.; Girshick, R.; Dollár, P. Rethinking Imagenet Pre-Training. In Proceedings of the IEEE/CVF International Conference on Computer Vision, Seoul, Republic of Korea, 27 October–2 November 2019; pp. 4918–4927. [CrossRef]
47. Isola, P.; Zhu, J.Y.; Zhou, T.; Efros, A.A. Image-to-Image Translation with Conditional Adversarial Networks. In Proceedings of the IEEE Conference on Computer Vision and Pattern Recognition, Honolulu, HI, USA, 21–26 July 2017; pp. 1125–1134. [CrossRef]
48. Wen, Y.; Chen, J.; Sheng, B.; Chen, Z.; Li, P.; Tan, P.; Lee, T.Y. Structure-Aware Motion Deblurring Using Multi-Adversarial Optimized Cyclegan. *IEEE Trans. Image Process.* **2021**, *30*, 6142–6155. [CrossRef] [PubMed]

49. Kupyn, O.; Budzan, V.; Mykhailych, M.; Mishkin, D.; Matas, J. Deblurgan: Blind Motion Deblurring Using Conditional Adversarial Networks. In Proceedings of the IEEE Conference on Computer Vision and Pattern Recognition, Salt Lake City, UT, USA, 18–22 June 2018; pp. 8183–8192. [CrossRef]
50. Cho, S.J.; Ji, S.W.; Hong, J.P.; Jung, S.W.; Ko, S.J. Rethinking Coarse-to-Fine Approach in Single Image Deblurring. In Proceedings of the IEEE/CVF International Conference on Computer Vision, Montreal, QC, Canada, 10–17 October 2021; pp. 4641–4650. [CrossRef]
51. Park, D.; Kang, D.U.; Kim, J.; Chun, S.Y. Multi-Temporal Recurrent Neural Networks for Progressive Non-Uniform Single Image Deblurring with Incremental Temporal Training. In Proceedings of the European Conference on Computer Vision, Glasgow, UK, 23–28 August 2020; Springer: Cham, Switzerland, 2020; pp. 327–343. [CrossRef]
52. Kingma, D.P.; Ba, J. Adam: A Method for Stochastic Optimization. *arXiv* **2014**, arXiv:1412.6980.
53. Loshchilov, I.; Hutter, F. SGDR: Stochastic Gradient Descent with Warm Restarts. *arXiv* **2016**, arXiv:1608.03983.
54. Loshchilov, I.; Hutter, F. Decoupled Weight Decay Regularization. *arXiv* **2017**, arXiv:1711.05101.
55. Li, X.; Wu, J.; Lin, Z.; Liu, H.; Zha, H. Recurrent Squeeze-and-Excitation Context Aggregation Net for Single Image Deraining. In Proceedings of the European Conference on Computer Vision (ECCV), Munich, Germany, 8–14 September 2018; pp. 254–269. [CrossRef]
56. Yasarla, R.; Patel, V.M. Uncertainty Guided Multi-Scale Residual Learning-Using a Cycle Spinning CNN for Single Image De-raining. In Proceedings of the IEEE/CVF Conference on Computer Vision and Pattern Recognition, Long Beach, CA, USA, 16–20 June 2019; pp. 8405–8414. [CrossRef]
57. Wei, W.; Meng, D.; Zhao, Q.; Xu, Z.; Wu, Y. Semi-Supervised Transfer Learning for Image Rain Removal. In Proceedings of the IEEE/CVF Conference on Computer Vision and Pattern Recognition, Long Beach, CA, USA, 16–20 June 2019; pp. 3877–3886. [CrossRef]
58. Fu, X.; Huang, J.; Zeng, D.; Huang, Y.; Ding, X.; Paisley, J. Removing Rain from Single Images via a Deep Detail Network. In Proceedings of the IEEE Conference on Computer Vision and Pattern Recognition, Honolulu, HI, USA, 21–26 July 2017; pp. 3855–3863. [CrossRef]
59. Zamir, S.W.; Arora, A.; Khan, S.; Hayat, M.; Khan, F.S.; Yang, M.H.; Shao, L. Multi-Stage Progressive Image Restoration. In Proceedings of the IEEE/CVF Conference on Computer Vision and Pattern Recognition, Online, 19–25 June 2021; pp. 14821–14831. [CrossRef]
60. Li, Y.; Tan, R.T.; Guo, X.; Lu, J.; Brown, M.S. Rain Streak Removal Using Layer Priors. In Proceedings of the IEEE Conference on Computer Vision and Pattern Recognition, Las Vegas, NV, USA, 26 June–1 July 2016; pp. 2736–2744. [CrossRef]
61. Wang, H.; Wu, Y.; Li, M.; Zhao, Q.; Meng, D. A Survey on Rain Removal from Video and Single Image. *arXiv* **2019**, arXiv:1909.08326. [CrossRef]
62. Wang, T.; Yang, X.; Xu, K.; Chen, S.; Zhang, Q.; Lau, R.W. Spatial Attentive Single-Image Deraining with a High Quality Real Rain Dataset. In Proceedings of the IEEE/CVF Conference on Computer Vision and Pattern Recognition, Long Beach, CA, USA, 16–20 June 2019; pp. 12270–12279. [CrossRef]
63. Luo, Y.; Xu, Y.; Ji, H. Removing Rain from a Single Image via Discriminative Sparse Coding. In Proceedings of the IEEE International Conference on Computer Vision, Santiago, Chile, 7–13 December 2015; pp. 3397–3405. [CrossRef]
64. Gu, S.; Meng, D.; Zuo, W.; Zhang, L. Joint Convolutional Analysis and Synthesis Sparse Representation for Single Image Layer Separation. In Proceedings of the IEEE International Conference on Computer Vision, Venice, Italy, 22–29 October 2017; pp. 1708–1716. [CrossRef]
65. Yang, W.; Tan, R.T.; Feng, J.; Guo, Z.; Yan, S.; Liu, J. Joint Rain Detection and Removal from a Single Image with Contextualized Deep Networks. *IEEE Trans. Pattern Anal. Mach. Intell.* **2019**, *42*, 1377–1393. [CrossRef]

sensors

MDPI

Article

A Novel Dynamic Bit Rate Analysis Technique for Adaptive Video Streaming over HTTP Support

Ponnai Manogaran Ashok Kumar [1], Lakshmi Narayanan Arun Raj [2], B. Jyothi [3], Naglaa F. Soliman [4,*], Mohit Bajaj [5] and Walid El-Shafai [6,7]

1 Department of Computer Science Engineering, Koneru Lakshmaiah Education Foundation, Vaddeswaram 522 302, India
2 Department of Computing Science and Engineering, B.S.A. Crescent Institute of Science and Technology, Vandalur, Chennai 600 048, India
3 Department of Electrical and Electronics Engineering, Koneru Lakshmaiah Educational Foundation, Vaddeswaram 522 302, India
4 Department of Information Technology, College of Computer and Information Sciences, Princess Nourah bint Abdulrahman University, P.O. Box 84428, Riyadh 11671, Saudi Arabia
5 Department of Electrical Engineering, Graphic Era (Deemed to Be University), Dehradun 248 002, India
6 Security Engineering Lab, Computer Science Department, Prince Sultan University, Riyadh 11586, Saudi Arabia
7 Department of Electronics and Electrical Communications Engineering, Faculty of Electronic Engineering, Menoufia University, Menouf 32952, Egypt
* Correspondence: nfsoliman@pnu.edu.sa

Citation: Ashok Kumar, P.M.; Arun Raj, L.N.; Jyothi, B.; Soliman, N.F.; Bajaj, M.; El-Shafai, W. A Novel Dynamic Bit Rate Analysis Technique for Adaptive Video Streaming over HTTP Support. *Sensors* **2022**, *22*, 9307. https://doi.org/10.3390/s22239307

Academic Editors: Kang Ryoung Park, Sangyoun Lee and Euntai Kim

Received: 26 September 2022
Accepted: 19 November 2022
Published: 29 November 2022

Abstract: Recently, there has been an increase in research interest in the seamless streaming of video on top of Hypertext Transfer Protocol (HTTP) in cellular networks (3G/4G). The main challenges involved are the variation in available bit rates on the Internet caused by resource sharing and the dynamic nature of wireless communication channels. State-of-the-art techniques, such as Dynamic Adaptive Streaming over HTTP (DASH), support the streaming of stored video, but they suffer from the challenge of live video content due to fluctuating bit rate in the network. In this work, a novel dynamic bit rate analysis technique is proposed to model client–server architecture using attention-based long short-term memory (A-LSTM) networks for solving the problem of smooth video streaming over HTTP networks. The proposed client system analyzes the bit rate dynamically, and a status report is sent to the server to adjust the ongoing session parameter. The server assesses the dynamics of the bit rate on the fly and calculates the status for each video sequence. The bit rate and buffer length are given as sequential inputs to LSTM to produce feature vectors. These feature vectors are given different weights to produce updated feature vectors. These updated feature vectors are given to multi-layer feed forward neural networks to predict six output class labels (144p, 240p, 360p, 480p, 720p, and 1080p). Finally, the proposed A-LSTM work is evaluated in real-time using a code division multiple access evolution-data optimized network (CDMA20001xEVDO Rev-A) with the help of an Internet dongle. Furthermore, the performance is analyzed with the full reference quality metric of streaming video to validate our proposed work. Experimental results also show an average improvement of 37.53% in peak signal-to-noise ratio (PSNR) and 5.7% in structural similarity (SSIM) index over the commonly used buffer-filling technique during the live streaming of video.

Keywords: adaptive video streaming; A-LSTM networks; bit rate measurement; client–server model; HTTP; reference metrics; video quality

1. Introduction

Adaptive media streaming through Hypertext Transfer Protocol (HTTP) is a widely used mechanism by the service provider. The main advantage is that it does not require any change in the underlying network layer to support streaming. The standard organization Moving Picture Experts Group (MPEG) and the 3rd Generation Partnership Project (3GPP) have standardized a method called Dynamic Adaptive Streaming over HTTP (DASH) to ensure interoperability [1]. In DASH implementation, the video is segmented, and each segment is stored with different video quality parameters, including spatial and temporal resolutions. The adaptation process at the server streams suitable segments targeted to match the link capacity of the client [2]. The present solution fails in case of fast changes in network bit carrying capacities leading to the video freezing and enhancing degradation of user satisfaction [3]. A new method can be integrated with the DASH technique to support live content streaming. In the buffer-based implementation of live video streaming [4,5], involving client observation of buffer threshold does not guarantee the quality since the variation in the bit rate depends on capturing and coding methods at the server. The scalable video coding (SVC) approach permits frame-level adaptation, but it requires switching different video layers during the session [6].

Another development in adaptive video streaming is the study of 3G/4G cellular networks offering Internet connection. Many times, the user equipment offered by a cellular operator to support Internet services fails to deliver the desired quality for many practical reasons. For example, Figure 1 shows the observation of a 4G dongle employing the CDMA20001xEVDO Rev-A technique and a 4G dongle based on the long-term evolution time-division duplex (LTE-TDD) category-3 system. The observed variation in bit rate depends on location and fluctuation with time. This clearly justifies the motivation behind the developing system, which can target wireless empowerment and offer the best performance in terms of user satisfaction.

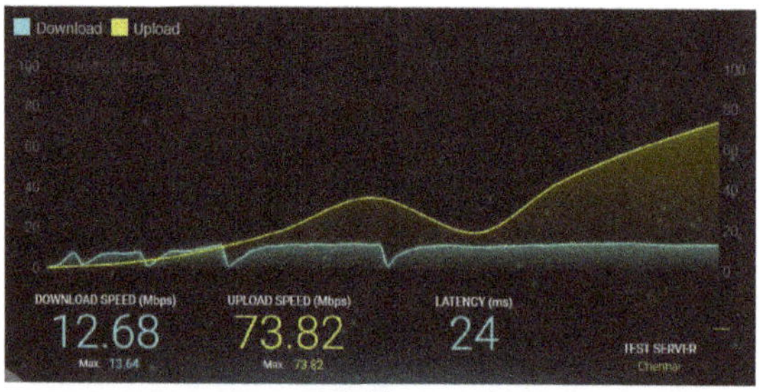

(a)

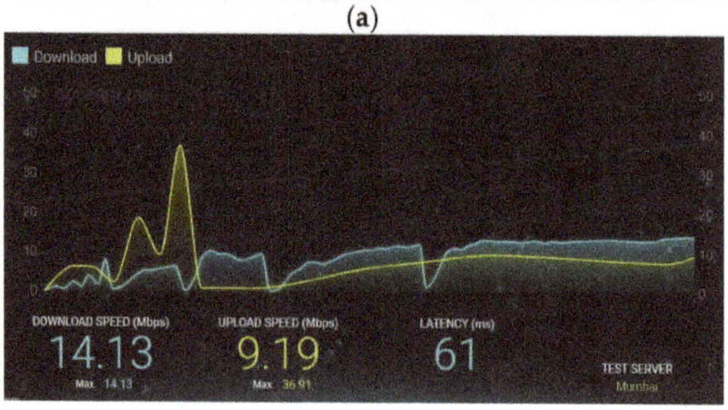

(b)

Figure 1. Download/Upload bit rate observed through different wireless Internet dongles (**a**) Reliance Netconnect+ (CDMA20001xEVDO Rev-A) 4G dongle (**b**) Airtel 4G Mobile Hotspot (LTE-TDD Category 3) dongle.

The objective performance of the proposed system is evaluated using standard metrics while meeting the design goal. The International Telecommunication Union Standardization Sector (ITU-T) recommends using full reference metrics when the original video is available at the receiver to test the individual system in a laboratory environment. The standard evaluation metrics can be applied to test video quality in different formats, including quarter common intermediate format (QCIF), common intermediate format (CIF), and Video Graphics Array (VGA). Table 1 lists the different parameters and corresponding values to test the large varying quality of the video.

Table 1. Test factors as per the ITU-T J.247 recommendation.

S. No.	Parameters	Values
1	Transmission	Errors with packet loss
2	Frame rate	5 fps to 30 fps
3	Video codec	H.264/AVC, VC-1, Windows Media 9, Real Video, MPEG4
4	Video resolution (QCIF, CIF, and VGA)	QCIF (6 to 320 Kbps) CIF (64 to 2000 Kbps) VGA (128–4000 Kbps)
5	Temporal errors (pausing with skipping)	Maximum of 2 s

The proposed work in this paper tries to use these parameters with their corresponding standard values in implementation and development. The existing link bit rate assessment method at the client involves sending a ping message to the server and computing the bit rate using the time spent by the packet to come back. However, this method lacks precision because of many external factors, for example, instantaneous congestion to the router can temporarily interrupt the incoming rate of a ping message. Thus, the best practice for dealing with this issue is to evaluate the capacity of the link in terms of bit rate to the receiver by analyzing the bit stream arrival on the fly.

The streaming that needs to be sampled at times is analyzed and sent a feedback message to the sender for performing remedial action on the outgoing stream so that the end user enjoys a better quality of experience during the entire viewing session. The schematic approach of the proposed architecture is shown in Figure 2, where the client applies a predefined algorithm to compute the arrival rate and forward the report to the server. The response action in the system loop needs to be proactive and stable to meet the satisfaction of the system's real-time streaming requirement. This provides the scope of additional intelligence for the link capacity estimation. In the proposed method, pattern matching is employed by the client to reduce the processing time and meet the requirements of live video streaming.

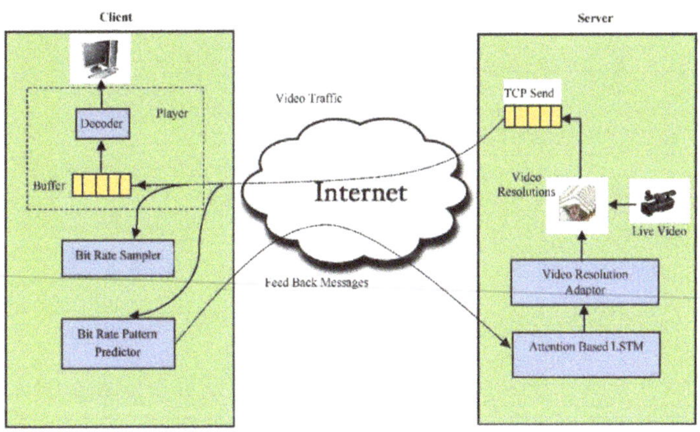

Figure 2. Schematic diagram of a client–server model for adaptive video streaming.

An analytical model is included to support the performance measure of the proposed work. The bit rate profile and performance measure are also presented in tabular form.

Although the proposed system is basically developed as on-the-top of HTTP (OTT), it has incorporated the inherent behavior of streamed data over the wireless network. The dynamics of the observed bit rate are due to the burst nature of the Internet traffic [7] and the time-varying nature of wireless signals in 3G/4G networks. Further, the system performance evaluated here corresponds to the Internet over 4G wireless networks (CDMA20001xEVDO Rev-A). Finally, the proposed work is compared against popular buffer-filling methods [8], the default Internet option to stream multimedia content.

We present a summary of our contributions:

1. We devised a novel feed-forward attention-based LSTM model using reinforcement learning to successfully integrate features from several layers of the LSTM network to solve sequential bit rate dependency problems and adaptive video streaming over HTTP.
2. We present a cost function for attention networks that maximizes video quality and minimizes re-buffering time.
3. Experimental results have revealed that the suggested A-LSTM technique performs better than the state-of-the-art buffer filling algorithms on the standard datasets.

The rest of the paper is organized as follows: video streaming, video buffering, and machine learning techniques in multimedia streaming-related works are summarized in Section 2. The architecture of the client–server model and A-LSTM is introduced first. Then, we present the suggested A-LSTM model employing reinforcement learning and feed forward attention-LSTM technique in Section 3. Section 4 presents the results of tests performed on standard datasets containing variant bit rates and buffer lengths with the current popular techniques. Lastly, we complete the paper with the hypotheses and future work in Section 5.

2. Related Work

The bit rate adaptation of video streaming involves many factors, including scheduling of segmented video, bit rate selection, bandwidth estimation, etc. Many commercially available services, such as Smooth Streaming, Akamai HD, Netflix, and Adobe OSMF, implement adaptive streaming through the Internet. Appropriate modeling and analysis of the key phase include switching of the multimedia data during streaming used by the service provider, which helps to refine the design of the system for improving performance in the feedback control loop [9].

In the HTTP based adaptive streaming (HAS), the quality of experience (QoE) depends on the appropriate selection of video segment and switching of bit stream based on client input [10]. The underflow probability of the media buffer is estimated during run time and is incorporated in the QoE framework while supporting the acceptable quality of the streaming video. The buffer stability is a vital parameter to maintain the quality of the video during play out, and this is implemented by estimating the buffer level during the streaming session of the client [11]. Furthermore, the estimation of buffer underflow probability can provide vital inputs in implementing layer switching of different video segments, i.e., adaptation of video content during the streaming process [12].

A new version of adaptive rate control algorithms [13] is proposed to improve the combined system performance of video play out smoothness and frame quality based on the feedback information of wireless network estimation, buffer content, and playback situation. However, their main disadvantage is the lack of adaptability of heterogeneous networks and noisy error data. To solve transmission errors, a novel error control coding technique is proposed [14] for video transmission over wireless network and to implement different error control techniques for video transmission. However, their main performance is not evaluated for real-time applications and does not consider the pixel intensity values.

To solve the pixel intensity problem, a novel algorithm [15] is presented for exploiting a general model of high-efficiency video coding (HEVC) technique with the help of decoding-energy fast compression (DEFC). This method does not consider routing parameters. A Novel Analytical framework [16] is proposed based on routing measure parameters to reduce distortion in wireless video traffic. A new hue saturation lightness (HSV), edge preserving, and Huffman-coding (HC)-based Huffman and differential pulse code modulation (DPCM) encodings algorithm [17] is proposed to increase the compression ratio of the video frames.

Dynamic Adaptive Streaming over HTTP (DASH) in the client–server environment has attracted worldwide attention for many reasons from researchers and developers [18]. There is a need to map the DASH layer with the scalable video coding (SVC) layer, not only to improve the throughput of streaming video with the help of HTTP overhead messages, but also to estimate the bit rates of media sessions [19]. A cross-layer method involving DASH and a physical (radio) layer can manage better scheduling and resource allocation [20] in the media accesses control layer to solve the throughput problem. To overcome the limitations of a single network, energy consumption of the end device and environmental factors are considered an important parameter by [21] to seamlessly transfer the requested video segments concurrently to mobile devices.

Further, a learning approach may help with streaming video through multilink. For example, the learning method can incorporate a Markov decision process with a finite state. The reward calculation in such implementations must include video quality of service (QoS) [21,22]. The estimation of network bandwidth as a Transport Control Protocol (TCP) throughput in the HAS system by the client may not be reliable when HAS traffic occupies a significant portion of the network traffic. The client encounters bottlenecks in the networks for supporting discrete characteristics of video bit rate while competing with other clients [23]. The physical layer information, i.e., statistics at the modem, can be passed to the application layer for fast identification and estimation of the wireless channel condition in a HAS system [24]. In [25], the physical layer throughput/goodput is used to adapt the rate of the HAS video client and improve the QoE of streaming video, but still, the system needs to consider the dynamic behavior of the wireless channel. At the physical layer, modem statistics can detect sharp variations in wireless link quality. Now, the HAS client can place a new request to the server based on its current state and the status of the existing request for the segment (as shown in Figure 2).

Another focus for adaptation of DASH/HAS video in cellular mobile systems can be enforced by the network operator based on the knowledge of cell load and channel conditions to optimize the content delivery. Further, this opens an avenue for joint optimization of resource allocation in multi-user networks and controlling video streaming for the DASH client [26]. Furthermore, assuming the proxy has the adaptive HAS content with multiple-bit rate encoding, it may eliminate the need for further processing of video, and this approach is desirable for on-the-top (OTT) streaming services, particularly when the DASH server is not present in the network operator's domain.

Understanding the time complexities and other quality of service (QoS) requirements of live video streaming, our work considers sampling of incoming bit rate alone and the buffer state for the client's decision-making. Furthermore, the proposed system is developed to cope with the fluctuation in bit rate due to the best-effort model of the Internet and the time-varying characteristics of a wireless channel. The combined network effect on the sampled data set (bit rate) tends to behave as random variables. Hence the Markov process-based decision-making is not suitable here [26]. Another novel contribution compared to the earlier work is the quick processing of link capacity estimation by using predefined pattern matching, as the system design is targeted to handle live video streams.

3. Proposed Methodology

3.1. System Architecture

The proposed system is demonstrated after the client–server architecture. The main function of the server side module is to receive the live or stored video for transcoding before streaming (Figure 2). The system architecture consists of two main modules, where the first module deals with transcoding and adaptive streaming of the content while the second module listens to the client feedback. A video-acquiring device capable of capturing high-definition content is attached to the server (Figure 3), and the H.264 video codec codes the resulting stream. The connection for live streaming is implemented on 4G wireless cellular networks. The system implementation at the client deals with (Figure 4) playing and analysis of the incoming video stream. After collecting N frames, the system simultaneously calls the bit rate estimation process and the media player task. The bit rate estimation algorithm defines the feedback category to be sent to the server.

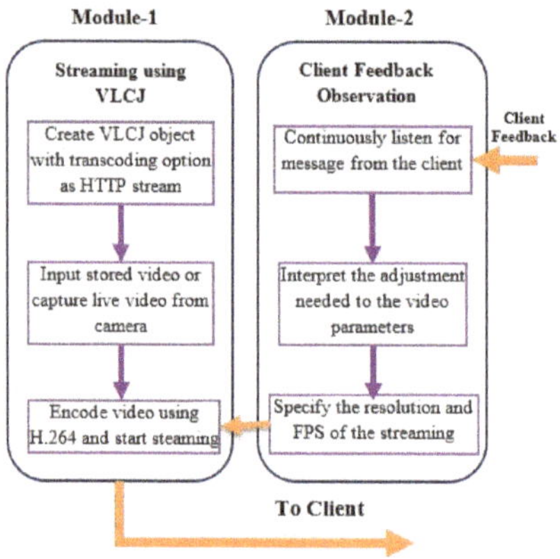

Figure 3. Modular flow diagram at server side representing transcoding, client feedback analysis, and adaptive video streaming operations.

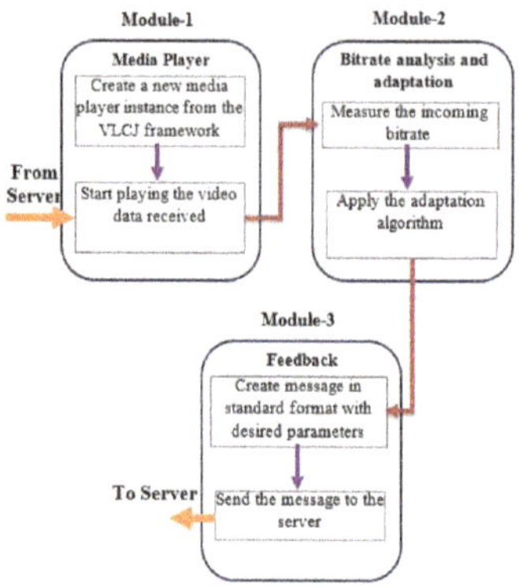

Figure 4. Modular flow diagram at client side performing bit rate analysis and giving feedback to the server.

The four unique patterns of variation in bit rate are defined (Figure 5) for comparison and analysis of streaming data. Category 1 denotes a progressive type of data flow where the bit rate increases in time. The bit rate fluctuation may gradually cease and finally stabilize (Category 2). Category 3 represents a case when the variation in bit rate diverges. If the decreasing trend of the bit rate continues, it represents a disturbing category of maintaining video quality (Category 4). Finally, the root mean square (RMS) method is adapted to dealing with unresolved categories.

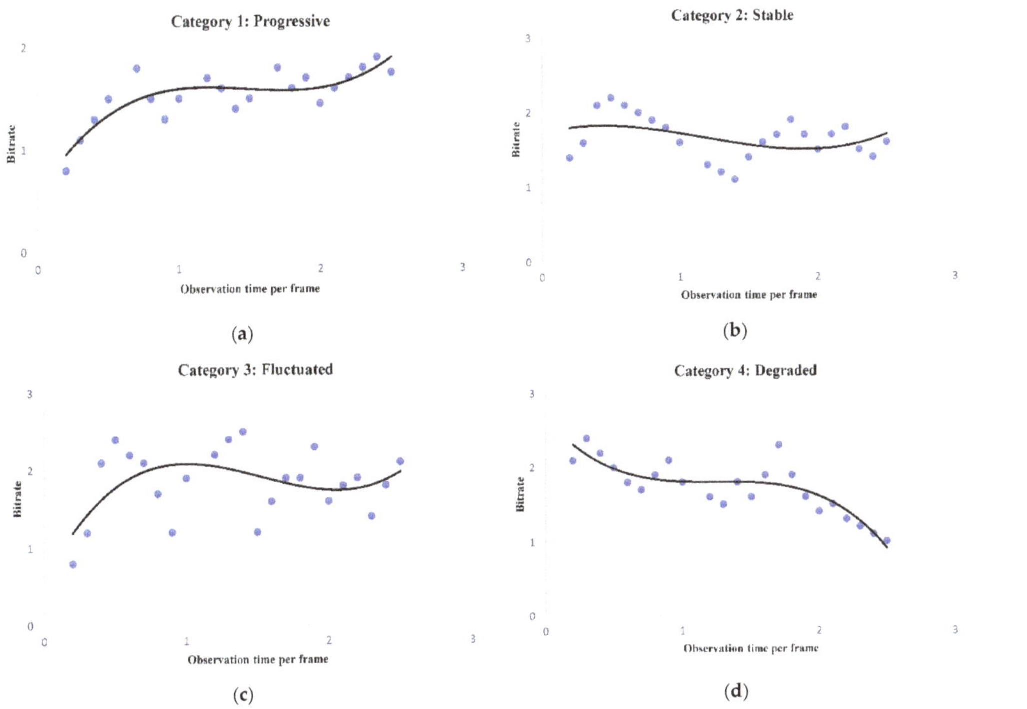

Figure 5. Different categories of bit rate arrived at the client side. (**a**) Progressive data pattern. (**b**) Fluctuated data pattern, (**c**) Stable data pattern, (**d**) Degraded data pattern.

3.1.1. Server Modules

The server's implementation procedure involves monitoring client feedback and assigning the appropriate parameter values to the ongoing video session. It consists of two main modules:

- The H.264 codec with VLCJ media framework captures the video and streams continuously through the HTTP port.
- The second module deals with listening to client feedback messages to adjust the video stream parameter, which includes resolution and frame rate.

3.1.2. Client Modules

The client periodically samples and analyzes the streaming data to estimate the dynamically changing bit rate trend. The client side implementation has three main modules:

- The first module consists of the VLCJ framework for playing streamed media data.
- The implementation of module 2 forms the core of the proposed system, which estimates the client's bit rate by applying a suitable bit pattern matching algorithm.
- The third module formats the feedback messages in a standard format than can be understood by the server.

3.2. Methodology

The proposed system at the client samples the incoming bit rate periodically at $(x_1, x_2, x_3, \ldots, x_n)$ and analyzes to find the trend of fluctuating data rate during the streaming, as shown in Algorithm 1. The fluctuating bit rate is categorized into predefined patterns (Figure 5) to simplify the estimation process. The proposed algorithm uses the theory of local maxima–minima in sampled bit rate to map the data arrival pattern into one of the four cases: progressive, stabilized, fluctuating, and degraded. When the system cannot resolve the streamed data into any four of these patterns, the status is declared as non-monotonic, and the system computes the data sample's RMS. By default, the system employs the RMS approach, which includes special cases such as monotonic flat patterns. Predicting the pattern depends on the values of α, β and γ, which are calculated based on analysis of startup, median, and endp for $(x_1, x_2, x_3, \ldots, x_n)$.

Algorithm 1. Client algorithm.

(1) Sample the data rate and put it in an array;
(2) Find maxima L_{max}:

(i) Input bit rate samples in pairs of three $(v_1,\ v_2,\ v_3)$;

(ii) If $(v_1 < v_2 > v_3)$, add v_2 to L_{max}, # find maximum bit rate;

(iii) Continue step 2 until N received frames;

(3) Find minima $L_{\min}$:

(i) Input bit rate samples in pairs of three $(v_1,\ v_2,\ v_3)$;

(ii) If $(v_1 > v_2 < v_3)$, add v_2 to $L_{\min}$, # find minimum bit rate;

(iii) Continue step 3 until N received frames;

(4) $Max = Analyse(L_{max})$, # Call procedure to acquire $\alpha,\ \beta$, and γ;

(5) $Min = Analyse(L_{min})$, # Call procedure to acquire $\alpha,\ \beta$, and γ;

(6) Find status:

(2)

- (i) If $(Max == \beta\ \&\ \&\ Min == \beta)$, set Status as **Progressive**;
- (ii) Else if $(Max == \alpha\ \&\ \&\ Min == \beta)$, set Status as **Stabilized**;
- (iii) Else if $(Max == \beta\ \&\ \&\ Min == \alpha)$, set Status as **Fluctuated**;
- (iv) Else if $(Max == \alpha\ \&\ \&\ Min == \alpha)$, set Status as **Degraded**;
- (v) If $(Max == \gamma\ ||\ Min == \gamma)$,

set Status as **non-monotonic** and call ***Find_rms* (*Bit rates*)**;
A. function Analyze (Bit rates):
Read start data point, *startp;*
Read end data point, *endp;*
Find median value, *medianp;*
If *(startp, medianp, endp)* tends to **monotonic increase**, return β;
Else if *(startp, medianp, endp)* tends to **monotonic decrease**, return α;
If *(startp, medianp and endp)* tends to neither **monotonic increase** nor **decrease**, return γ;
B. function Find_RMS (Bit rate):
Calculate the root mean square (RMS) of the samples;
$X_{rms} = \sqrt{1/n(x_1^2 + x_2^2 + \ldots + x_n^2)}$
Divide the N different samples into M segments (M = 3);
Continue Step1 to find the *RMS* values of each segments:
$\text{rms}_1, \text{rms}_2$, and rms_3;
Compute the difference among the overall RMS and
the RMS of the corresponding segments;
Calculate $diff_1 = RMS - rms_1$;
Calculate $diff_2 = RMS - rms_2$;
Calculate $diff_3 = RMS - rms_3$;
If $(diff_1 <= diff_2 <= diff_3)$, then return *1;*
Else if $(diff_1 >= diff_2 >= diff_3)$, then return *0;*
Else return 2.

The server side algorithm (Algorithm 2) decodes the client message and modifies the streaming video parameter accordingly. The execution time of the switching process from switching the current stream to the new stream is taken as one input parameter in the server's decision-making. If the client's algorithm wrongly classifies the arrival pattern (Ghuge, C. A et al., 2018), then it will lead to an improper action by the server, which may degrade the streaming video on the client side.

Algorithm 2. Server side algorithm.

Let S *and* T be the spatial and temporal resolution vector, respectively, given by:
$S = \{SR_1,\ SR_2,\ SR_3,\ SR_4, SR_5\}, T = \{TR_1,\ TR_2,\ TR_3,\ TR_4,\ TR_5\}$.

(1) Initially set SR_1 and TR_1 to a default value (S_{QCIF}, T_d);

(2) Continue:

Read feedback message $(status)$ from the client;
If (Status $== Stabilized$),
continue with the existing setup;
Else if (Status $== Progressive$),
call A-LSTM(status), ##Increase Spatial/Temporal resolution;
Else if (Status $== Fluctuated$),
find Switch_time (bit rate),
call A-LSTM(status), ##Update Spatial/Temporal resolution;
Else if (Status $== Degraded$),
call A-LSTM(status), ##decrease Spatial/Temporal resolution;
Else if (Status $== Non - monotonic$),
wait till next feedback message arrives;

(3) Continue till connection is terminated;

A. *Function Find_Switch_Time (bit rate):*
Calculate time for quality switch $T_{switch} = TQS_{k+1} - TQS_k$:
#where TQS_{k+1} is the time instant at the end of kth served quality switching,
request, and TQS_k is the present time instant attending previous quality,
switching request;
Set a timer when request for quality switch is received;
Wait for the next feedback message;
If $(T_{switch} > \mathrm{T}_{Fluctuation\_time})$,
discard the request for quality switching and wait for next client
feedback message;
Else,
serve the request for quality switching.

In this paper, we model the future bit-rate prediction for higher QoS as a time series prediction problem. Time series analysis, which involves analyzing past examples of bit rate in various network qualities to infer an optimal QoS, may be utilized to predict video bit rate. This time series analysis problem is learned in this work using attention-based LSTM (A-LSTM), an advanced variant of recurrent deep neural networks. The A-LSTM algorithm, which was trained using the back propagation through time (BPTT) algorithm, is more useful for learning long-duration dependencies.

To improve the quality of service (QoS) for individual consumers, we use a deep neural model consisting primarily of attention-based LSTM and reinforcement learning architecture (as shown in Figure 6). In reinforcement learning, an agent executes a task on an environment, and the environment responds with a reward based on the action performed. When the LSTM network's reinforcement learning (RL) agent receives the input state, it chooses an action that is equivalent to the bit rate of the next video sequence. The domain expert examines the performance of the proposed A-LSTM model using the reward function mentioned in Equation (2) based on the action (at) taken. The main goal of the proposed A-LSTM model is to choose an action class for the input state (S_t) that maximizes the overall video quality viewed by the end user. In Equation (1), cost function Q(t) has now been created to assess the total effectiveness of a video streaming session:

$$Q(t) = \sum_{n=1}^{N} q(R_n) - \mu \sum_{n=1}^{N} T_n \tag{1}$$

$$S_t = LSTM(S_{t-1}) \tag{2}$$

$$e_t = a(S_t) \tag{3}$$

$$\alpha_t = \frac{\exp(e_t)}{\sum\limits_{k=1}^{T} \exp(e_k)} \tag{4}$$

$$c = \sum_{t=1}^{T} \alpha_t S_t \tag{5}$$

where the first term $q(R_n)$ represents the video quality perceived by a user for N video sequence, S_t represents the output state, e_t represents the feed-forward attention network, α_t represents weighted attention, and c represents the weighted feature output.

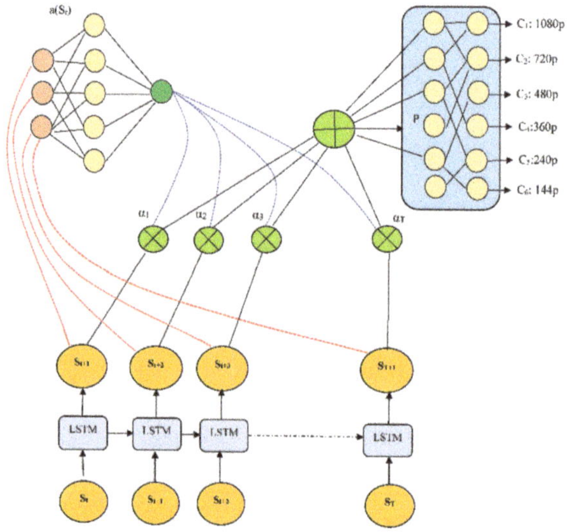

Figure 6. Architecture of the proposed attention based on LSTM at the server module.

The inputs to the proposed algorithm are bit rate b_t, current buffer size as bu_t, and output of the classification labels related to different spatial resolutions (144p, 240p, 360p, 480p, 720p, and 1080p). Furthermore, during training, attention networks are trained for parameters 'θ' with the help of rewards given by client feedback messages. Finally, these attention networks are responsible for maintaining efficient adaptive bit rate strategies for a particular video sequence.

The experiment uses the proposed attention-based model, which takes into account the video sequence, download times, and past (k = 16) bit rate measurements. As shown in Figure 6, an LSTM network receives these sequential inputs. The current buffer size as bt, the remaining video sequence as C_t, and the last chunk bit rate as b_t, are instantaneous inputs that are fed to a fully connected layer with 128 filters, each of size 4 and stride 1. The final layer, which is fully connected, chooses the state policies' action for state S_t using a softmax function. The softmax function's output is a selection of the bit rate for the following video segment with the highest probability, ensuring that the best bit rate is chosen for a corresponding state S_t. In the training phase, user feedback messages are reinforced to the attention network and LSTM network to obtain optimum parameters, using policy gradient strategy. Only the attention-LSTM network is used in predicting spatial resolutions in the testing phase.

For comparison purposes, we implemented a buffer filling algorithm based on the traditional adaptive streaming method [27] to analyze the performance of the proposed A-LSTM system. Implementation of the system at the client level involves monitoring the lower as well as the upper threshold of the media buffer. If the buffer reaches the upper threshold, it recommends slowing down the flow rate, but on the other hand, if the content arrival rate nears the lower threshold, it signals the server to increase the rate of content transfer. As a result, the server reduces or increases the stream bit rate by modifying the resolution of streaming video and/or reducing/increasing frames accordingly.

4. Results and Discussion

4.1. System Analysis

The streaming content can be treated as a chronological sequence of statistical data, which is sampled periodically for analysis and prediction. The error due to sampling and subsequent analysis need to be formulated and modeled to define the system's design objective and performance evaluation. The non-parametric approach [28] could be a better approach in micro-level system implementation.

A non-parametric prediction interval can be defined to include a simple maximum and minimum value in a sample set of a given population. Generally, for an exchangeable sequence of random variables, each sample qualifies as the maximum or minimum. In a bit rate sample set of $\{R_0, \ldots, R_n\}$, a sample R_i $(i = 0, 1, \ldots n)$ has the probability of $1/(n+1)$ being the maximum value, and probability of $1/(n+1)$ being the minimum value, while $(n-1)/(n+1)$ of probability, the sample R_i falls between the largest and smallest sample of $\{R_0, \ldots, R_n\}$. A sample maximum and minimum can be represented by L_{max} and L_{min}, respectively, and $(n-1)/(n+1)$ prediction interval of $[L_{max}, L_{min}]$.

For a given sample R_i, the error of the estimator $\hat{\theta}(R_i)$ can be denoted as:

$$e(R_i) = \hat{\theta}(R_i) - \theta_i \tag{6}$$

where θ_i is the parameter of estimation. Here, the error $e(R_i)$ depends on the process of estimation as well as on the sample value. The sampling deviation of the estimator $\hat{\theta}$ for a given sample R_i, is expressed as:

$$\begin{aligned} d(R_i) &= \hat{\theta}(R_i) - E(\hat{\theta}(R_i)) \\ &= \hat{\theta}(R_i) - E(\hat{\theta}) \end{aligned} \tag{7}$$

where $E(\hat{\theta}(R_i))$ is the expected value of the estimator. Like error of estimator, the sampling deviation R_i depends on the estimator as well the sample itself. The variance of $\hat{\theta}$ is computed as the expected value of the square of sampling deviations given by:

$$var(\hat{\theta}) = E[(\hat{\theta} - E(\hat{\theta}))^2] \tag{8}$$

The variance of the estimate indicates the distance from the expected value of the estimates. Sometimes, the distance between the average of the collection of estimates and the single parameter being estimated, called bias, need to be computed. The bias of θ can be denoted as:

$$bia(\hat{\theta}) = E(\hat{\theta}) - \theta \tag{9}$$

Further,

$$E(\hat{\theta}) - \theta = E(\hat{\theta}) \tag{10}$$

The mean squared error (MSE) can be expressed in terms of variance and bias as:

$$MSE(\hat{\theta}) = \text{var}(\hat{\theta}) + (bias(\hat{\theta}))^2 \tag{11}$$

If $\hat{R}$ denotes a set of n predicted values, and R the set of experiential values given as the input to the predictions, then the MSE of the predictor is computed as:

$$MSE = \frac{1}{n}\sum_{i=1}^{n}(\hat{R}_i - R_i)^2 \tag{12}$$

Since the full reference (FR) methods correspond to the objective evaluation of video quality and provide the most accurate result, it was used to evaluate the performance of the proposed system. The two widely used FR metrics are PSNR and SSIM. The paramount interest of the proposed solution is that the system response to the changing network resource should result in higher PSNR and SSIM while sustaining video communication.

4.1.1. Peak Signal-to-Noise Ratio (PSNR)

The PSNR provides information about the degradation of decoded video quality with respect to the original content. It is calculated on luminance components of the video (ITU-T recommendation), which can be formulated on a logarithmic scale as:

$$PSNR = 20\ \log_{10}\left(\frac{Max}{\sqrt{MSE(m)}}\right) \tag{13}$$

where $Max = 2^{no.\ of\ \text{bits}/(sample-1)}$, and for 8-bit word length, the luminance per sample is 255. The mean squared error (MSE (m)) is computed as the absolute difference between the original and the decoded video in the same frame (mth), denoted as:

$$MSE(m) = \frac{1}{M \times N}\sum_{i=1}^{M}\sum_{j=1}^{N}[X_{out}(i,j,m) - X_{in}(i,j,m)]^2 \tag{14}$$

The PSNR observation is basically an offline process that can be carried out on a few selected frames at the end of the experiment to ascertain the quality of the streaming system. In designing and developing a higher quality streaming system, achieving a minimum average PSNR of 30 dB may be desirable.

4.1.2. Structural Similarity (SSIM) Index

The SSIM [29] metrics measure the perceived degradation resulting from structural deformation at the frame level. In the real-world video, pixel positions exhibit temporal and spatial dependence between pixels. The spatial dependence information in a frame helps in estimating the structural similarity of the objects in decoded frames; therefore, SSIM is used as a perceptual measure of video quality.

The SSIM [30,31] metric is computed on three different components: luminance, contrast, and structure. It is defined by the Joint Video Team (JVT) of ISO/IEC MPEG and ITU-T VCEG as:

$$l(\mathbf{x},\mathbf{y}) = \frac{2\mu_x\mu_y + C_1}{\mu_x^2 + \mu_y^2 + C_1} \tag{15}$$

$$c(\mathbf{x},\mathbf{y}) = \frac{2\sigma_x\sigma_y + C_2}{\sigma_x^2 + \sigma_y^2 + C_2} \tag{16}$$

$$s(\mathbf{x},\mathbf{y}) = \frac{\sigma_{xy} + C_3}{\sigma_x\sigma_y + C_3} \tag{17}$$

where μ_x is the average of $\mathbf{x}$, μ_y is the average of $\mathbf{y}$, σ_x^2 is the variance of $\mathbf{x}$, σ_y^2 is the variance of $\mathbf{y}$, σ_{xy} is the covariance of $\mathbf{x}$ and $\mathbf{y}$. The constants C_1, C_2 and C_3 given by

$C_1 = (K_1\,L)^2$, $C_2 = (K_2\,L)^2 and\ C_3 = (K_3\,L)^2$ are used to stabilize the division operation while dealing with the weak denominator. L represents the dynamic range of the pixel values given by:

$L = 2^{no.\ of\ bit/pixel} - 1$ and $K_1 << 1$ and $K_2 << 1$ are two scalar constants.

Using these components, the SSIM is represented as:

$$SSIM(\mathbf{x},\mathbf{y}) = [l(\mathbf{x},\mathbf{y})]^{\alpha} \cdot [c(\mathbf{x},\mathbf{y})]^{\beta} \cdot [s(\mathbf{x},\mathbf{y})]^{\gamma} \tag{18}$$

where α, β, and γ state the different weightage assigned to each measure. The single-scale SSIM (Yue Wang et al., 2012) is now formulated as:

$$SSIM(x,y) = \frac{(2\mu_x\mu_y + c_1)(2\sigma_{xy} + C_2)}{\left(\mu_x^2 + \mu_y^2 + C_1\right)\left(\sigma_x^2 + \sigma_y^2 + C_2\right)} \tag{19}$$

An SSIM index of more than 0.95 represents a good decoded video, and it could be a design objective considering the requirement of the end user in high-quality video communication.

4.2. Experimental Setup

The proposed system was implemented in a client–server environment, where the server uses four standard video formats, namely common intermediate format (CIF), quarter CIF (QCIF), sub quarter CIF (SQCIF), and quarter Video Graphics Array (QVGA), to map the video quality dynamically corresponding to the feedback messages. The frame rate of 10, 18, 25, 30, and 35 fps was used to alter temporal resolution. The default value of the temporal resolution (in fps), and also during the initial setup, was assigned as 30. The streaming system at the server selects either one or a combination of spatial and temporal resolutions to achieve the predicted bit rate of the network by the client.

In the experimental setup, the wireless Internet connectivity was established by a 4G Internet-connect device, Reliance_Netconnect+, working on CDMA20001xRTT and 1xEV-DO Rev-A techniques. As per the specification provided by Reliance communication Ltd., the Reliance_Netconnect + device is designed to support a download and upload data rate of 3.1 Mbps and 1.8 Mbps, respectively, but a real-time measurement carried out using an online tool (SpeedOf.Me) had the average bit rate during uplink and downlink estimated at 0.54 Mbps and 0.45 Mbps, respectively, in the laboratory environment. Therefore, fluctuating throughput in data communication across the Internet connected by Reliance_Netconnect + provided us with the ideal real-time platform to test our proposed system.

The different modules of client and server were developed on Dell Inspiron N5010 personal computers having the Intel® Core ™ i7-3770 CPU @ 3.4 GHz processor and 8 GB RAM. The Window-7 Professional 32-bit operating system was installed to run the program for the client/server system. The streaming process was implemented over HTTP with the user datagram protocol as the transport protocol.

H.264 codec generates a variable bit rate for the input video depending on the scene change in visual contents. Figure 7 shows the bit rate vs. frame number for different frame size (176 × 144, 320 × 240, 640 × 480, and 800 × 600) observed during experimentation. The fluctuation in bit rate can be easily observed for all video resolution. This nature of video places additional constraints on system design, as the system proposed is based on best effort service model of the Internet, which operates in the wireless environment.

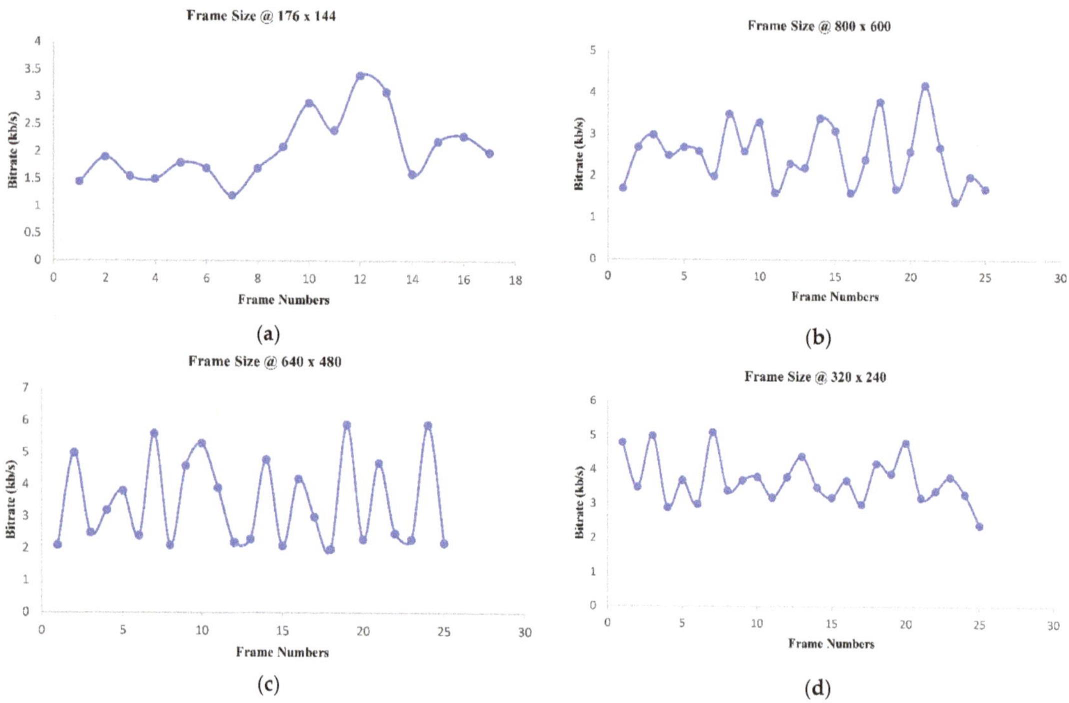

Figure 7. (**a**) Top-Left: 176 x 144, (**b**) Top-Right: 800 × 600, (**c**) Bottom-Left: 640 × 480, (**d**) Bottom-Right: 320 × 240.

VLC media player is used in a Java framework (VLCJ) at the sender as well as the receiver. VLCJ provides a higher level framework to cope with the complexities of VLC libraries. The VLC framework includes a variety of media formats through libavcodec library of codecs to the media player, which seamlessly plays the H.264 bit stream. Furthermore, JPCAP provides a library for packet capturing in-network applications using Java, which helps analyze real-time network traffic.

4.3. Results and Discussion

The proposed system was implemented using the open-source tool (VLCJ framework), and experimental results were obtained on live as well as stored video in wireless 4G CDMA networks. Since the intended application of this system is to support live video streaming over wireless on the top of HTTP, the result presented here corresponds to live-streamed video in the laboratory environment. Although the working of the system was successfully demonstrated many times, the numerical result analyzed here represents a single instance of experimentation.

4.3.1. Inter-Packet Arrival Delay

The variability of packet delay in the one-way end-to-end communication was observed while discarding packet loss during the experimentation of the proposed video streaming system. As shown in Figure 8, variation in the delay of packet arrival characterizes the inherent property of Internet traffic, which is attributed to the prevailing Internet traffic during the test. The measurement process includes additional delay in 4G wireless, which is used here as last-mile connectivity to the end user. Although the upper bound on packet delay as a design parameter was not directly incorporated in the proposed system, the server dynamically attuned the streaming bit rate to maximize visual eminence. The observed value, plotted in Figure 8, corresponds to an average inter-packet delay of 69 µs.

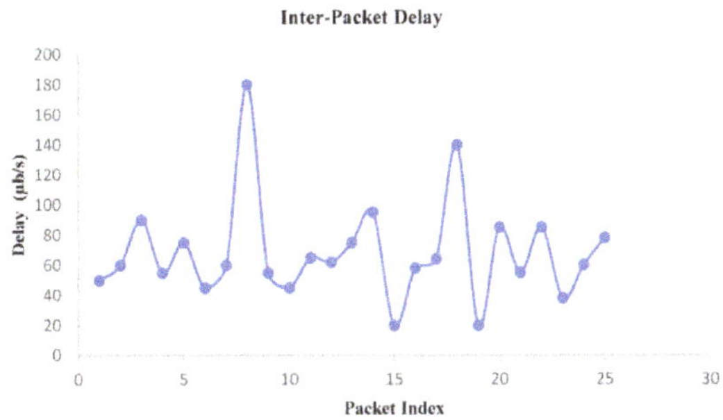

Figure 8. Observed packet arrival delay at the client with a graph drawn between delay and packet index.

4.3.2. PSNR Measurement

The PSNR observation (Figure 9) was performed offline based three methods: (i) without any adaptation, i.e., default existing mechanism in media streaming over the Internet, (ii) buffer filling algorithm, and (iii) the proposed adaptation method. The proposed algorithm achieves an average PSNR of 36.267 dB on video frames resulting from the live streaming, which is 37.53% higher than the buffer-filling algorithm. Further, the average improvement of PSNR is 74.37% higher than the default without an adaptation scheme. The augmented PSNR is accredited to the higher level of adaptation exhibited by the proposed technique, which continuously tries to deliver content at the maximum achievable quality. The observed PSNR for a few frames under different algorithms is also presented in Table 2.

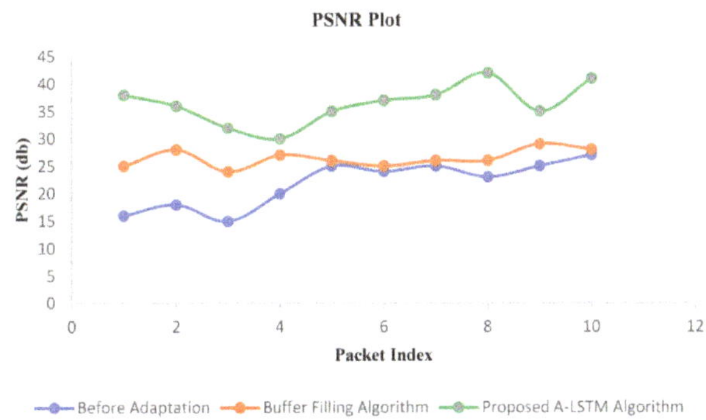

Figure 9. Performance comparison between before adaptation buffer filling algorithm and proposed A-LSTM technique using PSNR value.

Table 2. PSNR values for different frames.

Frame Number	Before Adaptation	Buffer Filling Algorithm	Proposed A-LSTM Algorithm
1	19.54	25.89	39.09
2	19.83	26.24	39.09
3	18.47	24.89	36.79
4	20.14	28.64	32.69
5	23.54	27.59	34.9
6	22.58	26.49	33.98
7	21.29	25.87	35.01
8	20.48	25.98	39.09
9	20.89	26.87	34.32
10	21.23	25.23	37.71
Average	**20.799**	**26.369**	**36.267**

4.3.3. SSIM Index

The SSIM was calculated offline using the same approach as that of PSNR. It is observed based on the data generated by three methods: default without adaptation scheme, existing buffer filling algorithm, and proposed adaptation algorithm resulting from the live streaming of video (Figure 10). The system implementing the proposed algorithm provides a 5.7% increase in average SSIM index than the existing buffer filling algorithm, and it achieves a much higher (11.44%) index than the method without adaptation. Due to the higher level of adaptation exhibited by the proposed system, the structural statistics were preserved, resulting in a higher value of the SSIM index. Although the design and implementation of the system do not directly deal with the retention of structural property during streaming, a higher SSIM index is an incentive for the system dynamics. Table 3 lists the perceived numerical ethics of SSIM under different adaptation algorithms.

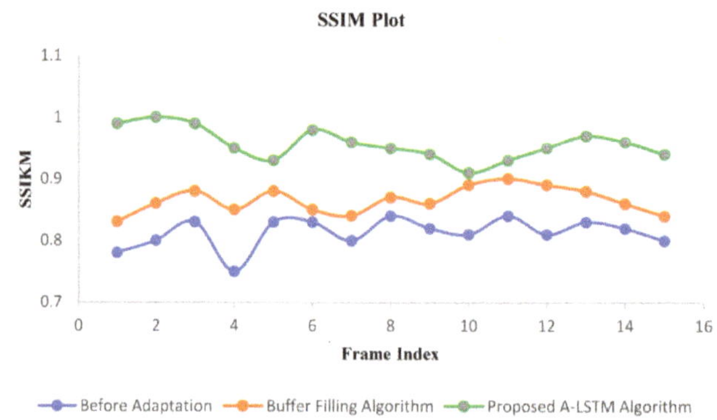

Figure 10. Performance comparison of before adaption, buffer filling algorithm, and proposed A-LSTM algorithm using SSIM index measurement.

Table 3. SSIM values for different frames.

Frame Number	Before Adaptation	Buffer Filling Algorithm	Proposed A-LSTM Algorithm
1	0.835	0.873	0.965
2	0.863	0.892	0.986
3	0.768	0.912	0.967
4	0.887	0.925	0.956
5	0.89	0.899	0.978
6	0.884	0.924	0.968
7	0.869	0.934	0.946
8	0.894	0.939	0.952
9	0.873	0.934	0.968
10	0.879	0.916	0.973
Average	**0.8689**	**0.9160**	**0.9683**

4.3.4. Selected Original and Decoded Frame

Figures 11 and 12 show the original and received decoded frames recorded during live streaming and stored foreman video during experimentation. Since the proposed system maintains an average PSNR of more than 36 dB and an SSIM index of 0.96, even a keen look at the received frames does not reveal a noticeable loss in quality of the decoded video, which is a requirement in developing a system for high-quality video applications. This also emphasizes the proposed method's importance over existing approaches and the default mechanism available on the Internet.

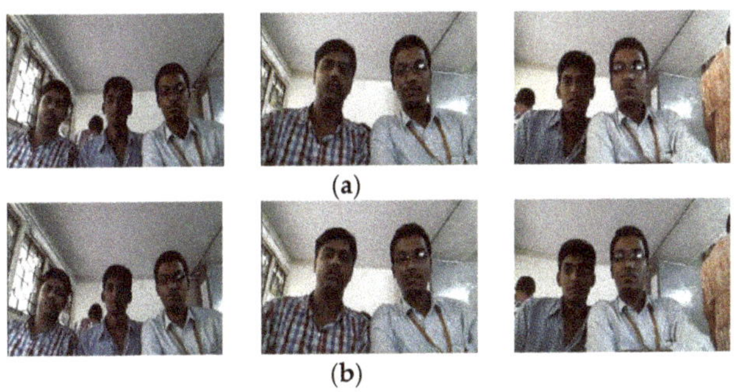

Figure 11. Quality comparison of sampled frames between (**a**) Original and (**b**) Received online video frames.

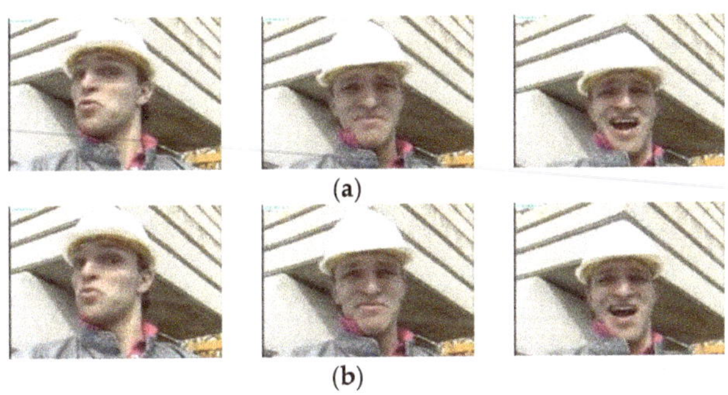

Figure 12. Quality comparison between (**a**) Original and (**b**) Received sampled stored foreman video frames.

5. Conclusions

The development of a mechanism to support adaptive streaming of video over HTTP in dealing with fluctuation in available bit rate on the Internet with last-mile connectivity as a wireless network is a rewarding approach. The proposed A-LSTM system adopted the theory of maxima–minima along with an RMS method, reinforcement techniques, along with attention-LSTM networks to compute and match the pattern of the bit stream. It also included the duration of network fluctuation as well as the time for switching excellence in effective decision-making, while switching between different video qualities. The proposed solution tackled the problem of inherent conduct of wireless and Internet traffic in a unified tactic. The link level quality of service parameters such as delay, jitter, and packet loss was not considered in the problem formulation as the system is developed on top of HTTP.

Although the proposed system is targeted toward the attainment of quality of the video, it can be used in many other video streaming applications. Abrupt congestion at the router on the Internet may cause extra delay in video streaming packets, and a suitable method is required to tackle it. Another future work includes supporting video streaming to hand-held devices (smart phones) connected to the Internet through a cellular network.

Author Contributions: Conceptualization, P.M.A.K., and L.N.A.R.; methodology, P.M.A.K., B.J., and N.F.S.; software, M.B.; validation, M.B., W.E.-S.; formal analysis, P.M.A.K., B.J. and N.F.S.; investigation, W.E.-S.; resources, N.F.S.; data curation, P.M.A.K.; writing—original draft preparation, P.M.A.K., M.B., W.E.-S., and B.J.; writing—review and editing, P.M.A.K., L.N.A.R., and W.E.-S.; visualization, M.B.; supervision, N.F.S.; project administration, L.N.A.R. and N.F.S.; funding acquisition, N.F.S., and W.E.-S. All authors have read and agreed to the published version of the manuscript.

Funding: This work is supported by Princess Nourah bint Abdulrahman University Researchers Supporting Project number (PNURSP2022R66), Princess Nourah bint Abdulrahman University, Riyadh, Saudi Arabia.

Data Availability Statement: The data used to support the findings of this study will be available with the corresponding author.

Acknowledgments: The authors would like to acknowledge the Princess Nourah bint Abdulrahman University Researchers Supporting Project number (PNURSP2022R66), Princess Nourah bint Abdulrahman University, Riyadh, Saudi Arabia. The authors would like to acknowledge the support of Prince Sultan University for paying the Article Processing Charges (APC) of this publication.

Conflicts of Interest: The authors declare no conflict of interest.

References

1. Krishna, Y.H.; Kumar, K.B.; Maharshi, D.; Amudhavel, J. Image processing and restriction of video downloads using cloud. *Int. J. Eng. Technol. (UAE)* **2018**, *7*, 327–330. [CrossRef]
2. Zhao, M.; Gong, X.; Liang, J.; Wang, W.; Que, X.; Cheng, S. QoE-driven cross-layer optimization for wireless dynamic adaptive streaming of scalable videos over HTTP. *IEEE Trans. Circuits Syst. Video Technol.* **2014**, *25*, 451–465. [CrossRef]
3. Han, S.; Go, Y.; Noh, H.; Song, H. Cooperative Server-Client HTTP Adaptive Streaming System for Live Video Streaming. In Proceedings of the 2019 International Conference on Information Networking (ICOIN), Malaysia, Malaysia, 9–11 January 2019.
4. Ghuge, C.A.; Ruikar, S.D.; Prakash, V.C. Query-specific distance and hybrid tracking model for video object retrieval. *J. Intell. Syst.* **2018**, *27*, 195–212. [CrossRef]
5. Reddy, K.S.; Prakash, B.L. HSV, edge preserved and huffman coding based intra frame high efficient video compression for multimedia communication. *Int. J. Eng. Technol. (UAE)* **2018**, *7*, 1090–1095. [CrossRef]
6. Chen, S.; Yang, J.; Ran, Y.; Yang, E. Adaptive Layer Switching Algorithm Based on Buffer Underflow Probability for Scalable Video Streaming Over Wireless Networks. *IEEE Trans. Circuits Syst. Video Technol.* **2016**, *26*, 1146–1160. [CrossRef]
7. Koganti, S.P.; Raja, K.H.; Sajja, S.; Sai Narendra, M. A study on volume, speed and lane distribution of mixed traffic flow by using video graphic technique. *Int. J. Eng. Technol. (UAE)* **2018**, *7*, 59–62.mel. [CrossRef]
8. El Meligy, A.O.; Hassan, M.S.; Landolsi, T. A Buffer-Based Rate Adaptation Approach for Video Streaming Over HTTP. In Proceedings of the Wireless Telecommunications Symposium (WTS), Washington, DC, USA, 22–24 April 2020.
9. De Cicco, L.; Mascolo, S. An Adaptive Video Streaming Control System: Modeling, Validation, and Performance Evaluation. *IEEE Trans. Netw.* **2014**, *22*, 526–539. [CrossRef]
10. Xu, Y.; Zhou, Y.; Chiu, D.M. Analytical QoE Models for Bit Rate Switching in Dynamic Adaptive Streaming Systems. *IEEE Trans. Mob. Comput.* **2014**, *13*, 2734–2748. [CrossRef]

11. Xing, M.; Xiang, S.; Cai, L. A Real-Time Adaptive Algorithm for Video Streaming over Multiple Wireless Access Networks. *IEEE Trans. Sel. Areas Commun.* **2014**, *32*, 795–805. [CrossRef]
12. Li, Z.; Zhu, X.; Gahm, J.; Pan, R.; Hu, H.; Begen, A.; Oran, D. Probe and Adapt Rate Adaptation for HTTP Video Streaming at Scale. *IEEE Trans. Sel. Areas Commun.* **2014**, *32*, 719–733. [CrossRef]
13. Anantharaj, B.; Balaji, N.; Sambasivam, G.; Basha, M.S.; Vengattaraman, T. EQVS: Enhanced Quality Video Streaming Distribution over Wired/Wireless Networks. In Proceedings of the 2017 International Conference on Technical Advancements in Computers and Communications (ICTACC), Melmaurvathur, India, 10–11 April 2017; pp. 148–153.
14. Nagageetha, M.; Mamilla, S.K.; Hasane Ahammad, S. Performance analysis of feedback based error control coding algorithm for video transmission on wireless multimedia networks. *J. Adv. Res. Dyn. Control Syst.* **2017**, *9*, 626–660.
15. Sripal Reddy, K.; Leelaram Prakash, B. Optimized lossless video compression analysis using decoding-energy fast compression. *J. Adv. Res. Dyn. Control Syst.* **2017**, *9*, 42–51.
16. Bulli Babu, R.; Shahid Afridi, S.K.; Satya Vasavi, S. A New enhancement to avoid video distortion in wireless multihop networks. *Int. J. Eng. Technol. (UAE)* **2018**, *7*, 326–330. [CrossRef]
17. Wankhede Vishal, A.; More, A.R.; Prasad, M.S.G. Suboptimal resource allocation scheme for scalable video multicast in integrated mobile WiMAX/WLANs network. *Int. J. Eng. Technol. (UAE)* **2018**, *7*, 69–76. [CrossRef]
18. Go, Y.; Kwon, O.C.; Song, H. An Energy efficient HTTP Adaptive Video Streaming with Networking Cost Constraint over Heterogeneous Wireless Networks. *IEEE Trans. Multimed.* **2015**, *17*, 1646–1657. [CrossRef]
19. Choi, W.; Yoon, J. SATE: Providing stable and agile adaptation in HTTP-based video streaming. *IEEE Access* **2019**, *7*, 26830–26841. [CrossRef]
20. El Essaili, A.; Schroeder, D.; Steinbach, E.; Staehle, D.; Shehada, M. QoE-based traffic and resource management for adaptive HTTP video delivery in LTE. *IEEE Trans. Circuits Syst. Video Technol.* **2014**, *25*, 988–1001. [CrossRef]
21. Zhou, C.; Lin, C.W.; Guo, Z. *m*DASH: A Markov decision-based rate adaptation approach for dynamic HTTP streaming. *IEEE Trans. Multimed.* **2016**, *18*, 738–751. [CrossRef]
22. Claeys, M.; Latre, S.; Famaey, J.; DeTurck, F. Design and Evaluation of a Self-Learning HTTP Adaptive Video Streaming Client. *IEEE Commun. Lett.* **2014**, *18*, 716–719. [CrossRef]
23. Sedano, I.; Brunnström, K.; Kihl, M.; Aurelius, A. Full-reference video quality metric assisted the development of no-reference bitstream video quality metrics for real-time network monitoring. *EURASIP J. Image Video Process.* **2014**, *2014*, 4. [CrossRef]
24. Wang, Y.; Jiang, T.; Ma, S.; Gao, W. Efficient Motion Weighted Spatial, Temporal Video SSIM Index. In Proceedings of the 2012 Visual Communications and Image Processing (VCIP 2012), San Diego, CA, USA, 27–30 November 2012.
25. Kumar, D.; Easwaran, N.K.; Srinivasan, A.; Shankar, A.M.; Raj, L.A. Adaptive video streaming over HTTP through 3G/4G wireless networks employing dynamic on the fly bitrate analysis. In Proceedings of the 2015 ITU Kaleidoscope: Trust in the Information Society (K-2015), Barcelona, Spain, 9–11 December 2015.
26. Ghuge, C.A.; Ruikar, S.D.; Prakash, V.C. Support vector regression and extended nearest neighbor for video object retrieval. *Evol. Intell.* **2018**, *15*, 837–850. [CrossRef]
27. Qasem, M.; Almohri, H.M.J. An Efficient Deception Architecture for Cloud-based Virtual Networks. *Kuwait J. Sci.* **2019**, *46*, 40–52.
28. ITU Telecommunication Standardization Sector. Objective perceptual multimedia video quality measurement in the presence of a full reference. *ITU-T Recomm. J.* **2008**, *247*, 18.
29. Jiang, J.; Sekar, V.; Zhang, H. Improving fairness, efficiency, and stability in http-based adaptive video streaming with festive. In Proceedings of the 8th International Conference on Emerging Networking Experiments and Technologies, Côte d'Azur, France, 10–13 December 2012; pp. 97–108.
30. Thang, T.C.; Ho, Q.D.; Kang, J.W.; Pham, A.T. Adaptive Streaming of Audiovisual Content using MPEG DASH. *IEEE Trans. Consum. Electron.* **2012**, *58*, 78–85. [CrossRef]
31. Zainab Mohammd Aljazzaf, Z. Modelling and measuring the quality of online services. *Kuwait J. Sci.* **2015**, *42*, 134–157.

sensors

MDPI

Article

A Hierarchical Spatial–Temporal Cross-Attention Scheme for Video Summarization Using Contrastive Learning

Xiaoyu Teng [1,2], Xiaolin Gui [1,2,*], Pan Xu [1,2], Jianglei Tong [1,2], Jian An [1,2], Yang Liu [3] and Huilan Jiang [4]

1 Department of Faculty of Electronic and Information Engineering, Xi'an Jiaotong University, Xi'an 710049, China
2 Shaanxi Province Key Laboratory of Computer Network, Xi'an Jiaotong University, Xi'an 710049, China
3 Medical College, Northwest Minzu University, Lanzhou 730030, China
4 ONYCOM Co., Ltd., Seoul 04519, Korea
* Correspondence: xlgui@mail.xjtu.edu.cn; Tel.: +86-157-2196-0091

Abstract: Video summarization (VS) is a widely used technique for facilitating the effective reading, fast comprehension, and effective retrieval of video content. Certain properties of the new video data, such as a lack of prominent emphasis and a fuzzy theme development border, disturb the original thinking mode based on video feature information. Moreover, it introduces new challenges to the extraction of video depth and breadth features. In addition, the diversity of user requirements creates additional complications for more accurate keyframe screening issues. To overcome these challenges, this paper proposes a hierarchical spatial–temporal cross-attention scheme for video summarization based on comparative learning. Graph attention networks (GAT) and the multi-head convolutional attention cell are used to extract local and depth features, while the GAT-adjusted bidirection ConvLSTM (DB-ConvLSTM) is used to extract global and breadth features. Furthermore, a spatial–temporal cross-attention-based ConvLSTM is developed for merging hierarchical characteristics and achieving more accurate screening in similar keyframes clusters. Verification experiments and comparative analysis demonstrate that our method outperforms state-of-the-art methods.

Keywords: video summarization; spatial–temporal features; cross-attention

Citation: Teng, X.; Gui, X.; Xu, P.; Tong, J.; An, J.; Liu, Y.; Jiang, H. A Hierarchical Spatial–Temporal Cross-Attention Scheme for Video Summarization Using Contrastive Learning. *Sensors* **2022**, *22*, 8275. https://doi.org/10.3390/s22218275

Academic Editors: Kang Ryoung Park, Sangyoun Lee and Euntai Kim

Received: 23 September 2022
Accepted: 25 October 2022
Published: 28 October 2022

1. Introduction

With the rapid development of multimedia information technology and intelligent terminal equipment, video data have emerged as a critical medium of information transmission due to its lack of reading threshold and high data-carrying capacity. However, the openness and informality of video production result in the accelerated growth of video data and several undesirable phenomena, such as widespread data redundancy [1], unclear content emphasis, and blurred video theme boundaries. Therefore, it is becoming vital to provide effective and efficient tools for the management, browsing, and retrieval of these videos. Video summarization, which uses a subset of the most informative frames to create a condensed version of the original video by removing redundant information [2–4], is an effective tool for addressing these issues.

Recent methods for video summarization rely heavily on the superior performance of deep learning, particularly in feature extraction. In addition, feature extraction is a fundamental component of video summarization algorithms that extract time series [5,6] or spatial–temporal features from video data [7,8]. From the perspective of a video feature, the performance of the video summary is dependent on the feature extraction technique. These deep learning video summarization algorithms constantly increase the depth and breadth of video feature extraction to improve its performance. The most important criterion for measuring video summarization performance is user satisfaction. User satisfaction is contingent upon their requirements for video summarization performance. Furthermore, user requirements can be translated into property constraints of algorithms [7]. These property constraints can be categorized as representativeness [5], content coverage [8], redundancy [3], diversity [5], interestingness [9], importance [10], etc. The variety of user requirements continues to expand, while their feature definitions are more hazy. Consequently, video summarization algorithms focusing on video salient characteristics extraction are incapable of satisfying the multi-source user requirements. In addition, with the rise in popularity of video terminal equipment and the evolution of multimedia technology, hand-held and fragmented time-created videos have become the predominant sources of new created video data. Certain more prominent properties of the new video production, such as significant redundancy, a lack of strong focus, and a fuzzy theme boundary, disrupt the video summarization's initial thinking mode based on video feature information and present it with new challenges. With the evolution of video characteristics and user requirements for video summarization, the demand for keyframe accuracy screening has increased. Some traditional methods are no longer applicable, such as clustering [11].

To be more precise, existing algorithms can meet a portion of the user-centered requirements and capture good summarization performance. However, the following challenges remain: contradiction between breadth extraction of salient video characteristics and multi-source of user diversified requirements; the contradiction between depth extraction of salient video characteristics and unbounded new video productions; the contradiction between similarity frames and more accurate keyframe screening.

To address the issues mentioned above, this paper proposes a hierarchical spatial–temporal cross-attention scheme based on contrastive learning, as shown in Figure 1. The central idea of this article is to extract features and relationships between frames that account for coarse and fine-grained, global and local, depth and breadth, to fuse hierarchical features while increasing the difference between similar frames, and then screen keyframes and generate summaries by evaluating their significance. From the perspective of video feature extraction, the solution to diverse user requirements for video summary lies in the extraction of the frame's own characteristics, relationship features between frames, and relationship features between frames and the entire video. This study uses DB-ConvLSTM and multi-head attention mechanisms to design multi-conv-attention cells and joint GAT to acquire the spatial–temporal connection of keyframes to extract fine-grained spatial–temporal feature information from video frames. The GAT adjusted DB-ConvLSTM to extract the global and breadth features. In addition, to amplify the difference of similar keyframes, a spatial–temporal cross-attention-based ConvLSTM is constructed for merging hierarchical characteristics. Finally, video summarization is generated by CB-ConvLSTM through possibility. Therefore, the major contributions of this work can be summarized as follows:

1. A hierarchical spatial–temporal video feature extraction approach is developed. The purpose is to ensure as much characteristic information as possible for generating video summarization;
2. A cross-attention cell that combines the local and global features information based on DB-ConvLSTM is proposed. It seeks to emphasize the difference between related frames and achieve more accurate screening in similar keyframes clusters for video summary generation;
3. Verification experiments and comparative analysis are performed on two benchmark datasets (TVSum and SumMe) for this paper's algorithm. The results demonstrate that the proposed algorithm is extremely rational, effective, and usable.

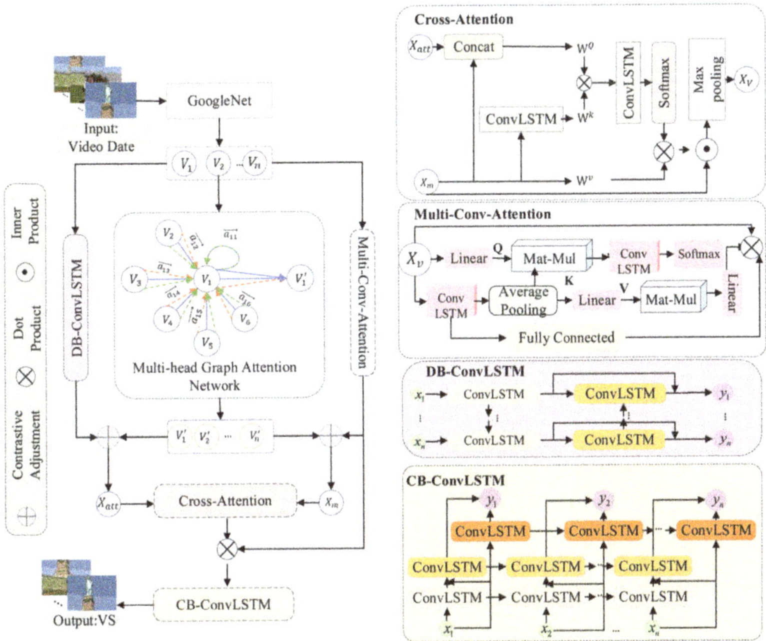

Figure 1. Overview of our approach.

2. Related Work

In this section, we briefly overview some state-of-the-art video summarization approaches and correlation techniques pertaining to our hierarchical spatial–temporal cross-attention scheme.

2.1. Video Summarization

Generally speaking, pre-processing, feature extraction, post-processing, and VS creation comprise the video summary generating procedures. The post-processing can be left out. In particular, feature extraction is the central stage of the algorithm. The initial algorithm is based on time series techniques such as vsLSTM/dppLSTM [5]. The initial method of similar keyframes decision is based on clustering [11]. Zhao et al. [12] develop an extended bidirectional LSTM (Bi-LSTM) for extracting both structure and information characteristics from video data. To acquire a more precise extraction of video features, refs. [3,13] offer a keyframe-selection strategy based on video spatial–temporal characteristics. In addition, graph neural networks are employed to implement this notion [1,6]. However, the aforementioned algorithms are all video-centric and lack comprehensive analysis of video topics and user demands. In [14], first-person (egocentric) videos-based models are proposed. A model of characterizing egocentric video frames uses a graph-based center-surround model. User requirements impose certain restrictions on the feature extraction results. The video summarization algorithms [15] are based on attention technologies, mimicking human keyframe filtering. Ji et al. [16] solve the problem of short-term contextual attention insufficiency and distribution inconsistency. Köprü [17] proposes two new architectures based on temporal attention (TA-AVSUM) and spatial attention (SA-AVSUM).

Additionally, for the video summarization algorithm, both video feature information and video frame relational are crucial [18]. Continuously improving the performance of the user-requirements-driven algorithm fundamentally necessitates more comprehensive and accurate feature extraction. This scheme is based on the concept of creating stereoscopic modeling using spatial–temporal feature information, relationship information, and other multi-elements.

2.2. Cross Attention

Refs. [19–22] have conducted substantial study on how to more properly and completely extract video features and the relationship features between video frames. Contextual information is vital in visual understanding problems [19] and is also applicable to generating video summarization. Huang et al. [19] proposes a Criss-Cross Network (CCNet) based on attention for obtaining video information in a more effective and efficient way. Lin et al. [20] presents a universal Cross-Attention Transformer (CAT) module for accurate and efficient semantic similarity comparison in one-shot object detection. In [22], the attention mechanism is incorporated at two main levels: a self-attention module leverages global interactions between encoder features, while cross-attention in the skip connections allows fine spatial recovery in the U-Net decoder by filtering out non-semantic features. It can be seen that cross-attention has the ability to simultaneously extract the depth and breadth characteristics of video data. This study uses cross-attention to merge the hierarchical spatial–temporal characteristics, and it aims to accentuate the distinctions between video frames.

2.3. Graph Attention Networks (GATs)

Veličković et al. [23] give a novel neural network architecture that operates on graph-structured data, leveraging masked self-attentional layers to address the shortcomings of prior methods based on graph convolutions or their approximations. GATs provide distinct weights to each neighbor based on their importance, effectively filtering the neighbors. Zhong et al. [1] build a method for video summarizing utilizing graph attention networks and Bi-LSTM. However, it does not take into account information loss throughout the confrontation process. This paper makes use of GATs to capture spatial–temporal relational attention between video frames and comparative-adjusting feature extraction.

3. Materials and Methods

Figure 1 shows an overview of our hierarchical spatial–temporal cross-attention scheme for video summarization. DB-ConvLSTM, multi-conv-attention, and multi-head attention GAT are all used for video feature extraction. The DB-ConvLSTM is employed to extract coarse-grained global spatial–temporal video characteristics. Effective fine-grained local features are extracted using multi-conv-attention networks and spatial–temporal relational feature extraction using multi-head attention GAT. This research derives hierarchical spatial–temporal feature information on the basis of cross-attention, taking into consideration both global and local characteristics and coarse-grained and fine-grained features. In particular, this scheme promotes comparative learning for acquiring local feature information for multi-conv-attention and GAT, and obtaining global feature knowledge for DB-ConvLSTM and GAT. The local and global characteristics are combined using spatial–temporal cross-attention. Finally, CB-ConvLSTM obtains the video summary.

Following the algorithm phases, this part elaborates the DB-ConvLSTM and CB-ConvLSTM, contrastive adjustment learning, and spatial–temporal cross-attention for the keyframes screening module. The contrastive adjustment learning is adjustment learning based on contrastive learning. Finally, we will introduce the loss function used in our framework.

3.1. DB-ConvLSTM and CB-ConvLSTM

Both DB-ConvLSTM and CB-ConvLSTM are founded on the technology of ConvLSTM. ConvLSTM is not only designed for extracting spatial–temporal information features but also for inferring saliency information concurrently. Then, suppose there are n frames in a video, the whole video can be written as $f = \{f_1, \cdots f_n\}$, c_t is the memory cell, f_t is the forget gate, and i_t is the input gate. From [24], we can obtain that the ConvLSTM is defined as:

$$\begin{aligned} i_t &= \sigma(W_i^X \times X_t + W_i^H \times H_{t-1}) \\ f_t &= \sigma(W_f^X \times X_t + W_f^H \times H_{t-1}) \\ o_t &= \sigma(W_o^X \times X_t + W_o^H \times H_{t-1}) \\ c_t &= f_t \circ c_{t-1} + i_t \circ \tanh(W_c^X \times X_t + W_c^H \times H_{t-1}) \\ H_t &= o_t \circ \tanh(c_t) \end{aligned} \quad (1)$$

3.1.1. DB-ConvLSTM

In the video information processing methods, the DB-ConvLSTM [25] network is suggested to extract spatial–temporal video characteristics more deeply and precisely. DB-ConvLSTM is a bidirectional two-layer architecture, one forward-oriented and one backward-oriented. The forward-oriented and backward-oriented have information interaction. The deeper layer is composed of backward-cells, its input is the output features of forward-cells, and the output is $\{Y_t\}_{t=1}^{t}$. The backward-ConvLSTM is defined as:

$$i_t^b = \sigma(W_i^{H^f} \times H_t^f + W_i^{H^b} \times H_{t+1}^b) \quad (2)$$

$$f_t^b = \sigma(W_f^{H^f} \times H_t^f + W_f^{H^b} \times H_{t+1}^b) \quad (3)$$

$$o_t^b = \sigma(W_o^{H^f} \times H_t^f + W_o^{H^b} \times H_{t+1}^b) \quad (4)$$

$$c_t^b = {f_t}^b \circ c_{t+1}^b + i_t^b \circ \tanh(W_c^{H^f} \times H_t^f + W_c^{H^b} \times H_{t+1}^b) \quad (5)$$

$$H_t^b = o_t^b \circ \tanh(c_t^b) \quad (6)$$

where W are the training parameters, denoting the learnable weights, H is the hidden state, σ is the activation function, $\times$ denotes the convolution operator, and $\circ$ denotes the hadamard product. In the VS algorithm, the DB-ConvLSTM can be written as:

$$Y_t = \tanh(W_y^{H^f} \times H_t^f + W_y^{H^b} \times H_{t-1}^b) \quad (7)$$

tanh is the activation function to normalize Y_t, and the loss function of training DB-ConvLSTM is distance minimization.

3.1.2. CB-ConvLSTM

CB-ConvLSTM is capable of extracting not only the characteristics of a single video frame but also the spatial–temporal relationships between different frames [7]. From [7], we can obtain the definition of CB-ConvLSTM, based on Equations (2)–(7), and replace the content in Equation (1) by ConvLSTM; then, CB-ConvLSTM is defined as follows:

$$H_t^f = ConvLSTM(X_t, H_{t-1}^f) \quad (8)$$

$$H_t^b = ConvLSTM(X_t \oplus H_{1,t}, H_{t+1}^b) \quad (9)$$

$\oplus$ is the operation of fusing two vectors, $H_{1,t}$ is the first hidden state, and the loss function of training CB-ConvLSTM is distance minimization. In this paper, the three layers in the network cell aim to extract and aggregate the features, and the final outputs are the possibility of whether a frame will be selected as a keyframe for video summarization.

3.2. *Contrastive Adjustment Learning*

Contrastive learning [26] introduces a novel idea of features derived from many perspectives: the learning algorithm does not have to concentrate on every element of the sample itself, as long as it learns enough traits to differentiate it from others. In our study, the use of contrastive learning serves three purposes: (1) to overcome the diversity theme of video, (2) to extract elastic traffic feature information, and (3) to increase feature extraction with surface breadth and detail while enlarging the difference between video frames. As shown in Figure 2, the specific application of our strategy is to use the GATs-obtained data as the primary line and generate positive and negative pairs from the results of DB-ConvLSTM and multi-conv-attention, respectively. D_m is supposed as the results of the two sections of the comparative learning. DDPG [27] is used to train the D_m adjusted DB-ConvLSTM, which is the same as [1].

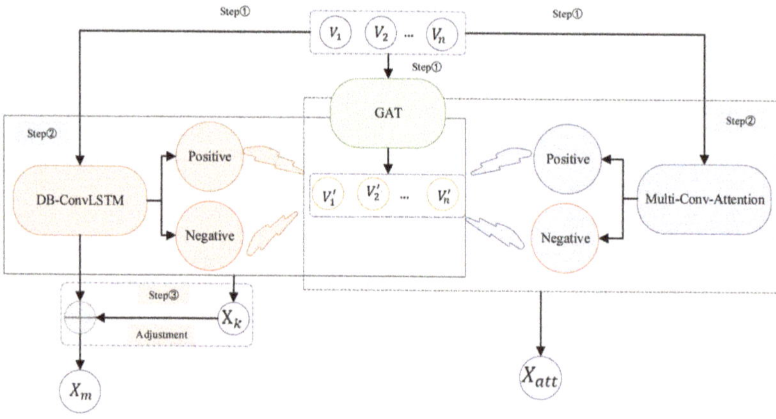

Figure 2. Step of contrastive adjustment learning.

x^+ is the positive sample, and x^- is the negative sample, S is the function for measuring the samples' similarity, and similar to [26], the rule for setting positive pairs is:

$$S(Y(x), Y(x^+)) \gg S(Y(x), Y(x^-)) \tag{10}$$

D_t is the video characteristics, which are extracted by multi-conv-attention and DB-ConvLSTM. D^+ is the keyframe sets, D^- is the non-keyframe sets, Q_j is feature mapping of the labeled data, and Q^+ is the annotated manually keyframe-sets. Then, the positive pairs include: $Y^A = \{D^+(x) \cap Q^+(x)\}$. Moreover, the loss function of a negative sample is InfoNCE in this paper, and it can be written as:

$$\mathcal{L}_{adj} = \sum_{x,x^+,x^-} \left[-log \left(\frac{e^{Y(x)^T Y(x^+)}}{e^{Y(x)^T Y(x^+)} + {}^{Y(x)^T Y(x^-)}} \right) \right] \tag{11}$$

3.3. *Multi-Conv-Attention and Cross-Attention*

3.3.1. Multi-Conv-Attention

The temporal, spatial, and multi-element video properties are all important parts of our approach. As a consequence, a new network cell is constructed using ConvLSTM and multi-head attention. It uses convolution to improve the attention mechanism's ability to get as much video information as possible. In our multi-conv-attention cell, we first adopt a set of projections to obtain query Q. Additionally, it employs ConvLSTM and average pooling to produce two sets of projections of key K and value V, enhancing the K and V dimensions of the attention mechanism while also boosting the performance and consistency of feature information extraction. Finally, the attention is calculated as:

$$M_c(Q, K, V) = Softmax(ConvLSTM(\frac{QK^T}{\sqrt{d_k}}))V \tag{12}$$

In this scheme, we employ $n = 8$, and $d_k = d_v = d_{model}/n = 64$.

3.3.2. Cross-Attention

The Cross-Attention module is shown in Figure 3. $F(\cdot)$ and $G(\cdot)$ are projections to align dimensions using interpolation function. Then, the module performs cross-attention between X_m and X_{att}, which can be expressed as

$$q = G(X_m) \cdot W^Q \tag{13}$$

$$k = ConvLSTM(G(X_{att})) \cdot W^k \tag{14}$$

$$v = F(X_m) \cdot W^v \tag{15}$$

Finally, calculate the cross-attention using Equation (12).

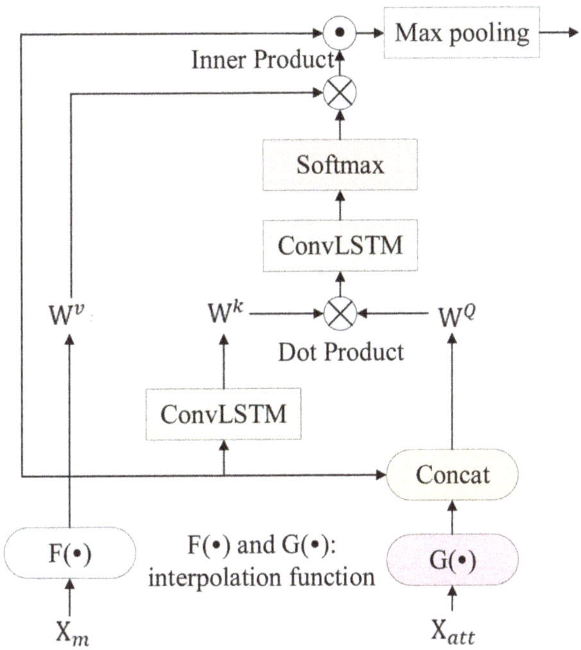

Figure 3. Spatial–temporal cross-attention cell.

3.4. Loss Function

The total loss is primarily made up of three components, and all the loss functions of these parts are based on cross-entropy. In items of supervised learning, the selection of keyframes is ultimately intended to decrease the discrepancy between predicted and background data. The cross-entropy is used to approximate the distribution of the learnt model to the background data. The lower the value is, the more similar the probability distributions of the anticipated and background data. p is the probability distribution of background data, and q is the predicted probability distribution, and the cross-entropy $H(p, q)$ is:

$$H(p,q) = \sum_{i=1}^{n} p_i log \frac{1}{q_i} = -\sum_{i=1}^{n} p_i log q_i \tag{16}$$

In our network, the softmax is used to normalize the cross-entropy, y_i is the output of network cells, $\hat{y}_i$ is the category i of background data, $\hat{y}_i \in \{0, 1\}$, and the loss function is:

$$\mathcal{L} = -\frac{1}{m}\left[\sum_{i=1}^{m} \hat{y}_i log \frac{e^{z_i}}{\sum_{i=1}^{k} e^{z_k}}\right] = -\frac{1}{m}\left[\sum_{m}^{i=1} \hat{y}_i log y_i\right] \tag{17}$$

$\mathcal{L}_{mat}$ is the loss function of the model of multi-conv-attention contrastive GAT. $\mathcal{L}_{dat}$ is the loss function of the model of DB-ConvLSTM contrastive adjustment GAT. $\mathcal{L}_{cro}$ is the loss function of cross-attention. Both $\mathcal{L}_{mat}$ and $\mathcal{L}_{dat}$ are cross-entropy, as defined by Equation (17). To resolve the centralization issue and reduce the ambiguity problem in key frame filtering, we use $\mathcal{L}_{cen}$ for centralization keyframe scores:

$$\mathcal{L}_{cen} = \lambda \cdot \frac{min(\mathcal{L}_{dat}, \mathcal{L}_{mat})}{max(\mathcal{L}_{dat}, \mathcal{L}_{mat})} \tag{18}$$

In Equation (17), λ balances the function of global and local domains. Formally, the objective function $\mathcal{L}_{obj}$ is written as

$$\mathcal{L}_{tol} = \mu \cdot \mathcal{L}_{cro} + \mathcal{L}_{cen} \tag{19}$$

μ balances the loss of cross-attention and multi-conv-attention.

4. Experiments Analysis

4.1. Datasets

Each database has its focus, so before the experiment, the two databases TVsum [28] and SumMe [29] should be analyzed, and the results are shown in Table 1. Additionally, we use two other public datasets, OVP (Open Video Project) [30] and YouTube [11], to augment the training sets.

Table 1. Analysis of TVsum and SumMe dataset.

Datasets	Description
TVsum	The title-based video summarization dataset contains 50 videos of various genres (e.g., news, documentary, egocentric) and 1000 annotations of shot-level importance scores (20 user annotations per video). The duration varies from 2 to 10 min.
SumMe	The SumMe dataset consists of 25 videos, each annotated with at least 15 human annotated summaries. The duration of videos varies from 1.5 to 6.5 min.

4.2. Evaluation Metrics

To facilitate a comparison study of the experimental influence on current research findings, the Precision, Recall, and F-score are used as measurement standards, similar to the literature [3]. S is the video summarization generated by the algorithm, G denotes the ground user-marked ground truth, and the following definitions apply to Precision, Recall, and F-score:

$$Precision = \frac{|S \cup G|}{|S|} \tag{20}$$

$$Recall = \frac{|S \cup G|}{|G|} \tag{21}$$

$$F - score = \frac{2 \times Precision \times Recall}{Precision + Recall} \tag{22}$$

As shown in [31], randomly generated video summaries may achieve equivalent performance when using the F-score measure. To avoid this problem, we evaluated our method as seen in Table 2. Under our comparison method, the comparing parameter is F-score. Furthermore, to be more precise, these datasets are randomly split into different training and testing sets five times, and the final measure is produced by averaging the five results.

Table 2. Datasets setting used for evaluation (C: Canonical; A: Augmented; T: Transfer).

Datasets	Setting	Training Phase	Testing Phase
TVSum	C	80% TVSum	The rest 20% of TVSum
	A	80% TVSum+SumMe+ OVP+YouTube	The rest 20% of TVSum
	T	SumMe+OVP+YouTube	TVSum
SumMe	C	80% SumMe	The rest 20% of SumMe
	A	TVSum+80% SumMe+ OVP+YouTube	The rest 20% of SumMe
	T	TVSum+OVP+YouTube	SumMe

4.3. Experimental Environment and Parameters Settings

The deep learning platform for operating our approach is Pytorch. The hidden states are with dimensionality of 256 for ConvLSTM, and other parameters settings are as follows: similar to other algorithms, we use the pool5 layer of GoogleNet to extract the visual features for each video frame. The number of ConvLSTMs hidden layers is 256, the learning rate initialized is $le-5$, the batch size is 5, the kernel size is set as (5,1), and the maximum training epoch is set as 100. Furthermore, considering that the training epochs are critical to summarization performance, after increasing for five epochs continuously, their influence on the validation set is plotted in Figures 4–6. The horizontal coordinate is the training epochs, and the vertical coordinates are the values the of F-scores, Recall, and Precision.

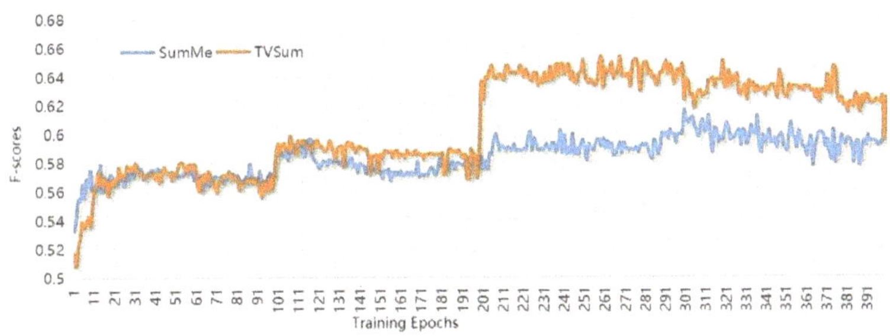

Figure 4. Plots show the influence of training epochs on the value of F-scores.

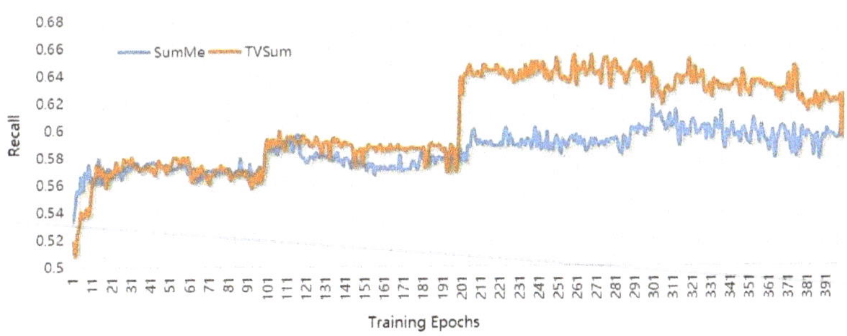

Figure 5. Plots show the influence of training epochs on the value of Recall.

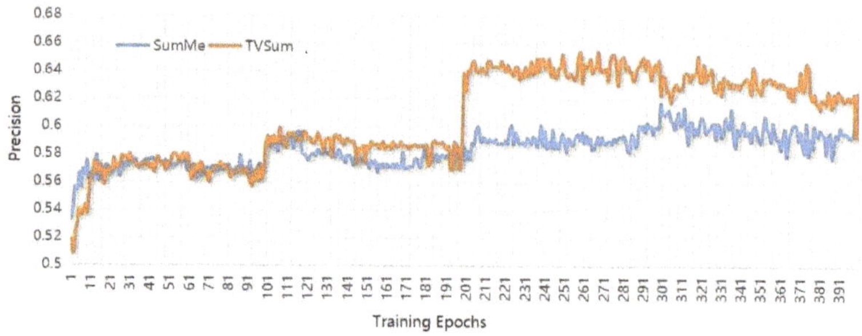

Figure 6. Plots show the influence of training epochs on the value of Precision.

4.4. Comparative Analysis of Schemes

This section verifies the feasibility and effectiveness of the proposed strategy through two ways: one is validation of the algorithm itself, and the other one is comparative analysis with state-of-the-art video summarization approaches.

4.4.1. Self-Verification

Before comparing the scheme to other state-of-the-art algorithms, it is vital to validate the scheme's performance itself. Table 3 and Figure 7 show the results of evaluating the performances of our methods on the SumMe and TVSum datasets. From the perspective of result stability, the test variation curves of C, T, and A on SumMe and TVSum datasets are shown in Figures 8 and 9. The horizontal coordinate of both figures is training epochs. Figure 7 gives an example of generating video summarization on the SumMe and TVSum datasets by our approach; the yellow lines show the annotation importance scores of ground truth summarization marked by the user, and the blue lines show the prediction score of our method. We clearly observe that our models achieve very competitive results against state-of-the-art methods.

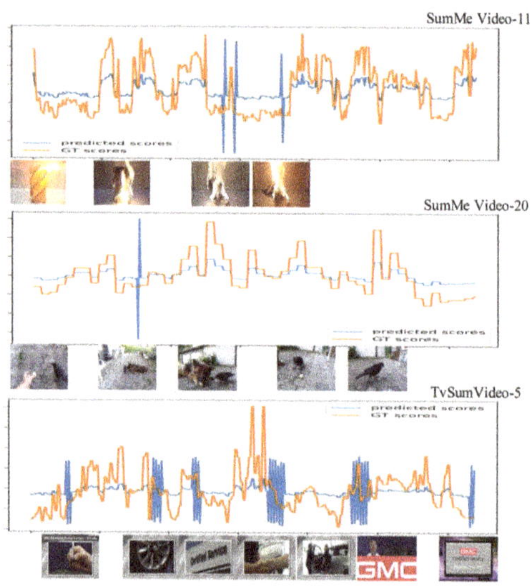

Figure 7. An example of generating video summarization on SumMe and TVSum datasets; the first two are samples from SumMe datasets and the last one is from TVSum datasets. The yellow lines show the annotation importance scores of ground truth summarization marked by the user, and the blue lines show the prediction score of our method.

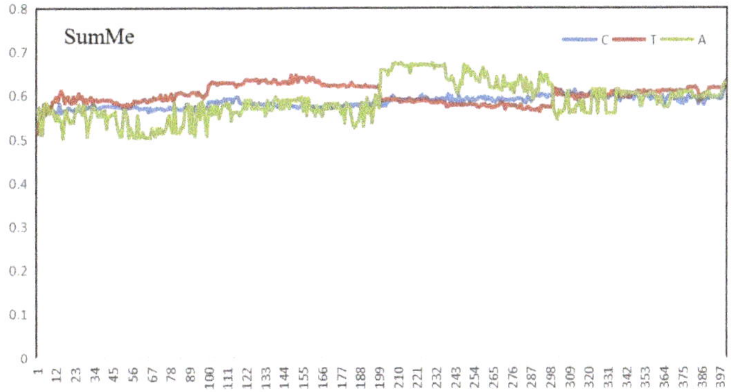

Figure 8. The A, C, and T results of SumMe.

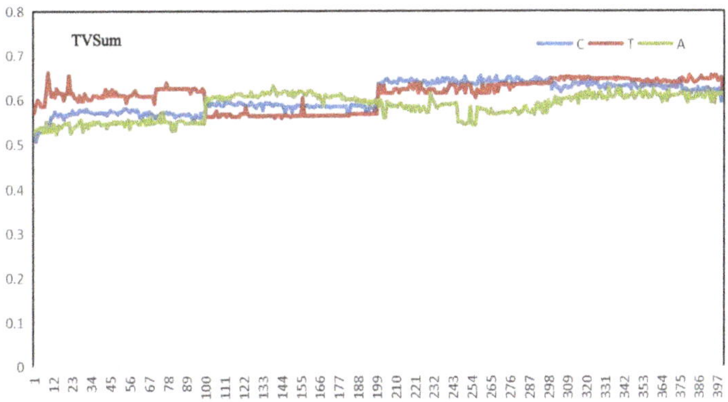

Figure 9. The A, C, and T results of TvSum.

Table 3. Performance analysis of self-verification (F-scores).

Data Sets	TVSum			SumMe		
Metric	C (%)	A (%)	T (%)	C (%)	A (%)	T (%)
MAX	65.3	67.4	66.2	61.6	63.1	64.5
MIN	50.8	50.2	55.9	53.3	52.4	51.3
AVERAGE	60.57	58.62	61.26	58.4	58.4	60.01

4.4.2. Comparative Analysis with Relative Approaches

The primary components of our algorithm consist of the attention mechanism, ConvLSTM, and GATs. In this section, we compared our approach with some state-of-the-art video summarization methods on SumMe and TvSum. Comparison methods can be classified into three categories: based on "LSTM+", based on "Attention+", and based on GATs methods.

(1) Comparison With "Bi-LSTM+" Methods

Due to the few research results on the summary algorithm based on ConvLSTM, this section compares our scheme to the Bi-LSTM based algorithms. Some classic algorithms are compared, as shown in Table 4.

Table 4. Performance analysis of methods based on "Bi-LSTM+".

Data Sets	TVSum			SumMe		
Metric	C (%)	A (%)	T (%)	C (%)	A (%)	T (%)
vsLSTM [5]	54.2	57.9	56.9	37.6	41.6	40.7
dppLSTM [5]	54.7	59.6	58.7	38.6	42.9	41.8
H-RNN [12]	57.9	61.9	–	42.1	43.8	–
HAS-RNN [32]	58.7	59.8	–	42.3	42.1	–
DHAVS [33]	60.8	61.2	57.5	45.6	46.5	43.5
Ours	65.3	67.4	66.2	58.4	58.4	60.01

H-RNN [12] and HAS-RNN [32] are based on hierarchical architecture. According to the findings of the comparison, we observe that our method outperforms state-of-the-art video summarization methods on both datasets.

(2) Comparison With "Attention+" Methods

Since the scheme in this paper involves not only the combination of ConvLSTM and attention but also the graph neural network, we will analyze it separately. The results of comparison with "Attention+" methods are shown in Table 5. SABTNet [15] is based on attention and a binary neural tree. Liang et al. [34] proposes a video summarization method based on dual-path attention, while Zhu et al. [35] is based on hierarchical attention. Table 5 demonstrates that the cross-attention method has clear benefits over the SumMe database.

Table 5. Performance analysis of methods based on "Attention+".

Data Sets	TVSum			SumMe		
Metric	C (%)	A (%)	T (%)	C (%)	A (%)	T (%)
M-AVS [36]	61.0	61.8	–	44.4	41.6	–
SABTNet [15]	61.0	–	–	51.7	–	–
[34]	61.58	61.2	58.9	51.7	52.1	44.1
[35]	61.5	62.8	56.7	51.1	52.1	45.6
Interp-SUM [2]	59.14	–	–	47.7	–	–
3DST-UNet [3]	58.3	58.9	56.1	47.4	49.9	47.9
Ours	65.3	67.4	66.2	58.4	58.4	60.01

(3) Comparison With "Graph Attention+" Methods

The extraction of spatial–temporal characteristics and frame–relationship features is facilitated by a graph neural network. Table 6 shows the results of comparing our method with some "Graph Attention+" video summarization methods including RSGN [13], GCAN [37], Bi-GAT [1] and SumGraph [38]. From the experimental results in Table 6, our method outperforms other approaches, which are based on "Graph Attention+".

Table 6. Performance analysis of methods based on "Graph Attention+".

Data Sets	TvSum (F-Score %)	SumMe (F-Score %)
RSGN [13]	60.1	45.0
GCAN [37]	60.1	53.0
Bi-GAT [1]	59.6	51.7
SumGraph [38]	63.9	51.4
Ours	65.36	58.48

4.4.3. Comparison Results

Following the comparison tests outlined above, it can be seen that the proposed method has certain advantages over existing approaches, most notably in the SumMe database set. Specifically, the hierarchical spatial–temporal cross-attention scheme in this research enhances the algorithm's stability, scalability, and other performance characteristics.

5. Conclusions

This paper proposes a hierarchical spatial–temporal cross-attention scheme for video summarization using contrastive learning. The scheme solves the contradictions of diversification user requirements, depth and breadth of features extraction and new creation videos. The hierarchical architecture is divided primarily into depth and breadth feature extraction and spatial–temporal cross-attention feature merging. This paper extracts local and depth features using a graph attention network and multi-head attention mechanism, and it extracts global and breadth features using a GAT adjusted DB-ConvLSTM. Furthermore, merging hierarchical characteristics via spatial–temporal cross-attention cells is used for more precise keyframe screening. Finally, video summarization is generated by CB-ConvLSTM. In practice, results from the TVSum and SumMe datasets indicate that the proposed algorithm is highly rational, effective, and usable. Nevertheless, the analysis of similarity keyframe screening is still insufficiently detailed.

Author Contributions: Conceptualization, X.T., X.G., P.X., J.T., J.A., Y.L. and H.J.; methodology, X.T. and X.G.; software, P.X. and J.T.; validation, X.T., X.G. and P.X.; formal analysis, J.A.; investigation, Y.L.; resources, H.J.; writing—original draft preparation, X.T.; writing—review and editing, X.T. and P.X.; project administration, X.G. and J.A.; funding acquisition, X.G. and J.A. All authors have read and agreed to the published version of the manuscript.

Funding: This paper was partially supported by the National Key Research and Development Project Under Grant (2018YFB1800304), National Natural Science Foundation of China (61472316), Fundamental Research Funds for the Central Universities (xzy012020112).

Institutional Review Board Statement: Not applicable.

Informed Consent Statement: Not applicable.

Data Availability Statement: Not applicable.

Conflicts of Interest: The authors declare no conflict of interest.

References

1. Zhong, R.; Wang, R.; Zou, Y.; Hong, Z.; Hu, M. Graph attention networks adjusted bi-LSTM for video summarization. *IEEE Signal Proc. Lett.* **2021**, *28*, 663–667. [CrossRef]
2. Yoon, U.-N.; Hong, M.-D.; Jo, G.-S. Interp-SUM: Unsupervised Video Summarization with Piecewise Linear Interpolation. *Sensors* **2021**, *21*, 4562. [CrossRef] [PubMed]
3. Liu, T.; Meng, Q.; Huang, J.-J.; Vlontzos, A.; Rueckert, D.; Kainz, B. Video summarization through reinforcement learning with a 3D spatio-temporal u-net. *IEEE Trans. Image Proc.* **2022**, *31*, 1573–1586. [CrossRef] [PubMed]
4. Li, W.; Pan, G.; Wang, C.; Xing, Z.; Han, Z. From coarse to fine: Hierarchical structure-aware video summarization. *ACM Trans. Mult. Comput. Commun. Appl. TOMM* **2022**, *18*, 1–16. [CrossRef]
5. Zhang, K.; Chao, W.-L.; Sha, F.; Grauman, K. Video summarization with long short-term memory. In Proceedings of the European Conference on Computer Vision (ECCV), Amsterdam, The Netherlands, 11–14 October 2016; pp. 766–782.
6. Zhao, B.; Li, H.; Lu, X.; Li, X. Reconstructive sequence-graph network for video summarization. *IEEE Trans. Patt. Anal. Mach. Intell.* **2021**, *44*, 2793–2801. [CrossRef] [PubMed]
7. Teng, X.; Gui, X.; Xu, P. A Multi-Flexible Video Summarization Scheme Using Property-Constraint Decision Tree. *Neurocomputing* **2022**, *506*, 406–417. [CrossRef]
8. Ji, Z.; Zhang, Y.; Pang, Y.; Li, X.; Pan, J. Multi-video summarization with query-dependent weighted archetypal analysis. *Neurocomputing* **2019**, *332*, 406–416. [CrossRef]
9. Rafiq, M.; Rafiq, G.; Agyeman, R.; Choi, G.S.; Jin, S.-I. Scene classification for sports video summarization using transfer learning. *Sensors* **2020**, *20*, 1702. [CrossRef]
10. Zhu, W.; Han, Y.; Lu, J.; Zhou, J. Relational Reasoning Over Spatial-Temporal Graphs for Video Summarization. *IEEE Trans. Image Proc.* **2022**, *31*, 3017–3031. [CrossRef]
11. De Avila, S.E.F.; Lopes, A.P.B.; da Luz, A., Jr.; de Albuquerque Araújo, A. VSUMM: A mechanism designed to produce static video summaries and a novel evaluation method. *Patt. Recognit. Lett.* **2011**, *32*, 56–68. [CrossRef]
12. Zhao, B.; Li, X.; Lu, X. Hierarchical recurrent neural network for video summarization. In Proceedings of the 25th ACM International Conference on Multimedia, New York, NY, USA, 23–27 October 2017; pp. 863–871.

13. An, Y.; Zhao, S. A Video Summarization Method Using Temporal Interest Detection and Key Frame Prediction. *arXiv* **2021**, arXiv:2109.12581.
14. Sahu, A.; Chowdhury, A.S. First person video summarization using different graph representations. *Patt. Recognit. Lett.* **2021**, *146*, 185–192. [CrossRef]
15. Fu, H.; Wang, H. Self-attention binary neural tree for video summarization. *Patt. Recognit. Lett.* **2021**, *143*, 19–26. [CrossRef]
16. Ji, Z.; Zhao, Y.; Pang, Y.; Li, X.; Han, J. Deep attentive video summarization with distribution consistency learning. *IEEE Trans. Neural Netw. Learn. Syst.* **2020**, *32*, 1765–1775. [CrossRef]
17. Köprü, B.; Erzin, E. Use of Affective Visual Information for Summarization of Human-Centric Videos. *arXiv* **2021**, arXiv:2107.03783.
18. Mi, L.; Chen, Z. Hierarchical Graph Attention Network for Visual Relationship Detection. In Proceedings of the IEEE/CVF Conference on Computer Vision and Pattern Recognition (CVPR), Seattle, WA, USA, 13–19 June 2020.
19. Huang, Z.; Wang, X.; Huang, L.; Huang, C.; Wei, Y.; Liu, W. Ccnet: Criss-cross attention for semantic segmentation. In Proceedings of the IEEE/CVF International Conference on Computer Vision, Seoul, Korea, 27 October–2 November 2019; pp. 603–612.
20. Lin, W.; Deng, Y.; Gao, Y.; Wang, N.; Zhou, J.; Liu, L.; Zhang, L.; Wang, P. CAT: Cross-Attention Transformer for One-Shot Object Detection. *arXiv* **2021**, arXiv:2104.14984.
21. Sanabria, M.; Precioso, F.; Menguy, T. Hierarchical multimodal attention for deep video summarization. In Proceedings of the 2020 25th International Conference on Pattern Recognition (ICPR), Milan, Italy, 10–15 January 2021; pp. 7977–7984.
22. Petit, O.; Thome, N.; Rambour, C.; Themyr, L.; Collins, T.; Soler, L. U-net transformer: Self and cross attention for medical image segmentation. In Proceedings of the International Workshop on Machine Learning in Medical Imaging, Strasbourg, France, 27 September 2021; pp. 267–276.
23. Veličković, P.; Cucurull, G.; Casanova, A.; Romero, A.; Lio, P.; Bengio, Y. Graph attention networks. *arXiv* **2017**, arXiv:1710.10903.
24. Shi, X.; Chen, Z.; Wang, H.; Yeung, D.-Y.; Wong, W.-K.; Woo, W.-c. Convolutional LSTM network: A machine learning approach for precipitation nowcasting. In Proceedings of the 28th International Conference on Neural Information Processing Systems, Montreal ON, Canada, 7–12 December 2015; Volume 28.
25. Song, H.; Wang, W.; Zhao, S.; Shen, J.; Lam, K.-M. Pyramid dilated deeper convlstm for video salient object detection. In Proceedings of the European Conference on Computer Vision (ECCV), Munich, Germany, 8–14 September 2018; pp. 715–731.
26. Gao, T.; Yao, X.; Chen, D. Simcse: Simple contrastive learning of sentence embeddings. *arXiv* **2021**, arXiv:2104.08821.
27. Lillicrap, T.P.; Hunt, J.J.; Pritzel, A.; Heess, N.; Erez, T.; Tassa, Y.; Silver, D.; Wierstra, D. Continuous control with deep reinforcement learning. In Proceedings of the International Conference on Learning Representations 2016, San Juan, Puerto Rico, 2–4 May 2016; pp. 1–15.
28. Song, Y.; Vallmitjana, J.; Stent, A.; Jaimes, A. Tvsum: Summarizing web videos using titles. In Proceedings of the IEEE Conference on Computer Vision and Pattern Recognition (CVPR), Salt Lake City, UT, USA, 18–23 June 2018; pp. 5179–5187.
29. Gygli, M.; Grabner, H.; Riemenschneider, H.; Van Gool, L. *Creating Summaries from User Videos*; Lecture Notes in Computer Science; Springer: Berlin/Heidelberg, Germany, 2014; pp. 505–520.
30. Open Video Project. Available online: https://open-video.org/ (accessed on 22 September 2022).
31. Otani, M.; Nakashima, Y.; Rahtu, E.; Heikkila, J. Rethinking the evaluation of video summaries. In Proceedings of the IEEE Conference on Computer Vision and Pattern Recognition (CVPR), Long Beach, CA, USA, 15–20 June 2019; pp. 7596–7604.
32. Zhao, B.; Li, X.; Lu, X. Hsa-rnn: Hierarchical structure-adaptive rnn for video summarization. In Proceedings of the IEEE Conference on Computer Vision and Pattern Recognition (CVPR), Salt Lake City, UT, USA, 18–23 June 2018; pp. 7405–7414.
33. Lin, J.; Zhong, S.-h.; Fares, A. Deep hierarchical LSTM networks with attention for video summarization. *Comput. Electr. Eng.* **2022**, *97*, 107618. [CrossRef]
34. Liang, G.; Lv, Y.; Li, S.; Wang, X.; Zhang, Y. Video summarization with a dual-path attentive network. *Neurocomputing* **2022**, *467*, 1–9. [CrossRef]
35. Zhu, W.; Lu, J.; Han, Y.; Zhou, J. Learning multiscale hierarchical attention for video summarization. *Patt. Recognit.* **2022**, *122*, 108–312. [CrossRef]
36. Ji, Z.; Zhao, Y.; Pang, Y.; Li, X.; Han, J. Video summarization with attention-based encoder–decoder networks. *IEEE Trans. Circ. Syst. Video Technol.* **2019**, *30*, 1709–1717. [CrossRef]
37. Li, P.; Tang, C.; Xu, X.Video summarization with a graph convolutional attention network. *Front. Inform. Technol. Electr. Eng.* **2021**, *22*, 902–913. [CrossRef]
38. Park, J.; Lee, J.; Kim, I.-J.; Sohn, K. Sumgraph: Video summarization via recursive graph modeling. In Proceedings of the European Conference on Computer Vision, Glasgow, UK, 23–28 August 2020; pp. 647–663.

sensors

MDPI

Article

Cross-Sensor Fingerprint Enhancement Using Adversarial Learning and Edge Loss

Ashwaq Alotaibi [1], Muhammad Hussain [1,*], Hatim AboAlSamh [1], Wadood Abdul [2] and George Bebis [3]

1 Department of Computer Science, CCIS, King Saud University, Riyadh 11451, Saudi Arabia
2 Department of Computer Engineering, King Saud University, Riyadh 11451, Saudi Arabia
3 Department of Computer Science and Engineering, University of Nevada, Reno, NV 89557, USA
* Correspondence: mhussain@ksu.edu.sa

Abstract: A fingerprint sensor interoperability problem, or a cross-sensor matching problem, occurs when one type of sensor is used for enrolment and a different type for matching. Fingerprints captured for the same person using various sensor technologies have various types of noises and artifacts. This problem motivated us to develop an algorithm that can enhance fingerprints captured using different types of sensors and touch technologies. Inspired by the success of deep learning in various computer vision tasks, we formulate this problem as an image-to-image transformation designed using a deep encoder–decoder model. It is trained using two learning frameworks, i.e., conventional learning and adversarial learning based on a conditional Generative Adversarial Network (cGAN) framework. Since different types of edges form the ridge patterns in fingerprints, we employed edge loss to train the model for effective fingerprint enhancement. The designed method was evaluated on fingerprints from two benchmark cross-sensor fingerprint datasets, i.e., MOLF and FingerPass. To assess the quality of enhanced fingerprints, we employed two standard metrics commonly used: NBIS Fingerprint Image Quality (NFIQ) and Structural Similarity Index Metric (SSIM). In addition, we proposed a metric named Fingerprint Quality Enhancement Index (FQEI) for comprehensive evaluation of fingerprint enhancement algorithms. Effective fingerprint quality enhancement results were achieved regardless of the sensor type used, where this issue was not investigated in the related literature before. The results indicate that the proposed method outperforms the state-of-the-art methods.

Keywords: biometrics; cross-sensor fingerprints; fingerprint enhancement; cGAN; adversarial learning; deep learning

Citation: Alotaibi, A.; Hussain, M.; AboAlSamh, H.; Abdul, W.; Bebis, G. Cross-Sensor Fingerprint Enhancement Using Adversarial Learning and Edge Loss. *Sensors* **2022**, 22, 6973. https://doi.org/10.3390/s22186973

Academic Editors: Euntai Kim, Sangyoun Lee and Kang Ryoung Park

Received: 2 August 2022
Accepted: 6 September 2022
Published: 15 September 2022

1. Introduction

The fingerprint is a biometric modality deployed mainly for human identification. Fingerprint recognition systems have several practical applications, including access control and criminal investigation [1].

Most available fingerprint systems compare data captured from the same sensor, where matching algorithms are designed to work on data obtained from a single sensor for enrollment and verification. Thus, the ability of these algorithms to work on data collected from multiple sensors is limited. It is known as the fingerprint sensor interoperability problem or the cross-sensor problem. In legacy databases, billions of fingerprints have been collected from different sensors based on diverse technologies. Every time the sensor of choice is changed, the re-enrollment of persons is a costly and substantial task. Moreover, due to the improvement in fingerprint sensors and the need to apply fingerprint recognition in devices such as those linked to the Internet of Things (IoT), the demand is high for an efficient fingerprint matching algorithm that can recognize fingerprints captured using different sensors. Therefore, the algorithms for the sensor interoperability problem, which improve the biometric system's ability to adapt to data obtained from several sensors, are highly needed and will significantly impact system usability [2].

The quality of fingerprints varies based on the sensor types used for capturing the fingerprint, even if the same sensing technology is employed (e.g., optical or capacitive). Additionally, the corresponding sets of features have high variability, which cannot be analyzed easily by a matching algorithm for accurate decisions. An example is shown in Figure 1, which shows the fingerprint of the same finger captured by nine different sensors [3].

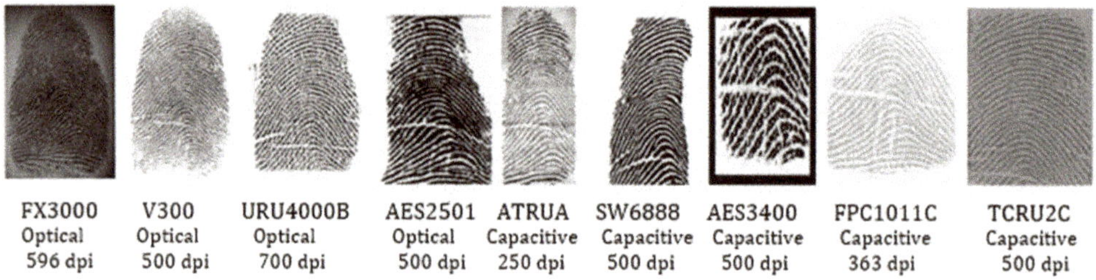

Figure 1. Fingerprints from the FingerPass database of the same finger that were captured by different sensors.

Differences in sensor technology and interaction type can cause significant variations in the quality of fingerprints. Thus, a considerable drop in the performance of the existing fingerprint recognition systems has been reported when different sensors are used for identification [2].

Moreover, the performance of cross-sensor matching algorithms is affected because of the variations in ridge patterns caused by the various types of noises and artifacts due to the difference in sensor technologies, as shown in Figure 1. There is a real need to enhance fingerprint images. However, this is challenging because fingerprints captured using various sensors include several kinds of texture patterns and noises [4].

A sample including a set of impressions taken from the MOLF dataset [5] is presented in Figure 2. These impressions were categorized into three subsets: DB1 comprises the flat dap (10) fingerprints captured by the Lumidigm Venus sensor; DB2 contains the fingerprints of the same fingers captured by the Secugen HamsterIV sensor; and DB3 consists of the dap fingerprints captured by CrossMatch L-Scan patrol sensor. Their quality was measured using the NFIQ (NBIS Fingerprint Image Quality) tool [6]. It is an open-source minutiae-based quality evaluation algorithm that provides a quality value {1, 2, 3, 4, 5}, with 1 representing the best quality and 5 denoting the worst one. Each row within the set stands for fingerprints captured by the same sensor. Each column, in turn, represents the same level of quality, in which the first column is excellent while the last column is poor. It can be noticed that DB1 has no images of the poor class. In addition, most of the ridge pattern information is unclear in the impressions belonging to classes poor and fair in DB2 and DB3.

In this paper, we present an efficient enhancement solution for the cross-sensor fingerprint problem. Specifically, motivated by the outstanding performance of deep learning-based techniques in various computer vision tasks such as image enhancement [7,8]. We designed an image-to-image mapping function $\mathcal{F}$ that receives a low-quality fingerprint and generates a high-quality one. We model $\mathcal{F}$ using Convolutional Neural Networks (CNN) based on encoder–decoder architecture. The learning of this kind of CNN is a challenging problem. Thus, we trained our method using two types of learning approaches i.e., the conventional end-to-end approach and using the adversarial learning (using a conditional GAN framework).

Adversarial learning generates fingerprints of higher quality than those produced by conventional learning, as demonstrated by comparing the outputs of the two methods using two frequent metrics: NFIQ and SSIM.

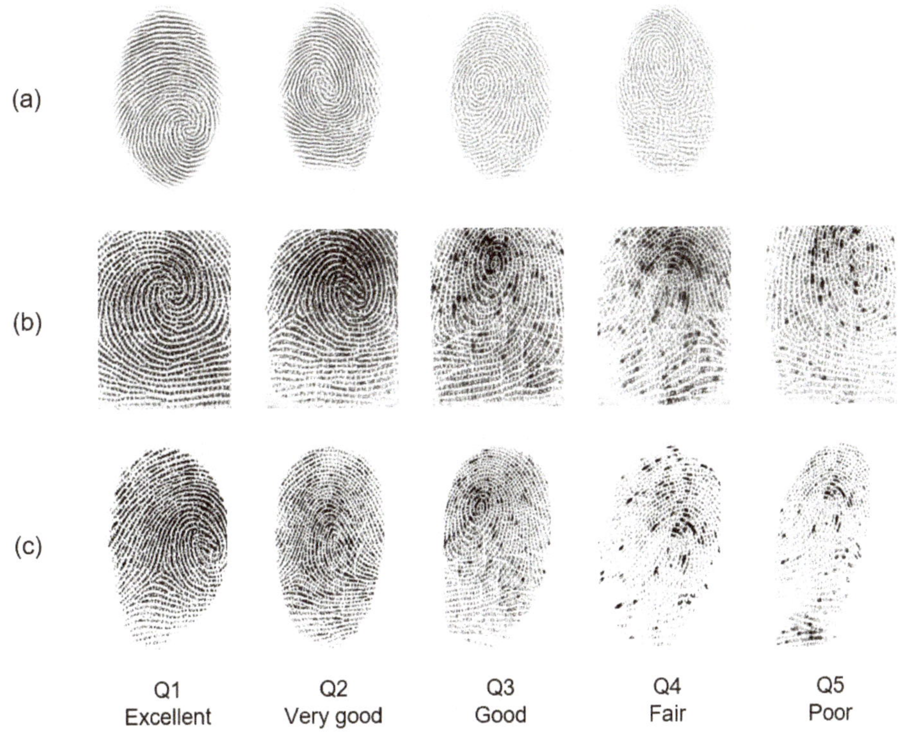

Figure 2. Quality variations $Q = \{1, 2, 3, 4,5\}$ per impression for the same subject across three sensors: (**a**) Lumidigm Venus, (**b**) Secugen Hamster-IV, and (**c**) CrossMatch L-Scan Patrol from MOLF dataset.

Our method was evaluated on two benchmark public datasets, FingerPass and MOLF. The results indicate that fingerprints are enhanced to higher quality regardless of the sensor type used.

To the best of our knowledge, this is the first work dealing with the problem of cross-sensor fingerprint enhancement using deep learning. Our contributions in this paper can be summarized as follow:

- We formulated the cross-sensor fingerprint enhancement problem as an image-to-image transformation problem and designed it using a CNN model with an encoder–decoder architecture that takes a low-quality fingerprint and produces an enhanced fingerprint. We trained the proposed CNN model using two different approaches: conventional learning and adversarial learning.
- Motivated by the success of adversarial learning in modeling image-to-image transformation [9], we learned the proposed image-to-image transformation (the CNN model) using a conditional GAN framework, where the proposed CNN model plays the role of a generator.
- To preserve the ridge patterns in the fingerprints, we incorporated the edge loss function [10] and $L1$ loss [9] into the adversarial loss [11]. This resulted in good quality enhanced fingerprints regardless of the type of sensor used to capture the fingerprints.
- For comprehensive evaluation of a fingerprint enhancement algorithm, we proposed a new metric called Fingerprint Quality Enhancement Index (FQEI). This metric yields a value between 1 and -1, where 1 represents the best enhancement and -1 represents the worst degradation.

The rest of this paper is structured as follows. Section 2 reviews previous enhancement methods, while Section 3 describes in detail the proposed method. Section 4 presents the training and testing stages of our model, while Section 5 gives details of the experiments. Section 6 discusses our results. Finally, Section 7 concludes the conducted work and suggests some directions for future work.

2. Related Work

In the last decade, various studies have been conducted to study the effect of reliable fingerprint enhancement for solving the matching problem assuming that the same sensor was used both for enrollment and verification.

A common technique is the HONG method proposed by Hong et al. [12], where fingerprints are enhanced using a bank of Gabor filters, which are adjusted to the orientation of the local ridges. Another state-of-the-art method is the CHIK method, which was proposed by Chikkerur et al. [13], where fingerprints are enhanced using the short-time Fourier transform (STFT). In this method, each fingerprint is initially divided into small overlapping windows, and the STFT is applied to each window. Next, the block energy, ridge orientation, and ridge frequency are estimated using the Fourier spectrum, and then contextual filtering is applied for fingerprint enhancement.

Other enhancement techniques focus on using off-line images, such as the latent fingerprint technique [14]. Researchers proposed an approach that employed a CNN model to predict ridge direction from a set of pre-trained ridge patterns. In [7], a direct end-to-end enhancement approach was proposed using the FingerNet architecture. This method relied on the use of a CNN within an encoder–decoder scheme. In [8], the authors employed a convolutional auto-encoder neural network to enhance the missing ridge pattern. A similar work was proposed in [15], where a method based on de-convolutional auto-encoders was developed to match sensor-scan and inked fingerprints.

All previous works have focused on using conventional learning only in the enhancement process, where CNNs learn to minimize the loss function. This process, however, requires a lot of manual effort. In contrast, the flexibility provided by Generative Adversarial Networks (GANs), which apply adversarial learning, allows for optimizing the objective function of the problem more effectively. It initially determines a single high-level goal, such as producing indistinguishable fake images from real images, and then learns to achieve such a goal automatically using a suitable loss function [9]. In the JOSHI method [16], a conditional GAN model was proposed based on an image-to-image translation to reconstruct the ridge structure of latent fingerprints. As discussed above, most previous enhancement methods have focused on matching latent fingerprints left unintentionally at a crime scene. Unlike previous methods, which deal with latent fingerprints, the proposed method addresses the problem of enhancing cross-sensor fingerprints. The problem of cross-sensor enhancement has been addressed in a few studies only. In [4,17], an adaptive histogram equalization method was proposed to enhance the contrast of contactless fingerprint ridges and valleys. To date, these are the only published studies concerning cross-sensor enhancement. No previous studies have addressed the cross-sensor enhancement problem using deep learning techniques.

3. Proposed Method

A critical issue when designing an effective cross-sensor fingerprint enhancement is preserving valleys, ridges, and other fingerprint features, such as minutiae. In view of this, we introduce a new method for cross-sensor fingerprint enhancement.

3.1. Problem Formulation

Fingerprint enhancement can be expressed as an image-to-image transformation problem. It aims to learn a mapping, denoted by $\mathcal{F}$, which transforms an input fingerprint $x \in \mathbb{R}^{mxn}$ to an enhanced fingerprint $\hat{y}$. This implies finding a mapping $\mathcal{F} : \mathbb{R}^{mxn} \to \mathbb{R}^{mxn}$ such that $\hat{y} = \mathcal{F}(x;\, \theta)$, where θ represents the transformation parameters. A critical question in this context is how to model the mapping function $\mathcal{F}$. From a practical standpoint, the application of both deep learning and CNNs has shown promising performance in pattern recognition problems, as indicated in various studies [4,14,15]. This, in turn, motivated us to model $\mathcal{F}$ using a CNN model. The learning method typically employed in CNNs is conventional learning, which is based on an objective function that minimizes the loss function between ground truth and the predicted labels. However, regardless of whether the learning process is automatic, several studies have sought to design more effective loss functions [9].

Another efficient learning approach is based on the Generative Adversarial Networks (GANs) framework. The learning method applied in GANs is adversarial learning, which is based on a min-max game and includes a specific loss function, where one agent tries to maximize while the other one tries to minimize [11].

3.2. *The Design of Mapping Function* ($\mathcal{F}$)

The design of the mapping function ($\mathcal{F}$) is a challenging problem since the captured fingerprints by different sensors have different texture patterns and noise [4]. The desired mapping must be developed to enhance fingerprints by preserving the underlying fingerprint features and removing possible corruption and noise. To address these issues and effectively learn $\mathcal{F}$, two learning frameworks were investigated: conventional learning and adversarial learning.

3.2.1. Conventional Learning Framework (One-Net)

In this case, $\mathcal{F}$ was designed using a CNN model following an encoder–decoder architecture [18]. It takes a low-quality fingerprint as input and produces a high-quality one as output. This architecture minimizes the loss between the target images and the predicted ones. This architecture was adopted from SegNet [19] with some modifications. SegNet comprises two networks: an encoder and a corresponding decoder, followed by a final pixel-wise classification layer.

SegNet has five encoders and corresponding five decoders. All the encoders include two consecutive layers and max pooling layers. Each convolutional layer consists of 64 filters with size 3×3, 1 padding and stride of 1 followed by batch normalization (BN) layer and then element-wise rectified linear non-linearity (ReLU). After that, 2×2 max pooling layer, with a stride of 2, is applied where the related max pooling indices (locations) are saved.

Each corresponding decoder up-samples its input using the recalled max-pooling indices using a 2×2 max unpooling layer with a stride of 2. Then, it convolves the input using two consecutive convolutional layers. Each convolutional layer contains 64 filters of size 3×3 and a stride of 1 followed by a batch normalization layer, then a ReLU layer. The final output is then fed into a multi-class soft-max classifier to compute class probabilities for each pixel independently.

This model has been specifically designed for segmentation purposes. However, since our goal is different and focuses on the enhancement task, the SegNet model was modified to fit the task of interest by receiving a low-quality, $300 \times 300 \times 1$ fingerprint and generating a same-size fingerprint with higher quality. Both the Softmax layer and the pixel-wise classification layer were removed. Since the target task is to produce a same-size fingerprint with a higher quality, a convolution layer with one filter of size 3×3, was also added, as shown in Figure 3.

The preservation of small and thin details is essential for fingerprint matching since they play an important role in determining the identity of each subject. Some of these details are the minutiae points formed mainly by ridge bifurcations and ridge endings. The ridge bifurcations are those points where ridges are divided into two ridges, whereas the ridge endings are those points where ridges end. The extraction of minutiae points is a difficult task in low-quality fingerprint images [1], see Figure 4.

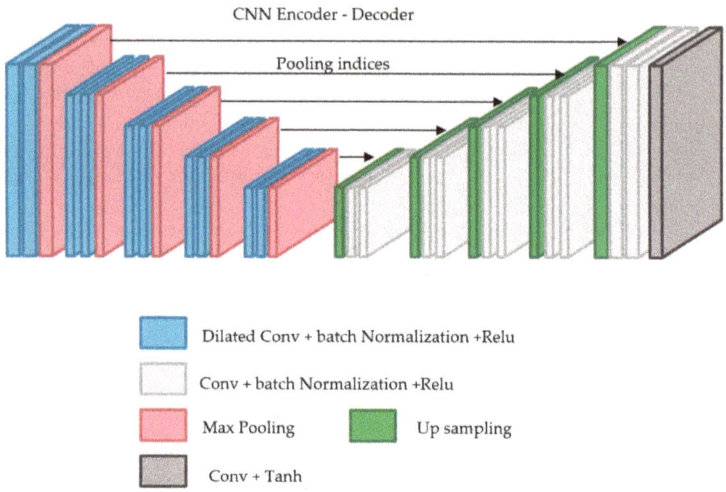

Figure 3. Conventional learning framework.

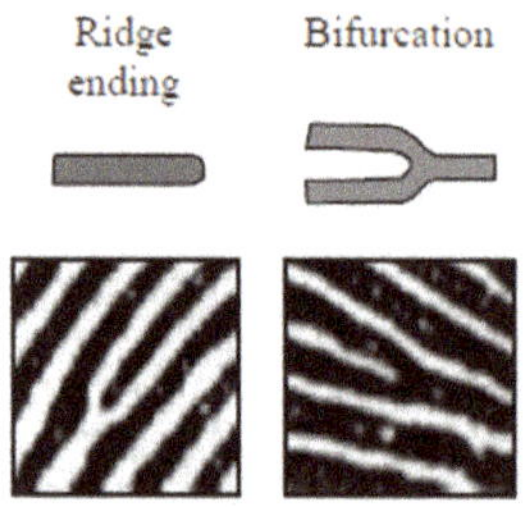

Figure 4. The two most common minutiae—ridge ending and bifurcation. Reprinted with permission from Ref. [1]. Copyright 2022, Springer Nature.

These small details should be considered when designing the target model. Convolutional networks are deployed to gradually reduce the image resolution until it is represented via tiny feature maps, where the spatial structure is not yet visible. However, this spatial acuity loss may restrict fingerprint enhancement. This issue can be addressed by dilated convolutions that can increase the output feature maps resolution without decreasing the individual neurons' receptive field. Thus, a second modification introduced to the SegNet model is adding dilated convolutions.

Generally, dilated convolution is a convolution having a wider kernel that is generated based on repeatedly adding spaces among the kernel elements [20]. Therefore, each convolution layer in the encoder was substituted by a dilated convolution layer using a different dilation factor in the range: 1, 1, 2, 2, 4, 4, 8, 8, 16, and 16. Our results illustrate that dilated convolution is appropriate for fingerprint enhancement since it enlarges the receptive field with no coverage or resolution loss.

In the decoder network, each decoder up-samples its input feature map(s) by deploying the memorized max-pooling indices related to its corresponding encoder's feature map(s). It should be noted that there is no conducted learning within the up-sampling stage. SegNet uses the max pooling indices to up-sample the feature map(s) and convolves them with a trainable decoder filter bank. Next, batch normalization is applied to each map. Subsequently, the high dimensional feature representation at the final decoder output is fed to a convolutional layer followed by a Tanh layer as shown in Table 1.

Table 1. Specifications of encoder and decoder models; FS represents the filter size; FN is number of filters and S represents the stride.

Encoder				Decoder			
Layer	FS	FN	S	Layer	FS	FN	S
Conv1_1	3	64	1	Conv1_1	3	64	1
Conv1_2	3	64	1	Conv1_2	3	64	1
Max Pooling 1	2	-	2	Max Un pooling 1	2	-	2
Dilated Conv 2_1	3	64	1	Conv 2_1	3	64	1
Dilated Conv 2_2	3	64	1	Conv 2_2	3	64	1
Max Pooling 2	2	-	2	Max Un pooling 2	2	-	2
Dilated Conv 3_1	3	64	1	Conv 3_1	3	64	1
Dilated Conv 3_2	3	64	1	Conv 3_2	3	64	1
Max Pooling 3	2	-	2	Max Un pooling 3	2	-	2
Dilated Conv 4_1	3	64	1	Conv 4_1	3	64	1
Dilated Conv 4_2	3	64	1	Conv 4_2	3	64	1
Max Pooling 4	2	-	2	Max Un pooling 4	2	-	2
Dilated Conv 5_1	3	64	1	Conv 5_1	3	64	1
Dilated Conv 5_2	3	64	1	Conv 5_2	3	64	1
Max Pooling 5	2	-	2	Max Un pooling 5	2	-	2
				Conv 6_1	3	1	1
				Tanh	-	-	-

3.2.2. The Adversarial Learning Framework (Two-Net)

This type of learning is based on the conditional generative adversarial network (cGAN) framework [9]. The cGAN framework consists of a generator and a discriminator. The role of the generator is to produce a transformed image from the input one. The discriminator determines if the input image is real or fake. In the training stage, both the generator and discriminator conduct a min-max game. For this task, $\mathcal{F}$ plays the role of the generator, which is to produce a high-quality fingerprint ($\hat{y}$) from a low-quality one (x). The enhanced high-quality fingerprint must have a clear ridge structure to preserve the valleys, ridges, and further fingerprint features, such as minutiae points. The discriminator differentiates real fingerprints from the generated ones, which helps to learn $\mathcal{F}$.

To effectively learn $\mathcal{F}$ via the cGANs framework, it is considered a generator that generates an enhanced image $\hat{y}$ from an input image x. To model $\mathcal{F}$, a dilated SegNet is deployed since both the input and output are images with the same size $300 \times 300 \times 1$, as explained in the first framework. The discriminator $\mathcal{D}$ is modeled using a patch GAN discriminator that was adopted from the paper [9]. The first convolution layer Conv contains 64 filters, stride 2, depth of 2, followed by a Leaky ReLU layer. The second Conv consists of 128 filters, stride 2, and the third contains 256 filters of stride 2; the fourth Conv contains 512 filters, stride 2; each of these layers is followed by a batch normalization layer and the Leaky ReLU. The last layer is a Conv layer consisting of one filter and stride of 1. All these Conv layers contain filters of size 4, as illustrated in Figure 5.

3.3. *Loss Functions and the Learning of* ($\mathcal{F}$)

For the first framework, $\mathcal{F}$ is learned through conventional learning based on taking a low-quality fingerprint x and producing a high-quality one. This model minimizes the gradient difference between the generated fingerprint and the ground truth y. We used two loss functions: $L1$ loss [9] and Edge Loss [10].

The first loss used is the $L1$ distance that is expressed as follows:

$$\mathcal{L}_{L1}(\mathcal{F}) = E_{x,y}[\| \mathrm{y} - \mathcal{F}(x) \|_1] \tag{1}$$

An ideal fingerprint image has valleys and ridges that flow in a locally regular direction. In this case, the detection of ridges is straightforward, and minutiae can be accurately located within the image. Nevertheless, skin conditions (e.g., dry/wet, bruises, and cuts), improper finger pressure, and sensor noise significantly impact fingerprint image quality.

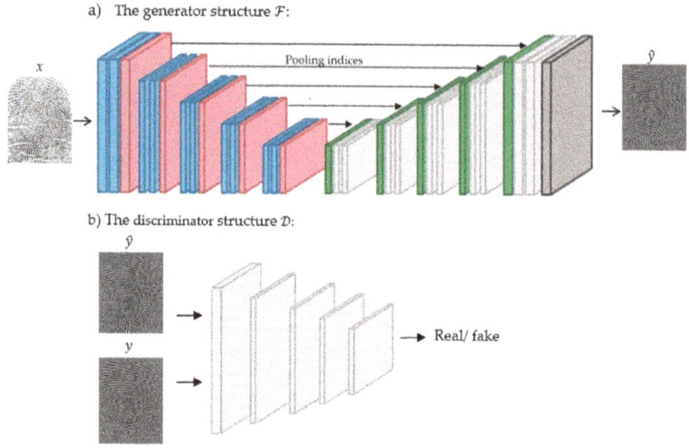

Figure 5. The adversarial learning framework. The blue color represents dilated Conv, BN and ReLU layers; the pink color represents Max-Pooling layer; the green color represents up sampling layer; light grey color represents Conv, BN and ReLU layers; the dark grey color represents Conv and Tanh layers.

Therefore, the edge loss function is added to improve the fingerprint ridge structures by calculating the edge direction. For this, the ridge pattern of the generated fingerprint and the corresponding ground truth fingerprint are initially computed, and then the loss is used to update the parameters of $\mathcal{F}$. The edge loss is denoted as L_{edge} and can be expressed as follows:

$$\mathcal{L}_{edge}(\mathcal{F}) = \sqrt{\| \Delta\mathcal{F}(x) - \Delta y \|^2 + \varepsilon^2} \quad (2)$$

where Δ represents the Laplacian of Gaussian operator, y denotes the ground truth fingerprint (high quality), and $\mathcal{F}(x)$ denotes the enhanced image. The parameter with constant ε empirically set to 10^{-3} as used in [10]. This loss is used to preserve edge features useful for improving ridge patterns.

The total loss

$$\mathcal{L}_{Conventional}(\mathcal{F}) = \mu\, \mathcal{L}_{L1}(\mathcal{F}) + \lambda \mathcal{L}_{edge}(\mathcal{F}). \quad (3)$$

In the second framework, $\mathcal{F}$ learning is inspired by the method [9]. Both $\mathcal{D}$ and $\mathcal{F}$ are learned using adversarial learning. The training dataset includes pairs of poor- and high-quality fingerprints. Such pairs are expressed as $(x_i;\ y_i)$, in which x_i stands for the poor-quality fingerprint image, while y_i stands for the corresponding high-quality one (ground truth).

A fingerprint x is fed into $\mathcal{F}$, which then maps it to an enhanced version $\hat{y}$. The channel-wise concatenation between the pairs $(x,\ y)$ and $(x, \hat{y})$ is then fed into $\mathcal{D}$ to classify them as real or generated fingerprints. The discriminator ensures that the generator effectively learns to preserve ridge structures of the generated enhanced fingerprints. The adversarial loss is given below:

$$\mathcal{L}_{GAN}(\mathcal{F},\ \mathcal{D}) = E_{(x,y)}[\log(\mathcal{D}(x,y) + E_x[\log(1 - \mathcal{D}(x, \mathcal{F}(x))\ ]]. \quad (4)$$

A custom training loop is deployed to train the model using the training dataset, in which the network weights are updated in each iteration. In the training stage, $\mathcal{F}$ produces a fingerprint that is hard to be classified as synthetic via $\mathcal{D}$. In contrast, $\mathcal{D}$ avoids being misled by $\mathcal{F}$ and increases the successful discrimination between the original and synthetic fingerprints by reducing the value of the loss function.

We combined the edge loss and $L1$ distance with adversarial learning. The final objective function is expressed below:

$$argmin\mathcal{F}max\mathcal{D}\ \mathcal{L}_{GAN}(\mathcal{F},\ \mathcal{D}) + \mu\ \mathcal{L}_{L1}(\mathcal{F}) + \lambda\mathcal{L}_{edge}(\mathcal{F}). \tag{5}$$

Figure 6 illustrates the training framework, which learns $\mathcal{F}$ to produce an enhanced fingerprint from an input one.

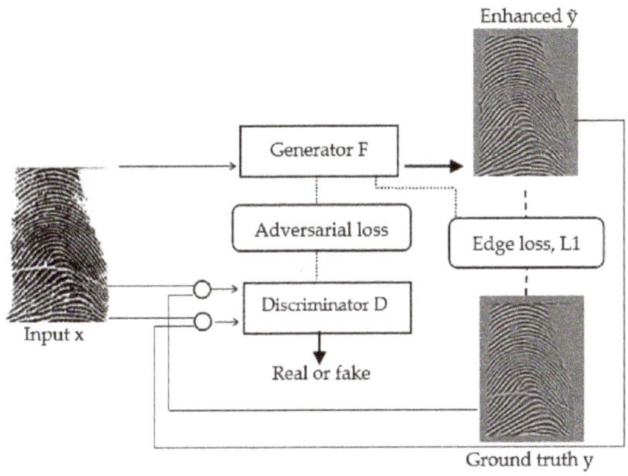

Figure 6. The learning procedure of $\mathcal{F}$ using adversarial learning. The thin arrows represent the input; the thick arrows represent the output; The dotted lines represent weights updating, the dashes represent the two fingerprints used to calculate the edge loss and $L1$ loss; and the circles represent the channel-wise concatenation.

3.4. *Assessing the Quality of the Enhancements*

Although both NFIQ [6] and SSIM [21] are popular and accurate metrics used widely to measure fingerprint quality, they do not offer a comprehensive description of what happens during enhancement. In these metrics, the number of enhanced or degraded images is not considered. A new metric has been designed to comprehensively describe each class's performance by analyzing the NFIQ results.

Fingerprint Quality Enhancement Index (FQEI)

The detail of the new metric for assessing the enhancement potential of an algorithm is given in the following paragraphs. A fingerprint can be assigned to one of five quality levels, i.e., $Q1$: excellent, $Q2$: very good, $Q3$: good, $Q4$: fair, or $Q5$: poor, based on the scores obtained from the NFIQ tool [6]. Using the quality levels of fingerprints before and after enhancement, we compute the Quality Confusion Matrix (QCM) as shown in Table 2, where Q_{jj} is the number of images with original quality Q_j have been enhanced to quality Q_i.

Table 2. The quality confusion matrix (QCM).

Q_{11}	Q_{12}	Q_{13}	Q_{14}	Q_{15}
Q_{21}	Q_{22}	Q_{23}	Q_{24}	Q_{25}
Q_{31}	Q_{32}	Q_{33}	Q_{34}	Q_{35}
Q_{41}	Q_{42}	Q_{43}	Q_{44}	Q_{45}
Q_{51}	Q_{52}	Q_{53}	Q_{54}	Q_{55}

To quantify the enhancement quality, each Q_{jj} in QCM is scaled according to the corresponding coefficient w_{ij} in the weight quality matrix (WQM), shown in Table 3.

Table 3. The weight quality matrix (WQM).

0	1	2	3	4
−1	0	1	2	3
−2	−1	0	1	2
−3	−2	−1	0	1
−4	−3	−2	−1	0

In WQM, (i) $w_{ii} = 0$ because there is no enhancement in the quality level of the fingerprints, (ii) w_{ij} ($i < j$) is 1, 2, 3, or 4 depending on enhancement levels, e.g., in case of Q_{13}, the quality of fingerprints after enhancement goes two levels up from Q_3 to Q_1, it must be weighted with $w_{13} = 2$, (iii) w_{ij} ($i > j$) is −1, −2, −3 or −4 depending on de-enhancement levels.

The enhancement score (E_s), which quantifies the quality of enhancement of fingerprints that were in a low-quality class before enhancement and assigned to a high-quality class after enhancement, can be expressed using QCM and WQM as follows:

$$E_s = \sum_{j=2}^{5} \sum_{j>i} Q_{ij} \times w_{ij} \tag{6}$$

The degradation score (D_s), which quantifies the quality of de-enhancement of fingerprints that were in a high-quality class before enhancement and assigned to a low-quality class after enhancement, can be expressed using QCM and WQM as follows:

$$D_s = \sum_{i=2}^{5} \sum_{j<i} Q_{ij} \times w_{ij} \tag{7}$$

In the ideal case (IS) scenario, all images are enhanced from low-quality class to excellent class. In other words, IS can be represented as a weighted sum of all images, except those of $Q1$ quality, using the following formula:

$$IS = (Q_{12} \times 1) + (Q_{13} \times 2) + (Q_{14} \times 3) + (Q_{15} \times 4) \tag{8}$$

where Q_{12} represents images from very good class that enhanced one degree up to be in class excellent, and so on.

However, in the worst-case (WS) scenario, all images move from the high-quality class to the poor-quality class, excluding the class poor since its images preserve their class. This means that WS can be expressed as a weighted sum of all images, except those in class poor, using the following formula:

$$WS = (Q_{51} \times -4) + (Q_{52} \times -3) + (Q_{53} \times -2) + (Q_{54} \times -1) \tag{9}$$

where Q_{51} represents images from excellent class that degraded four degrees down to be in poor class, and so on.

To measure the enhancement ratio (ER), the E_s computed using Equation (6), is divided by IS computed using Equation (8). Thus, the ER is expressed as follows:

$$ER = \frac{E_s}{IS} \tag{10}$$

In contrast, the degradation ratio can be measured by dividing the D_s by WS as follows:

$$DR = \frac{D_s}{WS} \tag{11}$$

The difference between the enhancement ratio and the degradation ratio is computed to determine the degree of enhancement for measuring the performance of an algorithm:

$$\text{FQEI} = ER - DR. \tag{12}$$

In the ideal case scenario FQEI = 1, and it is equal to −1 in the worst-case scenario.

The more the FQEI is close to one, the higher the enhancement is, and vice versa. An illustrative example is provided in the Appendix A.

4. Training and Testing

In this section, we discuss the training stage, which uses training data to learn the model, and the testing stage tests it using test data.

4.1. Training Details

The model constructed is a supervised generative one trained to generate high-quality fingerprint images from low-quality ones. Practically, a supervised model needs paired training data of low-quality fingerprints combined with their corresponding enhanced images. However, cross-sensor fingerprint datasets have low-quality fingerprints, and their related high-quality counterparts are not available. Moreover, cross-sensor fingerprint databases are not large enough with high-quality images. This results in training difficulties of deep neural network models. Therefore, there is a need to generate fingerprints with noise characteristics similar to those of real fingerprints, as shown in Figure 1, and their enhanced versions to train the enhancement model. The following subsections detail the datasets prepared for training the model.

FingerPass Database

The training data were fingerprints from the AES2501 sensor from the FingerPass dataset, which includes 8460 images of different qualities: excellent, very good, good, fair, and poor. To help the model learn how to enhance fingerprints with different quality levels, all fingerprints were enhanced using the HONG method [12], which were used as the target fingerprints.

The proposed method was trained using a minibatch SGD with Adam optimizer considering the following parameters: Momentum parameters $\beta 1$ = 0.5 and $\beta 2$ = 0.999, Learning rate 0.002, $\mu = 100$, and λ = 0.001.

4.2. Testing Details

The performance of the proposed method was tested using two benchmark public databases: FingerPass [3] and MOLF [5].

4.2.1. Multisensor Optical and Latent Fingerprint (MOLF) Dataset

This dataset includes images captured by using three different sensors, having the same sensor technology (optical sensors) and the same capturing method (press). Images in the database come from 100 subjects, where each one of the 10 fingerprints was captured in two sessions (two independent instances were captured in each session). Each sensor was used to capture 4000 images with 1000 fingerprint classes.

Live-scan images in the database are categorized into three subsets. DB1, DB2, and DB3. It can be noted from Figure 2 that those images are visually different due the acquisition sensor used and the capturing process applied.

4.2.2. FingerPass Database

FingerPass consists of images of the same eight fingers (thumb, index finger, middle finger, and ring finger of both hands) captured using nine sensors from 90 subjects; a sample is shown in Figure 1.

It includes two technological types (optical and capacitive sensors) and two capturing methods (in this case, press and sweep). Each subject was asked to take 12 impressions for each finger. Therefore, the database includes images of 720 fingers, where the total number of impressions for one sensor is $90 \times 8 \times 12$ = 8640 images.

Since our model is trained on fingerprints of size $300 \times 300 \times 1$, the fingerprints from the MOLF dataset and FingerPass are preprocessed to match the required size.

5. Experimental Results

In this section, we introduce the metric used to evaluate our results and present the outcome of the conducted experiments.

5.1. Fingerprint Image Quality Analysis

The NFIQ module of NBIS proposed in [6] was used to analyze the ability of the proposed enhancement algorithm to enhance the quality of cross-sensor fingerprints. The analysis offers a value between 1 and 5, where 1 represents the best quality while 5 represents the worst quality. The score distribution before and after applying the enhancement method was assessed using fingerprints from MOLF and FingerPass datasets to evaluate the performance. The results for MOLF enhancement using adversarial learning are shown in Table 4.

Table 4. NFIQ quality scores on the Original MOLF dataset and the enhanced dataset by our model (After E.). The up arrow represents better enhancement.

Quality	Q	DB1		DB2		DB3	
		Original	After E.	Original	After E.	Original	After E.
Excellent	1	2965	3796 ↑	1340	2255 ↑	2018	3303 ↑
Very good	2	985	183	1940	1724	985	646
Good	3	37	2	603	8	744	16
Fair	4	12	19	27	8	155	19
poor	5	0	0	89	5	97	18

It can be noticed from Table 4 that all images were enhanced, although different sensors were used to capture them. In DB1, there is a significant image quality enhancement, where 3796 images were enhanced out of 4000 to be in class excellent. The difference here is 204 images, which are enhanced compared to the original images.

Moreover, DB2 shows enhancement in class excellent results from 1340 to 2255 and a noticed reduction in a class fair and poor with 27 and 89 images before and 8 and 4 images after for each class. DB3 shows an increase in class excellent fingerprints by 1285 images and a reduction for all other classes; the number of fingerprints of class poor reduces from 97 to 18 after enhancement.

Two learning methods were applied: namely, conventional learning and adversarial learning. A single network was constructed with a loss function that aims to minimize the distance between the predicted and ground truth to test the impact of conventional learning, as described in Section 3.1. On the other hand, the impact of adversarial learning was tested using two networks: a generator and a discriminator, as described in Section 3.2. The results are shown in Table 5 on MOLF datasets.

Table 5. The effect of the learning approach on the quality of the MOLF database.

Quality Score	Q	DB1 Original	Conventional Learning (One Net)	Adversarial Learning (Two Net)
Excellent	1	2965	3827	3796
Very good	2	985	123	183
Good	3	37	12	2
Fair	4	12	36	19
poor	5	0	2	0
Quality Score	**Q**	**DB2 Original**	**Conventional Learning (One Net)**	**Adversarial Learning (Two Net)**
Excellent	1	1340	1915	2255
Very good	2	1940	2057	1724
Good	3	603	18	8
Fair	4	27	6	8

Table 5. *Cont.*

Quality Score	Q	DB3 Original	Conventional Learning (One Net)	Adversarial Learning (Two Net)
Excellent	1	2018	3206	3303
Very good	2	985	634	646
Good	3	744	39	16
Fair	4	155	86	19
poor	5	97	35	18

It can be noticed from Table 5 that the experiment based on adversarial learning offered better results than the conventional one, although the same network architecture was used to generate the fingerprints.

Comparison with the State-of-the-Art Method

There are various studies in the field of fingerprint enhancement, for example, the methods proposed in [7,8,14,15]. Although HONG and CHIK methods are a bit old, their performance is still better than the recent methods for cross-sensor fingerprint enhancement, and, due to this reason, they have been used in recent cross-sensor matching methods [4,22–26]. So, we compared our method with HONG and CHIK methods and a more recent method, i.e., JOSHI method [16].

Figures 7–9 illustrate the comparison results on DB1, DB2, and DB3.

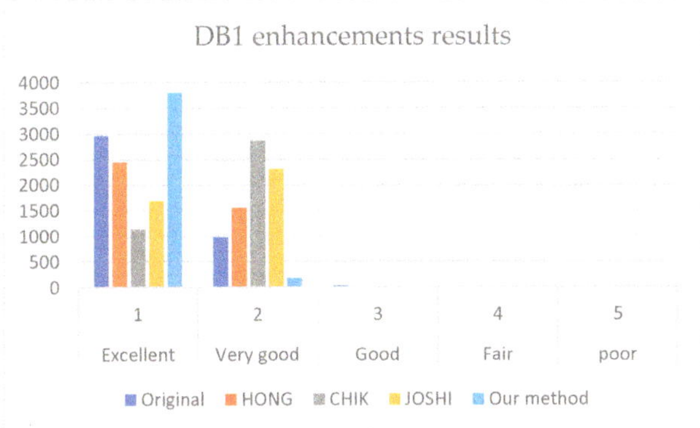

Figure 7. Comparison between the enhancement results of HONG [12], CHIK [13], JOSHI [16], and our method on DB1.

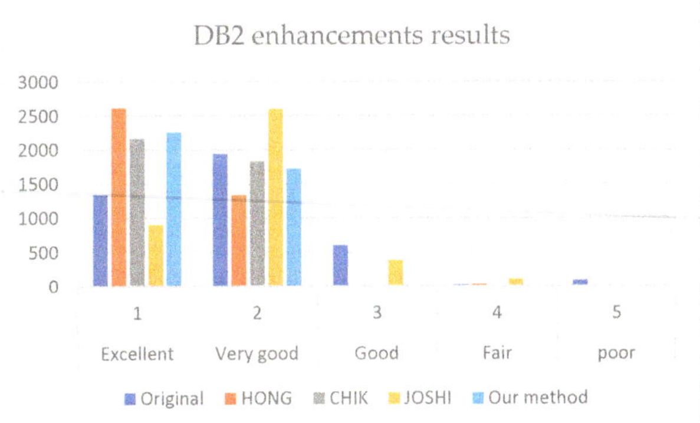

Figure 8. Comparison between the enhancement results of HONG [12], CHIK [13], JOSHI [16], and our method on DB2.

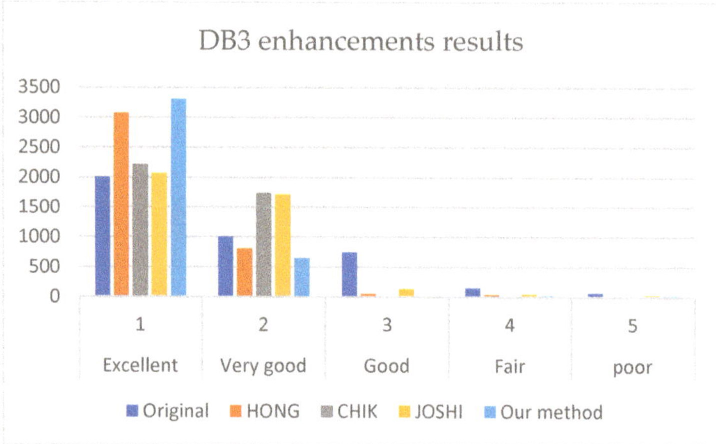

Figure 9. Comparison between the enhancement results of HONG [12], CHIK [13], JOSHI [16], and our method on DB3.

It is revealed from Figures 7–9 that our method outperforms HONG and CHIK methods in enhancing fingerprints to class excellent from DB1 and DB3. For DB2, the number of enhanced fingerprints to class excellent by HONG method is slightly higher than that by our method and CHIK.

5.2. Fingerprint Quality Enhancement Index (FQEI)

The FQEI metric was measured using MOLF datasets DB1, DB2, and DB3 by comparing three methods: HONG, CHIK, JOSHI [16], and our method, where obtained results are provided in Table 6. It can be clearly noticed that our method outperformed both HONG, CHIK, and JOSHI methods on DB1, DB2, and DB3.

Table 6. FQEI values computed for HONG method, CHIK method, JOSHI [16] method, and our method for the MOLF dataset.

The Enhancement Method	DB1	DB2	DB3
HONG [12]	0.2581	0.6342	0.7026
CHIK [13]	0.0231	0.5562	0.6508
JOSHI [16]	0.2012	0.1723	0.3270
Our method	0.8863	0.6760	0.8740

For DB1, the HONG method performance is 0.2581 since the E_s is 348, which is less than the D_s (-808). This means that the number of images above the diagonal is less than the images below the diagonal. The same case is for CHIK performance, where the E_s is 168, while the D_s is -1943 since a large number of fingerprints was degraded from excellent class to very good class. In contrast, our method has a higher enhancement score than the degradation score. Thus, our method outperformed both the HONG and CHIK methods on DB1, DB2, and DB3.

Tables 7–11 illustrate a comparison between the enhancement results obtained with HONG method, CHIK method, JOSHI method, and our method for FingerPass datasets using NFIQ and our metric FQEI.

Table 7. Analysis of the fingerprint quality scores measured by NFIQ of the FingerPass database before enhancement.

Quality	Q	AES2501	AES3400	ATRUA	FPC	FX3000	UPEK	V300	WS	URU4000B
Excellent	1	5519	0	28	40	4105	2016	7917	7395	4697
Very Good	2	2423	65	3149	508	4195	5472	637	895	3263
Good	3	662	7177	2356	5585	330	1142	76	304	647
Fair	4	32	0	0	0	0	0	10	42	33
Poor	5	4	1398	3107	2507	10	10	0	4	0

Table 8. Analysis of the fingerprint quality scores measured by NFIQ and FQEI of the FingerPass enhanced using HONG method [12].

NFIQ	Q	AES2501	AES3400	ATRUA	FPC	FX3000	UPEK	V300	WS	URU4000B
Excellent	1	6194	0	1136	13	4758	945	6381	6786	5245
Very Good	2	2443	202	6852	6020	3882	7693	2258	1853	3389
Good	3	2	8161	546	2596	0	2	1	0	0
Fair	4	1	0	1	0	0	0	0	1	0
Poor	5	0	277	105	10	0	0	0	0	6
FQEI		0.6146	0.1110	0.5868	0.4791	0.4497	0.1379	0.5065	0.5829	0.5912

Table 9. Analysis of the fingerprint quality scores measured by NFIQ and FQEI of the FingerPass enhanced using CHIK method [13].

NFIQ	Q	AES2501	AES3400	ATRUA	FPC	FX3000	UPEK	V300	WS	URU4000B
Excellent	1	4834	0	318	4	2253	838	3953	6217	545
Very Good	2	3806	124	7525	5743	6387	7800	4687	2423	8095
Good	3	0	7323	706	2848	0	2	0	0	0
Fair	4	0	0	0	0	0	0	0	0	0
Poor	5	0	1193	91	45	0	0	0	0	0
FQEI		0.5335	−0.0012	0.5410	0.4615	0.2562	0.1372	0.3202	0.5373	0.0683

Table 10. Analysis of the fingerprint quality scores measured by NFIQ and FQEI of the FingerPass enhanced using JOSHI method [16].

NFIQ	Q	AES2501	AES3400	ATRUA	FPC	FX3000	UPEK	V300	WS	URU4000B
Excellent	1	2216	197	4583	783	2418	2607	2874	2506	2160
Very Good	2	6497	0	3975	6394	6005	5672	1648	6126	6474
Good	3	27	3459	28	1453	189	359	2438	6	6
Fair	4	0	1877	3	8	22	1	464	2	0
Poor	5	0	3107	51	2	6	1	1216	0	0
FQEI		0.2291	−0.3792	0.7889	0.5660	0.1027	0.3114	−0.070	0.3096	0.1582

Table 11. Analysis of the fingerprint quality scores measured by NFIQ and FQEI of the FingerPass enhanced using our method.

NFIQ	Q	AES2501	AES3400	ATRUA	FPC	FX3000	UPEK	V300	WS	URU4000B
Excellent	1	5824	8192	8066	3958	4700	2924	4743	6779	8134
Very Good	2	2797	65	562	4680	2609	5716	241	1855	467
Good	3	2	82	6	1	813	0	3343	2	23
Fair	4	17	301	5	1	399	0	206	4	16
Poor	5	0	0	1	0	119	0	107	0	0
FQEI		0.5645	0.9388	0.9707	0.7836	0.3149	0.3407	0.2931	0.5825	0.9149

From Table 7 for FingerPass dataset before enhancement, it can be noticed that there are three sensors that have the highest number of images in poor class, including AES3400, ATRUA, and FPC sensors with 1398, 3107, and 2507 images, respectively.

Based on comparing the results of NFIQ for the three methods after enhancement, it can be noticed that our method offered the highest enhancement in these three sensors by extremely reducing it to zero poor images for the first sensor, one poor image for the second sensor and zero poor images for the third sensor. Moreover, it particularly enhanced the number of images in excellent class to more than 8000 images for the first sensors and the URU4000B sensor. In contrast, the HONG method revealed the highest enhancement for AES2501 sensor. There are also two sensors with the highest number of images in the excellent class: the WS and V300 sensors.

The overall results show that our method outperformed mostly in increasing the number of images in the excellent class. The CHIK method usually transforms fingerprints' quality to excellent and very good classes but with a noticeable reduction in the number of images in excellent class in most sensors. JOSHI method increases the number of poor fingerprints in two sensors: AES3400 and V3000.

In terms of FQEI metric, our method shows the highest results for five out of nine sensors. The results on AES3400, ATRUA, and URU4000B sensors are 0.9149, 0.9388, 0.9707, respectively, which are very close to 1, and hence a very high enhancement performance. However, a negative enhancement was achieved by JOSHI method in two sensors: AES3400 and V3000. On the other hand, CHIK method gave FQEI of −0.0012 for AES3400 sensor, where the minus sign means distortion in images, which can be obviously noticed by comparing it with the confusion matrix results as shown in Table 12, where most images preserved in good class without enhancement as well as a slight enhancement was revealed from poor class to good class.

Table 12. Quality Confusion Matrices for AES3400 sensor enhancements using: (**a**) Hong [12] (**b**) CHIK [13] (**c**) Our method.

	(a)					(b)					(c)				
	Q1	Q2	Q3	Q4	Q5	Q1	Q2	Q3	Q4	Q5	Q1	Q2	Q3	Q4	Q5
q1	0	0	0	0	0	0	0	0	0	0	0	61	6821	0	1310
q2	0	17	144	0	41	0	15	99	0	10	0	1	48	0	16
q3	0	47	6847	0	1267	0	46	6450	0	827	0	1	66	0	15
q4	0	0	0	0	0	0	0	0	0	0	0	2	242	0	57
q5	0	1	186	0	90	0	4	628	0	561	0	0	0	0	0

5.3. *Structural Similarity Index Metric (SSIM)*

Fingerprint enhancement algorithms are applied to improve fingerprints without changing the ridge structure. This feature can be assessed by computing the SSIM [21] on the generated fingerprints using anguli and their related ground truth, due to the lack of databases that include low-quality images and relative high-quality images. In other words, the higher the obtained SSIM value is, the higher the preserved structural similarity between the generated and ground truth is. Moreover, this denotes that the ridge structure is also maintained.

A comparison was conducted for fingerprints that were enhanced using HONG method, CHIK method, and our method. The test datasets contain two thousand synthetic fingerprints generated using anguli [27]. It is an open-source implementation from the fingerprint generator SFinGe [28] based on simulating synthetic live fingerprints having similar features, such as real-live fingerprints. Two thousand (2000) synthetic fingerprint images produced by Anguli are used to test the model with pattern types following the normal distribution, including the arch, right loop, left loop, whorl, and double loop. From those images generated using Anguli, other input images with lower quality were generated by adding Gaussian noise with morphological operations and blurring the filtering in the frequency domain.

Both the mean and standard deviation of SSIM were then computed as shown in Table 13.

Table 13. Mean and standard deviation (std) of SSIM.

The Enhancement Method	Mean of SSIM	Std
HONG [12]	0.4551	0.0482
CHIK [13]	0.4650	0.0460
JOSHI [16]	0.4125	0.0354
Our method	0.5127	0.0693

The mean of SSIM between the enhanced fingerprint generated using our model and the ground truth is 0.5127. It can be noticed that our method had the highest mean of SSIM, which means that the preservation of ridge patterns is the best in our method.

5.4. Computation Time

The average computation time needed to enhance the URU4000b sensor dataset was computed. All three methods were applied on the same environment (R2021b). The experiment was also applied using a laptop with an Intel Core i7-9750H CPU at 2.60 GHz -2.59 GHz, 32.0 GB RAM, Microsoft Windows 10 in the 64-bit operating system, and an x64-based processor. Our method is faster than HONG, CHIK, and JOSHI [16] methods as shown in Table 14.

Table 14. Comparison between the computation time for enhancement.

Method	Average Computation Time (in Seconds)
HONG [12]	0.63
CHIK [13]	0.48
JOSHI [16]	0.38
Our method	0.087

6. Discussion

The fingerprint sensor interoperability focuses on addressing how the fingerprint-matching system is able to compensate for the differences in the captured fingerprints for the same person by several sensors. The main causes of such variability in fingerprints are the differences in the used capturing technology of sensors, scanning area, sensor resolution, and interaction type.

In practice, each sensor generates its specific type of distortions. Hence, there is a need to enhance captured fingerprints by various sensors. To achieve this, a cross-sensor enhancement method was designed and trained using fingerprints from one sensor, which is the AES2501. On the other hand, this method revealed general enhancement results for other sensors in FingerPass and MOLF datasets. The learning approach considered is the adversarial learning one, which offers better enhancement than the conventional learning one. Moreover, it was found that there was no change in the global flow of ridge patterns within the captured fingerprints by different sensors. This proves its robustness to discrimination. Hence, the edge loss, $L1$ loss, and adversarial loss function were used as loss functions.

The use of dilation convolution offered better enhancement results than those measured using convolution only. This means that the small fingerprint details, considered important features for determining the identity, such as the minutia point and edges, were preserved. This is clearly illustrated in Table 15.

Table 15. The impact of using dilation operation and convolution operation for MOLF datasets.

FQEI	DB1	DB2	DB3
Convolution layer	0.8643	0.5208	0.7996
Dilation Convolution	0.8863	0.6760	0.8740

Based on comparing the results of our method with those of two state-of-art fingerprint methods: HONG and CHIK and a more recent method i.e., JOSHI method [16], using two metrics, our method outperformed both of them. However, the NFIQ metric does not offer a precise description for enhancement performance. Therefore, a new metric was designed, called FQEI. This metric gives one result value between 1 and −1 instead of the five classes results as in the NFIQ.

Figure 10 illustrates zoomed-in views of the fingerprints enhanced using the three methods. From the enhanced fingerprints examples shown in Figure 10, it can be noticed that the smoothed ridges related to the processed fingerprints by the HONG method were more enhanced than those of the CHIK method. On the other hand, our method enhanced fingerprints with preserving their original ridge pattern better than HONG and CHIK.

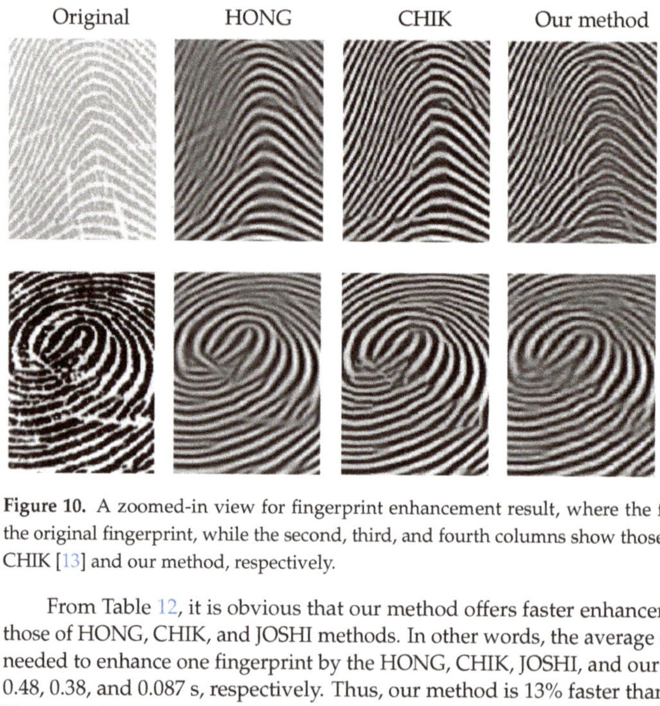

Figure 10. A zoomed-in view for fingerprint enhancement result, where the first column shows the original fingerprint, while the second, third, and fourth columns show those of the HONG [12], CHIK [13] and our method, respectively.

From Table 12, it is obvious that our method offers faster enhancement results than those of HONG, CHIK, and JOSHI methods. In other words, the average computation time needed to enhance one fingerprint by the HONG, CHIK, JOSHI, and our method was 0.63, 0.48, 0.38, and 0.087 s, respectively. Thus, our method is 13% faster than HONG method. However, there are two sensors FX3000 and V300 with less results than what was expected since the fingerprint nature is different than the original data.

7. Conclusions

It can be concluded that with the continuous developments in both fingerprint sensor technologies and the Internet of Things (IoT), the use of biometric fingerprint identification has been increasing over the years. Differences in sensor technologies and resolution can lead to different types of distortion, which affects fingerprint image quality. Therefore, fingerprints must be enhanced. On the other hand, there are no sufficient investigations of the cross-sensor enhancement problem in the related literature. Therefore, this paper proposed an efficient solution for this problem based on deep learning, in which cGAN framework is used for training the image-to-image transformation for fingerprint enhancement. It was demonstrated that the proposed method significantly enhanced the cross-sensor fingerprints regardless of the sensor type used. However, there is still space to achieve more enhancement. One of the suggested future works is to explore different loss functions to preserve and recover the ridge patterns.

Author Contributions: Conceptualization, A.A., M.H., H.A. and G.B.; methodology, A.A. and M.H.; software, A.A. and W.A.; validation, A.A., M.H. and W.A.; formal analysis, A.A. and M.H.; investigation, A.A. and M.H.; resources, M.H., W.A. and G.B.; data curation, A.A. and W.A.; writing—original draft preparation, A.A. and M.H.; writing—review and editing, A.A., W.A. and G.B.; visualization, A.A. and W.A.; supervision, M.H. and H.A.; project administration, M.H. and H.A.; funding acquisition, M.H. and H.A. All authors have read and agreed to the published version of the manuscript.

Funding: This Project was funded by the National Plan for Science, Technology and Innovation (MAARIFAH), King Abdulaziz City for Science and Technology, Kingdom of Saudi Arabia, under Project no. 13-INF946-02.

Institutional Review Board Statement: Not applicable.

Informed Consent Statement: Not applicable.

Data Availability Statement: The MOLF dataset is available at: http://research.iiitd.edu.in/groups/iab/molf.html, and the FingerPass (accessed on 22 May 2021) dataset is available at: http://www.fingerpass.csdb.cn/ (accessed on 22 May 2021).

Conflicts of Interest: The authors declare no conflict of interest.

Appendix A

To clarify FQEI metric, the following example is provided: For a small dataset of 40 fingerprint images having different qualities, the table below represents the NFIQ degrees before (original) and after the enhancement. The QCM is then computed, where the first column represents the total number of images before enhancement, while the first row represents the total number of images after enhancement and so on. Both the IS and WS are then computed as follows:

$$IS = 10 \times 1 + 10 \times 2 + 5 \times 3 + 10 \times 4$$
$$WS = 5 \times -4 + 10 \times -3 + 10 \times -2 + 5 \times -1$$

Table A1. NFIQ quality score of the example before and after enhancements.

NFIQ Values	Q1	Q2	Q3	Q4	Q5
Original	5	10	10	5	10
The enhanced	20	4	14	1	1

Table A2. Computing the (QCM).

	Q1	Q2	Q3	Q4	Q5
q1	1	8	5	2	4
q2	0	1	2	0	1
q3	3	0	3	3	5
q4	1	0	0	0	0
q5	0	1	0	0	0

Table A3. Computing the WCM by Multiplying QCM with WQM.

	Q1	Q2	Q3	Q4	Q5
q1	0	8	10	6	16
q2	0	0	2	0	3
q3	−6	0	0	3	10
q4	−4	0	0	0	0
q5	0	−3	0	0	0

Table A4. Calculating the FQEI.

E_s	D_s	IS	WS	ER	DR	FQEI
58	−13	85	−75	0.6823	−0.1733	0.509

References

1. Maltoni, D.; Maio, D.; Jain, A.K.; Prabhakar, S. *Handbook of Fingerprint Recognition*; Springer Science & Business Media: Berlin/Heidelberg, Germany, 2009.
2. Ross, A.; Jain, A. Biometric Sensor Interoperability: A Case Study in Fingerprints. In *Biometric Authentication*; Maltoni, D., Jain, A.K., Eds.; Lecture Notes in Computer Science 3087; Springer: Berlin/Heidelberg, Germany, 2004; pp. 134–145.
3. Jia, X.; Yang, X.; Zang, Y.; Zhang, N.; Tian, J. A Cross-Device Matching Fingerprint Database from Multi-Type Sensors. In Proceedings of the 21st International Conference on Pattern Recognition (ICPR2012), Tsukuba, Japan, 11–15 November 2012; pp. 3001–3004.
4. Lin, C.; Kumar, A. A CNN-based framework for comparison of contactless to contact-based fingerprints. *IEEE Trans. Inf. Forensics Secur.* **2018**, *14*, 662–676. [CrossRef]
5. Sankaran, A.; Vatsa, M.; Singh, R. Multisensor Optical and Latent Fingerprint Database. *IEEE Access* **2015**, *3*, 653–665. [CrossRef]
6. NIST. NIST Biometric Image Software (NBIS). 2021. Available online: https://www.nist.gov/services-resources/software/nist-biometric-image-software-nbis (accessed on 9 May 2022).
7. Li, J.; Feng, J.; Kuo, C.-C.J. Deep convolutional neural network for latent fingerprint enhancement. *Signal Process. Image Commun.* **2018**, *60*, 52–63. [CrossRef]
8. Svoboda, J.; Monti, F.; Bronstein, M.M. Generative convolutional networks for latent fingerprint reconstruction. In Proceedings of the 2017 IEEE International Joint Conference on Biometrics (IJCB), Denver, CO, USA, 1–4 October 2017; pp. 429–436.
9. Isola, P.; Zhu, J.-Y.; Zhou, T.; Efros, A.A. Image-to-image translation with conditional adversarial networks. In Proceedings of the IEEE Conference on Computer Vision and Pattern Recognition, Honolulu, HI, USA, 21–26 July 2017; pp. 1125–1134.
10. Zamir, S.W.; Arora, A.; Khan, S.; Hayat, M.; Khan, F.S.; Yang, M.H.; Shao, L. Multi-stage progressive image restoration. In Proceedings of the IEEE/CVF Conference on Computer Vision and Pattern Recognition, Nashville, TN, USA, 20–25 June 2021; pp. 14821–14831.
11. Goodfellow, I.; Pouget-Abadie, J.; Mirza, M.; Xu, B.; Warde-Farley, D.; Ozair, S.; Courville, A.; Bengio, Y. Generative Adversarial nets. In *Advances in Neural Information Processing Systems, Proceedings of the 27th International Conference on Neural Information Processing Systems, Montreal, QC, Canada, 8–13 December 2014*; MIT Press: Cambridge, MA, USA, 2014; pp. 2672–2680.
12. Hong, L.; Wan, Y.; Jain, A. Fingerprint image enhancement: Algorithm and performance evaluation. *IEEE Trans. Pattern Anal. Mach. Intell.* **1998**, *20*, 777–789. [CrossRef]
13. Chikkerur, S.; Wu, C.; Govindaraju, V. A systematic approach for feature extraction in fingerprint images. In *Biometric Authentication*; Zhang, D., Jain, A.K., Eds.; Lecture Notes in Computer Science 3072; Springer: Berlin/Heidelberg, Germany, 2004; pp. 344–350.
14. Wong, W.J.; Lai, S.-H. Multi-Task CNN for Restoring Corrupted Fingerprint Images. *Pattern Recognit.* **2020**, *101*, 107203. [CrossRef]
15. Schuch, P.; Schulz, S.; Busch, C. De-convolutional auto-encoder for enhancement of fingerprint samples. In Proceedings of the 2016 Sixth International Conference on Image Processing Theory, Tools and Applications (IPTA), Oulu, Finland, 12–15 December 2016; IEEE: New York, NY, USA, 2016. [CrossRef]
16. Joshi, I.; Anand, A.; Vatsa, M.; Singh, R.; Roy, S.D.; Kalra, P. Latent fingerprint enhancement using generative adversarial networks. In Proceedings of the 2019 IEEE Winter Conference on Applications of Computer Vision (WACV), Waikoloa, HI, USA, 7–11 January 2019; pp. 895–903.
17. Lin, C.; Kumar, A. Improving cross sensor interoperability for fingerprint identification. In Proceedings of the 2016 23rd International Conference on Pattern Recognition (ICPR), Cancun, Mexico, 4–8 December 2016; IEEE: New York, NY, USA, 2016. [CrossRef]
18. Hinton, G.E.; Salakhutdinov, R.R. Reducing the Dimensionality of Data with Neural Networks. *Science* **2006**, *313*, 504–507. [CrossRef] [PubMed]
19. Badrinarayanan, V.; Kendall, A.; Cipolla, R. Segnet: A deep convolutional encoder-decoder architecture for image segmentation. *IEEE Trans. Pattern Anal. Mach. Intell.* **2017**, *39*, 2481–2495. [CrossRef] [PubMed]
20. Hassani, I.K.; Pellegrini, T.; Masquelier, T. Dilated convolution with learnable spacings. *arXiv* **2021**, arXiv:2112.03740.
21. Wang, Z.; Bovik, A.; Sheikh, H.; Simoncelli, E.P. Simoncelli, Image quality assessment: From error visibility to structural similarity. *IEEE Trans. Image Process.* **2004**, *13*, 600–612. [CrossRef] [PubMed]
22. Zang, Y.; Yang, X.; Jia, X.; Zhang, N.; Tian, J.; Zhu, X. A coarse-fine fingerprint scaling method. In Proceedings of the 2013 International Conference on Biometrics (ICB), Madrid, Spain, 4–7 June 2013; IEEE: New York, NY, USA, 2013. [CrossRef]
23. Alshehri, H.; Hussain, M.; Aboalsamh, H.A.; Al Zuair, M.A. Cross-sensor fingerprint matching method based on orientation, gradient, and gabor-hog descriptors with score level fusion. *IEEE Access* **2018**, *6*, 28951–28968. [CrossRef]
24. AlShehri, H.; Hussain, M.; AboAlSamh, H.; AlZuair, M. A large-scale study of fingerprint matching systems for sensor interoperability problem. *Sensors* **2018**, *18*, 1008. [CrossRef] [PubMed]

25. Alshehri, H.; Hussain, M.; Aboalsamh, H.A.; Emad-Ul-Haq, Q.; AlZuair, M.; Azmi, A.M. Alignment-Free Cross-Sensor Fingerprint Matching Based on the Co-Occurrence of Ridge Orientations and Gabor-HoG Descriptor. *IEEE Access* **2019**, *7*, 86436–86452. [CrossRef]
26. Alrashidi, A.; Alotaibi, A.; Hussain, M.; AlShehri, H.; AboAlSamh, H.; Bebis, G. Cross-Sensor Fingerprint Matching Using Siamese Network and Adversarial Learning. *Sensors* **2021**, *21*, 3657. [CrossRef] [PubMed]
27. Ansari, A.H. Generation and Storage of Large Synthetic Fingerprint Database. Master's Thesis, Indian Institute of Science, Bangalore, India, 2011.
28. Cappelli, R.; Maio, D.; Maltoni, D. SFinGe: An approach to synthetic fingerprint generation. In Proceedings of the International Workshop on Biometric Technologies (BT'04), Prague, Czech Republic, 15 May 2004; pp. 147–154.

sensors

MDPI

Article

Learning Gait Representations with Noisy Multi-Task Learning

Adrian Cosma and Emilian Radoi *

Faculty of Automatic Control and Computer Science, University Politehnica of Bucharest, 006042 Bucharest, Romania
* Correspondence: emilian.radoi@upb.ro

Abstract: Gait analysis is proven to be a reliable way to perform person identification without relying on subject cooperation. Walking is a biometric that does not significantly change in short periods of time and can be regarded as unique to each person. So far, the study of gait analysis focused mostly on identification and demographics estimation, without considering many of the pedestrian attributes that appearance-based methods rely on. In this work, alongside gait-based person identification, we explore pedestrian attribute identification solely from movement patterns. We propose DenseGait, the largest dataset for pretraining gait analysis systems containing 217 K anonymized tracklets, annotated automatically with 42 appearance attributes. DenseGait is constructed by automatically processing video streams and offers the full array of gait covariates present in the real world. We make the dataset available to the research community. Additionally, we propose GaitFormer, a transformer-based model that after pretraining in a multi-task fashion on DenseGait, achieves 92.5% accuracy on CASIA-B and 85.33% on FVG, without utilizing any manually annotated data. This corresponds to a +14.2% and +9.67% accuracy increase compared to similar methods. Moreover, GaitFormer is able to accurately identify gender information and a multitude of appearance attributes utilizing only movement patterns. The code to reproduce the experiments is made publicly.

Keywords: gait recognition; self-supervised learning; pose estimation; multi-task learning; weakly-supervised learning

Citation: Cosma, A.; Radoi, E. Learning Gait Representations with Noisy Multi-Task Learning. *Sensors* **2022**, 22, 6803. https://doi.org/10.3390/s22186803

Academic Editors: Euntai Kim, Sangyoun Lee and Kang Ryoung Park

Received: 12 August 2022
Accepted: 6 September 2022
Published: 8 September 2022

1. Introduction

Technologies relying on facial and pedestrian analysis play a crucial role in intelligent video surveillance and security systems. Facial and pedestrian analysis systems have become the norm in video intelligence, such systems being deployed ubiquitously. However, appearance-based pedestrian re-identification [1] and facial recognition models [2] invariably suffer from extrinsic factors related to camera viewpoint and resolution, and to the change in a person's appearance such as different clothing, hairstyles and accessories. Moreover, due to the proliferation of privacy laws such as GDPR, it is increasingly difficult to deploy appearance-based solutions for video-intelligence. Human movement is highly correlated with many internal and external aspects of a particular individual including age, gender, body mass index, clothing, carrying conditions, emotions and personality [3]. The manner of walking is unique to each person, it does not significantly change in short periods of time [4] and cannot be easily faked to impersonate another person [5]. Gait analysis has gained significant attention in recent years [6,7], due to solving many of the problems of appearance-based technologies without relying on the direct cooperation of subjects. However, compared to appearance-based methods, gait analysis is intrinsically harder to perform with reliable accuracy, due to the influence of many confounding factors that affect the manner of walking. This problem is tackled in literature in two major ways, either by building specialized neural architectures that are invariant to walking variations [8–10], or by creating large-scale and diverse datasets for training [11–15].

One of the first attempts of building a large-scale gait recognition dataset is OU-ISIR [14], which is comprised of 10,307 identities that walk in a straight line for a short duration of time. Such a dataset is severely limited by its lack of walking variability, having only viewpoint change as a confounding factor. Building sufficiently large datasets that account for all the walking variations imply an immense annotation effort. For example, the GREW benchmark [12] for gait-based identification, reportedly took 3 months of continuous manual annotation by 20 workers. In contrast, automatic, weakly annotated datasets are much easier to gather by leveraging existing state-of-the-art models—UWG [11], a comparatively large dataset of individual walking tracklets proved to be a promising new direction in the field. Increasing the dataset size is indeed correlated with performance on downstream gait recognition benchmarks [11], even though no manual annotations are provided. One limitation of these datasets is that they are annotated with attributes per individual only sparsely, and not addressing the problem of pedestrian attribute identification (PAI), currently performed only through appearance-based methods [16–18]. Walking pedestrians are often annotated only with their gender, age, and camera viewpoint [8,12,14,15]. Even though gait-based demographic identification is a viable method for pedestrian analysis [19], it is also severely limited by the lack of data. Also, many attributes from PAI networks such as gender, age and body type have a definite impact on walking patterns [20–22], and we posit that they can be identified with a reasonable degree of accuracy using only movement patterns and not utilizing appearance information.

We propose DenseGait, the largest gait dataset for pretraining to date, containing 217 K anonymized tracklets in the form of skeleton sequences, automatically gathered by processing real-world surveillance streams through state-of-the-art models for pose estimation and pose tracking. An ensemble of PAI networks was used to densely annotate each skeleton sequence with 42 appearance attributes such as their gender, age group, body fat, camera viewpoint, clothing information and apparent action. The purpose of DenseGait is to be used for pretraining networks for gait recognition and attribute identification, it is not suitable for evaluation since it is annotated automatically and does not contain manual, ground-truth labels. DenseGait contains walking individuals in real scenarios, it is markerless, non-treadmill, and avoids unnatural and constrictive laboratory conditions, which have been shown to affect gait [23]. It practically contains the full array of factors that are present in real world gait patterns.

The dataset is fully anonymized, and any information pertaining to individual identities is removed, such as the time, location and source of the video stream, and the appearance and height information of the person. DenseGait is a gait analysis dataset primarily intended for pretraining neural models—using it to explicitly identify the individuals within it is highly unfeasible, requiring extensive external information about the individuals, such as personal identifying information (i.e., their name or ID) and a baseline gait pattern. According to GDPR (https://eur-lex.europa.eu/eli/reg/2016/679/oj, accessed on 1 July 2022) legislation, data used for research purposes can be used if anonymized. Moreover, anonymized data does not conform to the rigors of personal data and can be processed without explicit consent. Nevertheless, *any attempt to use of DenseGait to explicitly identify individuals present in it is highly discouraged.*

We chose to utilize only skeleton sequences for gait analysis, as current appearance-based methods that rely on silhouettes are not privacy preserving, potentially allowing for identification based only on the person's appearance, rather than their movement [24]. Skeleton sequences encode only the movement of the person, abstracting away any visual queues regarding identity and attributes. Moreover, skeleton-based solutions have the potential to generalize across tasks such as action recognition, allowing for a flexible and extensible computation.

DenseGait, compared to other similar datasets [11], contains 10× more sequences and is automatically annotated with 42 appearance attributes through a pretrained PAI ensemble (Table 1). In total, 60 h of video streams were processed, having a cumulative walking duration of pedestrians of 410 h. We release the dataset under open credentialized access, for research purposes only, under CC-BY-NC-ND-4.0 (https://creativecommons.org/licenses/by-nc-nd/4.0/legalcode, accessed on 1 July 2022) License.

Table 1. List of attributes extracted by each network in the PAI ensemble. Each network is trained on a different dataset, with a separate set of attributes. After coalescing similar attributes and eliminating appearance-only attributes, we obtain 42 appearance attributes.

PA100k	PETA	RAP
Female, AgeOver60, Age18–60, AgeLess18, Front, Side, Back, Hat, Glasses, HandBag, ShoulderBag, Backpack, HoldObjectsInFront, ShortSleeve, LongSleeve, UpperStride, UpperLogo, UpperPlaid, UpperSplice, LowerStripe, LowerPattern, LongCoat, Trousers, Shorts, Skirt & Dress, Boots	Age16–30, Age31–45, Age46–60, AgeAbove61, Backpack, CarryingOther, Casual lower, Casual upper, Formal lower, Formal upper, Hat, Jacket, Jeans, LeatherShoes, Logo, LongHair, Male, Messenger Bag, Muffler, No accessory, No carrying, Plaid, PlasticBags, Sandals, Shoes, Shorts, Short Sleeve, Skirt, Sneaker, Stripes, Sunglasses, Trousers, TShirt, UpperOther, V-Neck	Female, AgeLess16, Age17–30, Age31–45, BodyFat, BodyNormal, BodyThin, Customer, Clerk, BaldHead, LongHair, BlackHair, Hat, Glasses, Muffler, Shirt, Sweater, Vest, TShirt, Cotton, Jacket, Suit-Up, Tight, ShortSleeve, LongTrousers, Skirt, ShortSkirt, Dress, Jeans, TightTrousers, LeatherShoes, SportShoes, Boots, ClothShoes, CasualShoes, Backpack, SSBag, HandBag, Box, PlasticBags, PaperBag, HandTrunk, OtherAttchment, Calling, Talking, Gathering, Holding, Pusing, Pulling, CarryingbyArm, CarryingbyHand

We also propose GaitFormer, a multi-task transformer-based architecture [25] that is pretrained on DenseGait in a self-supervised manner, being able to perform exceptionally well in zero-shot gait recognition scenarios on benchmark datasets, achieving 92.5% identification accuracy from direct transfer on the popular CASIA-B dataset, without using any manually annotated data. Moreover, it obtains good results on demographic and pedestrian attribute identification from walking patterns, with no manual annotations. GaitFormer represents the first use of a plain transformer encoder architecture in gait skeleton sequence processing, without relying on hand-crafted architectural modifications as in the case of graph neural networks [26,27].

This paper makes the following contributions:

1. We release DenseGait, the largest dataset of skeleton walking sequences, densely annotated with appearance information, for use in pretraining neural architectures that can be further fine-tuned on specific gait analysis tasks. The dataset can be found at https://bit.ly/3SLO8RW, under open credentialized access, for research purposes only.
2. We propose GaitFormer, a multi-task transformer that is pretrained on the DenseGait dataset and achieves exceptional results in zero-shot gait recognition scenarios on benchmark datasets, achieving 92.52% accuracy on CASIA-B and 85.33% on FVG, without training on any manually annotated data (+14.2% and +9.67% increase compared to similar methods [11]). The code is made publicly available at: https://github.com/cosmaadrian/gaitformer
3. We explore the performance of GaitFormer on other gait analysis tasks, such as gait-based gender estimation and attribute identification.

2. Related Work

2.1. Gait Analysis

Video gait analysis encompasses research efforts dedicated to automatically estimate and predict various aspects of a walking person. Research has been mostly dedicated into gait-based person recognition, with many benchmark datasets [8,11,12,14,28–31] available for training and testing models. Moreover, there have been improvements in areas such as estimating demographics information [19,32], emotion detection [33] and ethnicity estimation [34] from only movement patterns. Sepas-Moghaddam and Etemad [35] proposed a taxonomy to organize the existing works in the field of gait recognition. In this work, we focus mainly on body representation, as we made a deliberate choice of providing DenseGait with only movement information for anonymization. Broadly, works in gait analysis can be divided into two major approaches in terms of body representation: silhouette-based and skeleton-based.

2.1.1. Silhouette-Based Solutions

Silhouette-based approaches make use of silhouettes of walking individuals estimated either through background subtraction methods or through instance segmentation and tracking. Silhouettes are used in various forms, either in a condensed representation [36–38], or as a sequences, as it is the norm in more modern methods [8,39–41]. Most notably, GaitSet [39] processes the silhouettes as a set, as opposed to preserving the temporal information present in a sequence. As such, the authors can include silhouettes from multiple videos of the same walking subjects, achieving good invariance to walking variations. GaitPart [40] processes the temporal variation of each individual body part separately in a Micro-motion Capture Module (MCM), taking inspiration from model-based approaches. Each body part exhibits different visual queues and temporal variation and the authors propose to combine the each feature part to construct the final gait representation. Recently, Lin et al. [41], advance the construction of neural architectures for processing silhouette sequences by proposing a Global-Local Feature Extractor (GLFE), which obtains good results on benchmark datasets. Zhang et al. [8] propose GaitNet, a model which directly makes use of the appearance of the individual and is able to output invariant feature representations for gait recognition. Moreover, they also propose FVG, a dataset with 226 individuals, only from the front-view angle, one of the more challenging angles in gait analysis, due to the lack of perceived variation in limb movements.

2.1.2. Skeleton-Based Solutions

Skeleton-based approaches, on the other hand, avoid making use of appearance information in the form of silhouettes, and instead focus on the moving anatomical skeleton of the person, effectively processing only movement patterns. Approaches typically imply processing walking sequences with a pose estimation [42] model, and processing the resulting skeletons with a neural network, either by adapting conventional CNN modules [43], or with an LSTM [44,45]. More modern approaches make use of graph neural networks to model the relationships between human joints [46,47]. Liao et al. [44] make use of a combined CNN and LSTM architecture to model 2D skeleton sequences. A later improvement makes use of 3D skeletons [45] to further improve results. Li et al. [46] propose a graph-based convolutional architecture to process skeleton sequences, and a Joints Relationship Pyramid Mapping to map spatio-temporal gait features into a discriminative feature space. Li and Zhao [47] propose CycleGait, a graph-based approach that incorporates multiple walking paces in the augmentation procedure and obtains robust results in gait recognition on CASIA-B. In contrast to these approaches, we opted to take a data-driven approach, instead of an algorithmic approach, and use a standard transformer architecture and pretrain it on a large amount of weakly-labelled data. Recently, Cosma and Radoi [11] proposed an approach called WildGait to skeleton-based gait recognition, in which they automatically mine surveillance streams and pretrain a ST-GCN [26] model in a self-supervised manner. Through fine-tuning, good results are obtained in recognition on CASIA-B and FVG. Similarly to WildGait, we also process publicly available surveillance streams, but increase the DenseGait dataset size by an order of magnitude. Moreover, we densely annotate each skeleton sequence with 42 appearance attributes for use in zero-shot attribute identification scenarios.

However, model-based approaches still lag behind methods utilizing appearance (i.e., silhouettes). This is most likely due to the imperfect extraction of skeletons by modern pose estimators, which struggle to accurately detect fine-grained movements at a distance. Moreover, using appearance-based methods is fundamentally easier, since a single silhouette can contain identifying information about a subject. For instance Xu et al. [48] obtained reasonable results for gait recognition using a single silhouette, which cannot be considered gait, as no temporal movement is being processed at all. This implies that recognition is performed through "shortcuts" in the form of appearance features (i.e., body composition, height, haircut, side-profile etc). For this reason, a more privacy-aware approach is to process only movement patterns, which constitutes the motivation for releasing DenseGait with only anonymized skeleton sequences, and disregarding silhouettes.

2.2. Transformers and Self-Supervised Learning

In recent years, there has been a insurgence of research in the area of self-supervised learning, mostly due to the extremely high performance obtained in natural language processing with models such as BERT [49] and GPT [50]. Self-supervised learning presumes training models using aspects of the data itself as a supervisory signal. While initial efforts in computer vision relied on creating artificial pretext tasks [51–53], the field is moving towards contrastive-based approaches [29,54,55]. Methods such as SimCLR [29], Barlow Twins [54] and Dino [55] obtaining almost similar performance to direct supervision. Moreover, the transformer has proven to be a flexible architecture, capable of handling a multitude of modalities such as text [49], images [56], video [57], speech [58], and highly benefit from large-scale pretraining [59]. Taking inspiration from related efforts to process non-textual data with transformers [56], we construct GaitFormer by processing flattened skeletons as input "tokens". In this manner, any human bias related to hand-crafted graph relationships between the body joints is eliminated. Moreover, as opposed to graph networks such as ST-GCN [27], training a similarly large transformer encoder make more efficient use of computational resources, significantly reducing training time.

3. Method

3.1. Dataset Construction

For building the DenseGait dataset, we made use of public video streams (e.g., street cams), and processed them with AlphaPose [42], a modern, state-of-the-art multi-person pose estimation model. AlphaPose's raw output is comprised of skeletons with (x, y, c) coordinates for each of the 18 joints of the COCO skeleton format Lin et al. [60], corresponding to 2D coordinates in the image plane and a prediction confidence score. We performed intra-camera tracking for each skeleton with on SortOH [61]. SortOH is based on the SORT [62] algorithm, which relies only on coordinate information and not on appearance information. As opposed to DeepSORT [63] which makes use of person re-identification models, SortOH is only using coordinates and bounding box size for faster computation time while having comparably similar performance. SortOH ensures that tracking is not significantly affected by occlusions.

To ensure that the skeleton sequences can be properly processed by a deep learning model, we performed extensive data cleaning. We have filtered low confidence skeletons by computing the average confidence of each of the 18 joints, and in each sequence, skeletons with an average confidence of less than 0.5 were removed. Furthermore, skeletons with feet confidence less than 0.4 were removed. This step guarantees that the feet are visible and confidently detected—leg movement is one of the most important signals for gait analysis. In our processing, we chose a period length T of 48 frames, which corresponds to approximately 2 full gait cycles on average [64]. Surveillance streams do not have the same frame rate between them, which makes the sequences have different paces and durations. As such, we filtered short tracklets which have a duration of less than $\frac{T*fps}{24}$. We consider 24 FPS to be real-time video speed, and each video was processed according to its own frame rate. Moreover, skeletons are linearly interpolated such that the pace and duration is unified across video streams.

Similar to [11], we further normalized each skeleton by centering at the pelvis coordinates (x_{pelvis}, y_{pelvis}) and scaling vertically by the distance between the head and the hips $(y_{neck} - y_{pelvis})$ and horizontally by the distance between the shoulders $((x_{R.shoulder} - x_{L.shoulder}))$. This procedure is detailed in Equations (1) and (2). The normalization procedure aligns the skeleton sequences in a similar manner to the alignment step in face recognition pipelines [65]. This step eliminated the height and body type information about the subject, ensuring that the person cannot be directly identified.

$$x_{joint} = \frac{x_{joint} - x_{pelvis}}{|x_{R.shoulder} - x_{L.shoulder}|} \tag{1}$$

$$y_{joint} = \frac{y_{joint} - y_{pelvis}}{|y_{neck} - y_{pelvis}|} \tag{2}$$

However, body type information should be preserved through the analysis of the walking patterns. Moreover, normalization obscures the human position in the frame, to prevent identification of the source video stream.

Finally, we filtered standing/non-walking skeletons in each sequence by computing the average movement speed of the legs, which is indicative of the action the person is performing. As such, if the average leg speed is less than 0.0015 and higher than 0.09, the sequence was removed. The thresholds were determined through manual inspection of the sequences. This eliminated both standing skeleton sequences as well as sequences with erratic leg movement, which is most probably due to poor pose estimation output in that case.

DenseGait is fully anonymized. Any information regarding the identity of particular individuals in the dataset is eliminated, including appearance information (by keeping only movement information in the form of skeleton sequences), height and body proportions (through normalization), and the time, location, and source of the video stream. Identifying individuals in DenseGait is highly unfeasible, as it requires external information (i.e., name, email, ID, etc.) and specific collection of gait patterns.

The final dataset contains 217 K anonymized tracklets, with a combined length of 410 h. DenseGait is currently the largest dataset of skeleton sequences for use in pretraining gait analysis models. Table 2 showcases a comparison between DenseGait and other popular gait recognition datasets. Since the skeleton sequences are collected automatically through pose tracking, it is impossible to quantify exactly the number of different identities in the dataset, as, in some cases, tracking might be lost due to occlusions. However, DenseGait contains a significantly larger number of tracklets compared to other available datasets while also being automatically densely annotated with 42 appearance attributes. In the case of UWG [11] and DenseGait, the datasets do not contain explicit covariates for each identity, but rather covariates in terms of viewing angle, carrying conditions, clothing change, and apparent action are present across the tracklet duration, similar to GREW [12].

Similarly to UWG [11], DenseGait does not contain multiple walks per person, rather each tracklet is considered a unique identity. Compared to other large-scale datasets, DenseGait tracks individuals for a longer duration, which makes it suitable for use in self-supervised pretraining, as longer tracked walking usually contains more variability for a single person. Figure 1 shows boxplots with a five-number summary descriptive statistics for the distribution of track durations in each dataset. DenseGait has a mean tracklet duration of 162 frames, which is significantly larger (z-test $p < 0.0001$) compared to other datasets: CASIA-B [15]—83 frames, FVG [8]—97 frames, GREW [12]—98 frames, UWG [11]—136 frames). Due to potential loss of tracking information, the dataset is noisy, and can be used only for self-supervised pretraining.

3.2. *Annotations with Appearance Attributes*

Appearance attributes are essential for pretraining for tasks such as gender estimation [19], age estimation [66] and pedestrian attribute identification [16–18]. To ensure that the dataset is densely annotated with appearance attributes, we made use of an ensemble of pretrained PAI networks, each trained on different popular PAI datasets. Specifically, we employed three InceptionV3 [67] networks trained on RAP [68], PETA [69] and PA100k [16], respectively. Figure 2 showcases the annotation procedure.

Table 2. Comparison of popular datasets for gait recognition. DenseGait is an order of magnitude larger, has more identities in terms of skeleton sequences (highlighted in **bold**), and each sequence is annotated with 42 appearance attributes. * Approximate number given by pose tracker. † Implicit covariates across tracking duration.

Dataset	# IDs	Sequences	Covariates	Views	Env.
USF HumanID [31]	122	1870	Y	2	Outdoor
TUM-GAID [28]	305	3370	Y	1	Outdoor
FVG [8]	226	2857	Y	1	Outdoor
CASIA-B [15]	124	13,640	Y	11	Indoor
OU-ISIR [14]	10,307	144,298	N	14	Indoor
GREW [12]	26,000	128,000	Y	-	Outdoor
UWG [11]	38,502 *	38,502	Y †	-	Outdoor
DenseGait (**ours**)	**217,954** *	**217,954**	Y †	-	Outdoor

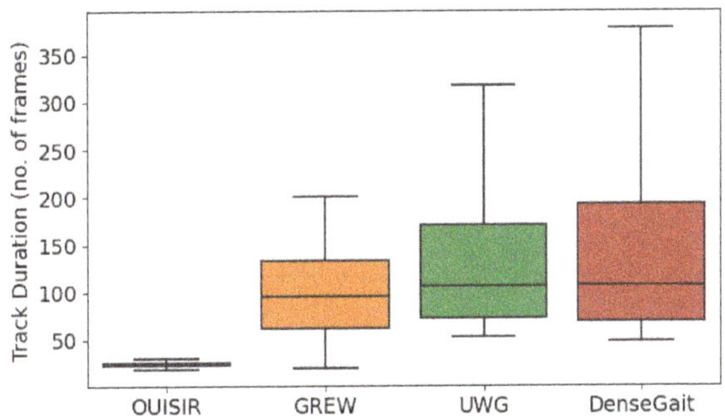

Figure 1. Comparison between existing large-scale skeleton gait databases and DenseGait in terms of distributions of tracklet duration. DenseGait is an order of magnitude larger than the next largest skeleton database, while having a longer average duration (136 frames UWG vs 162 frames DenseGait).

Since each dataset has a different set of pedestrian attributes, we averaged similar classes (e.g., *AgeLess16* and *AgeLess18* into *AgeChild*), coalesced similar classes (e.g., *Formal* and *Suit-Up* into *FormalWear*) and removed attributes that cannot evidently be estimated from movement patterns (e.g., *BaldHead, Hat, V-Neck, Glasses, Plaid* etc.).

For a particular sequence, we take the cropped image of the pedestrian at every T frames (where T is the period length), and randomly augment it $k = 4$ times (e.g., random horizontal flips, color jitter and small random rotation). For each crop, each augmented version is then processed by a PAI network and the results are averaged such that the output is robust to noise [70]. Finally, to have a unified prediction for the walking sequence, results are averaged according to the size of the bounding box relative to the image, similar to Catruna et al. [19]. Predictions on larger crops have a higher weight, with the assumption that the pedestrian appearance is more clearly distinguishable when closer to the camera.

Figure 3 showcases the final list of attributes, and their distribution across the dataset. We have a total of 42 attributes, split into 8 groups: *Gender, Age Group, Body Type, Viewpoint, Carry Conditions, Clothing, Footwear* and *Apparent Action*. For the final annotations, we chose to keep the soft-labels and not round them, as utilizing soft-labels for model training was shown to be a more robust approach when dealing with noisy data [70].

Figure 4 showcases selected examples of attribute predictions from the PAI ensemble. Since surveillance cameras usually have low resolution and the subject might be far away from the camera, some pedestrian crops are blurry and might affect prediction by the PAI ensemble. For gender, age group, body composition and viewpoint, the models are confidently identifying these attributes. However, for specific pieces of clothing (i.e., footwear: Sandals/LeatherShoes), predictions are not always reliable, due to the low resolution of some of the crops, but the errors are negligible when taking into account the scale of the dataset.

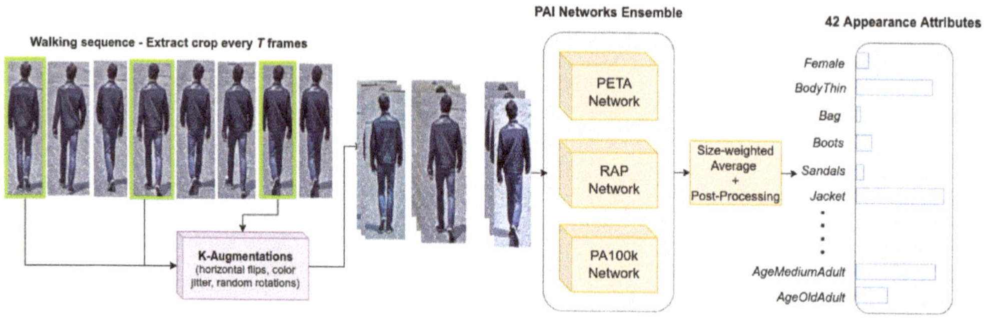

Figure 2. Overview of the automatic annotation procedure for the 42 appearance attributes. To robustly annotate attributes, an ensemble of pretrained networks is used in conjunction with multiple augmentations of the same crop. Predictions across the sequence are averaged according to their bounding-box area.

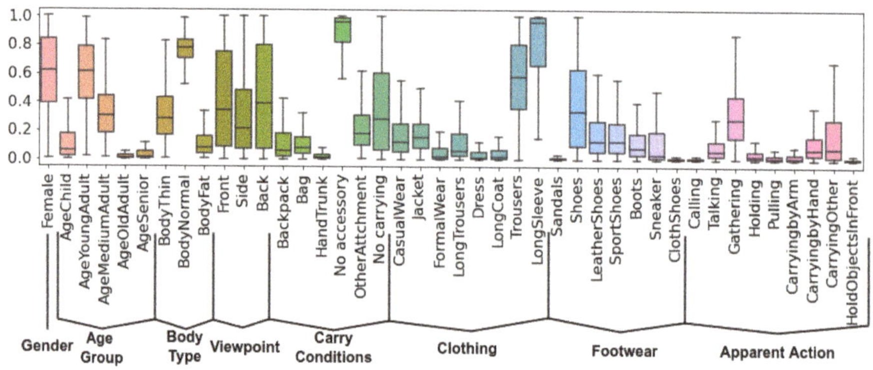

Figure 3. Distribution of the 42 appearance attributes in DenseGait. The dataset is annotated in a fine-grained manner with attributes ranging from internal aspects of the person (Gender, Age Group, Body Type) to appearance only labels (Clothing, Footwear).

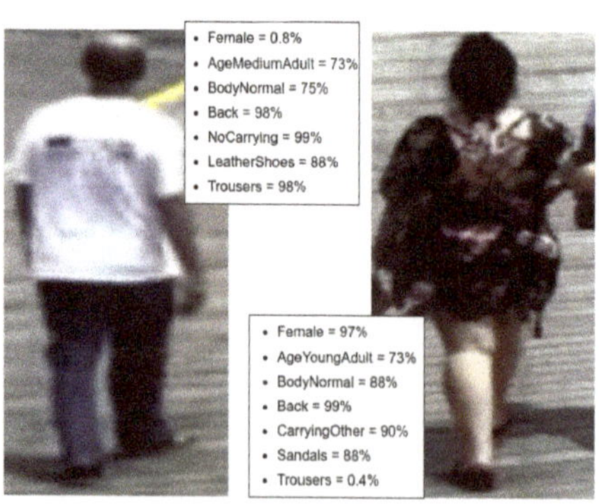

Figure 4. Qualitative examples for selected attributes from the PAI ensemble. The networks correctly identify gender, age group and viewpoint. However, in some cases, clothing and, more specifically, footwear are more difficult to estimate in low resolution scenarios.

3.3. Description of Model Architecture

For pretraining on the DenseGait dataset for the tasks of gait-based recognition and attribute identification, we chose to adapt the popular transformer encoder architecture [25] to handle skeleton sequences. Initially, transformers were immensely successful in handling sequential data in the form of text, effectively replacing LSTM [71] networks, the de facto approach for these problems. However, lately, transformers have been used in a variety of problems, being able to handle images [56], video [57] and multi-modal data [72]. Moreover, transformer architectures in particular highly benefit from large-scale, self-supervised pretraining [49,50,55], allowing models to be effectively fine-tuned on more specific datasets with small amounts of annotated data.

To handle skeleton sequences, we abstain from making any hand-crafted architectural modifications, as in the case of Plizzari et al. [27], which uses a hybrid approach by combining graph computation on the skeleton and using multi-head attention on the extracted features. Instead, we take inspiration from ViT [56], which processes images as a sequence of flattened patches that are fed into a standard transformer encoder network. Figure 5 showcases the training procedure for GaitFormer in the multi-task training regime. Each skeleton is flattened into a 54 dimensional vector and is linearly projected with a standard learnable feed-forward layer into a 256 dimensional space. Each skeleton projection is then fed into a transformer encoder network. We opted for learnable positional embedding that is added to each projection instead of concatenated, to avoid increasing the dimensionality. After the transformer encoder, representations for each skeleton are averaged, and a final linear feed-forward layer of 256 elements is used as the final embedding. Further, as described in SimCLR [29], we used an additional 128-dimension linear layer for training with a supervised contrastive objective [73]. Additionally, a linear layer is used as appearance head to estimate the pedestrian attributes that is trained using a standard binary-crossentropy loss.

We used three different model sizes for the transformer encoder in our experiments, with 4 encoder layers (*SM*), 8 encoder layers (*MD*) and 12 encoder layers (*XL*). In all types of architectures, 8 attention heads were used, and the internal feed-forward dimensionality was 256 [25].

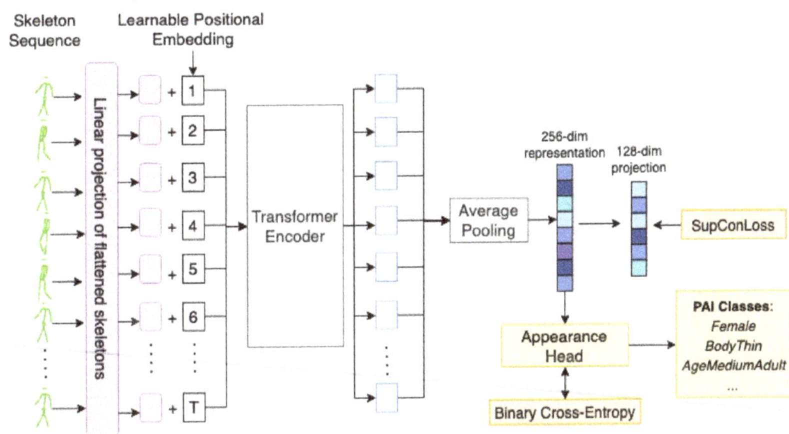

Figure 5. Overview of GaitFormer (Multi-Task) training procedure. Flattened skeletons are linearly projected using a standard feed-forward layer and fed into a transformer encoder. The vectorized representations are average pooled and the resulting 256-dimensional vector is used for estimating the identity and to estimate the 42 appearance attributes through the "Appearance Head". The contrastive objective (SupConLoss) is applied to a lower 128-dimensional linear projection, similar to the approach in SimCLR [29].

3.4. Training Details

For training on DenseGait, we chose to use contrastive learning [29] as a supervisory signal. By design, contrastive methods work by attracting representations belonging to the same class, while simultaneously repelling samples from different classes. This paradigm is identical to the objective for recognition problems, which constitues one of the main tasks in gait analysis. Specifically, we used SupConLoss [73], with a temperature of $\tau = 0.001$, alongside a two-view sampler for each skeleton in the batch. SupConLoss assumes a multi-viewed batch, with multiple augmentations for the same sample. Each view of a skeleton sequence is randomly augmented by the standard suite of augmentations for this data modality: random sequence crops of fixed length of $T = 48$, random flips with 50% probability, random paces [53], and random gaussian noise added to joints coordinates. Let $i \in I \equiv \{1 \ldots 2N\}$ be the index of an arbitrary augmented sample. SupConLoss is defined as:

$$\mathcal{L}^{sup} = \sum_{i \in I} \frac{-1}{|P(i)|} \sum_{p \in P(i)} \log \frac{\exp(z_i \cdot z_p / \tau)}{\sum_{a \in A(i)} \exp(z_i \cdot z_a / \tau)} \tag{3}$$

In Equation (3), $z_l = Enc(\tilde{x}_l)$ denotes the embedding of a skeleton sequence x_l, "$\cdot$" denotes the dot product operation and $A(i) \equiv I \setminus \{i\}$. Moreover, $P(i) \equiv \{p \in A(i) : \tilde{y}_p = \tilde{y}_i\}$ is the set of indices of all positives in the multi-viewed batch distinct from i. In our case, the positive pairs are constructed by two different augmentations of the same skeleton sequence. The variability of the two augmentations is higher if the skeleton is tracked for a longer duration of time, as the walking individual might change direction.

As suggested in Chen et al. [29], the supervisory signal given by SupConLoss is applied to a lower dimensional embedding (128 dimensions) to avoid the curse of dimensionality.

For predicting appearance attributes, which is a multi-label problem, we used a standard binary-crossentropy loss between each appearance label (p_i) and its corresponding prediction (y_i) (Equation (4)). As previously mentioned, we keep the soft labels as a supervisory signal, to prevent the network from overfitting and be more robust to noisy or incorrect labels [74]. Moreover, since learning appearance labels can regarded as a knowledge distillation problem between the PAI ensemble and the transformer network, soft labels help improve the distillation process [75].

$$\mathcal{L}_{appearance} = -(y_i \log(p_i) + (1 - y_i) \log(1 - p_i))) \tag{4}$$

In multi-task (MT) training scenarios, we used a combination of the two losses, with a weight penalty of $\lambda = 0.5$ on the appearance loss $\mathcal{L}_{appearance}$. We chose $\lambda = 0.5$ empirically, such that the two losses have similar magnitudes. The final loss function is defined as:

$$\mathcal{L}_{final} = \mathcal{L}_{SupCon} + \lambda \mathcal{L}_{appearance} \tag{5}$$

In plain contrastive training scenarios, we employ only the SupConLoss, without predicting attributes (i.e., $\mathcal{L}_{final} = \mathcal{L}_{SupCon}$).

The motivation for pre-training the network in a multi-task setting is that the network not only learns to cluster walking sequences by their identity, but also to take appearance attributes into account. For instance, predicting the gender and age, even if they are not completely reliable, could prove useful for gait recognition, as demographics can be considered soft-biometrics, allowing the network to automatically filter identities by these attributes. On the other hand, in contrastive-only scenario, the network is under a classical self-supervised regime.

We used a batch size of 1024 across our experiments, with a cyclical learning rate [76] ranging from 0.0001 and 0.001 across 20 epochs. We trained all models for 400 epochs.

4. Experiments and Results

This section explores the performance of GaitFormer on gait-based recognition, gender identification and pedestrian attribute identification. We are primarily interested in evaluating the model in scenarios with low amounts of annotated data and we opted to use the two popular benchmark datasets originally constructed for gait recognition: CASIA-B [15] and FVG [8]. For gender estimation, we manually annotated the gender information for each identity in the two datasets and constructed CASIA-gender and FVG-gender. We briefly describe each dataset below.

We chose CASIA-B to compare with other skeleton-based gait recognition models, since it is one of the most popular gait recognition datasets in literature. It contains 124 subjects walking indoors in a straight line, captured with 11 synchronized cameras with three walking variations—normal walking (NM), clothing change (CL) and carry conditions (BG). According to Yu et al. [15], the first 62 subjects are used for training and the rest for evaluation. CASIA-gender consists of manually annotated the subjects in CASIA-B with gender information, having a split of 92 males and 32 females. We maintain the training and validation splits from the recognition task, using the first 62 subjects for training (44 males and 18 females) and the rest for validation (48 males and 14 females). We use FVG to evaluate the robustness of GaitFormer, as it contains different covariates than CASIA-B such as varying degrees of walking speed, the passage of time and cluttered background. Moreover, FVG only contains walks from the front-view angle, which is more difficult for gait processing due to lower perceived limb variation. According to Zhang et al. [8], from the 226 identities present in FVG, the first 136 are used for training and the rest for testing. Similarly, FVG-gender contains manual annotations with gender information, obtaining 149 males and 77 females. We maintain the training and validation splits from the recognition task, utilizing the first 136 individuals for training (83 males and 53 females) and the rest for validation (66 males and 24 females).

4.1. Recognition

We initially trained GaitFormer under two regimes: (i) contrastive only and (ii) multi-task (MT), which implies training with SupConLoss [73] on the tracklet ID while simultaneously estimating the appearance attributes (Figure 5). We experiment with three models sizes: SM—4 encoder layers (2.24M parameters), MD—8 encoder layers (4.35M parameters) and XL—12 encoder layers (6.46M parameters).

We pretrain GaitFormer on the DenseGait dataset under the mentioned conditions and directly evaluate recognition performance in terms of accuracy on CASIA-B and FVG, without fine-tuning. In all experiments we perform a deterministic crop in the middle of the skeleton sequences of T = 48 frames, and use no test-time augmentations. For each cropped skeleton sequence, features are extracted using the 256-dimensional representation and are normalized with the l_2 norm. In Table 3 we present results on the walking variations for each model size and training regime. For CASIA-B, we show mean accuracy where the gallery set contains all viewpoints except the probe angle, in the three evaluation scenarios: normal walking (NM), change in clothing (CL) and carry bag (CB). For FVG, we show accuracy results based on the evaluation protocols mentioned by Zhang et al. [8], corresponding to different walking scenarios (walk speed (WS), change in clothing (CL), carrying bag (CB), cluttered background (CBG) and ALL). Results show that unsupervised pretraining on DenseGait is a viable way to perform gait recognition, achieving an accuracy of 92.52% on CASIA-B and 85.33% on FVG, without any manually annotated data available. Notably, multi-task learning on appearance attributes provides a consistent positive gap in the downstream performance.

Model size in terms of number of layers does not seem to considerably affect performance on benchmark datasets. GaitFormerMD (8 layers) fairs consistently better than GaitFormerXL (12 layers), while being similarly close to GaitFormerSM (4 layers).

Figure 6 compares GaitFormerMD pretrained on DenseGait in the two training regimes (contrastive only—Cont. and Multi-Task—MT) and GaitFormerMD randomly initialized. The networks were fine-tuned on progressively larger samples of the corresponding datasets: for CASIA-B, we sampled multiple runs for the same identity (from 1 to 10 runs per ID), and for FVG, we randomly sampled a percentage of runs per each identity. Models were fine-tuned using Layer-wise Learning Rate Decay (LLRD) [77], which implies a higher learning rate for top layers and a progressively lower learning rate for bottom layers. The learning rate was decreased linearly from 0.0001 to 0, across 200 epochs. The results show that unsupervised pretraining has a substantial effect on downstream performance especially in low data scenarios (direct transfer and 10% of available data). Moreover, pretraining the model in Multi-Task learning regime, in which the network was tasked to estimate appearance attributes from movement alongside with the identity, provides a consistent increase in performance.

Table 3. GaitFormer direct transfer performance on gait recognition on CASIA-B and FVG datasets. We highlight in **bold** the best overall result for each dataset.

		CASIA-B			FVG			
Size	**Training**	**NM**	**CL**	**CB**	**WS**	**CB**	**CL**	**CBG**	**ALL**
SM	Contrastive	89.00	22.36	61.88	77.33	81.82	54.27	86.75	77.33
MD	Contrastive	90.18	23.46	60.78	78.33	72.73	49.15	83.33	78.33
XL	Contrastive	91.79	21.11	63.12	76.33	69.70	48.29	87.61	76.33
SM	MT	92.52	22.73	**67.16**	84.67	81.82	**59.40**	**91.45**	84.67
MD	MT	**92.52**	**23.31**	65.10	**85.33**	**87.88**	53.42	88.89	**85.33**
XL	MT	90.69	20.75	60.34	85.00	81.82	51.71	91.03	85.00

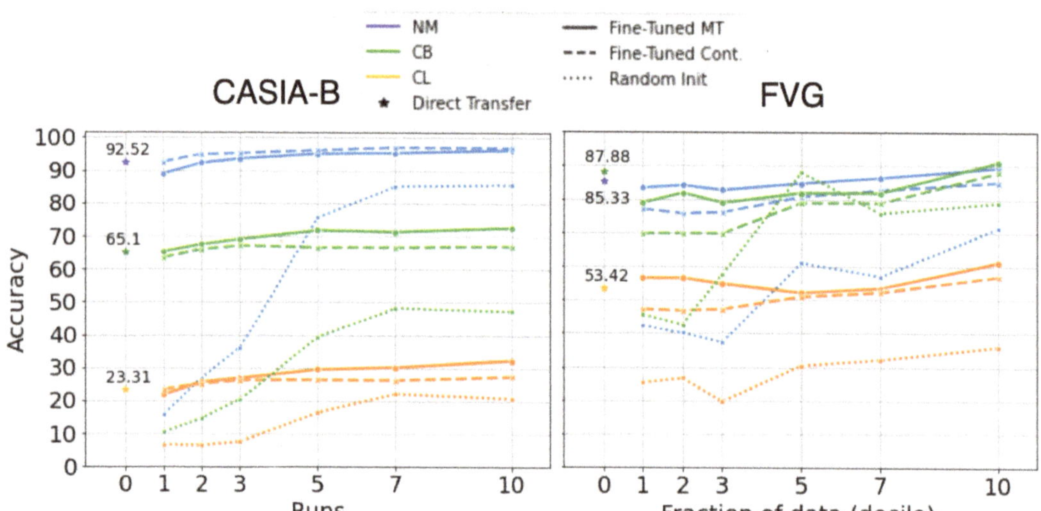

Figure 6. Fine-tuning results on gait recognition on CASIA-B and FVG, on progressively larger number of runs per identity. Compared to the same network randomly initialized, pretraining on DenseGait offers substantial improvements, even in the direct transfer regime. A consistent performance increase is obtained when also estimating attributes.

Table 4 presents state-of-the-art results compared with other skeleton-based gait recognition models. We showcase the results of GaitFormerSM trained in the Multi-Task (MT) regime, without fine-tuning (direct) and tuned with all the available training data in CASIA-B. For comparison, we include WildGait [11] with and without fine-tuning, as this model is also pretrained on a large dataset of skeleton sequences. We also compare with our implementation of GaitGraph Teepe et al. [78]—a multi-branch ST-GCN which processes joint coordinates, velocities and bone angles, achieving great results on CASIA-B—and with a ST-GCN pretrained on DenseGait.

It is clear that the fine-tuned GaitFormerSM has very good results even without fine-tuning, achieving comparable results with the state of the art. Fine-tuning marginally increases the performance, achieving 96.2% accuracy on normal walking (NM) and 72.5% performance in carry bag (CB).

Table 4. GaitFormer comparison to other skeleton-based gait recognition methods on CASIA-B dataset. In all methods the gallery set contains all viewpoints except the proble angle. In **bold** and underline we highlight the best and second best results for a particular viewpoint and walking condition.

	Method	**0°**	**18°**	**36°**	**54°**	**72°**	**90°**	**108°**	**126°**	**144°**	**162°**	**180°**	**Mean**
	GaitGraph	79.8	89.5	91.1	92.7	87.9	89.5	94.35	95.1	92.7	93.5	80.6	89.7
	ST-GCN (DenseGait)	89.5	89.5	95.1	87.9	81.4	68.5	64.5	89.5	88.7	84.6	82.2	83.8
	WildGait—direct	72.6	84.6	90.3	83.8	63.7	62.9	66.1	83.0	86.3	84.6	83.0	78.3
NM	PoseFrame	66.9	90.3	91.1	55.6	89.5	**97.6**	**98.4**	97.6	89.5	69.4	68.5	83.1
	GaitFormerSM—MT—direct	94.3	97.5	99.2	98.4	79.8	80.6	89.5	**100.0**	94.3	95.1	88.7	92.5
	WildGait—tuned	86.3	96.0	97.6	94.3	**92.7**	94.3	94.3	98.4	97.6	91.1	83.8	93.4
	GaitFormerSM—MT—tuned	**96.7**	**99.2**	**100.0**	**99.2**	91.9	91.9	95.1	98.4	**96.7**	**97.6**	**91.1**	**96.2**
	GaitGraph	27.4	33.0	40.3	37.1	33.8	33.0	35.4	33.8	34.6	21.7	17.7	31.6
	ST-GCN (DenseGait)	18.5	22.5	25.0	21.7	13.7	18.5	21.7	31.4	21.7	21.7	16.9	21.2
	WildGait—direct	12.1	33.0	25.8	18.5	12.9	11.3	21.7	24.2	20.1	26.6	16.1	20.2
CL	PoseFrame	13.7	29.0	20.2	19.4	28.2	**53.2**	**57.3**	**52.4**	25.8	26.6	21.0	31.5
	GaitFormerSM/MT—direct	12.9	21.7	29.0	25.8	16.1	18.5	22.5	29.0	27.4	26.6	20.1	22.7
	WildGait—tuned	29.0	32.2	**35.5**	**40.3**	26.6	25.0	38.7	38.7	31.4	34.6	**31.4**	**33.0**
	GaitFormerSM/MT—tuned	**35.5**	**35.5**	33.8	33.8	20.9	30.6	31.4	31.4	28.2	**42.7**	29.8	32.2
	GaitGraph	64.5	69.3	70.1	62.9	61.2	58.8	59.6	58.0	57.2	55.6	45.9	60.3
	ST-GCN (DenseGait)	78.2	68.5	71.7	60.4	59.6	45.9	46.7	58.0	58.0	58.0	51.6	59.7
	WildGait—direct	67.7	60.5	63.7	51.6	47.6	39.5	41.1	50.0	52.4	51.6	42.7	51.7
BG	PoseFrame	45.2	66.1	60.5	42.7	58.1	**84.7**	**79.8**	**82.3**	65.3	54.0	50.0	62.6
	GaitFormerSM/MT—direct	78.2	71.7	**84.7**	**74.2**	56.4	50.0	57.2	66.1	69.3	70.9	59.6	67.1
	WildGait—fine-tuned	66.1	70.1	72.6	65.3	56.4	64.5	65.3	67.7	57.2	66.1	52.4	64.0
	GaitFormerSM/MT—tuned	**82.2**	**80.6**	83.8	72.6	**62.9**	69.3	68.5	70.1	**69.3**	**77.4**	**60.4**	**72.5**

4.2. Comparison with ST-GCN and Other Pretraining Datasets

In Table 5, we compare GaitFormer with ST-GCN [26] under different pretraining datasets. Reported results are mean accuracy across all angles for CASIA-B, under normal walking (NM) scenario, and accuracy under ALL scenario for FVG. The networks were not fine-tuned on these datasets; we present direct transfer performance after pretraining. We chose to pretrain on OU-ISIR [13], as this dataset is one of the most popular, large-scale datasets for gait recognition. However, OU-ISIR lacks data diversity, as all individuals are walking on a treadmill for a short duration, which is not the case for DenseGait. We also chose to pretrain on GREW [12], as it is also a diverse dataset collected in the wild, but contains fewer identities that walk for a comparably shorter duration of time.

Results show that, as a pretraining dataset, DenseGait is consistently outperforming GREW and OU-ISIR across the two architectures. These results are consistent with the insights in Figure 1, in which we posit that longer tracking duration for the individuals imply larger data diversity when pretraining in a contrastive self-supervised fashion, which directly improves performance.

4.3. Gait-Based Gender Detection

Table 6 presents results for direct transfer (zero-shot) performance for gender estimation on CASIA-gender and FVG-gender. In this case, we compared different sizes of GaitFormer trained on DenseGait in two manners: i) only estimating attributes, without a constrastive objective (Attributes Only), and ii) estimating attributes and identity using a constrastive objective (MT). Similarly to the case of gait recognition, the Multi-Task networks consistently outperforms the other training regime. Moreover, the networks achieved reasonable performance in terms of F_1 score (76.18% for CASIA-gender and

86.81% for FVG-gender), considering that the networks were not exposed to any manually annotated data.

Table 5. Comparison between GaitFormer and ST-GCN pretrained with Supervised Contrastive on GREW [12], OU-ISIR [13] and our proposed DenseGait. Performance is directly correlated with mean tracklet duration on each dataset as shown in Figure 1. We highlight in **bold** the best results for each architecture and dataset.

Backbone	Pretraining Data	CASIA-B (NM)	FVG (ALL)
ST-GCN	OU-ISIR	55.65	63.33
	GREW	61.14	56.67
	DenseGait **(ours)**	**83.80**	**75.28**
GaitFormer **(ours)**	OU-ISIR	25.73	51.34
	GREW	65.40	64.04
	DenseGait **(ours)**	**89.0**	**77.33**

Table 6. GaitFormer direct transfer performance on gait-based gender estimation on CASIA-gender and FVG-gender. We highlight in **bold** the best overall results for each dataset.

		CASIA-Gender			FVG-Gender		
Size	Training	Prec.	Recall	F_1	Prec.	Recall	F_1
SM	Attributes	94.54	62.46	72.10	85.87	84.9	85.10
MD	Attributes	94.48	63.66	73.07	82.03	79.87	80.27
XL	Attributes	94.59	62.43	72.08	84.36	84.43	84.26
SM	MT	**94.84**	63.61	73.00	86.47	86.34	86.33
MD	MT	94.71	61.50	71.31	86.96	**86.87**	**86.81**
XL	MT	94.72	**67.67**	**76.18**	**87.06**	86.21	86.38

Figure 7 presents the performance under fine-tuning of GaitFormerXL on CASIA-gender and FVG-gender, in similar conditions to the recognition task. All networks are trained with a binary-crossentropy objective on the gender estimation task, without taking the person identity into account at training time. GaitFormerXL under Multi-Task training regime is consistently superior to a network initialized from random weights, achieving an F_1 score of 93.09% on CASIA-gender and of 91.51% on FVG-gender. The pretrained models significantly benefit from fine-tuning when small amounts of training data is available. Performance slightly increases with the availability of more training data.

4.4. Gait-Based Pedestrian Attribute Identification

For pedestrian attribute identification, we process a 10-h surveillance stream, corresponding to 10,733 tracklets, and use it for testing. For evaluation, we use the attribute pseudo-labels annotated automatically by the PAI ensemble. Figure 8 showcases R^2 score results for GaitFormerMD trained with a multi-task objective. This score is computed relative to the soft pseudo-labels estimated by the PAI ensemble. We emphasize that the model only uses movement information to estimate these labels, and has no information regarding appearance. Using a skeleton-based model for pedestrian attribute identification is useful in situations where the appearance of the person is unavailable (i.e., in privacy-critical scenarios). The model is effectively distilling external appearance into movement representations.

The model obtains good results in categories such as *Gender*, *AgeGroup*, *BodyType* and *Viewpoint*. The model is able to obtain better than average performance on categories such as *Footwear*, and some types of clothing. However, some clothing categories have proven to be very difficult to model, especially *LongCoat* and *Trousers*. We hypothesize that such pieces of clothing negatively affect the accuracy of the pose estimation model, resulting in low quality extracted skeletons.

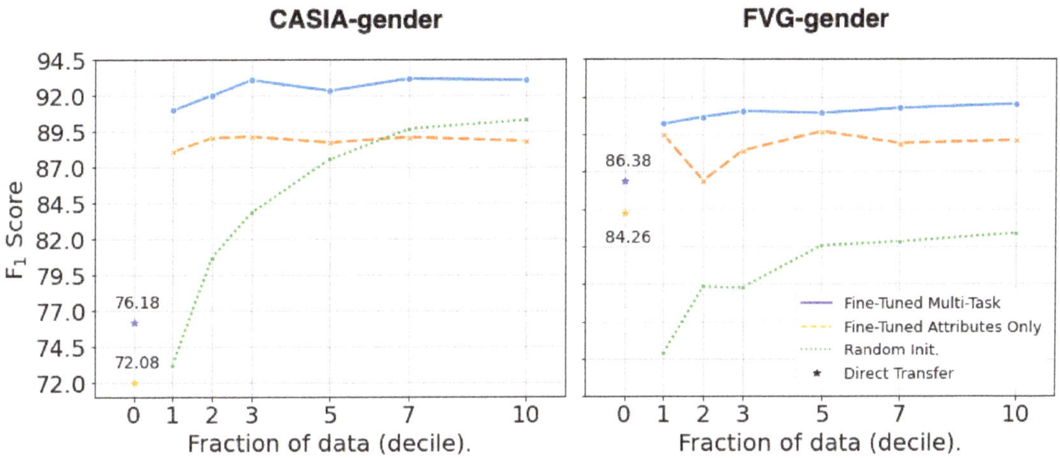

Figure 7. Fine-tuning results for GaitFormer on CASIA-gender and FVG-gender, trained on progressively larger samples of the datasets. Compared to a randomly initialized network, GaitFormer benefits significantly from fine-tuning in extremely low data regimes (e.g., 10% of available annotated data). Compared to only pretraining on predicting attributes (Attributes Only), the Multi-Task network has consistently better performance across all fractions of the datasets.

These are promising results which show that external appearance and movement are intrinsically linked together. This is evident in the more explicit relationship between, for example, footwear and gait, in which, intuitively, gait is severely affected by the walker's choice of shoes. Clothing, accessories, and actions while walking can be regarded as "distractor" attributes, which affect gait only temporarily. However, there are more subtle information cues which are present in gait, related to the developmental aspects of the person (e.g., gender, age, body composition, mental state etc). These attributes are more stable in time, and can provide insights into the internal workings of the walker. We posit that, in the future, works in gait analysis will tackle more rigorously the problem of estimating the internal state of the walker (i.e., personality/mental issues) through specialized datasets and methods.

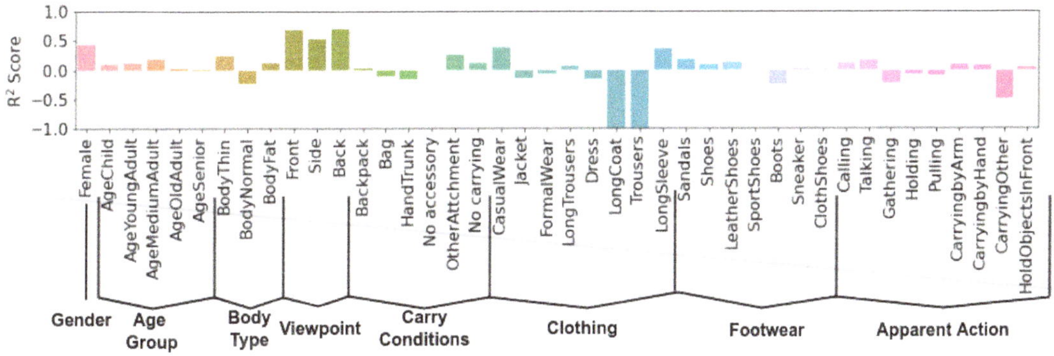

Figure 8. GaitFormerMD performance in terms of R^2 score. GaitFormerMD was trained with the multi-task objective. The model uses only movement information to predict attributes, and no information regarding the appearance of the individual.

4.5. *Inference Time*

Using transformer architectures for processing gait has other advantages besides a noticeable increase in downstream performance. Transformers have been shown to be more efficient in terms of inference time when compared to convolutional networks [56]. This effect is not directly correlated with the number of parameters, but is rather more influenced by the network structure [79].

In Figure 9, we show a comparison between multiple sizes of GaitFormer, a plain transformer module minimally adapted for processing skeleton sequences, with the ST-GCN network, a popular architecture for skeleton action recognition [26] and gait analysis [46]. We computed the inference time across multiple period lengths (from 12 frames to 96 frames) to evaluate the scalability when processing shorter/longer sequences. For each period length, we run 100 experiments with a batch size of 512 and show the mean inference time in seconds, along with the standard deviation. All experiments were run on a NVIDIA RTX 3060 GPU. Even with comparable and exceeding number of parameters (ST-GCN from Cosma and Radoi [11] has 3.11M parameters), the transformer architecture clearly outperforms graph-convolutional models for processing gait sequences across multiple sequence lengths.

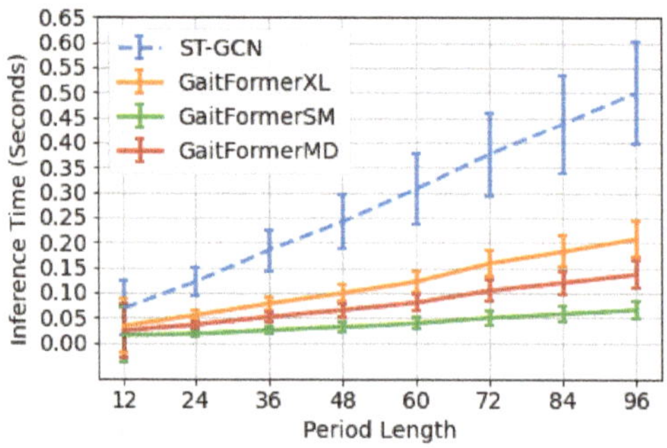

Figure 9. Inference times across processed walking duration length (period length) for ST-GCN and the various sizes of GaitFormer. We report the mean and stardard deviation across 100 runs, for each period length.

5. Conclusions

In this work, we presented DenseGait, currently the largest dataset for pretraining gait analysis models, consisting of 217 K anonymized skeleton sequences. Each skeleton sequence is automatically annotated with 42 appearance attributes by making use of an ensemble of pretrained PAI networks. We make DenseGait available to the research community, under open credentialized access, to promote further advancement in the skeleton-based gait analysis field. We proposed GaitFormer, a transformer that is pretrained on DenseGait in a self-supervised and multi-task fashion. The model obtains 92.5% accuracy on CASIA-B and 85.3% accuracy on FVG, without processing any manually annotated data, achieving higher performance even compared to fully supervised methods. GaitFormer represents the first application of plain transformer encoders for skeleton-based gait analysis, without any hand-crafted architectural modifications. We explored pedestrian attribute identification based solely on movement, without utilizing appearance information. GaitFormer achieves good results in gender, age body type, and clothing attributes.

Author Contributions: Conceptualization, A.C and E.R.; Formal analysis, A.C.; Funding acquisition, E.R.; Methodology, A.C. and E.R.; Project administration, E.R.; Resources, E.R.; Software, A.C.; Supervision, E.R.; Validation, A.C.; Writing—original draft, A.C.; Writing—review & editing, E.R. All authors have read and agreed to the published version of the manuscript.

Funding: This work was partly supported by CRC Research Grant 2021, with funds from UEFISCDI in project CORNET (PN-III 1/2018) and by the Google IoT/Wearables Student Grants.

Institutional Review Board Statement: Not applicable.

Informed Consent Statement: Not applicable.

Data Availability Statement: The dataset analyzed in this study are made publicly available. This data can be found here: https://bit.ly/3SLO8RW, accessed on 1 July 2022.

Conflicts of Interest: The authors declare no conflict of interest.

References

1. Ye, M.; Shen, J.; Lin, G.; Xiang, T.; Shao, L.; Hoi, S.C. Deep Learning for Person Re-identification: A Survey and Outlook. *IEEE Trans. Pattern Anal. Mach. Intell.* **2021**, *44*, 2872–2893. [CrossRef] [PubMed]
2. Wang, M.; Deng, W. Deep face recognition: A survey. *Neurocomputing* **2021**, *429*, 215–244. [CrossRef]
3. Nixon, M.S.; Tan, T.N.; Chellappa, R. *Human Identification Based on Gait (The Kluwer International Series on Biometrics)*; Springer: Berlin/Heidelberg, Germany; 2005.
4. McGibbon, C.A. Toward a Better Understanding of Gait Changes With Age and Disablement: Neuromuscular Adaptation. *Exerc. Sport Sci. Rev.* **2003**, *31*, 102–108. [CrossRef] [PubMed]
5. Kumar, R.; Isik, C.; Phoha, V.V. Treadmill Assisted Gait Spoofing (TAGS) An Emerging Threat to Wearable Sensor-based Gait Authentication. *Digit. Threat. Res. Pract.* **2021**, *2*, 1–17. [CrossRef]
6. Singh, J.P.; Jain, S.; Arora, S.; Singh, U.P. Vision-based gait recognition: A survey. *IEEE Access* **2018**, *6*, 70497–70527. [CrossRef]
7. Makihara, Y.; Nixon, M.S.; Yagi, Y. Gait recognition: Databases, representations, and applications. In *Computer Vision: A Reference Guide*; Springer: Berlin/Heidelberg, Germany, 2020; pp. 1–13.
8. Zhang, Z.; Tran, L.; Liu, F.; Liu, X. On Learning Disentangled Representations for Gait Recognition. *IEEE Trans. Pattern Anal. Mach. Intell.* **2020**, *44*, 345–360. [CrossRef]
9. Sepas-Moghaddam, A.; Etemad, A. View-invariant gait recognition with attentive recurrent learning of partial representations. *IEEE Trans. Biom. Behav. Identity Sci.* **2020**, *3*, 124–137. [CrossRef]
10. Thapar, D.; Nigam, A.; Aggarwal, D.; Agarwal, P. VGR-net: A view invariant gait recognition network. In Proceedings of the 2018 IEEE 4th International Conference on Identity, Security, and Behavior Analysis (ISBA), Singapore, 11–12 January 2018; pp. 1–8.
11. Cosma, A.; Radoi, I.E. WildGait: Learning Gait Representations from Raw Surveillance Streams. *Sensors* **2021**, *21*, 8387. [CrossRef]
12. Zhu, Z.; Guo, X.; Yang, T.; Huang, J.; Deng, J.; Huang, G.; Du, D.; Lu, J.; Zhou, J. Gait Recognition in the Wild: A Benchmark. In Proceedings of the IEEE International Conference on Computer Vision (ICCV), Montreal, BC, Canada, 11–17 October 2021.
13. Makihara, Y.; Mannami, H.; Tsuji, A.; Hossain, M.; Sugiura, K.; Mori, A.; Yagi, Y. The OU-ISIR Gait Database Comprising the Treadmill Dataset. *IPSJ Trans. Comput. Vis. Appl.* **2012**, *4*, 53–62. [CrossRef]
14. Xu, C.; Makihara, Y.; Ogi, G.; Li, X.; Yagi, Y.; Lu, J. The OU-ISIR Gait Database Comprising the Large Population Dataset with Age and Performance Evaluation of Age Estimation. *IPSJ Trans. Comput. Vis. Appl.* **2017**, *9*, 1–14. [CrossRef]
15. Yu, S.; Tan, D.; Tan, T. A framework for evaluating the effect of view angle, clothing and carrying condition on gait recognition. In Proceedings of the 18th International Conference on Pattern Recognition (ICPR'06), Hong Kong, China, 20–24 August 2006; Volume 4, pp. 441–444.
16. Liu, X.; Zhao, H.; Tian, M.; Sheng, L.; Shao, J.; Yi, S.; Yan, J.; Wang, X. Hydraplus-net: Attentive deep features for pedestrian analysis. In Proceedings of the IEEE International Conference on Computer Vision, Venice, Italy, 22–29 October 2017; pp. 350–359.
17. Tang, C.; Sheng, L.; Zhang, Z.; Hu, X. Improving pedestrian attribute recognition with weakly-supervised multi-scale attribute-specific localization. In Proceedings of the IEEE/CVF International Conference on Computer Vision, Seoul, Korea, 27–28 October 2019; pp. 4997–5006.
18. Jian, J.; Houjing, H.; Wenjie, Y.; Xiaotang, C.; Kaiqi, H. Rethinking of pedestrian attribute recognition: Realistic datasets with efficient method. *arXiv* **2020**, arXiv:2005.11909.
19. Catruna, A.; Cosma, A.; Radoi, I.E. From Face to Gait: Weakly-Supervised Learning of Gender Information from Walking Patterns. In Proceedings of the 2021 16th IEEE International Conference on Automatic Face and Gesture Recognition (FG 2021), Jodhpur, India, 15–18 December 2021; pp. 1–5.
20. uk Ko, S.; Tolea, M.I.; Hausdorff, J.M.; Ferrucci, L. Sex-specific differences in gait patterns of healthy older adults: Results from the Baltimore Longitudinal Study of Aging. *J. Biomech.* **2011**, *44*, 1974–1979. [CrossRef]
21. Ko, S.u.; Hausdorff, J.M.; Ferrucci, L. Age-associated differences in the gait pattern changes of older adults during fast-speed and fatigue conditions: results from the Baltimore longitudinal study of ageing. *Age Ageing* **2010**, *39*, 688–694. [CrossRef]
22. Choi, H.; Lim, J.; Lee, S. Body fat-related differences in gait parameters and physical fitness level in weight-matched male adults. *Clin. Biomech.* **2021**, *81*, 105243. [CrossRef]
23. Takayanagi, N.; Sudo, M.; Yamashiro, Y.; Lee, S.; Kobayashi, Y.; Niki, Y.; Shimada, H. Relationship between daily and in-laboratory gait speed among healthy community-dwelling older adults. *Sci. Rep.* **2019**, *9*, 3496. [CrossRef]
24. Liu, Z.; Malave, L.; Osuntogun, A.; Sudhakar, P.; Sarkar, S. Toward understanding the limits of gait recognition. In Proceedings of the Biometric Technology for Human Identification, Orlando, FL, USA, 12–13 April 2004; Volume 5404, pp. 195–205.

25. Vaswani, A.; Shazeer, N.; Parmar, N.; Uszkoreit, J.; Jones, L.; Gomez, A.N.; Kaiser, Ł.; Polosukhin, I. Attention is All you Need. In *Proceedings of the Advances in Neural Information Processing Systems;* Guyon, I., Luxburg, U.V., Bengio, S., Wallach, H., Fergus, R., Vishwanathan, S., Garnett, R., Eds.; Curran Associates, Inc.: Red Hook, NY, USA, 2017; Volume 30.
26. Yan, S.; Xiong, Y.; Lin, D. Spatial temporal graph convolutional networks for skeleton-based action recognition. In Proceedings of the Thirty-Second AAAI Conference on Artificial Intelligence, New Orleans, LA, USA, 2–7 February 2018.
27. Plizzari, C.; Cannici, M.; Matteucci, M. Spatial temporal transformer network for skeleton-based action recognition. In Proceedings of the International Conference on Pattern Recognition, Virtual Event, 10–15 January 2021; pp. 694–701.
28. Hofmann, M.; Geiger, J.; Bachmann, S.; Schuller, B.; Rigoll, G. The tum gait from audio, image and depth (gaid) database: Multimodal recognition of subjects and traits. *J. Vis. Commun. Image Represent.* **2014**, *25*, 195–206. [CrossRef]
29. Chen, T.; Kornblith, S.; Norouzi, M.; Hinton, G. A simple framework for contrastive learning of visual representations. In Proceedings of the International Conference on Machine Learning, PMLR, Virtual Event, 13–18 July 2020; pp. 1597–1607.
30. Shutler, J.D.; Grant, M.G.; Nixon, M.S.; Carter, J.N. On a large sequence-based human gait database. In *Applications and Science in Soft Computing;* Springer: Berlin/Heidelberg, Germany, 2004; pp. 339–346.
31. Sarkar, S.; Phillips, P.J.; Liu, Z.; Vega, I.R.; Grother, P.; Bowyer, K.W. The humanid gait challenge problem: Data sets, performance, and analysis. *IEEE Trans. Pattern Anal. Mach. Intell.* **2005**, *27*, 162–177. [CrossRef]
32. Do, T.D.; Nguyen, V.H.; Kim, H. Real-time and robust multiple-view gender classification using gait features in video surveillance. *Pattern Anal. Appl.* **2020**, *23*, 399–413. [CrossRef]
33. Xu, S.; Fang, J.; Hu, X.; Ngai, E.; Guo, Y.; Leung, V.; Cheng, J.; Hu, B. Emotion recognition from gait analyses: Current research and future directions. *arXiv* **2020**, arXiv:2003.11461.
34. Zhang, D.; Wang, Y.; Bhanu, B. Ethnicity classification based on gait using multi-view fusion. In Proceedings of the 2010 IEEE Computer Society Conference on Computer Vision and Pattern Recognition-Workshops, San Francisco, CA, USA, 13–18 June 2010; pp. 108–115.
35. Sepas-Moghaddam, A.; Etemad, A. Deep gait recognition: A survey. *IEEE Trans. Pattern Anal. Mach. Intell.* **2022**. [CrossRef]
36. Han, J.; Bhanu, B. Individual Recognition Using Gait Energy Image. *IEEE Trans. Pattern Anal. Mach. Intell.* **2006**, *28*, 316–322. [CrossRef]
37. Wang, C.; Zhang, J.; Pu, J.; Yuan, X.; Wang, L. Chrono-Gait Image: A Novel Temporal Template for Gait Recognition. In *Proceedings of the Computer Vision—ECCV 2010;* Daniilidis, K.; Maragos, P., Paragios, N., Eds.; Springer: Berlin/Heidelberg, Germany, 2010; pp. 257–270.
38. Bashir, K.; Xiang, T.; Gong, S. Gait recognition using Gait Entropy Image. In Proceedings of the 3rd International Conference on Imaging for Crime Detection and Prevention (ICDP 2009), London, UK, 3 December 2009; pp. 1–6. [CrossRef]
39. Chao, H.; He, Y.; Zhang, J.; Feng, J. Gaitset: Regarding gait as a set for cross-view gait recognition. In Proceedings of the AAAI Conference on Artificial Intelligence, Honolulu, HI, USA, 27 January–1 February 2019; Volume 33, pp. 8126–8133.
40. Fan, C.; Peng, Y.; Cao, C.; Liu, X.; Hou, S.; Chi, J.; Huang, Y.; Li, Q.; He, Z. Gaitpart: Temporal part-based model for gait recognition. In Proceedings of the IEEE/CVF Conference on Computer Vision and Pattern Recognition, Seattle, WA, USA, 14–19 June 2020; pp. 14225–14233.
41. Lin, B.; Zhang, S.; Yu, X. Gait recognition via effective global-local feature representation and local temporal aggregation. In Proceedings of the IEEE/CVF International Conference on Computer Vision, Montreal, QC, Canada, 11–17 October 2021; pp. 14648–14656.
42. Li, J.; Wang, C.; Zhu, H.; Mao, Y.; Fang, H.S.; Lu, C. Crowdpose: Efficient crowded scenes pose estimation and a new benchmark. In Proceedings of the IEEE/CVF Conference on Computer Vision and Pattern Recognition, Long Beach, CA, USA, 15–20 June 2019; pp. 10863–10872.
43. Cosma, A.; Radoi, I.E. Multi-Task Learning of Confounding Factors in Pose-Based Gait Recognition. In Proceedings of the 2020 19th RoEduNet Conference: Networking in Education and Research (RoEduNet), Bucharest, Romania, 11–12 December 2020; pp. 1–6. [CrossRef]
44. Liao, R.; Cao, C.; Garcia, E.B.; Yu, S.; Huang, Y. Pose-Based Temporal-Spatial Network (PTSN) for Gait Recognition with Carrying and Clothing Variations. In Proceedings of the Biometric Recognition, Shenzhen, China, 28–29 October 2017; pp. 474–483.
45. An, W.; Liao, R.; Yu, S.; Huang, Y.; Yuen, P.C. Improving Gait Recognition with 3D Pose Estimation. In Proceedings of the CCBR, Urumqi, China, 11–12 August 2018.
46. Li, N.; Zhao, X.; Ma, C. JointsGait: A model-based Gait Recognition Method based on Gait Graph Convolutional Networks and Joints Relationship Pyramid Mapping. *arXiv* **2020**, arXiv:2005.08625.
47. Li, N.; Zhao, X. A Strong and Robust Skeleton-based Gait Recognition Method with Gait Periodicity Priors. *IEEE Trans. Multimed.* **2022**. [CrossRef]
48. Xu, C.; Makihara, Y.; Li, X.; Yagi, Y.; Lu, J. Gait recognition from a single image using a phase-aware gait cycle reconstruction network. In Proceedings of the European Conference on Computer Vision, Virtual Event, 23–28 August 2020; pp. 386–403.
49. Devlin, J.; Chang, M.W.; Lee, K.; Toutanova, K. BERT: Pre-training of Deep Bidirectional Transformers for Language Understanding. In *Proceedings of the 2019 Conference of the North American Chapter of the Association for Computational Linguistics: Human Language Technologies, Volume 1 (Long and Short Papers);* Association for Computational Linguistics: Minneapolis, MN, USA, 2019; pp. 4171–4186. [CrossRef]

50. Brown, T.; Mann, B.; Ryder, N.; Subbiah, M.; Kaplan, J.D.; Dhariwal, P.; Neelakantan, A.; Shyam, P.; Sastry, G.; Askell, A.; et al. Language Models are Few-Shot Learners. In *Proceedings of the Advances in Neural Information Processing Systems*; Larochelle, H., Ranzato, M., Hadsell, R., Balcan, M.F., Lin, H., Eds.; Curran Associates, Inc.: Red Hook, NY, USA, 2020; Volume 33, pp. 1877–1901.
51. Gidaris, S.; Singh, P.; Komodakis, N. Unsupervised Representation Learning by Predicting Image Rotations. In Proceedings of the International Conference on Learning Representations, Vancouver, Canada, 30 April–3 May 2018.
52. Doersch, C.; Gupta, A.; Efros, A.A. Unsupervised visual representation learning by context prediction. In Proceedings of the IEEE International Conference on Computer Vision, Santiago, Chile, 7–13 December 2015; pp. 1422–1430.
53. Wang, J.; Jiao, J.; Liu, Y.H. Self-supervised video representation learning by pace prediction. In Proceedings of the European Conference on Computer Vision, Virtual Event, 23–28 August 2020; pp. 504–521.
54. Zbontar, J.; Jing, L.; Misra, I.; LeCun, Y.; Deny, S. Barlow twins: Self-supervised learning via redundancy reduction. In Proceedings of the International Conference on Machine Learning, PMLR, Virtual Event, 18–24 July 2021; pp. 12310–12320.
55. Caron, M.; Touvron, H.; Misra, I.; Jégou, H.; Mairal, J.; Bojanowski, P.; Joulin, A. Emerging properties in self-supervised vision transformers. In Proceedings of the IEEE/CVF International Conference on Computer Vision, Virtual Event, 11–17 October 2021; pp. 9650–9660.
56. Dosovitskiy, A.; Beyer, L.; Kolesnikov, A.; Weissenborn, D.; Zhai, X.; Unterthiner, T.; Dehghani, M.; Minderer, M.; Heigold, G.; Gelly, S.; et al. An Image is Worth 16 × 16 Words: Transformers for Image Recognition at Scale. In Proceedings of the International Conference on Learning Representations, Virtual Event, 26–30 April 2020.
57. Zhang, H.; Hao, Y.; Ngo, C.W. Token shift transformer for video classification. In Proceedings of the 29th ACM International Conference on Multimedia, Chengdu, China, 20–24 October 2021; pp. 917–925.
58. Dong, L.; Xu, S.; Xu, B. Speech-Transformer: A No-Recurrence Sequence-to-Sequence Model for Speech Recognition. In Proceedings of the 2018 IEEE International Conference on Acoustics, Speech and Signal Processing (ICASSP), Calgary, Canada, 15–20 April 2018; pp. 5884–5888. [CrossRef]
59. Beal, J.; Wu, H.Y.; Park, D.H.; Zhai, A.; Kislyuk, D. Billion-Scale Pretraining with Vision Transformers for Multi-Task Visual Representations. In Proceedings of the IEEE/CVF Winter Conference on Applications of Computer Vision, Waikoloa, HI, USA, 4–8 January 2022; pp. 564–573.
60. Lin, T.Y.; Maire, M.; Belongie, S.; Hays, J.; Perona, P.; Ramanan, D.; Dollár, P.; Zitnick, C.L. Microsoft coco: Common objects in context. In Proceedings of the European Conference on Computer Vision, Zurich, Switzerland, 6–12 September 2014; pp. 740–755.
61. Nasseri, M.H.; Moradi, H.; Hosseini, R.; Babaee, M. Simple online and real-time tracking with occlusion handling. *arXiv* **2021**, arXiv:2103.04147.
62. Bewley, A.; Ge, Z.; Ott, L.; Ramos, F.; Upcroft, B. Simple online and realtime tracking. In Proceedings of the 2016 IEEE International Conference on Image Processing (ICIP), Phoenix, AZ, USA, 25–28 September 2016; pp. 3464–3468.
63. Wojke, N.; Bewley, A.; Paulus, D. Simple Online and Realtime Tracking with a Deep Association Metric. In Proceedings of the 2017 IEEE International Conference on Image Processing (ICIP), Beijing, China, 17–20 September 2017; pp. 3645–3649. [CrossRef]
64. Murray, M.P.; Drought, A.B.; Kory, R.C. Walking Patterns of Normal Men. *JBJS* **1964**, *46*, 335–360. [CrossRef]
65. Xu, X.; Meng, Q.; Qin, Y.; Guo, J.; Zhao, C.; Zhou, F.; Lei, Z. Searching for alignment in face recognition. In Proceedings of the AAAI Conference on Artificial Intelligence, Virtual Event, 19 August 2021; Volume 35, pp. 3065–3073.
66. Li, X.; Makihara, Y.; Xu, C.; Yagi, Y.; Ren, M. Gait-based human age estimation using age group-dependent manifold learning and regression. *Multimed. Tools Appl.* **2018**, *77*, 28333–28354. [CrossRef]
67. Szegedy, C.; Vanhoucke, V.; Ioffe, S.; Shlens, J.; Wojna, Z. Rethinking the inception architecture for computer vision. In Proceedings of the IEEE Conference on Computer Vision and Pattern Recognition, Las Vegas, NV, USA, 26 June–1 July 2016; pp. 2818–2826.
68. Li, D.; Zhang, Z.; Chen, X.; Huang, K. A richly annotated pedestrian dataset for person retrieval in real surveillance scenarios. *IEEE Trans. Image Process.* **2019**, *28*, 1575–1590. [CrossRef] [PubMed]
69. Deng, Y.; Luo, P.; Loy, C.C.; Tang, X. Pedestrian attribute recognition at far distance. In Proceedings of the 22nd ACM International Conference on Multimedia, Orlando, FL, USA, 3–7 November 2014; pp. 789–792.
70. Sohn, K.; Berthelot, D.; Carlini, N.; Zhang, Z.; Zhang, H.; Raffel, C.A.; Cubuk, E.D.; Kurakin, A.; Li, C.L. Fixmatch: Simplifying semi-supervised learning with consistency and confidence. *Adv. Neural Inf. Process. Syst.* **2020**, *33*, 596–608.
71. Hochreiter, S.; Schmidhuber, J. Long short-term memory. *Neural Comput.* **1997**, *9*, 1735–1780. [CrossRef]
72. Gabeur, V.; Sun, C.; Alahari, K.; Schmid, C. Multi-modal transformer for video retrieval. In Proceedings of the European Conference on Computer Vision, Virtual Event, 23–28 August 2020; pp. 214–229.
73. Khosla, P.; Teterwak, P.; Wang, C.; Sarna, A.; Tian, Y.; Isola, P.; Maschinot, A.; Liu, C.; Krishnan, D. Supervised Contrastive Learning. In *Proceedings of the Advances in Neural Information Processing Systems*; Larochelle, H., Ranzato, M., Hadsell, R., Balcan, M.F., Lin, H., Eds.; Curran Associates, Inc.: Red Hook, NY, USA, 2020; Volume 33, pp. 18661–18673.
74. Müller, R.; Kornblith, S.; Hinton, G.E. When does label smoothing help? *Adv. Neural Inf. Process. Syst.* **2019**, *32*, 4694–4703.
75. Hinton, G.; Vinyals, O.; Dean, J. Distilling the Knowledge in a Neural Network. In Proceedings of the NIPS Deep Learning and Representation Learning Workshop, Montreal, QC, Canada, 11–12 December 2015.
76. Smith, L.N. Cyclical learning rates for training neural networks. In Proceedings of the 2017 IEEE Winter Conference on Applications of Computer Vision (WACV), Santa Rosa, CA, USA, 24–31 March 2017; pp. 464–472.
77. Zhang, T.; Wu, F.; Katiyar, A.; Weinberger, K.Q.; Artzi, Y. Revisiting Few-sample BERT Fine-tuning. In Proceedings of the International Conference on Learning Representations, Virtual Event, 26–30 April 2020.

78. Teepe, T.; Khan, A.; Gilg, J.; Herzog, F.; Hörmann, S.; Rigoll, G. GaitGraph: Graph Convolutional Network for Skeleton-Based Gait Recognition. In Proceedings of the 2021 IEEE International Conference on Image Processing (ICIP), Anchorage, AK, USA, 19–22 September 2021; pp. 2314–2318. [CrossRef]
79. Langerman, D.; Johnson, A.; Buettner, K.; George, A.D. Beyond Floating-Point Ops: CNN Performance Prediction with Critical Datapath Length. In Proceedings of the 2020 IEEE High Performance Extreme Computing Conference (HPEC), Waltham, MA, USA, 22–24 September 2020; pp. 1–9. [CrossRef]

sensors

MDPI

Article

Occluded Pedestrian-Attribute Recognition for Video Sensors Using Group Sparsity

Geonu Lee [1], Kimin Yun [2] and Jungchan Cho [1,*]

[1] College of Information Technology, Gachon University, Seongnam 13120, Korea
[2] Artificial Intelligence Research Laboratory, Electronics and Telecommunications Research Institute, Daejeon 34129, Korea
* Correspondence: thinkai@gachon.ac.kr; Tel.: +82-31-750-5328

Abstract: Pedestrians are often obstructed by other objects or people in real-world vision sensors. These obstacles make pedestrian-attribute recognition (PAR) difficult; hence, occlusion processing for visual sensing is a key issue in PAR. To address this problem, we first formulate the identification of non-occluded frames as temporal attention based on the sparsity of a crowded video. In other words, a model for PAR is guided to prevent paying attention to the occluded frame. However, we deduced that this approach cannot include a correlation between attributes when occlusion occurs. For example, "boots" and "shoe color" cannot be recognized simultaneously when the foot is invisible. To address the uncorrelated attention issue, we propose a novel temporal-attention module based on group sparsity. Group sparsity is applied across attention weights in correlated attributes. Accordingly, physically-adjacent pedestrian attributes are grouped, and the attention weights of a group are forced to focus on the same frames. Experimental results indicate that the proposed method achieved 1.18% and 6.21% higher F_1-scores than the advanced baseline method on the occlusion samples in DukeMTMC-VideoReID and MARS video-based PAR datasets, respectively.

Keywords: deep learning; group-sparsity loss; temporal attention module; video-based pedestrian-attribute recognition

Citation: Lee, G.; Yun, K.; Cho, J. Occluded Pedestrian-Attribute Recognition for Video Sensors Using Group Sparsity. *Sensors* **2022**, *22*, 6626. https://doi.org/10.3390/s22176626

Academic Editors: Euntai Kim, Sangyoun Lee and Kang Ryoung Park

Received: 21 July 2022
Accepted: 28 August 2022
Published: 1 September 2022

1. Introduction

Pedestrian-attribute recognition (PAR) is a task that predicts various attributes of pedestrians detected by surveillance vision sensors, e.g., CCTV. It is a human-searchable semantic description and can be adopted in soft biometrics for visual surveillance [1]. Several studies have been conducted on this subject [2–8], owing to the importance of its applications, such as in finding missing persons and criminals. A few studies have focused on occlusion situations for pedestrian detection [9] and person re-identification [10–12] based on visual sensors. However, the occlusion problem in the field of PAR remains an open problem.

Due to the fact that other objects and persons obstruct pedestrians, it is impossible to resolve this challenge based on a single image. However, a video sensor contains more pedestrian information than an image, thus allowing a model to leverage information from multiple frames. Consider a case in which the lower body of a pedestrian is occluded in some frames but the other frames contain a visible lower-body appearance of the same pedestrian. In this case, we must use only the information obtained from the frame with the visible lower body rather than the one in which the lower body is occluded. Recently, Chen et al. [13] proposed a video-based PAR method that calculates temporal attention probabilities to focus on frames that are important for attribute recognition. However, this method concentrates on incorrect frames when a pedestrian is occluded by other objects or people.

Recent studies are yet to comprehensively consider occlusion analysis. In this study, we propose a novel method for improving PAR performance in occlusion cases. As an intuitive idea, to avoid concentrating on the frame with the occlusion, we select a frame that can best estimate each attribute. Therefore, one solution adopts the sparsity regularization [14] of temporal attention weights. In other words, sparse attention maximizes relevant information in the other weighted frames. However, our experimental results indicate that adding this simple sparsity constraint to the baseline method [13] does not accurately handle occlusion. This is because the method proposed in [13] employs multiple independent branches for multi-attribute classification. Sparsity-constrained temporal attention cannot understand the relationships between the attributes. However, pedestrian attributes are closely related to each other. In particular, semantically adjacent attributes exhibit more significant relationships, as illustrated in Figure 1. Therefore, the relationship between attributes is key to finding meaningless frames, and we formulate this relationship as temporal attention based on group sparsity.

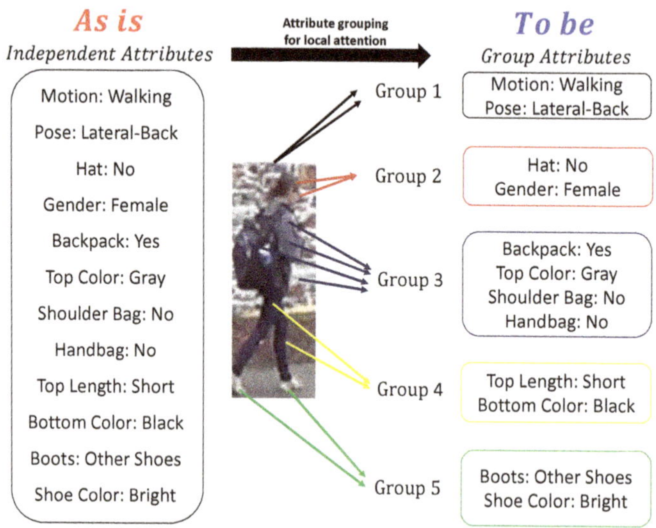

Figure 1. Attribute grouping for local attention. Physically-adjacent pedestrian attributes are grouped into one group. Group 1 is for attributes related to the entirety of a pedestrian. Groups 2, 3, 4, and 5 are for attributes related to the head, upper body, lower body, and feet of a pedestrian, respectively. The network focuses on the semantic information of the pedestrian such that it helps in recognizing pedestrian attributes occluded by obstacles.

Group sparsity [15] is a more advanced method than sparsity; it can gather the related attention of the attributes into a single group. For instance, in Figure 1, information regarding "boots" and "shoe color" is destroyed at the same time an obstacle occludes the feet of a pedestrian. In this case, group sparsity categorizes the "boots" and "shoe color" together into one group. Then, their attention weights are simultaneously suppressed. Therefore, the group constraint achieves more improved results for occlusion situations than those of the sparsity method. Figure 2 presents an overview of the proposed method comprising a shared feature extractor, multiple attribute-classification branches, and a temporal attention module based on group sparsity across multiple branches.

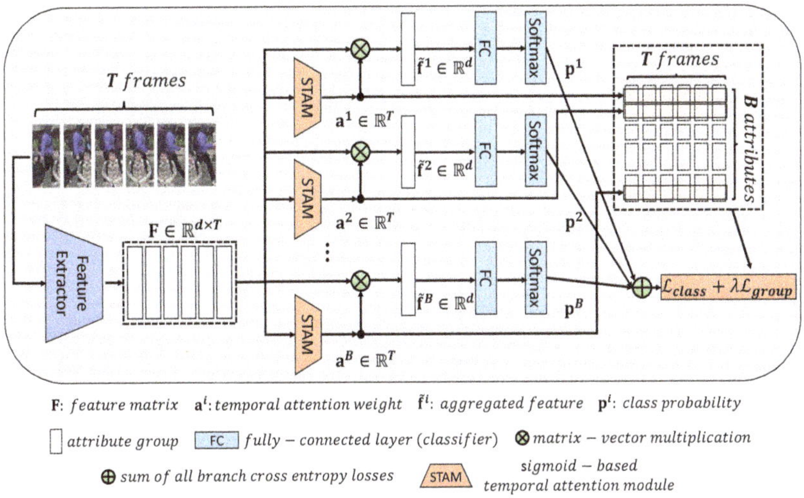

Figure 2. Overview of the network architecture of the proposed method. It comprises a feature extractor, Sigmoid-based temporal attention modules, and attribute classifiers. Due to the fact that the attributes of the pedestrians are closely related to each other, the attention weights for semantically adjacent attributes have similar values to each other, i.e., temporal frame attentions are not independent. To reflect this point, we formulate a temporal attention module based on the group-sparsity constraint. In the $T \times B$ block, the attention weights of the related attributes are grouped by the L_2 norm in each frame.

Extensive experiments were conducted to demonstrate the improvement of the proposed method in its effectiveness against occlusion. The proposed method outperformed the advanced methods on the DukeMTMC-VideoReID [13,16,17] and MARS [13,18] benchmark datasets. In particular, the proposed method achieved 1.18% and 6.21% higher F_1-scores than those of the advanced baseline method on occlusion samples. We also validated the proposed method on additional occlusion scenarios with synthetic data, demonstrating that the proposed method consistently outperformed the advanced baseline method with a maximum F_1-score of 6.26%.

Our main contributions are summarized as follows.

- The proposed temporal attention module is designed to reflect the temporal sparsity of useful frames in a crowded video. Our model is guided to not pay attention to the occluded frame, but rather to the frame where relevant attributes are visible.
- When a pedestrian is occluded owing to obstacles, information on several related attributes is difficult to infer simultaneously. Therefore, we propose a novel group-sparsity-based temporal attention module. This module allows a model to robustly pay attention to meaningful frames to recognize the group attributes of a pedestrian.
- Extensive experiments provide performance analysis of PAR methods on various occlusion scenarios, where the proposed method outperformed the state-of-the-art methods.

The remainder of this paper is organized as follows. First, we introduce sparsity and group-sparsity regularizations, as well as other related work in Section 2.2, and then the proposed method is described in Section 3. Subsequently, Section 4 presents details on the implementation and experimental results. Finally, we discuss and conclude the paper in Section 5.

2. Preliminaries

2.1. Sparsity and Group-Sparsity Regularizations

In deep learning, training a classifier model f is an under-determined problem due to finite datasets [19]. A regularization term R is used to impose prior knowledge on parameters $\mathbf{w}$ as

$$\min_{\mathbf{w}} \sum_{i=1}^{n} L(f(\mathbf{x}_i; \mathbf{w}), y_i) + \lambda R(\mathbf{w}), \tag{1}$$

where $\mathbf{x}_i$, L, and λ represents the i-th training example, a loss function between predicting results $f(\mathbf{x}_i; \mathbf{w})$ and their ground truths y_i, and a hyper-parameter that controls the importance of the regularization term, respectively.

Sparsity Regularization is adopted to induce the model to be sparse. The feasible constraint for sparsity is to reduce the number of nonzero parameter elements, defined as L_0 norm $R(\mathbf{w}) = \|\mathbf{w}\|_0$. However, because the L_0 norm solution is NP-hard problem, the L_1 norm $R(\mathbf{w}) = \|\mathbf{w}\|_1 = \sum_j |w_j|$ is used to approximate L_0 norm in several deep learning problems [20].

Group-sparsity regularization is employed to introduce the K-group structure into the leaning problem as $R(\mathbf{w}) = ||\mathbf{w}||_{2,1} = \sum_{k=1}^{K} \|\mathbf{w}^k\|_2$, where $\|\mathbf{w}^k\|_2 = \sqrt{\sum_{j=1}^{|\mathcal{G}^k|} (w_j^k)^2}$. This is interpreted as imposing an L_2 norm regularizer on members of each group, $\mathbf{w}^k \in \mathbb{R}^{|\mathcal{G}^k|}$, and then inducing an L_1 norm over groups [21,22].

Applications: Nguyen et al. [20] proposed a sparse temporal pooling network for action localization in a video. Unlike the sparsity loss method that adjusts each value, the group-sparsity loss method simultaneously controls the values associated with each other [21–24]. We propose a method that simultaneously adjusts the attention weights of pedestrian attributes by designing the group-sparsity constraint.

2.2. Pedestrian-Attribute Recognition

Video-based PAR: Chen et al. [13] proposed an attention module that indicates the extent to which the model pays attention to each frame for each attribute. They designed branches and classifiers for each attribute in the video. Specker et al. [25] employed global features before temporal pooling to utilize the different pieces of information from various frames. However, existing video-based PAR methods are yet to comprehensively consider the occlusion problem. In this study, we focus on the occlusion handling of video-based PAR.

Image-based PAR: Liu et al. [2] proposed the HydraPlus-Net network that utilizes multi-scale features. Tang et al. [26] proposed an attribute localization module (ALM) that learns specific regions for each attribute generated from multiple levels. Furthermore, Ji et al. [27] proposed a multiple-time-steps attention mechanism that considers the current, previous, and next time steps to understand the complex relationships between attributes and images. Jia et al. [28] proposed Spatial and Semantic Consistency Regularizations (SSC_{soft}). The spatial consistency regularization understands the regions related to each attribute. In addition, they proposed a semantic consistency regularization to extract the unique semantic features of each attribute.

With image-based PAR, it is difficult to achieve accurate attribute recognition for various situations, such as occlusion situations. On the other hand, videos contain more information than images; recently, the number of video-based studies has been increasing.

3. Proposed Method

3.1. Problem Formulation

Figure 3 presents examples of occluded pedestrian images from two video PAR datasets (DukeMTMC-VideoReID and MARS [13]). Typically, pedestrian images obtained from surveillance cameras in the real world are often obscured by crowds of people, cars, and buildings. In addition, the instability of pedestrian tracking results in distorted pedestrian images. Therefore, it is important to correctly recognize pedestrian attributes in occlusion situations; however, occluded pedestrian images make it impossible to obtain single-image-based PAR. This study attempts to achieve improved PAR using multiple frames, i.e., video-based PAR.

(**a**) (**b**)

Figure 3. (**a**,**b**) represent the occlusion types in MARS and DukeMTMC-VideoReID datasets, respectively. Various occlusion types exist, such as a lower body or head of a pedestrian occluded by other pedestrians, tracking failure, and so forth.

3.2. Overview

The proposed method comprises a feature extractor, attention modules, and attribute classifiers. In addition, the inputs are a set of T frames, as illustrated in Figure 2.

First, any feature-extraction network can be used. Here, we utilize the same feature extractor employed in our baseline [13], which comprises a ResNet [29] and two convolution modules, to extract two types of features according to their relevance to the identification (for more details, please refer to [13]). It returns a feature matrix $\mathbf{F} \in \mathbb{R}^{d \times T}$ that contains a set of d-dimensional feature vectors corresponding to T frames as $\mathbf{F} = [\mathbf{f}_1, \mathbf{f}_2, \ldots, \mathbf{f}_T]$. However, the body parts of a pedestrian are often occluded, owing to obstacles and other pedestrians in actual videos. Therefore, the information required to recognize pedestrian attributes differs for each frame, even in the same video.

Second, the proposed network includes a temporal attention module for aggregating multiple frames that is implemented by multiplying the feature matrix $\mathbf{F}$ as

$$\tilde{\mathbf{f}}^i = \mathbf{F}\mathbf{a}^i = \sum_{t=1}^{T} a_t^i \cdot \mathbf{f}_t^i, \tag{2}$$

where $\tilde{\mathbf{f}}^i \in \mathbb{R}^d$ is an aggregated feature vector, while $\mathbf{a}^i$ is an attention-weight vector obtained by the temporal attention module in Section 3.3. The superscript i indicates the i-th attribute type (e.g., hat, backpack, shoe type, and color).

Finally, multi-branch classifiers are employed for multi-labeled attribute classifications as depicted in Figure 2. Notably, unlike the existing work [13], which trains multiple attribute classifiers by solely adopting independent classification losses, the proposed method reliably trains multiple classifiers using feature vectors constrained by a temporal attention module based on group sparsity.

In the following sections, we will explain the novel temporal attention module based on group sparsity.

3.3. Temporal-Attention-Module-Based Classification

Chen et al. [13] designed the temporal attention as a Softmax-based probabilistic temporal attention module ($PTAM$) that calculates important probabilities for frames in the temporal direction and returns an attention weight vector $\mathbf{a} \in \mathbb{R}^T$. However, $PTAM$ comprises Conv-ReLU-Conv-ReLU-Softmax. ReLU [30] converts all the negative values to 0 as illustrated in Figure 4a, while Softmax normalizes the sum of the attention weights of the T frame equal to 1, i.e., $Softmax(\mathbf{a}) = [\frac{e^{a_1}}{\sum_{j=1}^{T} e^{a_j}}, \frac{e^{a_2}}{\sum_{j=1}^{T} e^{a_j}}, \ldots, \frac{e^{a_T}}{\sum_{j=1}^{T} e^{a_j}}]$. This makes it difficult to obtain attention weights that reflect the sparsity constraints [20]. In other words, if the weight of a particular frame becomes 1, the weight of the rest of the frame becomes 0. This is not optimal, as the weights of several frames should have high values. To address this issue, Ref. [20] adopted the Sigmoid-based attention module. Inspired by [20], we use a Sigmoid-based temporal attention module ($STAM$) configured with Conv-ReLU-Conv-Sigmoid. The Sigmoid after Conv allows any frame to have a weight close to 0 or 1, as illustrated in Figure 4b.

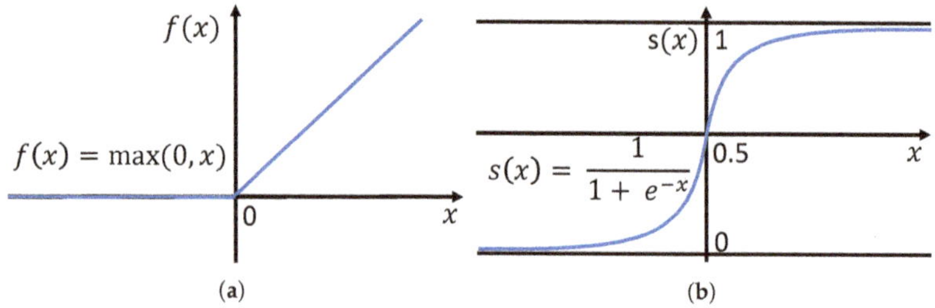

Figure 4. Activation functions. (**a**) ReLU function; (**b**) Sigmoid function.

In multi-branch cases, a temporal-attention-weight vector for the i-th attribute type, $\mathbf{a}^i \in \mathbb{R}^T$, can be obtained as

$$\mathbf{a}^i = STAM^i(\mathbf{F}). \quad (3)$$

Finally, an aggregated feature vector for the i-th attitude classification, $\tilde{\mathbf{f}}^i \in \mathbb{R}^d$, is obtained by Equation (2). Subsequently, we pass $\tilde{\mathbf{f}}^i$ to the i-th linear attribute classifier, and a prediction vector $\mathbf{p}^i$ is obtained for each attribute as:

$$\mathbf{p}^i = Softmax(\mathbf{W}^i\tilde{\mathbf{f}}^i), \quad (4)$$

where $\mathbf{W}^i \in \mathbb{R}^{c_i \times d}$ represents a weight matrix of a fully connected layer for the i-th attribute classification branch, and c_i denotes the number of classes of the branch. The classification loss $\mathcal{L}_{class}$ is the sum of the cross-entropy (CE) [31] of the attributes.

$$\mathcal{L}_{class} = \sum_{i=1}^{B} \beta^i CE(\mathbf{p}^i), \quad (5)$$

where B denotes the number of branches for each attribute in Figure 2. β^i is a balancing hyperparameter for the i-th attribute classification. It is set as a reciprocal of the number of classes in each attribute, because each attribute classification has a different number of classes.

3.4. Limitation of Sparsity Constraint on STAM

The temporal attention weight $\mathbf{a}^i$ in Equation (2) is an indicator that represents the importance of each frame. The sparsity constraint for the attention weight is used to improve the importance indication of frames and is computed by the L_1 norm on $\mathbf{a}^i$.

$$\mathcal{L}_{sparsity} = \sum_{i=1}^{B} \|\mathbf{a}^i\|_1, \quad (6)$$

where B denotes the number of branches of each attribute. The sparsity loss is the operation of the L_1 norm per branch of each attribute. From the formulation, the sparsity constraint is expected to have the effect of selecting frames that are not occluded from T frames independently for each branch.

However, compared with the baselines, our experimental results, presented in Section 4, indicate that the sparsity constraint on the $STAM$ fails to assign importance to the correct frame, thereby degrading the PAR performance sometimes.

Why does the sparsity constraint fail to improve the overall performance?

As illustrated on the left-hand side of Figure 5, the sparsity constraint on $STAM$ is independently applied to the temporal attention weights by the L_1 norm for each branch; hence, the attention weights of each branch solely depend on the temporal information in each attribute. This implies that the sparsity constraint does not help a model understand the relationship between each attribute. However, pedestrian attributes are closely related to each other. As presented in Figure 3, information about some attributes, such as the type and color of the bottom and shoe of a pedestrian, respectively, is damaged simultaneously if a the lower body or feet of the pedestrian is/are occluded. Therefore, another constraint is required to guide the model to understand the relationship between pedestrian attributes, which is important for achieving improved performance, by considering various occlusion situations. In the next section, we design the attribute relationships as attribute groups and formulate the group constraints of these attributes.

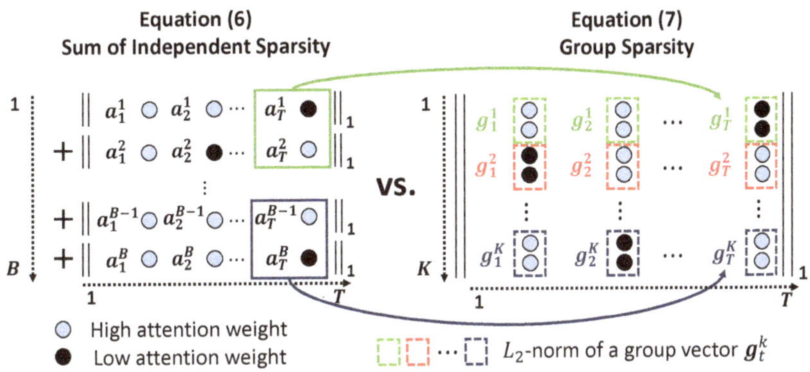

Figure 5. Comparison between the sparsity- and group-sparsity-based constraints. Unlike the sparsity-based method that adjusts each value independently, the group-sparsity-based method simultaneously controls the values associated with each other.

3.5. Group-Sparsity Constraint on STAM

Group sparsity extends and generalizes how to learn the correct sparsity regularization by which prior assumptions on the structure of the input variables can be incorporated [15]. Regarding the attributes of an occluded pedestrian, the prior assumption is that these attributes can be partitioned into K groups based on their relevance, i.e., $\mathcal{G}^k$ where $k = 1, 2, \ldots, K$, as illustrated in Figure 1. Accordingly, the attention weights in the same group at time t, $\{a_t^i | i \in \mathcal{G}^k\}$, can be constrained by considering the group structure.

The method for grouping multiple attribute weights at time t involves introducing a new vector at time t using each attribute group, i.e., $\mathbf{g}_t^k \in \mathbb{R}^{|\mathcal{G}^k|}$, as presented on the right-hand side of Figure 5. By summing the L_2 norm of a group vector $\mathbf{g}_t^k$, we can define two sparsity constraints on attributes and time as

$$\mathcal{L}_{group} = \sum_{t=1}^{T} \sum_{k=1}^{K} \gamma_k \|\mathbf{g}_t^k\|_2, \tag{7}$$

where $\|\mathbf{g}_t^k\|_2$ always has positive values; hence, the sum of these values has the same effect as the L_1 norm [21–23]. γ_k is a balancing hyperparameter for the k-th group in the sum of all the group-sparsity loss functions. It is set as a reciprocal of the number of attributes in each group, because each group has a different number of attributes.

The $\mathcal{L}_{group}$ constraint on $STAM$ simultaneously increases or decreases the attention weights of specific groups in particular frames. This helps a model understand the frames that are more important for each group, including the groups that are recognizable in the same frame. This constraint is consistent with the prior assumption that groups exist between attributes. In addition, it does not employ explicit local patches in frames for the recognition of specific attributes. It adopts implicit attention via attribute groups, thereby enabling improved attribute recognition for pedestrian appearance distortions due to tracking failures.

Finally, the total loss function comprises $\mathcal{L}_{class}$ and $\mathcal{L}_{group}$, described above, as follows:

$$\mathcal{L}_{total} = \mathcal{L}_{class} + \lambda\mathcal{L}_{group}, \tag{8}$$

where λ represents a weight factor that combines the classification and group-sparsity losses.

4. Experiments

4.1. Implementation Details

Table 1 presents the attribute groups of the group sparsity for the experiments. We employed the same feature extractor as [13], pretrained on the ImageNet dataset [32]. The initial learning rate was set to 3×10^{-4} and multiplied by 0.3 at 100 epochs. The weight decay was set to 5×10^{-4} for the Adam optimizer [33]. For the input, the width and height of the frame were resized to 112 and 224 pixels, respectively. The weight factor λ in Equation (8) was set to 0.02. The batch size for training was set to 64. The model was trained for 200 epochs, and the best results among the measurements were reported every 20 epochs. The sequence length T of the consecutive and non-overlapping frames for training was set to 6, according to a previous study [13]. In the test phase, we divided the trajectory of a pedestrian into segments comprising 6 frames. The divided segments were independently inferred, and the results were averaged for PAR. In other words, performance was measured using one prediction per trajectory, according to [13]. We utilized a single NVIDIA Titan RTX GPU for both training and inference. Regarding our experimental setting, in the absence of an additional explanation, we follow the process detailed in the baselines [13] for a fair comparison. The random seed for the experiments was fixed deterministically.

Table 1. Attribute groups for DukeMTMC-VideoReID and MARS datasets.

Group	DukeMTMC-VideoREID	MARS
Whole	motion, pose	motion, pose
Head	hat, gender	age, hat, hair, gender
Upper Body	backpack, top color, shoulder bag, handbag	backpack, top color, shoulder bag, handbag, top length
Lower Body	top length, bottom color	bottom length, bottom color, type of bottom
Foot	boots, shoe color	-

4.2. Evaluation Metrics and Datasets

We evaluated the proposed method using the average accuracy and F_1-score that decrease when the algorithm fails to recognize the correct pedestrian attributes. For the extensive experiments, we used two video-based PAR datasets: DukeMTMC-VideoReID and MARS [13], which were derived from the reidentification datasets, DukeMTMC-VideoReID [16] and MARS [18], respectively. Chen et al. [13] reannotated them for the video-based PAR datasets.

4.2.1. DukeMTMC-VideoReID Dataset

The DukeMTMC-VideoReID dataset contains 12 types of pedestrian-attribute annotations. Eight of these attributes are binary types: backpack, shoulder bag, handbag, boots, gender, hat, shoe color, and top length. The other four attributes are multi-class types: motion (walking, running, riding, staying, various), pose (frontal, lateral-frontal, lateral, lateral-back, back, various), bottom color (black, white, red, gray, blue, green, brown, complex), and top color (black, white, red, purple, gray, blue, green, brown, complex). The attributes were annotated per trajectory and the total number of trajectories was 4832. We excluded four trajectories with fewer frames than the segment length T, while the remaining 4828 trajectories were adopted in the experiments. For the training, 2195 trajectories were used, 413 of which contained occlusions, as illustrated in Figure 3b. For the test, 2633 trajectories were employed, 449 of which contained occlusions. The average length of the trajectories was approximately 169 frames.

4.2.2. MARS Dataset

The MARS dataset contains 14 types of pedestrian-attribute annotations. Ten of these attributes are binary types: shoulder bag, gender, hair, bottom type, bottom length, top length, backpack, age, hat, and handbag. The other four attributes are multi-class types: motion (walking, running, riding, staying, various), pose (frontal, lateral-frontal, lateral, lateral-back, back, various), top color (black, purple, green, blue, gray, white, yellow, red, complex), and bottom color (white, purple, black, green, gray, pink, yellow, blue, brown, complex). The attributes were also annotated per trajectory, and the total number of trajectories was 16,360. We also excluded five trajectories with fewer frames than the segment length T, and the remaining trajectories were 16,355. For the training, 8297 trajectories were used, 35 of which contained occlusions, as illustrated in Figure 3a. For the test, 8058 trajectories were used, 30 of which contained occlusions. The average length of the trajectories was approximately 60 frames.

4.3. Comparisons with State-of-the-Art Methods

The proposed method was compared with five baselines: Chen et al. [13], 3D-CNN [34], CNN-RNN [35], ALM [26], and SSC_{soft} [28]. The Chen et al. [13] method is a state-of-the-art video-based PAR method. CNN-RNN and 3D-CNN are video-based PAR methods compared in [13]. ALM [26] and SSC_{soft} [28] are two state-of-the-arts for image-based PAR. For fair comparisons, we adopted the average values for each image of trajectories to evaluate the ALM and SSC_{soft} methods on video-based datasets. We retrained the ALM [26] using the officially published code. For SSC_{soft} [28], we re-implemented it because there is no official code. In the case of ALM [26] and SSC_{soft} [28], the image batch size was set to 96, and the learning rate was adjusted to 7.5×10^{-5}, according to [36].

To evaluate the improvement of the proposed method in occlusion situations, we compared its performance with those of the baselines by only adopting the occlusion samples. Table 2 presents the results on the DukeMTMC-VideoReID and MARS datasets. To ensure accurate evaluation, we excluded the "hat" and "handbag" attributes of the MARS dataset when evaluating all methods, because the ground truth of both attributes for all occlusion samples was the same, i.e., "no". As presented in Table 2, the proposed method outperformed the baselines in all cases and achieved average accuracies of 88.36% and 71.94%, including average F_1-scores of 70.21% and 61.88% on the occlusion samples of the DukeMTMC-VideoReID and MARS datasets, respectively. In particular, the proposed method achieves superior performance over the state-of-the-art ALM [26] and SSC_{soft} [28] methods, which extended to video using multi-frame averages. This shows that the image-based PAR methods have limitations in effectively using multiple frames when extended to video. In the real world, pedestrians are often occluded by various environments, so performance improvement of the proposed method in occlusive situations is not trivial.

Table 2. Comparisons of the results obtained for the occlusion samples of the DukeMTMC-VideoReID and MARS datasets. The **bold** indicates the best result.

Dataset	Method	Average Accuracy (%)	Average F_1-Score (%)
DukeMTMC-VideoReID	Chen et al. [13]	88.33	69.03
	3DCNN [34]	84.41	61.38
	CNN-RNN [35]	87.94	68.12
	ALM [26]	86.99	65.87
	SSC_{soft} [28]	86.86	65.01
	Ours	**88.36**	**70.21**
MARS	Chen et al. [13]	66.39	55.67
	3DCNN [34]	60.83	46.16
	CNN-RNN [35]	65.83	53.79
	ALM [26]	67.50	55.73
	SSC_{soft} [28]	68.89	57.44
	Ours	**71.94**	**61.88**

To verify that the proposed method does not have severe negative effects on non-occlusion samples, we also evaluated its performance using total samples, including the occlusion and non-occlusion samples. Table 3 presents the performances of the methods on the total samples of the DukeMTMC-VideoReID and MARS datasets, where the proposed method outperformed the baselines. The Chen et al. [13] method exhibited a slightly better average accuracy in just one case, in the DukeMTMC-VideoReID dataset. However, because the measure of average accuracy did not consider a data imbalance, the difference was negligible. For instance, if there are 90 negative samples and 10 positive samples among the 100 total samples, the model can obtain high accuracy by predicting most of the samples as being negative, e.g., when true negative, true positive, false negative, and false positive are 90, 1, 9, and 0, respectively, the accuracy is 91%, and the F_1-score is 18.18%. Therefore, the average F_1-score is a better measure than the average accuracy for imbalanced datasets.

Table 3. Comparisons of the results for the total samples of the DukeMTMC-VideoReID and MARS datasets. The **bold** indicates the best result.

Dataset	Method	Average Accuracy (%)	Average F_1-Score (%)
DukeMTMC-VideoReID	Chen et al. [13]	**89.12**	71.58
	3DCNN [34]	85.38	64.66
	CNN-RNN [35]	88.80	71.73
	ALM [26]	88.13	69.66
	SSC_{soft} [28]	87.52	68.71
	Ours	88.98	**72.30**
MARS	Chen et al. [13]	86.42	69.92
	3DCNN [34]	81.96	60.39
	CNN-RNN [35]	86.49	69.89
	ALM [26]	86.56	68.89
	SSC_{soft} [28]	86.01	68.15
	Ours	**86.75**	**70.42**

4.4. *Ablation Study*

4.4.1. Effects of the Weight Factor λ

We compared the experimental results according to the weight factor λ in Equation (8). The weight factor λ is a parameter that adjusts sparsity. As presented in Table 4, the proposed method exhibits higher F_1-scores than those of the baseline methods, regardless of the λ values, and the best results were obtained with $\lambda = 0.02$.

Table 4. Analysis of the group-sparsity loss for the occlusion samples of the DukeMTMC-VideoReID and MARS datasets. The **bold** indicates the best result.

Dataset	Method	Average Accuracy (%)	Average F_1-Score (%)
DukeMTMC -VideoReID	Chen et al. [13]	88.33	69.03
	$\lambda = 0.005$	**88.38**	69.85
	$\lambda = 0.03$	88.16	69.62
	$\lambda = 0.02$	88.36	**70.21**
MARS	Chen et al. [13]	66.39	55.67
	$\lambda = 0.005$	68.06	55.07
	$\lambda = 0.03$	70.00	58.89
	$\lambda = 0.02$	**71.94**	**61.88**

4.4.2. Comparisons Between PTAM and STAM

We analyzed $PTAM$ and $STAM$ by applying them along with each method. Table 5 demonstrates that sparsity has the worst performance for occlusion samples in terms of both accuracy and F_1-scores. As explained in Section 3.4, the sparsity constraint cannot help a model understand the relationship between attributes. However, the proposed method using the group-sparsity-constrained $STAM$, which understands the relationship between each attribute, exhibited the best performance among the other methods.

Table 5. Comparisons between the sparsity-based and group-sparsity-based (ours) constraints for the occlusion samples of the DukeMTMC-VideoReID and MARS datasets. The **bold** indicates the best result.

Dataset	Method	PTAM	STAM	Average Accuracy (%)	Average F_1-Score (%)
DukeMTMC -VideoReID	Chen et al. [13]	✓	-	88.33	69.03
	Sparsity	✓	-	87.99	69.05
	Group sparsity	✓	-	88.23	**70.24**
	Chen et al. [13]	-	✓	87.94	69.26
	Sparsity	-	✓	87.68	67.52
	Group sparsity	-	✓	**88.36**	70.21
MARS	Chen et al. [13]	✓	-	66.39	55.67
	Sparsity	✓	-	70.00	57.76
	Group sparsity	✓	-	**71.94**	61.70
	Chen et al. [13]	-	✓	66.94	55.92
	Sparsity	-	✓	69.17	57.80
	Group sparsity	-	✓	**71.94**	**61.88**

4.5. Qualitative Results

We visualized the temporal attention weight vector with various segment frames to analyze the improvement of the proposed method in occlusion situations. Figure 6 presents the temporal attention vectors and PAR results of the method presented by Chen et al. [13] and that of our method for all the groups in the DukeMTMC-VideoReID dataset. The values of the baseline method are similar in all the frames. Thereby, the baseline method failed to recognize the "shoe color" attribute. In contrast, the values of the proposed method are different in each frame. Moreover, the values of the occlusion frames are lower than those of the general frames. The attention weights of the bottom- and top-length attributes are simultaneously controlled, because they belong to the same group. For the same reason, the attention weights of the "shoe color" and "boot" attributes are also simultaneously adjusted. Consequently, the proposed method accurately predicted all attributes. It shows that the proposed group-sparsity constraint helps STAM accurately focus on non-occlusion frames.

Attribute name	GT	Baseline	Ours
Motion	Wakling	Wakling	Wakling
Pose	Lateral-Frontal	Lateral-Frontal	Lateral-Frontal
Backpack	Yes	Yes	Yes
Shoulder Bag	No	No	No
Handbag	No	No	No
Boots	Other Shoes	Other Shoes	Other Shoes
Gender	Male	Male	Male
Hat	No	No	No
Shoe Color	Bright	Dark	Bright
Top Length	Short	Short	Short
Bottom Color	Brown	Brown	Brown
Top Color	Black	Black	Black

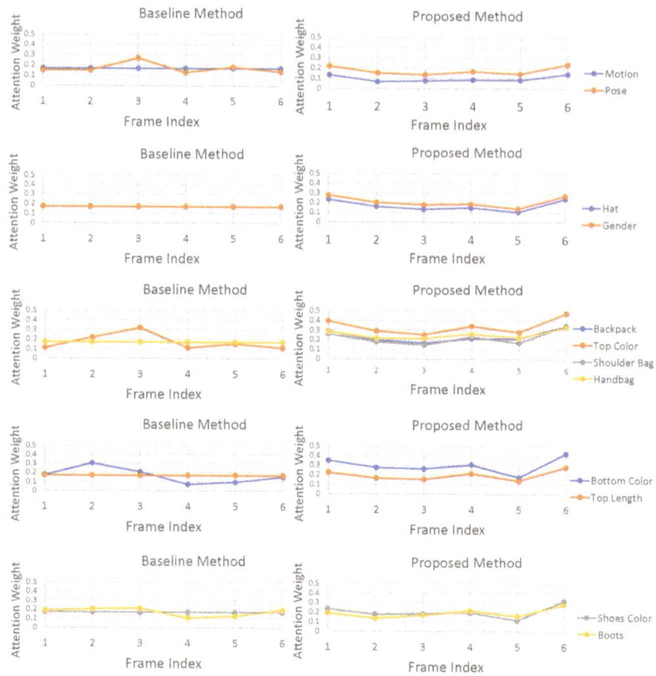

Figure 6. Qualitative results for the DukeMTMC-VideoReID dataset. It presents the attention weights of the group attributes and PAR results. For the groups related to the lower body, the proposed method has low attention weights in the occluded frames. However, the attention weights of the baseline method (Chen et al. [13]) are almost the same in all the frames.

4.6. Evaluation of Additional Occlusion Scenarios

We designed two synthetic occlusion scenarios, as illustrated in Figure 7, to validate the improvement of the proposed method on several occlusion samples. These two occlusion scenarios are designed to analyze the impact on recognition performance if a part of the appearance of pedestrian frames is distorted by blurring, low illumination, or an object such as another pedestrian or car.

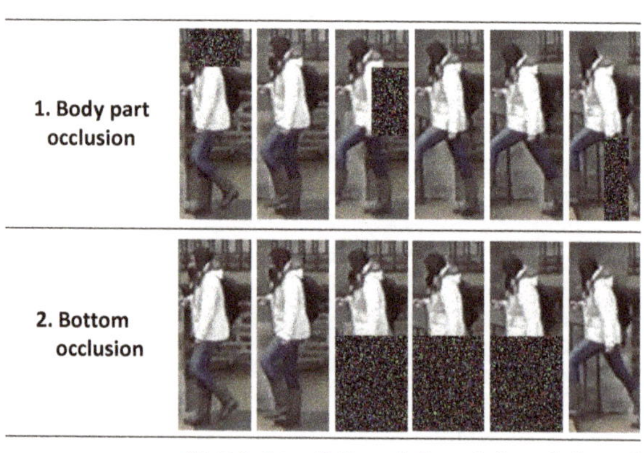

Figure 7. Examples of two occlusion scenarios.

The first scenario was a body-part occlusion. In this scenario, we randomly selected three frames among the segment frames. Subsequently, the head of a pedestrian and the left and right sides of their upper and lower body, respectively, were randomly occluded. The second scenario was the bottom occlusion scenario that simulated a situation in which cars and bicycles passed through and occluded the lower body of the pedestrian. We randomly selected three consecutive frames.

In the process of constructing the scenarios, we did not apply the additional occlusion situations to real occlusion samples in the datasets. The number of test samples for each scenario was 2633 and 8058 for the DukeMTMC-VideoReID and MARS datasets, respectively, which are the same as the total number of test samples in the original datasets.

We did not retrain the baseline and proposed methods to prevent the models from learning the tendency of synthetic occlusion. We used the same models in Sections 4.3 and 4.4 and evaluated them on two scenario samples. Table 6 presents the results for the body-part and bottom occlusion scenarios. In all cases, the proposed method achieved better results than those of the baseline methods. Table 7 shows the average F_1-scores according to the number of consecutive occlusion frames on the bottom occlusion scenario samples of the DukeMTMC-VideoReID and MARS datasets. As the number of consecutive occlusion frames increases, the amount of information for recognizing attributes decreases, and, thus, the performances of all methods were degraded. Nevertheless, the proposed method consistently achieved better average F_1-scores in comparison to those of the baselines as the number of consecutive occlusion frames increased. The obtained experimental results indicate that the proposed method is effective in handling occlusions, regardless of the scenario. Accordingly, we can conclude that the proposed method is more suitable for real-world scenarios with many occlusions than the compared methods.

Table 6. Comparisons of the results for the two occlusion scenarios of the DukeMTMC-VideoReID and MARS datasets. The **bold** indicates the best result.

Dataset	Method	Body Part		Bottom	
		Average Accuracy (%)	**Average F_1-Score (%)**	**Average Accuracy (%)**	**Average F_1-Score (%)**
DukeMTMC-VideoReID	Chen et al. [13]	88.67	70.94	87.03	66.85
	3DCNN [34]	85.31	63.99	82.28	58.40
	CNN-RNN [35]	88.73	71.17	88.50	70.00
	ALM [26]	88.08	69.45	87.17	66.98
	SSC_{soft} [28]	87.60	67.87	86.60	65.64
	Ours	**88.95**	**71.97**	**88.59**	**70.66**
MARS	Chen et al. [13]	85.97	68.34	82.79	62.55
	3DCNN [34]	81.64	59.42	79.05	55.58
	CNN-RNN [35]	86.42	69.49	85.95	68.34
	ALM [26]	86.32	67.87	85.77	65.96
	SSC_{soft} [28]	85.34	65.18	84.61	63.95
	Ours	**86.73**	**70.05**	**86.08**	**68.81**

Table 7. Comparisons of the average F_1-scores (%) for according to the number of consecutive occluded frames on the bottom occlusion scenario of the DukeMTMC-VideoReID and MARS datasets. The **bold** indicates the best result.

Dataset	# Consecutive Occlusion Frames	Chen et al. [13]	3DCNN [34]	CNN-RNN [35]	ALM [26]	SSC$_{soft}$ [28]	Ours
DukeMTMC -VideoReID	1	70.79	63.46	71.15	69.18	68.00	**72.04**
	2	69.63	61.15	70.38	68.37	66.98	**71.61**
	3	66.85	58.40	70.00	66.98	65.64	**70.66**
	4	61.28	55.77	67.44	64.78	63.68	**68.16**
	5	55.94	54.22	**63.77**	62.03	61.16	63.54
MARS	1	68.37	59.13	69.59	68.40	67.28	**70.19**
	2	66.11	57.38	69.09	67.62	65.50	**69.69**
	3	62.55	55.58	68.34	65.96	63.95	**68.81**
	4	57.48	54.14	67.01	64.30	61.84	**67.62**
	5	51.50	52.97	65.04	62.18	59.13	**65.67**

5. Conclusions and Future Work

This study proposed a novel video-based PAR method to improve PAR in various occlusion situations. The proposed method was formulated as a group sparsity to consider the relationship between pedestrian attributes. In addition to improving the temporal attention weights for non-occluded frames, it exhibited the effect of simultaneously excluding multiple occluded attributes by understanding the relationship between each attribute within the frame. In other words, the proposed method focused more on information about attributes that were not occluded and related to each other in the specific frames.

The proposed method was designed to improve PAR in occlusion situations; however, only a few datasets contained sufficient occlusion samples. To address this limitation, the proposed method was also validated on additional scenarios with synthetic samples. The results obtained from extensive experiments demonstrate that the proposed method consistently outperformed most of the baselines. In the future, we will study how to generate an extensive and natural occlusion situation. Furthermore, we will investigate a one-stage method that can detect and track pedestrians and better recognize pedestrian attributes in an extensive occlusion situation.

Author Contributions: G.L. conceived the idea, and he designed and performed the experiments. K.Y. refined the idea and the experiments. J.C. conceived and refined the idea, and he refined the experiments. G.L., K.Y. and J.C. wrote the paper. All authors have read and agreed to the published version of the manuscript.

Funding: This work was supported by the Institute of Information & Communications Technology Planning & Evaluation (IITP) grant funded by the Korea government (MSIT) (No. 2014-3-00123, Development of High Performance Visual BigData Discovery Platform for Large-Scale Realtime Data Analysis).

Institutional Review Board Statement: Not applicable.

Informed Consent Statement: Not applicable.

Data Availability Statement: The dataset information. DukeMTMC-VideoReID: https://github.com/Yu-Wu/DukeMTMC-VideoReID MARS: http://zheng-lab.cecs.anu.edu.au/Project/project_mars.html (accessed on 20 July 2022).

Conflicts of Interest: The authors declare no conflict of interest.

References

1. Wang, X.; Zheng, S.; Yang, R.; Zheng, A.; Chen, Z.; Tang, J.; Luo, B. Pedestrian-attribute recognition: A survey. *Pattern Recognit.* **2022**, *121*, 108220.
2. Liu, X.; Zhao, H.; Tian, M.; Sheng, L.; Shao, J.; Yi, S.; Yan, J.; Wang, X. Hydraplus-net: Attentive deep features for pedestrian analysis. In Proceedings of the IEEE International Conference on Computer Vision, Venice, Italy, 22–29 October 2017.
3. Zhao, X.; Sang, L.; Ding, G.; Han, J.; Di, N.; Yan, C. Recurrent attention model for pedestrian-attribute recognition. In Proceedings of the AAAI Conference on Artificial Intelligence, Honolulu, HI, USA, 27 January–1 February 2019.

4. Li, Y.; Huang, C.; Loy, C.C.; Tang, X. Human attribute recognition by deep hierarchical contexts. In Proceedings of the European Conference on Computer Vision, Amsterdam, The Netherlands, 8–16 October 2016.
5. Han, K.; Wang, Y.; Shu, H.; Liu, C.; Xu, C.; Xu, C. Attribute aware pooling for pedestrian-attribute recognition. In Proceedings of the International Joint Conference on Artificial Intelligence, Macao, China, 10–16 August 2019.
6. Liu, P.; Liu, X.; Yan, J.; Shao, J. Localization guided learning for pedestrian-attribute recognition. In Proceedings of the British Machine Vision Conference, Newcastle upon Tyne, UK, 3–6 September 2018.
7. Li, Y.; Xu, H.; Bian, M.; Xiao, J. Attention based CNN-ConvLSTM for pedestrian-attribute recognition. *Sensors* **2020**, *20*, 811. [CrossRef] [PubMed]
8. Li, Q.; Zhao, X.; He, R.; Huang, K. Visual-semantic graph reasoning for pedestrian-attribute recognition. In Proceedings of the AAAI Conference on Artificial Intelligence, Honolulu, HI, USA, 27 January–1 February 2019.
9. Zou, T.; Yang, S.; Zhang, Y.; Ye, M. Attention guided neural network models for occluded pedestrian detection. *Pattern Recognit. Lett.* **2020**, *131*, 91–97. [CrossRef]
10. Zhou, S.; Wu, J.; Zhang, F.; Sehdev, P. Depth occlusion perception feature analysis for person re-identification. *Pattern Recognit. Lett.* **2020**, *138*, 617–623. [CrossRef]
11. Chen, Y.; Yang, T.; Li, C.; Zhang, Y. A Binarized segmented ResNet based on edge computing for re-identification. *Sensors* **2020**, *20*, 6902. [CrossRef] [PubMed]
12. Yang, Q.; Wang, P.; Fang, Z.; Lu, Q. Focus on the visible regions: Semantic-guided alignment model for occluded person re-identification. *Sensors* **2020**, *20*, 4431. [CrossRef] [PubMed]
13. Chen, Z.; Li, A.; Wang, Y. A temporal attentive approach for video-based pedestrian attribute recognition. In Proceedings of the Chinese Conference on Pattern Recognition and Computer Vision, Xi'an, China, 8–11 November 2019.
14. Hastie, T.; Tibshirani, R.; Friedman, J.H.; Friedman, J.H. *The Elements of Statistical Learning: Data Mining, Inference, and Prediction*; Springer: New York, NY, USA, 2009; Volume 2.
15. Yuan, M.; Lin, Y. Model selection and estimation in regression with grouped variables. *J. R. Stat. Soc. Ser. B* **2006**, *68*, 49–67. [CrossRef]
16. Wu, Y.; Lin, Y.; Dong, X.; Yan, Y.; Ouyang, W.; Yang, Y. Exploit the unknown gradually: One-shot video-based person re-identification by stepwise learning. In Proceedings of the IEEE Conference on Computer Vision and Pattern Recognition, Salt Lake City, UT, USA, 18–22 June 2018.
17. Ristani, E.; Solera, F.; Zou, R.; Cucchiara, R.; Tomasi, C. Performance measures and a data set for multi-target, multi-camera tracking. In Proceedings of the European Conference on Computer Vision, Amsterdam, The Netherlands, 8–16 October 2016.
18. Zheng, L.; Bie, Z.; Sun, Y.; Wang, J.; Su, C.; Wang, S.; Tian, Q. Mars: A video benchmark for large-scale person re-identification. In Proceedings of the European Conference on Computer Vision, Amsterdam, The Netherlands, 8–16 October 2016.
19. Carter, B.; Jain, S.; Mueller, J.W.; Gifford, D. Overinterpretation reveals image classification model pathologies. In Proceedings of the Advances in Neural Information Processing Systems, Virtual, 6–14 December 2021.
20. Nguyen, P.; Liu, T.; Prasad, G.; Han, B. Weakly supervised action localization by sparse temporal pooling network. In Proceedings of the IEEE Conference on Computer Vision and Pattern Recognition, Salt Lake City, UT, USA, 18–22 June 2018.
21. Scardapane, S.; Comminiello, D.; Hussain, A.; Uncini, A. Group sparse regularization for deep neural networks. *Neurocomputing* **2017**, *241*, 81–89. [CrossRef]
22. Yoon, J.; Hwang, S.J. Combined group and exclusive sparsity for deep neural networks. In Proceedings of the International Conference on Machine Learning, Sydney, NSW, Australia, 6–11 August 2017.
23. Cho, J.; Lee, M.; Chang, H.J.; Oh, S. Robust action recognition using local motion and group sparsity. *Pattern Recognit.* **2014**, *47*, 1813–1825. [CrossRef]
24. Gao, Z.; Zhang, H.; Xu, G.P.; Xue, Y.B.; Hauptmann, A.G. Multi-view discriminative and structured dictionary learning with group sparsity for human action recognition. *Signal Process.* **2015**, *112*, 83–97. [CrossRef]
25. Specker, A.; Schumann, A.; Beyerer, J. An evaluation of design choices for pedestrian-attribute recognition in video. In Proceedings of the IEEE International Conference on Image Processing, Virtual, Abu Dhabi, United Arab Emirates, 25–28 October 2020.
26. Tang, C.; Sheng, L.; Zhang, Z.; Hu, X. Improving pedestrian-attribute recognition with weakly-supervised multi-scale attribute-specific localization. In Proceedings of the IEEE International Conference on Computer Vision, Seoul, Korea, 27 October– 2 November 2019.
27. Ji, Z.; Hu, Z.; He, E.; Han, J.; Pang, Y. Pedestrian-attribute recognition based on multiple time steps attention. *Pattern Recognit. Lett.* **2020**, *138*, 170–176. [CrossRef]
28. Jia, J.; Chen, X.; Huang, K. Spatial and Semantic Consistency Regularizations for Pedestrian Attribute Recognition. In Proceedings of the IEEE Conference on Computer Vision and Pattern Recognition, Virtual, 19–25 June 2021.
29. He, K.; Zhang, X.; Ren, S.; Sun, J. Deep residual learning for image recognition. In Proceedings of the IEEE Conference on Computer Vision and Pattern Recognition, Las Vegas, NV, USA, 26 June–1 July 2016.
30. Nair, V.; Hinton, G.E. Rectified linear units improve restricted boltzmann machines. In Proceedings of the International Conference on Machine Learning, Haifa, Israe, 21–24 June 2010.
31. Goodfellow, I.; Bengio, Y.; Courville, A. *Deep Learning*; MIT Press: Cambridge, MA, USA, 2016.
32. Deng, J.; Dong, W.; Socher, R.; Li, L.J.; Li, K.; Li, F.-F. Imagenet: A large-scale hierarchical image database. In Proceedings of the IEEE Conferenceon Computer Vision and Pattern Recognition, Miami, FL, USA, 20–25 June 2009.

33. Kingma, D.P.; Ba, J. Adam: A method for stochastic optimization. In Proceedings of the International Conference on Learning Representations, San Diego, CA, USA, 7–9 May 2015.
34. Ji, S.; Xu, W.; Yang, M.; Yu, K. 3D convolutional neural networks for human action recognition. *IEEE Trans. Pattern Anal. Mach. Intell.* **2012**, *35*, 221–231. [CrossRef]
35. McLaughlin, N.; Del Rincon, J.M.; Miller, P. Recurrent convolutional network for video-based person re-identification. In Proceedings of the IEEE Conference on Computer Vision and Pattern Recognition, Las Vegas, NV, USA, 26 June–1 July 2016.
36. Goyal, P.; Dollár, P.; Girshick, R.; Noordhuis, P.; Wesolowski, L.; Kyrola, A.; Tulloch, A.; Jia, Y.; He, K. Accurate, large minibatch sgd: Training imagenet in 1 hour. *arXiv* **2017**, arXiv:1706.02677.

sensors

MDPI

Article

Content Swapping: A New Image Synthesis for Construction Sign Detection in Autonomous Vehicles

Hongje Seong, Seunghyun Baik, Youngjo Lee, Suhyeon Lee and Euntai Kim *

School of Electrical and Electronic Engineering, Yonsei University, Seoul 03722, Korea;
hjseong@yonsei.ac.kr (H.S.); shbaik104@yonsei.ac.kr (S.B.); lzozo95@yonsei.ac.kr (Y.L.);
hyeon93@yonsei.ac.kr (S.L.)
* Correspondence: etkim@yonsei.ac.kr; Tel.: +82-2-2123-7729

Abstract: Construction signs alert drivers to the dangers of abnormally blocked roads. In the case of autonomous vehicles, construction signs should be detected automatically to prevent accidents. One might think that we can accomplish the goal easily using the popular deep-learning-based detectors, but it is not the case. To train the deep learning detectors to detect construction signs, we need a large amount of training images which contain construction signs. However, collecting training images including construction signs is very difficult in the real world because construction events do not occur frequently. To make matters worse, the construction signs might have dozens of different construction signs (i.e., contents). To address this problem, we propose a new method named content swapping. Our content swapping divides a construction sign into two parts: the board and the frame. Content swapping generates numerous synthetic construction signs by combining the board images (i.e., contents) taken from the in-domain images and the frames (i.e., geometric shapes) taken from the out-domain images. The generated synthetic construction signs are then added to the background road images via the cut-and-paste mechanism, increasing the number of training images. Furthermore, three fine-tuning methods regarding the region, size, and color of the construction signs are developed to make the generated training images look more realistic. To validate our approach, we applied our method to real-world images captured in South Korea. Finally, we achieve an average precision (AP_{50}) score of 84.98%, which surpasses that of the off-the-shelf method by 9.15%. Full experimental results are available online as a supplemental video. The images used in the experiments are also released as a new dataset CSS138 for the benefit of the autonomous driving community.

Keywords: construction sign detection; image synthesis; cut-and-paste; perspective transformation

Citation: Seong, H.; Baik, S.; Lee, Y.; Lee, S.; Kim, E. Content Swapping: A New Image Synthesis for Construction Sign Detection in Autonomous Vehicles. *Sensors* **2022**, 22, 3494. https://doi.org/10.3390/s22093494

Academic Editor: Son Lam Phung

Received: 5 April 2022
Accepted: 1 May 2022
Published: 4 May 2022

1. Introduction

The misdetection of a construction sign may lead to accidents by unexpectedly entering blocked roads. Therefore, the reliable detection of construction signs is quite important in realizing autonomous driving. With the recent progress in object detection based on deep learning [1–6], one might think that we can accomplish the reliable detection of construction signs easily, but it is not true. To train the deep learning detector, we need large-scale training images including construction signs for robust and high-quality results. Unfortunately, construction signs appear infrequently on roads. Thus, collecting large amounts of training data for construction sign detection is required, but it is time-consuming and expensive. To address this problem, we propose a new method for learning to detect construction signs on roads. The main idea of the proposed method is to synthesize training images using a small number of construction sign images. To synthesize training images, we follow the cut-and-paste mechanism [7–9], which cuts an instance from the source image (i.e., construction sign region in an image) and pastes it into a background image (i.e., road image). The cut-and-paste method enables a model to avoid overfitting on a small number of backgrounds in source images, but it cannot generalize limited instances.

A construction sign can be divided into two parts: the board and frame. The content of the sign is contained in a rectangular board, and the board is supported by a frame. The frame can be shared for any sign. Using this characteristic, we effectively generate new construction sign images by swapping the contents in the rectangular board between two different construction sign images, as shown in Figure 1. Our content swapping synthesizes numerous synthetic construction signs by combining the board images (i.e., contents) taken from the in-domain images and the frames (i.e., geometric shapes) taken from the out-domain images. This approach allows us to obtain new $N_I N_O$ images from N_I in-domain construction sign images and N_O out-domain construction sign images. Although in-domain sign images need to be collected using the same camera setting as in the test set, out-domain images can be collected from the Internet. Therefore, we can synthesize a large-scale training dataset with only a small number of in-domain sign images and train a detector on them.

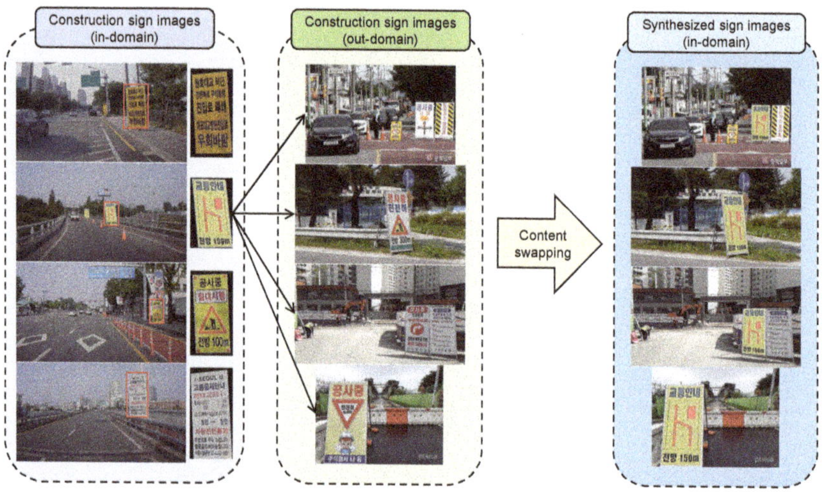

Figure 1. Content swapping. With N_I in-domain and N_O out-domain construction sign images, we synthesize $N_I N_O$ in-domain construction sign images via perspective transformation. The synthesized sign images are used as source images for cut-and-paste.

We also develop three fine-tuning methods to improve the quality of synthetic training images. The three methods deal with the (1) pasted region, (2) instance size, and (3) color difference of the synthesized images, respectively. The first method guides us to paste the synthetic construction sign image on the drivable region. Because the construction sign cannot be placed on the sky, car, or other objects, it should be placed only on the drivable region for realistic purposes. The second method helps us to select the size of the instance based on the location where the sign is to be pasted. If we assume that the construction sign is always pasted on the road and the road is flat, then we can automatically predict the size of the instance in the image. The prediction not only avoids making construction images either too large or too small but also resizes the images to match nearby objects, thereby improving global consistency. Finally, we blend the synthesized construction signs with the training image to reduce the gap between the source and background images. The blending also reduces the domain gap between in-domain and out-domain construction sign images in content swapping. To our best knowledge, no other research has been conducted to detect the construction signs. To validate the effectiveness of the proposed methods, we collect the CSS138 (Construction Signs in Seoul with 138 images) dataset for training and testing construction sign detection. All the images are captured in Seoul, Korea. The CSS138 dataset can be downloaded at https://github.com/Hongje/content-swapping, (accessed on 5 April 2022). In the experiment, we synthesize a large-scale training dataset with only

12 in-domain sign images and achieve a robust and accurate result with an AP_{50} score of 84.98% for CSS138. Our result surpasses off-the-shelf cut-and-paste by 9.15% in the AP_{50} score. Full experimental results are available online: https://youtu.be/us_qso6C5pw, (accessed on 5 April 2022).

The main contributions are summarized as follows:

- This is the first paper which deals with the *construction* sign detection.
- We propose a new image synthesis method, content swapping, to avoid overfitting on limited instances in source images.
- We further present three fine-tunning methods for creating realistic construction images on roads.
- To demonstrate the efficacy of the proposed method, we construct a new dataset, CSS138, for construction sign detection.
- Finally, we achieve an AP_{50} score of 84.98%, creating a gap of 9.15% from the naive cut-and-paste method.

The remainder of this paper is organized as follows. Previous works related to this study are discussed in Section 2. The proposed method for synthesizing construction images is described in Section 3. The experimental results for CSS138 and the analysis are presented in Section 4. Finally, the conclusions are presented in Section 5.

2. Related Work

2.1. Sign Detection

Early methods designed models for detecting signs heuristically. Specifically, Prince et al. [10] design a sign detection algorithm based on a geometrical analysis of the edges and groups of the sign image features. Escalera et al. [11] segment images using color thresholding and then analyze the shape to detect signs. Fang et al. [12] formulate three types of shapes—circular, triangular, and octagonal—to extract the color features of the signs. Shadeed et al. [13] convert the RGB color space to HSV and YUV color spaces and then defined a heuristic algorithm. Loy et al. [14] exploit the symmetric nature and the pattern of the edge of the triangular, square, and octagonal shapes to predict the shape of the sign image. Bahlmann et al. [15] propose a joint color and shape information modeling approach using a set of Haar wavelet features.

Recently, state-of-the-art approaches have used convolutional neural network (CNN)-based supervised models. Shao et al. [16] train CNNs with simplified Gabor filters. Cao et al. [17] use shallow CNNs to classify the traffic signs. Zhang et al. [18] propose a new cascaded R-CNN architecture that includes multiscale attention and imbalanced samples. Liu et al. [19] propose TSingNet, which is based on feature pyramid networks and includes several attention-based modules. Ahmed et al. [20] propose a new DNN-based framework that is robust in detecting traffic signs, even under challenging weather conditions. Zeng et al. [21] propose an improved YOLOv3 architecture for real-time traffic-sign detection. All previous methods considered only traffic or road signs.

The basic difference between general sign detection and construction sign detection is how much training samples are provided. Differently from the large amount of training images in general sign detection, only dozens of training images are given in construction sign detection. Furthermore, collecting the training images for construction signs is much more difficult. The key idea of our method is how to augment the training images and train a detector on them effectively. The purpose to detect and recognize construction signs is to alert the unplanned situations made by road construction. Understandably, commercial autonomous vehicles can handle not only the planned situations but also the unplanned situations. The typical example of the unplanned situation might be the road construction. In this case, the autonomous vehicle may not have to obey the traffic law. For example, our vehicle may have to cross the road following policeman's hand signal, ignoring the traffic sign. The goal of our paper is to handle that kind of unplanned abnormal situation.

Our construction sign detection can also be considered as a special kind of class imbalance problem. We are dealing with only a single class (i.e., construction sign) and the instances of the class are highly imbalanced with the background instances such as buildings, roads, or pedestrians. The key idea of the paper is to tackle the serious imbalance problem by augmenting the training samples.

2.2. Image Synthesis for Network Training

Several studies [9,22] have synthesized training images with a focus on realism. Furthermore, task-specific image synthesis has also been extensively studied. Dwibedi et al. [7] propose a simple yet effective training image synthesis method that uses cut-and-paste for object detection. Lee et al. [8] propose content transfer, which transfers tail-class content from source to target to address the class imbalance problem in unsupervised domain-adaptive semantic segmentation. Leon et al. [9] synthesize training images by rendering that does not require real-world images. In this paper, we propose methods for synthesizing construction sign images for sign detection. The key idea of our image synthesis is that the contents of the board are taken from in-domain images, whereas the frame is taken from the out-domain (and in-domain) images. Since the frames includes only the geometrical shape of the sign board, they can be collected from any images (out-domain images) without affecting the detection performance. However, since the board images have their own style, the construction sign images taken only from the in-domain images are used to facilitate the synthesis onto the background road images.

3. Method

3.1. Overview

An overview of the proposed method for synthesizing training data is shown in Figure 2. The entire process of synthesizing the training images comprised four main steps. In the first step, we prepared images by collecting construction sign images and road images. As acquiring construction sign images is difficult, we could only prepare a limited number of sign images. Therefore, we collected additional out-domain construction sign images from the Internet. In the second step, the four corners of the content and segmentation mask were labeled in the construction sign images. In the third step, content swapping was performed using these labels. Finally, the training images were generated via the cut-and-paste mechanism using the proposed realistic transformations.

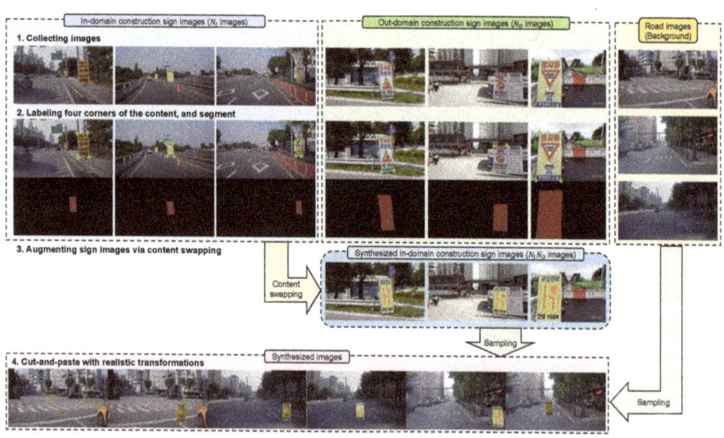

Figure 2. An overview of training data synthesis. The entire process was divided into four steps. First, we collected three types of images: in-domain construction sign images, out-domain construction sign images, and road images. Then, we labeled four corners of the contents and segmented the construction sign images. The labels were then used for content swapping. Finally, a pair of a construction image and a road image was randomly sampled and synthesized via the cut-and-paste mechanism with proposed realistic transformations. The synthesized images are used for training networks.

For a clearer explanation, we provide a pseudo-code of the proposed method in Algorithm 1. Each step in Algorithm 1 matches Figure 2. In the following subsections, we describe the details of each step.

Algorithm 1 Pseudo-code of the proposed method.

Step1: Collecting construction sign and road images

In-domain construction sign image: I_{In}

Out-domain construction sign image: I_{Out}

Road image: I_{Road}

Step2: Labeling bounding box, segment, and four corners of the board

Bounding box labels: $Bbox_{In}$, $Bbox_{Out}$

Segment labels: M_{In}, M_{Out}

Four corners of the board: $([\ x_I^1\ \ y_I^1\ ][\ x_I^2\ \ y_I^2\ ][\ x_I^3\ \ y_I^3\ ][\ x_I^4\ \ y_I^4\ ])$, $([\ x_O^1\ \ y_O^1\ ][\ x_O^2\ \ y_O^2\ ][\ x_O^3\ \ y_O^3\ ][\ x_O^4\ \ y_O^4\ ])$

Step3: Content swapping

Randomly select content image (source): $S \in In$

Randomly select frame image (target): $T \in [\ In \quad Out\ ]$

Set content region mask of target image using four corners label: C_T

Compute transformation matrix $\mathcal{T}$: ▷ Section 3.4

$$\mathcal{T} = \begin{bmatrix} x_S^1 & y_S^1 & 1 & 0 & 0 & 0 & -x_S^1 x_T^1 & -y_S^1 x_T^1 \\ 0 & 0 & 0 & x_S^1 & y_S^1 & 1 & -x_S^1 y_T^1 & -y_S^1 y_T^1 \\ x_S^2 & y_S^2 & 1 & 0 & 0 & 0 & -x_S^2 x_T^2 & -y_S^2 x_T^2 \\ 0 & 0 & 0 & x_S^2 & y_S^2 & 1 & -x_S^2 y_T^2 & -y_S^2 y_T^2 \\ x_S^3 & y_S^3 & 1 & 0 & 0 & 0 & -x_S^3 x_T^3 & -y_S^3 x_T^3 \\ 0 & 0 & 0 & x_S^3 & y_S^3 & 1 & -x_S^3 y_T^3 & -y_S^3 y_T^3 \\ x_S^4 & y_S^4 & 1 & 0 & 0 & 0 & -x_S^4 x_T^4 & -y_S^4 x_T^4 \\ 0 & 0 & 0 & x_S^4 & y_S^4 & 1 & -x_S^4 y_T^4 & -y_S^4 y_T^4 \end{bmatrix}^{-1} \begin{bmatrix} x_T^1 \\ y_T^1 \\ x_T^2 \\ y_T^2 \\ x_T^3 \\ y_T^3 \\ x_T^4 \\ y_T^4 \end{bmatrix}$$

Swap content: $I_T^{CS} = \mathcal{T}(I_S) \odot C_T + I_T \odot (1 - C_T)$

Step4: Cut-and-paste with realistic transformations

Randomly select road image (background): $B \in Road$

Compute pasteable region: P_B ▷ Section 3.5.1

Randomly select bottom point of the sign: $\mathbf{p}_1 = [\ p_1^x \quad p_1^y\ ] \in P_B$

Compute top point of the sign: ▷ Section 3.5.2

$$\mathbf{p}_2 = [\ p_2^x \quad p_2^y\ ] = \left[\ p_1^x \quad \left(\tan^{-1}\left(\frac{\tan(\alpha \cdot p_1^y + \beta)}{1-h}\right) - \beta\right) \Big/ \alpha\ \right]$$

Cut sign image I_T^{CS} and paste to road image I_B: $I_T^{CP} = \text{Cut} - \text{and} - \text{Paste}(I_T^{CS}, M_T, I_B, \mathbf{p}_1, \mathbf{p}_2)$

Transform segment label to $\mathbf{p}_1$ and $\mathbf{p}_2$: $M_T^{CP} = \text{Transform\_mask}(M_T, \mathbf{p}_1, \mathbf{p}_2)$

Transform bounding box label to $\mathbf{p}_1$ and $\mathbf{p}_2$: $Bbox_T^{CP} = \text{Transform\_box}(Bbox_T, \mathbf{p}_1, \mathbf{p}_2)$

Reduce color difference: $I_T^{GP} = \text{GP} - \text{GAN}(I_T^{CS}, M_T^{CS})$ ▷ Section 3.5.3

Output: Synthesized training image: I_T^{GP};
Synthesized training label: $Bbox_T^{CP}$

3.2. Collecting Images

To collect construction sign and road images, we used the FHD390C-USB(D) (Autonomous A2Z, Gyeongsan, South Korea) camera model. This model captures full HD images (1080p) in 30 frames per second. It has a field of view of 60 degrees. We built a data-collecting platform using this camera model, as shown in Figure 3. The camera was installed at a height of 1500mm from the ground and was positioned in front of a platform so that we could collect front-view images of the roads. In total, we collected 138 construction sign images, of which 12 were used for training and the remaining 126 were used for testing. The collected construction sign images were used as in-domain images. In addition, we collected 992 road images that did not contain any construction signs. All the images were captured in Seoul, Korea. Twelve images were used to collect the contents of the construction signs. We also collected an additional 24 construction sign images from the Internet. They were out-domain construction sign images, and they were used to capture the frame of the construction sign boards.

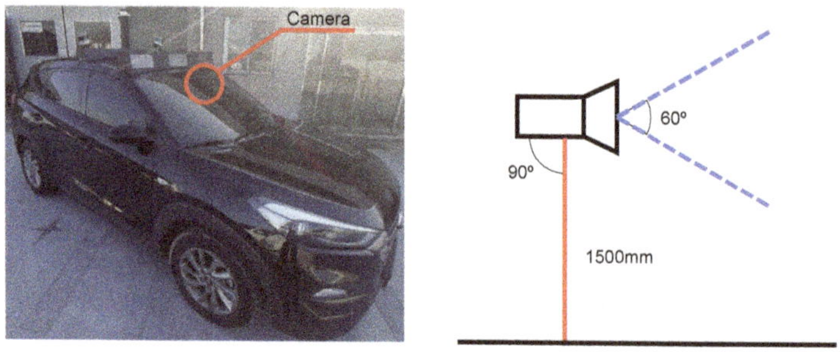

(a) Camera setting in real image (b) Camera setting details

Figure 3. The camera setting in the data-collecting platform.

We collected 12 construction signs using our platform. Thus, we had 12 kinds of construction signs (12 in-domain images) for the board region. We also gathered 24 construction sign images from the Internet, making 36 kinds of construction signs (12 in-domain + 24 out-domain images) for the frame region. The collected 12 in-domain construction signs are shown in Figure 4.

Figure 4. Collected in-domain construction signs.

3.3. Labeling

We annotate three types of labels in the construction sign images. First, we annotate the bounding box for all the collected construction sign images. Bounding box annotations are needed to compute the loss during training and evaluate the detection quality during testing. Second, we annotate the four corners of the board in the training set of construction sign images. Corner annotation is required to calculate the transformation matrix between two construction sign images. Third, we annotate the per-pixel label of the construction sign. Pixel-level annotations are used for both content swapping (detailed in Section 3.4) and cut-and-paste (detailed in Section 3.5).

3.4. Content Swapping

To overcome the lack of the training image, we synthesize training images using a cut-and-paste [7–9] mechanism, as shown in Figure 1. The cut-and-paste effectively helps to prevent the networks from overfitting on the limited backgrounds of the training images. However, the cut-and-paste method cannot augment the content of the training images. This means that only background images can be diversified, and the contents of the construction signs are still limited. We address this problem using content swapping.

The construction sign can be divided into two parts: a rectangular board and frame, as shown in Figure 5. Therefore, we can reuse the frame for other constructions by replacing only the board. To replace the board in the target sign image with the source sign image, we need to formulate the transformation function between the source image and the target image. Thankfully, because the shape of the board is rectangular, replacing the content is possible with four pairs of corner points on the board using perspective transformation, as follows:

$$\begin{bmatrix} wx_T \\ wy_T \\ w \end{bmatrix} = \mathcal{T}\begin{bmatrix} x_S \\ y_S \\ 1 \end{bmatrix} = \begin{bmatrix} p_{11} & p_{12} & p_{13} \\ p_{21} & p_{22} & p_{23} \\ p_{31} & p_{32} & 1 \end{bmatrix}\begin{bmatrix} x_S \\ y_S \\ 1 \end{bmatrix} \tag{1}$$

where $\begin{bmatrix} x_S & y_S \end{bmatrix}^T$ and $\begin{bmatrix} x_T & y_T \end{bmatrix}^T$ are the source and target points of the construction sign images, respectively, and $\mathcal{T} = \begin{bmatrix} p_{11} & p_{12} & p_{13} \\ p_{21} & p_{22} & p_{23} \\ p_{31} & p_{32} & 1 \end{bmatrix}$ is the perspective transformation matrix with 8 parameters. Here, the eight parameters in the perspective transformation matrix $\mathcal{T}$ are unknown. Then, we can unfold the equations as:

$$x_T = (p_{11}x_S + p_{12}y_S + p_{13})/w, \tag{2}$$

$$y_T = (p_{21}x_S + p_{22}y_S + p_{23})/w, \tag{3}$$

$$w = p_{31}x_S + p_{32}y_S + 1. \tag{4}$$

By substituting Equation (4) into Equations (2) and (3), we can join a parameter w into x_T and y_T as:

$$x_T = \frac{p_{11}x_S + p_{12}y_S + p_{13}}{p_{31}x_S + p_{32}y_S + 1}, \tag{5}$$

$$y_T = \frac{p_{21}x_S + p_{22}y_S + p_{23}}{p_{31}x_S + p_{32}y_S + 1}. \tag{6}$$

To easily formulate each unknown parameter in $\mathcal{T}$ into a matrix form, we can rearrange Equations (5) and (6) into:

$$x_T = p_{11}x_S + p_{12}y_S + p_{13} - p_{31}x_Sx_T - p_{32}y_Sx_T, \tag{7}$$

$$y_T = p_{21}x_S + p_{22}y_S + p_{23} - p_{31}x_Sy_T - p_{32}y_Sy_T, \tag{8}$$

respectively. Here, there are eight unknown parameters (i.e., $\begin{bmatrix} p_{11} & p_{12} & \cdots & p_{32} \end{bmatrix}$). Therefore, to estimate the eight parameters' values, we need eight different formulas. With Equations (7) and (8), we can make eight different formulas using four known pairs of corresponding points ($[\, x_S^1 \;\; y_S^1 \,] \cdots [\, x_S^4 \;\; y_S^4 \,]$ for source points and $[\, x_T^1 \;\; y_T^1 \,] \cdots [\, x_T^4 \;\; y_T^4 \,]$ for target points), and then we can write them into a matrix as follows:

$$\begin{bmatrix} x_T^1 \\ y_T^1 \\ x_T^2 \\ y_T^2 \\ x_T^3 \\ y_T^3 \\ x_T^4 \\ y_T^4 \end{bmatrix} = \begin{bmatrix} x_S^1 & y_S^1 & 1 & 0 & 0 & 0 & -x_S^1 x_T^1 & -y_S^1 x_T^1 \\ 0 & 0 & 0 & x_S^1 & y_S^1 & 1 & -x_S^1 y_T^1 & -y_S^1 y_T^1 \\ x_S^2 & y_S^2 & 1 & 0 & 0 & 0 & -x_S^2 x_T^2 & -y_S^2 x_T^2 \\ 0 & 0 & 0 & x_S^2 & y_S^2 & 1 & -x_S^2 y_T^2 & -y_S^2 y_T^2 \\ x_S^3 & y_S^3 & 1 & 0 & 0 & 0 & -x_S^3 x_T^3 & -y_S^3 x_T^3 \\ 0 & 0 & 0 & x_S^3 & y_S^3 & 1 & -x_S^3 y_T^3 & -y_S^3 y_T^3 \\ x_S^4 & y_S^4 & 1 & 0 & 0 & 0 & -x_S^4 x_T^4 & -y_S^4 x_T^4 \\ 0 & 0 & 0 & x_S^4 & y_S^4 & 1 & -x_S^4 y_T^4 & -y_S^4 y_T^4 \end{bmatrix} \begin{bmatrix} p_{11} \\ p_{12} \\ p_{13} \\ p_{21} \\ p_{22} \\ p_{23} \\ p_{31} \\ p_{32} \end{bmatrix}. \tag{9}$$

The objective is to estimate eight unknown parameters in $\mathcal{T}$. Therefore, we can finally obtain the transformation matrix $\mathcal{T}$ by computing the inverse of the 8×8 matrix in Equation (9) and performing matrix multiplication as follows:

$$\begin{bmatrix} x_S^1 & y_S^1 & 1 & 0 & 0 & 0 & -x_S^1 x_T^1 & -y_S^1 x_T^1 \\ 0 & 0 & 0 & x_S^1 & y_S^1 & 1 & -x_S^1 y_T^1 & -y_S^1 y_T^1 \\ x_S^2 & y_S^2 & 1 & 0 & 0 & 0 & -x_S^2 x_T^2 & -y_S^2 x_T^2 \\ 0 & 0 & 0 & x_S^2 & y_S^2 & 1 & -x_S^2 y_T^2 & -y_S^2 y_T^2 \\ x_S^3 & y_S^3 & 1 & 0 & 0 & 0 & -x_S^3 x_T^3 & -y_S^3 x_T^3 \\ 0 & 0 & 0 & x_S^3 & y_S^3 & 1 & -x_S^3 y_T^3 & -y_S^3 y_T^3 \\ x_S^4 & y_S^4 & 1 & 0 & 0 & 0 & -x_S^4 x_T^4 & -y_S^4 x_T^4 \\ 0 & 0 & 0 & x_S^4 & y_S^4 & 1 & -x_S^4 y_T^4 & -y_S^4 y_T^4 \end{bmatrix}^{-1} \begin{bmatrix} x_T^1 \\ y_T^1 \\ x_T^2 \\ y_T^2 \\ x_T^3 \\ y_T^3 \\ x_T^4 \\ y_T^4 \end{bmatrix} = \begin{bmatrix} p_{11} \\ p_{12} \\ p_{13} \\ p_{21} \\ p_{22} \\ p_{23} \\ p_{31} \\ p_{32} \end{bmatrix}. \tag{10}$$

Using the estimated transformation matrix $\mathcal{T}$, we warped the board from the source image to the target image, which is called content swapping.

With content swapping, we can effectively augment in-domain construction sign images using out-domain construction sign images. Given the N_I in-domain and N_O out-domain construction sign images, we can synthesize in-domain images by content swapping from in-domain sign images to out-domain images, resulting in $N_I N_O$ pairs. Therefore, although we obtained only 12 in-domain construction sign images for training, 288 in-domain images can be obtained using 24 out-domain sign images. Furthermore, we use the frame region in the in-domain sign images for content swapping, which resulted in 432 construction sign images.

Figure 5. Visualization of the two parts of the construction sign. In the first column, we show a construction sign image. In the second column, we denote the board and frame regions with green and red, respectively.

3.5. *Cut-and-Paste with Realistic Transformations*

We synthesize training images by cutting a construction sign image and then pasting it onto the background road images. Here, naively cutting and pasting would result in unrealistic synthetic images, which may lead to performance degradation. We address this problem by proposing three fine-tunning methods. They are developed from three perspectives: pasteable region, instance size, and color difference. Detailed explanations of each fine-tuning methods are provided below.

3.5.1. Pasteable Region

The construction sign cannot fly and is never placed on a car. Therefore, we set the pasteable region as the road. To find road regions in the background image, we used two independent pre-trained networks: semantic segmentation and depth estimation. For the semantic segmentation network, we used DeepLab v3+, trained on Cityscapes https://www.cityscapes-dataset.com, (accessed on 5 April 2022). Because the road class is included in the Cityscapes dataset, the predicted score of the road is used directly. For the depth estimation network, we use the off-the-shelf depth prediction network, MiDaS [23]. The estimated depth is used to filter the noise by thresholding. Thus, the regions that are predicted as roads and with estimated depths lower than the predefined threshold are defined as pasteable regions.

3.5.2. Instance Size

Close objects look large and far objects look small. This property is also preserved in the images. Using this property, we adjust the instance size of the construction sign according to the pasted position. In the image, we first randomly select a pixel within the pasteable region ($\mathbf{p}_1 = [\ p_1^x \ \ p_1^y\ ]$). The selected pixel is the bottom point of the construction sign. In real-world coordinates, we compute the distance between the camera and the sign (d), under the assumption that the road is flat, as follows:

$$d = H_{cam} \tan \theta_1 \tag{11}$$

where H_{cam} denotes the height of the camera from the road, and θ_1 is the angle between the line from the camera to the road and the line from the camera to the bottom of the construction sign line. The angle θ_1 is proportional to p_1^y:

$$\theta_1 = \alpha \cdot p_1^y + \beta \tag{12}$$

where α and β are constants. Given the computed distance d, we can calculate the angle θ_2, which is the angle between the line from the camera to the road and the line from the camera to the top of the construction sign, as follows:

$$\theta_2 = \tan^{-1}\left(\frac{d}{H_{cam} - H_{sign}}\right) \tag{13}$$

where H_{sign} denotes the height of the construction sign, and $H_{sign} < H_{cam}$. For simplification, we assume that all construction signs have the same height H_{sign} and stand perpendicular to the road. Then, in the image coordinates, we compute the top point of the construction sign ($\mathbf{p}_2 = [\ p_2^x \ \ p_2^y\ ]$) using the proportionality between θ_2 and p_2^y as follows:

$$p_2^x = p_1^x, \tag{14}$$

$$
\begin{aligned}
p_2^y &= (\theta_2 - \beta)/\alpha \\
&= \left(\tan^{-1}\left(\frac{d}{H_{cam} - H_{sign}}\right) - \beta\right) \Big/ \alpha \\
&= \left(\tan^{-1}\left(\frac{H_{cam}\tan\theta_1}{H_{cam} - H_{sign}}\right) - \beta\right) \Big/ \alpha \\
&= \left(\tan^{-1}\left(\frac{H_{cam}\tan\left(\alpha \cdot p_1^y + \beta\right)}{H_{cam} - H_{sign}}\right) - \beta\right) \Big/ \alpha.
\end{aligned}
\tag{15}
$$

In Equation (15), we divide the denominator and numerator by H_{cam} as:

$$
\begin{aligned}
p_2^y &= \left(\tan^{-1}\left(\frac{\tan\left(\alpha \cdot p_1^y + \beta\right)}{1 - (H_{sign}/H_{cam})}\right) - \beta\right) \Big/ \alpha. \\
&= \left(\tan^{-1}\left(\frac{\tan\left(\alpha \cdot p_1^y + \beta\right)}{1 - h}\right) - \beta\right) \Big/ \alpha.
\end{aligned}
\tag{16}
$$

where h denotes the ratio of the height of the sign to the camera. Using Equations (14) and (16), the top point of the construction sign can be directly computed from the bottom point. We empirically set the parameters α, β, and h to $\pi/3888$, $\pi/3$, and 0.75, respectively. The overall process for computing the size of the construction signs is summarized in Figure 6.

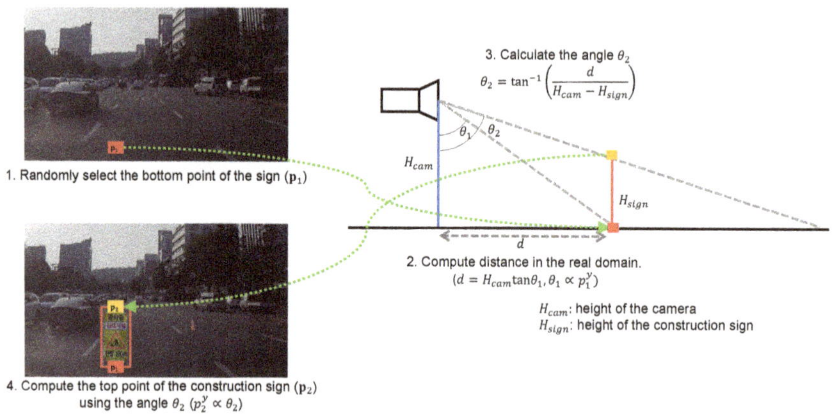

Figure 6. Step-by-step processes for computing the size of the construction sign. (1) We randomly select the bottom position of the construction sign in the image. The position is selected only within the pasteable region. (2) From the randomly selected point in the image, we estimate the angle θ_1 and compute the distance between the camera and the sign in the real domain. (3) We estimate the angle θ_2 by assuming that all construction signs have the same height H_{sign} and stand perpendicular to the road. (4) Using the estimated angle θ_2, we compute the point of the top of the construction sign in the image.

3.5.3. Color Difference

One of the main reasons for the artifacts in the synthesized images, which is made using cut-and-paste, is the color difference between the two images. As shown in Figure 7c, the color difference is caused by differences in illumination, weather, and environment. To match the color difference between the construction sign image and road image, we blend the synthesized image using an off-the-shelf model, GP-GAN [24]. By blending, we can reduce the artifacts of the synthesized image, as shown in Figure 7d.

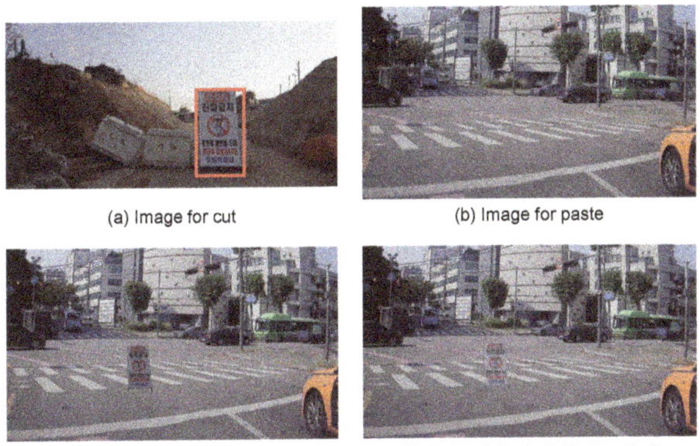

(a) Image for cut (b) Image for paste

(c) Cut-and-Paste (without blending) (d) Cut-and-Paste (with blending)

Figure 7. Effect of blending. Using the construction sign image in (**a**) and the road image in (**b**), we synthesize the training image via cut-and-paste. As shown in (**c**), however, the artifact seems prominent because of the color difference between the construction sign and the road. This problem is mitigated by blending, as shown in (**d**).

4. Experiments

4.1. Implementation Details

We conduct some experiments using our collected construction sign detection dataset, CSS138. We use YOLOv3 [1] as a construction sign detector. Basically, we follow the training and inference details in the original YOLOv3 paper [1]. We use Darknet-53 [1] as a backbone network. Darknet-53 consists of 53 convolutional layers and 23 residual connections. Darknet-53 outputs three different sizes of features, which have 1/8, 1/16, and 1/32 resolutions with respect to the input image. To detect construction signs from encoder's feature, a decoder is used. The decoder takes three outputs of Darknet-53, and outputs detection results at three different resolutions, i.e., 1/8, 1/16, and 1/32 resolutions with respect to the input image. Each output predicts five values: four for coordinates of the bounding box and one for objectness. Unlike vanilla YOLOv3, which predicts the class of the object, we do not predict the object class because we have only a single object class, construction sign, in this paper.

Additionally, we apply our method to YOLOv3-tiny to see the effectiveness of our method in other networks. YOLOv3-tiny uses Darknet-19 [25] as a backbone network. Darknet-19 has 19 convolutional layers without residual connections. YOLOv3 has 61.5M parameters, while YOLOv3-tiny has 8.7M parameters. These parameter numbers are comparable with state-of-the-art object detectors: Faster R-CNN [26] has 52.7M parameters, FPN [27] has 60.6M parameters, and RetinaNet [28] has 56.9M parameters. To train YOLOv3, we use an RGB image as an input.

A total of 9920 RGB images are synthesized for training using the CSS138 training set. The training set includes 992 road images and 36 construction sign images. Among 36 sign images, 12 images are in-domain and 24 are out-domain. We randomly crop 640×640 patches for the training. The learning rate is initially set to 1×10^{-2}, and we decrease the learning rate to 1×10^{-3} using the cosine decay schedule. The network is trained in 375,000 iterations with a mini-batch size of 16. The entire training process takes approximately 60 h using a single NVIDIA Titan V GPU. During inference, we obtain multiple detection results at three different resolutions. To select accurate results and dismiss overlapped noisy results, we use non-maximal suppression with an IoU threshold of 0.45.

4.2. Quantitative Results

We validate our approach on the CSS138 validation set. The CSS138 validation set includes 126 images containing at least one construction sign. For quantitative evaluation, we measure the average precision with Intersection over Union (IoU) thresholds of 0.5, and we denote it as AP_{50}. Following the recent object detection benchmark https://cocodataset.org/#detection-eval, (accessed on 5 April 2022), we additionally measure AP, which is calculated by computing 10 average precision values with IoU thresholds of {0.5, 0.55, ...0.9, 0.95} and then averaging them. To demonstrate the superiority of our approach, we set a baseline that synthesizes 9920 training images by using a naive cut-and-paste method. From the baseline, we add the proposed methods in a step-by-step manner. The experimental results for the CSS138 validation set are listed in Table 1. As shown in Table 1, our method achieves AP and AP_{50} scores of 70.36% and 84.98%, respectively, whereas the baseline achieves scores of 60.53% and 75.84%, respectively. We surpass the baseline by >+9% for both the AP and AP_{50} scores. Table 1 shows the contributions of each step of the proposed method. Each step improves the performance by >+2% for both AP and AP_{50}. This demonstrates that all our approaches are effective in synthesizing training images for construction sign detection.

Table 1. Experimental results on CSS138 validation set.

Method	AP	AP_{50}
A. Baseline (cut-and-paste)	60.53	75.84
B. + Pasteable region	62.74	77.30
C. + Instance size	65.57	82.44
D. + Content swapping	68.51	82.73
E. + Color difference	70.36	84.98

In Table 2, we additionally validate the efficacy of our instance size adjustment method. As described in Section 3.5.2, we resize the instance by projecting it to real-world coordinates. We can compare it with Fixed, which uses the original scale of the construction sign image. We can further compare it with Random, which uses a randomly sampled value for scaling construction sign images and was used in cut-and-paste [7]. As shown in Table 2, we significantly surpass Fixed and Random by 5% and 2%, respectively, in terms of the AP_{50} score. The results demonstrate the superiority of our instance size adjustment method.

Table 2. Analysis experiment on instance size.

Instance Size	AP	AP_{50}
Fixed	62.74	77.30
Random [7]	65.14	80.12
Ours	65.57	82.44

In Table 3, we conduct an ablation study using a different backbone. In the ablation study, we use DarkNet-19 in YOLOv3-tiny. As shown in Table 3, our proposed method improves the detection quality of both YOLOv3-tiny and YOLOv3 networks. This result demonstrates that our proposed method is effective in various networks.

Table 3. Ablation study with various backbone networks.

Method	YOLOv3-tiny (DarkNet-19)		YOLOv3 (DarkNet-53)	
	AP	AP_{50}	AP	AP_{50}
Baseline	53.40	70.67	60.53	75.84
Proposed	54.95	75.57	70.36	84.98

4.3. *Grad-CAM Result*

In this subsection, we analyze the effectiveness of our proposed method using Grad-CAM. In Figure 8, we visualize the Grad-CAM [29] results of YOLOv3. To extract Grad-CAM, we compute the gradient of the score for objectness at three different resolution outputs. Then, we average the three activations in each last layer of the decoder. To validate the efficacy of our proposed method, we compare two methods for synthesizing training images. One is to synthesize images simply using naive cut-and-paste method (baseline), and the other one is to synthesize the images using our proposed method. As shown in Figure 8, YOLOv3 trained using a baseline often cannot detect construction signs (second and third rows), while YOLOv3 trained with our proposed method gives accurate activation maps. This result demonstrates that our proposed method helps to learn the discriminative features for construction sign detection.

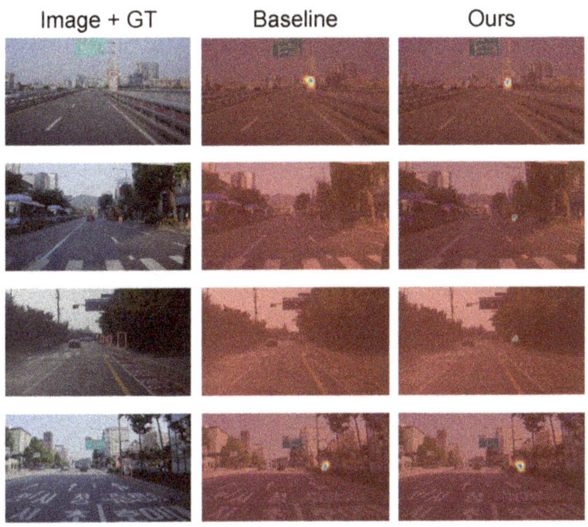

Figure 8. Grad-CAM result. In the first column, input images and corresponding ground truth bonding boxes of the construction sign are shown. In the second and third columns, Grad-CAM results are given. In the Grad-CAM results, high activation values are visualized in blue, while low activation values are visualized in red.

4.4. *Effect of Daylight*

In this subsection, we analyze the effect of daylight and whether on the performance. We build a hierarchical structure in our CSS138 training set by splitting it into two parts: one is captured under sufficient daylight (i.e., outdoor scene), and the other one is captured under low daylight (i.e., tunnel scene). Among 992 road images, 796 images were taken outdoors and 196 images were taken in tunnel. Examples of outdoor and tunnel scenes are given in Figure 9.

Figure 9. An example of synthesized training set in outdoor and tunnel scenes.

With this split, we train detection networks, and the results for the two different daylight conditions are given in Table 4. As shown in the table, daylight significantly contributed to the performance. Specifically, the performance difference between outdoors and the tunnel is about 30%, and the training set captured under sufficient daylight is more effective than the one under low daylight in improving detection performance. Therefore, daylight and weather are crucial for construction sign detection.

Table 4. Results on two different daylight conditions.

Split	YOLOv3-tiny (DarkNet-19)		YOLOv3 (DarkNet-53)	
	AP	AP_{50}	AP	AP_{50}
Outdoor	56.34	76.81	69.32	84.88
Tunnel	16.20	32.29	39.91	55.53

4.5. Qualitative Analysis

The synthesized training images are shown in Figure 10. Baseline (*A*) often pastes the construction sign on the sky, which never occurs in real-world scenarios. After considering the pasteable region (*B*), the construction sign is placed on the road, but the scale seems very unfamiliar. Our instance size adjustment method (*C*) could address this problem, but the problem of limited sign images remained. Our content swapping (*D*) effectively augments the construction sign images, preventing overfitting. Finally, the color difference (*E*) between the background road image and foreground construction sign image is adjusted to create a realistic image.

Figure 10. Synthesized training images. For each method, we sampled from the same five background images.

Figure 11 shows the qualitative results of the proposed methods on the CSS138 validation set, as well as the results of the baseline. As shown in the figure, our method finds small instances (first, second, and third rows) and precisely determines the bounding box of the construction sign (fourth row).

In Figure 12, we present some failure cases, and they show some limitation of our method. The first two rows present false negatives, while the last row present false positive. In the first two rows, construction signs are often missed when they are occluded by other objects such as traffic cones. The last row is the example of the false positive. As can be seen, a rectangular shape object is sometimes detected as a construction sign. We expect that this problem can be solved by various methods, e.g., pre-designing sign shape [10,12], hard example mining [30,31], or learning with strong generalization [32,33].

Figure 11. Qualitative results on CSS138 validation set. From left to right, each row shows ground truth (GT), the baseline (naive cut-and-paste), and Ours. For each result on the baseline and ours, we denote the number of missed construction signs.

Figure 12. Limitations: if the construction sign is severely occluded, we cannot detect it accurately (first and second rows). A rectangle shape can be detected as a construction sign (third row).

5. Conclusions

In this paper, we have presented a new approach for synthesizing training images for construction sign detection and trained a deep learning detector on them. Since this is the first paper which deals with the construction sign detection, there is not a benchmark set, and we have applied our method to real-world images. Our approach is effective, even when only a few construction sign images are available. Furthermore, our main proposal, content swapping, allows us to use out-domain construction sign data, effectively alleviating the problem of data hunger. To demonstrate the efficacy of our approach, we collected road and construction sign images in person and collected out-domain construction sign images from the Internet. The images used in our experiments are gathered as a dataset CSS138, and we made the dataset available online for the benefit of our community. Even though our method was tested only on the dataset gathered in Seoul, South Korea, we firmly believe that our methods will be applied to other countries and other similar sign-related tasks successfully. Since our content swapping allows us to train networks with a few images, it has the potential to be applied to the few-shot learning field. In this paper, we applied our method only to images, but our proposed method can be extended to videos by applying content swapping and realistic transformations smoothly over time. In addition, our method can be extended to stereo-camera by modeling a construction sign in 3D and projecting it into stereo-view. In addition, a laser scanner sensor can also be considered to measure the distance between the vehicle and the construction sign. The measured distance can improve the quality of the realistic transformations. Furthermore, the future direction of this work would be deciding the action of the autonomous vehicles, after detecting construction signs.

Author Contributions: Conceptualization, H.S., S.L., and E.K.; methodology, H.S., S.L., and E.K.; software, S.B. and Y.L.; validation, H.S., S.B., and Y.L.; formal analysis, H.S., S.L., and E.K.; investigation, H.S., S.L., and E.K.; resources, S.B. and Y.L.; data curation, S.B.; writing—original draft preparation, H.S.; writing—review and editing, S.L. and E.K.; visualization, H.S.; supervision, E.K.; project administration, E.K.; funding acquisition, E.K. All authors have read and agreed to the published version of the manuscript.

Funding: This work was supported by the Technology Innovation Program, (No. 20014124, Development of multipurpose mid-size bus platform technology for automated driving based on predefined route) funded By the Ministry of Trade, Industry and Energy (MOTIE), Korea.

Institutional Review Board Statement: Not applicable.

Informed Consent Statement: Not applicable.

Data Availability Statement: The raw data supporting the conclusions of this article will be made available by the corresponding author upon reasonable request.

Conflicts of Interest: The authors declare no conflict of interest.

References

1. Redmon, J.; Farhadi, A. YOLOv3: An Incremental Improvement. *arXiv* **2018**, arXiv:1804.02767.
2. Ahmed, M.; Hashmi, K.A.; Pagani, A.; Liwicki, M.; Stricker, D.; Afzal, M.Z. Survey and Performance Analysis of Deep Learning Based Object Detection in Challenging Environments. *Sensors* **2021**, *21*, 5116. [CrossRef] [PubMed]
3. Zhao, H.; Zhou, Y.; Zhang, L.; Peng, Y.; Hu, X.; Peng, H.; Cai, X. Mixed YOLOv3-LITE: A Lightweight Real-Time Object Detection Method. *Sensors* **2020**, *20*, 1861. [CrossRef] [PubMed]
4. Charouh, Z.; Ezzouhri, A.; Ghogho, M.; Guennoun, Z. A Resource-Efficient CNN-Based Method for Moving Vehicle Detection. *Sensors* **2022**, *22*, 1193. [CrossRef] [PubMed]
5. Miller, D.; Sünderhauf, N.; Milford, M.; Dayoub, F. Uncertainty for Identifying Open-Set Errors in Visual Object Detection. *IEEE Robot. Autom. Lett.* **2021**, *7*, 215–222. [CrossRef]
6. Jiang, L.; Nie, W.; Zhu, J.; Gao, X.; Lei, B. Lightweight object detection network model suitable for indoor mobile robots. *J. Mech. Sci. Technol.* **2022**, *36*, 907–920. [CrossRef]
7. Yun, W.H.; Kim, T.; Lee, J.; Kim, J.; Kim, J. Cut-and-Paste Dataset Generation for Balancing Domain Gaps in Object Instance Detection. *IEEE Access* **2021**, *9*, 14319–14329. [CrossRef]

8. Lee, S.; Hyun, J.; Seong, H.; Kim, E. Unsupervised Domain Adaptation for Semantic Segmentation by Content Transfer. In Proceedings of the Thirty-Fifth AAAI Conference on Artificial Intelligence, virtual, 2–9 February 2021.
9. Eversberg, L.; Lambrecht, J. Generating Images with Physics-Based Rendering for an Industrial Object Detection Task: Realism versus Domain Randomization. *Sensors* **2021**, *21*, 7901. [CrossRef] [PubMed]
10. Prince, S.; Bergevin, R. Road sign detection and recognition using perceptual grouping. In Proceedings of the International Symposium on Automotive Technology & Automation, Florence, Italy, 16–19June 1997.
11. De La Escalera, A.; Moreno, L.E.; Salichs, M.A.; Armingol, J.M. Road Traffic Sign Detection and Classification. *IEEE Trans. Ind. Electron.* **1997**, *44*, 848–859. [CrossRef]
12. Fang, C.Y.; Chen, S.W.; Fuh, C.S. Road-Sign Detection and Tracking. *IEEE Trans. Veh. Technol.* **2003**, *52*, 1329–1341. [CrossRef]
13. Shadeed, W.; Abu-Al-Nadi, D.I.; Mismar, M.J. Road traffic sign detection in color images. In Proceedings of the 10th IEEE International Conference on Electronics, Circuits and Systems, 2003. ICECS 2003. Proceedings of the 2003, Sharjah, United Arab Emirates, 14–17 December 2003; Voume 2, pp. 890–893.
14. Loy, G.; Barnes, N. Fast shape-based road sign detection for a driver assistance system. In Proceedings of the 2004 IEEE/RSJ International Conference on Intelligent Robots and Systems (IROS)(IEEE Cat. No. 04CH37566), Sendai, Japan, 28 September–2 October 2004; Volume 1, pp. 70–75.
15. Bahlmann, C.; Zhu, Y.; Ramesh, V.; Pellkofer, M.; Koehler, T. A system for traffic sign detection, tracking, and recognition using color, shape, and motion information. In Proceedings of the EE Proceedings. Intelligent Vehicles Symposium, Las Vegas, NV, USA, 6–8 June 2005; pp. 255–260.
16. Shao, F.; Wang, X.; Meng, F.; Rui, T.; Wang, D.; Tang, J. Real-time traffic sign detection and recognition method based on simplified Gabor wavelets and CNNs. *Sensors* **2018**, *18*, 3192. [CrossRef] [PubMed]
17. Cao, J.; Song, C.; Peng, S.; Xiao, F.; Song, S. Improved traffic sign detection and recognition algorithm for intelligent vehicles. *Sensors* **2019**, *19*, 4021. [CrossRef] [PubMed]
18. Zhang, J.; Xie, Z.; Sun, J.; Zou, X.; Wang, J. A cascaded R-CNN with multiscale attention and imbalanced samples for traffic sign detection. *IEEE Access* **2020**, *8*, 29742–29754. [CrossRef]
19. Liu, Y.; Peng, J.; Xue, J.H.; Chen, Y.; Fu, Z.H. TSingNet: Scale-aware and context-rich feature learning for traffic sign detection and recognition in the wild. *Neurocomputing* **2021**, *447*, 10–22. [CrossRef]
20. Ahmed, S.; Kamal, U.; Hasan, M.K. DFR-TSD: A deep learning based framework for robust traffic sign detection under challenging weather conditions. *IEEE Trans. Intell. Transp. Syst.* **2021**, 1–13. [CrossRef]
21. Zeng, H. Real-Time Traffic Sign Detection Based on Improved YOLO V3. In Proceedings of the 11th International Conference on Computer Engineering and Networks, Beijing, China, 9–11 December 2022; pp. 167–172.
22. Frolov, V.; Faizov, B.; Shakhuro, V.; Sanzharov, V.; Konushin, A.; Galaktionov, V.; Voloboy, A. Image Synthesis Pipeline for CNN-Based Sensing Systems. *Sensors* **2022**, *22*, 2080. [CrossRef] [PubMed]
23. Ranftl, R.; Lasinger, K.; Hafner, D.; Schindler, K.; Koltun, V. Towards Robust Monocular Depth Estimation: Mixing Datasets for Zero-shot Cross-dataset Transfer. *IEEE Trans. Pattern Anal. Mach. Intell.* **2020**, *44*, 1623–1637. [CrossRef] [PubMed]
24. Wu, H.; Zheng, S.; Zhang, J.; Huang, K. Gp-gan: Towards realistic high-resolution image blending. In Proceedings of the ACM Multimedia 2019 Conference, Nice, France, 21–25 October 2019; pp. 2487–2495.
25. Redmon, J.; Farhadi, A. YOLO9000: better, faster, stronger. In Proceedings of the 2017 IEEE Conference on Computer Vision and Pattern Recognition, Honolulu, HI, USA, 22–25 July 2017; pp. 7263–7271.
26. Ren, S.; He, K.; Girshick, R.; Sun, J. Faster r-cnn: Towards real-time object detection with region proposal networks. *NIPS* **2015**, *28*, 91–99. [CrossRef] [PubMed]
27. Lin, T.Y.; Dollár, P.; Girshick, R.; He, K.; Hariharan, B.; Belongie, S. Feature pyramid networks for object detection. In Proceedings of the 2017 IEEE Conference on Computer Vision and Pattern Recognition, Honolulu, HI, USA, 22–25 July 2017; pp. 2117–2125.
28. Lin, T.Y.; Goyal, P.; Girshick, R.; He, K.; Dollár, P. Focal loss for dense object detection. In Proceedings of the 2017 IEEE International Conference on Computer Vision, Venice, Italy, 22–29 October 2017; pp. 2980–2988.
29. Selvaraju, R.R.; Cogswell, M.; Das, A.; Vedantam, R.; Parikh, D.; Batra, D. Grad-CAM: Visual Explanations from Deep Networks via Gradient-Based Localization. *Int. J. Comput. Vis.* **2020**, *128*, 336–359. [CrossRef]
30. Shrivastava, A.; Gupta, A.; Girshick, R. Training region-based object detectors with online hard example mining. In Proceedings of the 6 IEEE Conference on Computer Vision and Pattern Recognition, Negas, NV, USA, 27–30 June 2016; pp. 761–769.
31. Lee, S.; Seong, H.; Lee, S.; Kim, E. Correlation Verification for Image Retrieval. In Proceedings of the 2022 Conference on Computer Vision and Pattern Recognition, New Orleans, LA, USA, 19–24 June 2022.
32. Zhou, K.; Yang, Y.; Qiao, Y.; Xiang, T. Domain Generalization with MixStyle. In Proceedings of the 9th International Conference on Learning Representations, Virtual, 3–7 May 2021.
33. Lee, S.; Seong, H.; Lee, S.; Kim, E. WildNet: Learning Domain Generalized Semantic Segmentation from the Wild. In Proceedings of the 2022 Conference on Computer Vision and Pattern Recognition, New Orleans, LA, USA, 19–24 June 2022.

sensors

MDPI

Article

Towards Enhancing Traffic Sign Recognition through Sliding Windows

Muhammad Atif [1], Tommaso Zoppi [1], Mohamad Gharib [2] and Andrea Bondavalli [1,*]

[1] Department of Mathematics and Informatics, 50142 Florence, Italy; muhammad.atif@unifi.it (M.A.); tommaso.zoppi@unifi.it (T.Z.)
[2] Institute of Computer Science, University of Tartu, 51009 Tartu, Estonia; mohamad.gharib@ut.ee
* Correspondence: andrea.bondavalli@unifi.it

Abstract: Automatic Traffic Sign Detection and Recognition (TSDR) provides drivers with critical information on traffic signs, and it constitutes an enabling condition for autonomous driving. Misclassifying even a single sign may constitute a severe hazard, which negatively impacts the environment, infrastructures, and human lives. Therefore, a reliable TSDR mechanism is essential to attain a safe circulation of road vehicles. Traffic Sign Recognition (TSR) techniques that use Machine Learning (ML) algorithms have been proposed, but no agreement on a preferred ML algorithm nor perfect classification capabilities were always achieved by any existing solutions. Consequently, our study employs ML-based classifiers to build a TSR system that analyzes a sliding window of frames sampled by sensors on a vehicle. Such TSR processes the most recent frame and past frames sampled by sensors through (i) Long Short-Term Memory (LSTM) networks and (ii) Stacking Meta-Learners, which allow for efficiently combining base-learning classification episodes into a unified and improved meta-level classification. Experimental results by using publicly available datasets show that Stacking Meta-Learners dramatically reduce misclassifications of signs and achieved perfect classification on all three considered datasets. This shows the potential of our novel approach based on sliding windows to be used as an efficient solution for TSR.

Keywords: traffic sign recognition; sliding windows; meta learning; deep learning; classification

Citation: Atif, M.; Zoppi, T.; Gharib, M.; Bondavalli, A. Towards Enhancing Traffic Sign Recognition through Sliding Windows. *Sensors* **2022**, *22*, 2683. https://doi.org/10.3390/s22072683

Academic Editors: Euntai Kim, Sangyoun Lee and Kang Ryoung Park

Received: 11 February 2022
Accepted: 29 March 2022
Published: 31 March 2022

1. Introduction

Intelligent transportation systems are nowadays of utmost interest for researchers and practitioners as they aim at providing advanced and automatized functionalities, such as obstacle detection, traffic sign recognition, car plate recognition, and automatic incident detection or stopped vehicle detection systems. Particularly, Traffic Sign Detection and Recognition (TSDR) systems aim at detecting (TSD) and recognizing (TSR) traffic signs from images or frames sampled by sensors [1–3] installed on vehicles (e.g., webcams). Those systems synergize with the human driver, who may misinterpret or miss an important traffic sign, potentially leading to accidents that may generate safety-related hazards [4]. When integrated into intelligent vehicles [5,6], in terms of Advanced Driver-Assistance Systems (ADAS) [2,7–9], TSDR can automatically provide drivers with actionable warnings or even trigger reaction strategies (e.g., automatic reduction of speed, braking) that may be crucial to avoid or reduce the likelihood of accidents [3,10].

Humans are expected to naturally miss or misinterpret a traffic sign occasionally because of being distracted [11]. Similarly, to humans, TSDR systems are also subject to error as they may misinterpret or miss a traffic sign. This could happen due to various reasons, such as unsatisfactory road situations, imperfect traffic sign state, adverse environmental conditions (e.g., foggy weather [12]) or imperfect analysis processes. Nevertheless, researchers and practitioners are trying to minimize misclassifications at the automatic TSDR side, which is expected to increase safety by providing drivers with accurate and timely notifications.

Available TSD systems can precisely extract areas of an image or a frame, which are supposed to contain a traffic sign. Thereto, TSR systems that embed Machine Learning (ML) algorithms [13–16], process features that are extracted from those images through feature descriptors (e.g., Histogram of Oriented Gradients (HOG) [17], Local Binary Pattern (LBP) [16] to recognize traffic sign categories [18,19]. Alternatively, deep learning algorithms, such as AlexNet, googLeNet [20] can directly process images coming from sensors and classify them according to internal representation learning processes, which are orchestrated through multiple convolutional and fully connected layers. Throughout the years, many studies tackled TSR [21–23] using different feature descriptors and ML-based classifiers. Different combinations of such classifiers and features have been proven to generate heterogeneous classification scores [15,19,24–27] motivating the need for comparisons to discover the optimal classifier for a given TSR problem [3,28,29].

Regardless of the outcomes of comparison studies, most of the existing solutions for TSR process a single image or frame and output a classification result. Instead, vehicles gradually approach traffic signs during their road trips, generating sequences of images: the closer the vehicle is to the traffic sign, the better the quality of the image, even under slightly different environmental conditions. Therefore, the problem of TSR naturally scales to knowledge extraction from a set or sequence of images that potentially contain traffic signs. Consequently, the classification process should not depend only on a single frame to make a decision; instead, it should build on the knowledge acquired as the vehicle moves forward, i.e., the sequence of images.

This study considers a sliding window of images to commit classification rather than classifying frames individually. First, we process each image with the most effective single-frame classifier for TSR: then, we combine classification scores assigned to images in the sliding window to provide a unified and improved classification result. Such a combination is performed by appropriate Meta-Learners [30], which suit model combination, and therefore, show potential to be applied in such a context.

We conduct an experimental evaluation by processing three public datasets, namely, (i) German Traffic Sign Recognition Benchmark (GTSRB) [31] (ii) BelgiumTSC [32], and (iii) the Dataset of Italian Traffic Signs (DITS) [33], which report on sequences or unordered sets of images of traffic signs. From each image, we extracted 12 different feature sets, which use handcrafted features (HOG [17], LBP [16]), deep features (from AlexNet [34], ResNet-18 [35]), and their combinations, to debate their impact in TSR. Those features were fed to supervised classifiers as Decision Trees [36], Random Forests [37], k-th Nearest Neighbour (K-NN, [13]), Linear Discriminant Analysis Classifier (LDA) [38], Support Vector Machines (SVMs) [14], and AdaBoost [39]. We also exercised single-frame classifiers that do not rely on feature descriptors as deep learners, namely Inceptionv3 [40], MobileNet-v2 [41] and AlexNet [34]. We used the classifiers above both as single-frame classifiers and as base-level learners of a Stacking meta-learner, which aggregates individual classification scores into sliding windows. The meta-level classifier for Stacking was experimentally chosen out of supervised (non-deep) classifiers, the Majority Voting [42] and Discrete Hidden Markov Model (DHMM, [43]).

Additionally, we compare the classification performance of those meta-learners with Long Short-Term Memory (LSTM) networks, which naturally deal with sequences of data coming at different time instants. We trained those LSTM networks on the same sliding windows of images processed through Stacking. Results show how single-frame classifiers achieve 100% accuracy on the GTRSB, 99.72% on BelgiumTSC and 96.03% on DITS datasets. Then, we applied our approach based on sliding windows by using LSTM networks and stacking meta-learners, finding that both approaches greatly improve the accuracy of TSR: particularly, specific stacking meta-learners achieved perfect accuracy (i.e., no misclassifications at all) on the three datasets by using a sliding window of two or three images.

Summarizing the contribution and novelty of the paper mainly lies in the following items:

- a deep literature review about ML-based TSR;

- presentation of an approach based on sliding windows of frames to be processed either by meta-learners or LSTM;
- an experimental campaign that relies on heterogeneous and public datasets of traffic signs; and finally
- a discussion of results that clearly shows how a sliding window of at least two items, deep base-level classifiers and K-NN as stacking meta-learner allow achieving perfect TSR on all datasets considered in the study, dramatically improving the state of the art.

The rest of the paper is organized as follows: Section 2 elaborates on related works and a review of existing TSR systems. Section 3 expands on our approach based on sliding windows. Section 4 reports on our experimental setup and methodology, classifiers, and feature sets to compare different TSR systems. Finally, Section 5 discusses and comments on those experimental results, letting Section 6 conclude the paper.

2. Background on Traffic Sign Recognition

2.1. Classifiers for TSR

In the last decade, researchers, practitioners, and companies devised automatic TSR systems to be integrated into ADAS. Amongst all the possible approaches, most TSR systems rely on the same main blocks, namely: (i) Dataset creation/identification, (ii) pre-processing (e.g., resizing, histogram equalization), (iii) Feature extraction and supervised model learning, or (iv) model learning through deep classifiers (i.e., deep learners).

As depicted in Figure 1, these building blocks interact with each other sequentially. Each image in the dataset is pre-processed to make feature extraction easier. These features are then fed into the classifier, either for training or for testing (right of Figure 1) if the model was already learned. Alternatively (see bottom left of Figure 1) we could rely on deep learning algorithms, which—unlike traditional supervised classifiers—embed representation learning, and therefore, do not require feature extraction.

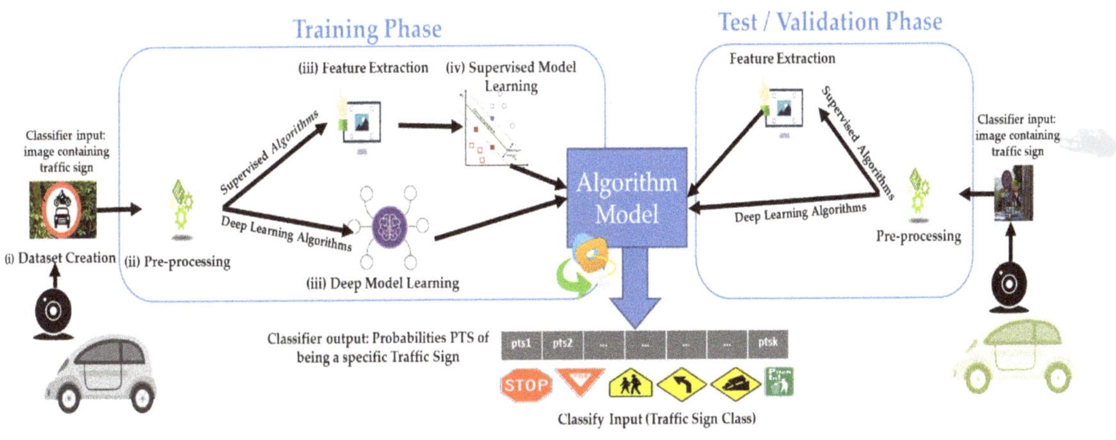

Figure 1. Block diagram of traffic sign recognition through deep learners or supervised classifiers.

Regardless of their type, classifiers output Probabilities of Traffic Sign categories (PTS), or rather, assign probabilities belonging to any known class of traffic signs to each image. The category of a traffic sign corresponds to the highest probability in PTS which defines the predicted class of a given image.

2.2. Related Works on Single-Frame TSR

Feature extractors and supervised classifiers have been arranged differently to minimize misclassifications in a wide variety of domains. Soni et al. [24] processed the Chinese traffic sign dataset through SVM, trained on the HOG or LBP after Principal Component

Analysis (PCA), reaching an accuracy of 84.44%. A similar setup was used by Manisha and Liyanage [21], who achieved 98.6% accuracy on vehicles moving at 40–45 km/h. Moreover, Matoš et al. [22] used an SVM trained on HOG features and achieved recognition of 93.75% accuracy on the GTSRB dataset. The same dataset was used in [44], where Extreme Learning Machine (ELM) improved accuracy to 96%. Agrawal and Chaurasiya [45] extracted HOG features from the traffic signs of the GTSRB dataset and applied PCA for the dimensionality reduction to obtain an accuracy of 73.99%, 66.46%, 91.86% on denial, mandatory and danger traffic sign categories. Similar studies as [15,46,47] processed the same datasets with different feature sets and algorithms, obtaining similar scores.

Deep learners and the Viola–Jones framework allowed the authors of [8] to enhance classification on the GTSRB dataset with up to 90% of accuracy. Li et al. [28] proposed a new Convolutional Neural Network and trained on the GTSRB and BelgiumTSC datasets. The proposed architecture achieved an accuracy of 98.1% and 97.4% on BelgiumTSC and GTSRB datasets, respectively. In [48], the fifteen-layer WAF-LeNet network reached a detection accuracy of 96.5% on GTSRB. The authors of [49] proposed an approach for TSDR using SegU-Net and a modified Tversky loss function With L1-Constraint that achieved 94.60% and 80.21% precision and recall, respectively, on the CURE-TSD dataset. Liu et al. [50] proposed traffic sign recognition and detection approaches, which first extract the region of interest and after verification of the traffic sign through an SVM classifier, it classifies the traffic sign into traffic sign categories. The proposed approach achieves the highest accuracy 94.81%.

Another study [51] used the Inceptionv3 model trained with transfer learning on the BelgiumTSC dataset, obtaining an accuracy of 99.18%. In [52], the authors found that the Tiny-YOLOv2 network is fast but outperformed by YOLOv2 or YOLOv3 deep learners. While the authors of [53] introduced real time image enhancement CNN and achieved an accuracy of 99.25% for the BelgiumTSC, 99.75% for GTSRB, and 99.55% for Croatian Traffic Sign (rMASTIF). Authors of [54], developed a real time TSR by using the You Only Look Once (YOLO) algorithm to train the model for Malaysian traffic sign recognition and tested it on five types of warning traffic signs. In [55] authors propose a lightweight CNN architecture for the recognition of the traffic sign GTSRB dataset, and they achieved 99.15% accuracy. In one another study [56], a novel semi supervised classification technique is adopted for TSR with weakly-supervised learning and self-training. An ensemble of CNN was used for the recognition of the traffic signs and achieved higher than 99% accuracy for the circular traffic signs of the German and BelgiumTSC datasets [57]. Lu et al. [58] use multi-modal tree structure embedded multitask learning for the GTSRB dataset and achieved an overall accuracy of 98.27%. In [59], the authors improved the VGG-16 deep model by removing some redundant convolutional layers and adding Batch Normalization and global average pooling layer to improve the performance of the network, while [60] proposed a hybrid 2D-3D CNN. In [61], the authors proposed a traffic sign recognition system that learns learning hierarchical features based on multi-scale CNNs. In one another study [62], the authors proposed a real-time TSDR for Chinese and German roads. In [63] authors proposed a robust custom feature extraction method and multilayer artificial neural network for the recognition of traffic signs in real time.

Additionally, a few works perform classification depending on multiple frames. In a study [64], authors considered the sequences of frames of the street view images and achieved an 87.03% evaluation score, i.e., the ratio of true positive and true positive + false positive + false negative. In another study, Yuan et al. [65] proposed a video based traffic sign detection and recognition mechanism to fuse the result of all frames for final classification. They utilized a multi-class SVM with two different fusion strategies, i.e., equal weighting and a scale based weighting scheme which achieved 99.48% accuracy on the TS2010 dataset.

In the literature, there are many studies [66–69] focusing on single frame TSR, and very few studies [64,65] that process multiple frames. According to our knowledge based on the literature review, there is no study available that considers the sliding windows approach for traffic sign recognition.

2.3. Background on Comparative Studies

Only a few comparative studies have been proposed in the literature. For example, Jo [15] trained different supervised classifiers on HOG features extracted from the GTSRB dataset. Similarly, Schuszter [70] reported on experiments with the BelgiumTSC dataset [32], where HOG features were extracted from images and then fed to the SVM to classify one of the six basic traffic sign subclasses. Yang et al. [19] provide a comparison of different classifiers, such as the K-NN, SVM, Random Forest and AdaBoost trained by using combinations of features. This study reported the highest accuracy by using Random Forest with the combination of LBP and HOG features. Another study [29] compared traditional supervised classifiers and deep learning models on three datasets, i.e., GTSRB, BelgiumTSC and DITS considering three broad categories of traffic signs, i.e., red circular, blue circular and red triangular. Noticeably, both traditional supervised classifiers and deep classifiers achieved perfect accuracy on GTSRB. Moreover, the authors of [18] trained different classifiers for traffic sign recognition. They considered the GTSRB dataset and extracted HOG features to train LDA and Random Forest. Additionally, they used the committee of Convolutional Neural Networks (CNN) and multiscale-scale CNN. While in the study [31] authors organized a competition to classify GTSRB dataset traffic signs. These traffic signs were categorized by human and ML algorithms and an accuracy of 98.98% was achieved which is comparable to human performance on this dataset.

3. Sliding Windows to Improve TSR

TSR naturally fits the analysis of sequences of images being collected as the vehicle approaches the traffic sign. Therefore, we organize a complex classifier that processes sliding windows of frames

As shown in Figure 2, a sliding window of size s contains (i) the most recent frame sampled by the sensors on the vehicle plus (ii) the s-1 most recent frames. The figure represents how sliding windows of size s = 2 and s = 3 evolve as time passes by as the vehicle approaches a speed limit sign. Intuitively, the closer the vehicle gets to the traffic sign, the more visible and clearer the traffic sign gets. On the other hand, the sooner the TSR correctly classifies a traffic sign, the better it is for the ADAS, e.g., it may provide more time for emergency braking, whenever needed.

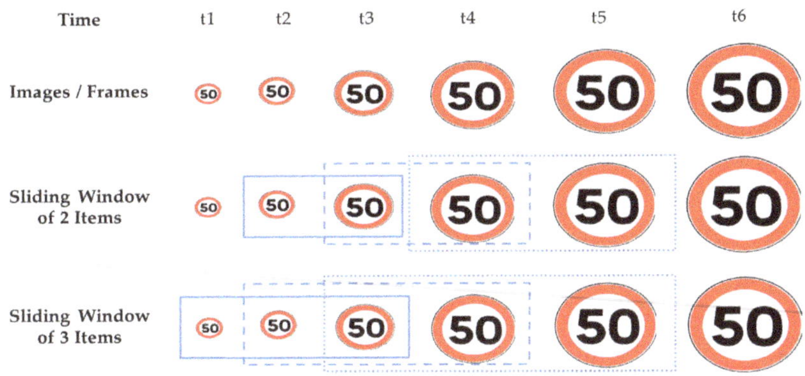

Figure 2. Example of Sliding Windows. Dotted, dashed and solid boxes show sliding windows, respectively, at t5, t4, t3.

3.1. Sliding Windows and Meta-Learning

Adopting sliding windows of s images calls for a rework of the TSR system. In particular, classification should be carried out using s subsequent classifications, which contribute to the final decision on the traffic sign. Those single-frame classifications for subsequent frames have to be combined by utilizing an independent strategy that delivers the result of this ensemble of single-frame classifiers.

Such a combination is usually orchestrated through meta-learning [30,71], which uses knowledge acquired during base-learning episodes, i.e., meta-knowledge, to improve classification capabilities. More specifically [72], a base-learning process starts feeding images into one or more base classifiers to create meta-data at the first stage. Results of those base learners, i.e., meta-data are provided alongside other features to the meta-level classifier as input features, which in turn provides the classification result of the whole meta-learner.

The paradigm of meta-learning can be adapted to TSR as shown in Figure 3. Let k be the number of different categories of traffic signs (i.e., classes), and let s be the size of the sliding window. Starting from the left of the figure, frames are processed by means of single-frame base-level classifiers, which provide k probabilities $PTS_i = \{pts_{i1}, \ldots pts_{ik}\}$ to classify each frame. Depending on the current time t_j, we create a sliding window of at most s×k items, namely swsj = $\{PTS_j, PTS_{j-1}, \ldots PTS_{j-s-1}\}$, which builds the meta-data to be provided to the meta-level classifier. On the right side of Figure 3, the meta-level classifier processes such meta-data and provides the k probabilities PTS_{final}, which will constitute the classification result of the whole sequence within the sliding window. As time moves on, we will have newly captured images and the sliding window will process the most recent s × k items. Note, that the sliding window sw_{sj} may contain less than s × k items when j < s (e.g., the window of size 3 at time t2 in Figure 2). In those cases, the TSR system will decide based on a single-frame classification of the most recent image.

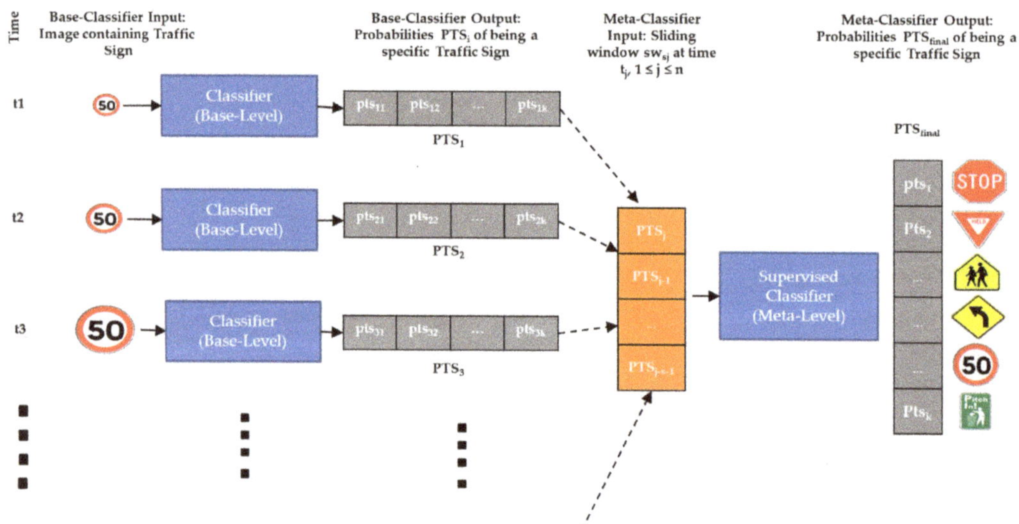

Figure 3. Diagram representing TSR which uses sliding windows. Blue solid boxes represent single-frame classifier in Figure 1.

3.2. A Stacking Meta-Learner

The structure of the meta-learner we described previously is traditionally referred to as Stacking. Stacking [73] builds a base-level of different classifiers as base learners. Base-learners can be trained with the exact same training set or with different training sets, mimicking Bagging [74]. Each of the n base-learners generates meta-features (PTS_i, $1 \leq i \leq n$ in Figure 3) that are fed to another independent classifier, the meta-level classifier, which calculates and delivers the final output (PTS_{final} in Figure 3).

In our instantiation of the Stacker, we use the same base-level classifier, which can either be a deep learner or a traditional supervised classifier but feed each base-learner with a different image. The meta-level classifier is necessarily a supervised (non-deep) classifier as it has to process numeric features contained in sw_{sj} rather than images.

3.3. Long Short-Term Memory Networks (LSTM)

As an alternative to stacking, we plan the usage of LSTM networks [75,76]. An LSTM network is a Recurrent Neural Network that learns the long-term dependencies between time steps of sequence data by orchestrating two layers. Those networks do not have a meta-learning structure as a stacker: however, they perfectly fit the analysis of sliding windows of traffic signs as they are intended to be used for the classification of sets or sequences by directly processing multiple frames. The first layer contains a sequence of inputs, which are then forwarded to the LSTM fully connected layer, and finally, the output layer shows the classification result.

4. Methodology, Inputs and Experimental Setup

This section describes the methodology, inputs, and experimental setup to compare single-frame classifiers and approaches built upon sliding windows, such as Stacking and LSTM networks. Results will be presented, analyzed, and discussed in Section 5.

4.1. Methodology for a Fair Comparison of TSR Systems

We orchestrate our experimental methodology as follows:

- **Datasets and Pre-processing.** Images go through a pre-processing phase to resize them to the same scale and enhance the contrast between background and foreground through histogram equalization.
- **Feature Extraction** (Section 4.3). Then, each pre-processed image is analyzed to extract features: these will be used with traditional supervised classifiers, while deep learners will be directly fed with pre-processed images.
- **Classification Metrics** (Section 4.4). Before exercising classifiers, we select metrics to measure the classification capabilities of ML algorithms which apply both to single-frame classifiers and to others based on sliding windows.
- **Single-Frame Classification.** Both supervised (Section 4.5) classifiers and deep learners (Section 4.6) will be trained and tested independently, executing grid searches to identify proper values for hyper-parameters.
- **Sliding Windows with Stacking Meta-Learners** (Section 4.7). Results of single-frame classifiers will then be used to build Stacking learners as described in Section 3.2 and by adopting different meta-level classifiers.
- **Sliding Windows with LSTM** (Section 4.8). Furthermore, sliding windows will be used to exercise LSTM networks as described in Section 3.3.

Exercising such methodology with its inputs required approximately 6 weeks of execution. The experiments were conducted on an Intel(R) Core (TM) i5-8350U CPU@1.7 GHz 1.9 GHz running MATLAB. MATLAB implementations of Deep Learners also use our NVIDIA Quadro RTX 5000 GPU.

4.2. TSR Datasets and Traffic Sign Categories

We conducted extensive research to identify commonly used labeled datasets reporting on sequences of traffic signs with overlapping categories. We selected three public datasets which report on sequences of images of traffic signs, namely: (i) the BelgiumTSC dataset [32], (ii) the GTSRB dataset [31], and (iii) the DITS [33]. Details about their structure and the categories of traffic signs are in Tables 1 and 2, respectively.

Table 1. Details of the three datasets used in this study.

Dataset	Train Images	Test Images	Images per Sequence	Training Sequences	Test Sequences
GTSRB	39210	12570	30	1307	419
DITS	7500	1159	15	500	123
BelgiumTSC	4581	2505	3	1527	835

Table 2. Categorization of Traffic Signs into 9 categories based on their shape, color, and content.

Category	**1**	**2**	**3**	**4**	**5**	**6**	**7**	**8**	**9**
Traffic Signs							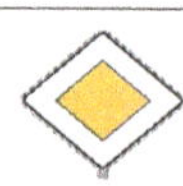		
GTSRB								✗	✓
BelgiumTSC	✓	✓	✓	✓	✓	✓	✓	✓	✗
DITS								✓	✓

4.2.1. German Traffic Signs Recognition Benchmark Dataset

The German Traffic Signs Recognition Benchmark (GTSRB [31]) dataset is widely used in the literature [15,18,19,31] as it reports on images of traffic signs belonging to eight categories with heterogeneous illumination, occlusion and distance from the camera. The dataset contains sequences of 30 images for each traffic sign, which were gathered as the vehicle was approaching it. The authors made available 1307 training and 419 testing sequences of images for a total of 51,780 images contained in the dataset. Table 2 depicts examples of traffic signs for each category of traffic sign contained in this dataset. Importantly, the rectangular traffic signs we mapped into category 8 in the table do not appear in the GTSRB dataset but appear in other datasets considered in this study.

4.2.2. BelgiumTSC Dataset

The BelgiumTSC dataset [32] is another dataset of traffic signs which was extensively used in the last decade [32,70]. The BelgiumTSC contains eight categories of traffic signs, shown from category 1 to category 8 in Table 2. The dataset is smaller than the GTSRB: the BelgiumTSC contains only 2362 sets of three images taken with different cameras from different viewpoints. It follows that this dataset reports triple images for each traffic sign which are all taken at the same time and thus are not time- ordered: this requires a dedicated discussion that we expand on in Section 5.4.

4.2.3. Dataset of Italian Traffic Signs Dataset

The Dataset of Italian Traffic Signs (DITS) dataset is considered more challenging than others in the literature [33] as it contains traffic signs images that were taken under non-optimal lighting conditions, e.g., day, night-time, foggy weather. The DITS contains 623 sequences containing a varying, time-ordered, number of frames. We point out that DITS is the only dataset in this study that contains all the nine categories of traffic signs reported in Table 2 and as such, it provides a complete view of all potential traffic signs. The dataset contains 500 training sequences and 123 testing sequences of varying lengths as summarized in Table 1.

4.3. Feature Descriptors

In this study, we extract features from images by means of *handcrafted*, i.e., HOG, LBP and *deep*, i.e., AlexNet and ResNet, feature descriptors, as described below.

- **Histogram of Oriented Gradients (HOG)** mostly provides information about key points in images. The process partitions an image into small squares and computes the normalized HOG histogram for each key point in each square [17].
- **Local Binary Patterns (LBP)** encode local textures [16] by partitioning each image into non-overlapping cells. Then, LBP isolates local binary patterns and uses small gray-scale discrepancies to identify specific features. Its behavior is invariant to the monotonic transformation of grayscale.
- **AlexNet Features (AFeat)** are extracted through a pre-trained AlexNet [34], composed of five convolutional layers and three fully connected layers. Convolutional layers are basically extracting deep features from RGB images of size 227 × 227. We extract a feature vector of 4096 items by fetching data at the fully connected layer "fc7".
- **ResNet Features (RFeat)** are extracted from a ResNet-18 [35], a convolutional neural network with 18 hidden layers. Convolutional layers are extracting deep features from RGB images of size 224 × 224 Similarly to AlexNet, we extract 512 features by extracting data at the global average pooling layer "pool5".

In addition, we combine hand-crafted and deep feature descriptors that are consequently fed simultaneously to classifiers: couples as {HOG ∪ LBP}, {AFeat ∪ HOG}, {AFeat ∪ LBP}, {RFeat ∪ HOG}, {RFeat ∪ LBP}, {AFeat ∪ RFeat}, and triples of {AFeat ∪ HOG ∪ LBP} and {RFeat ∪ HOG ∪ LBP}.

4.4. Classification Metrics

The performance of classifiers for TSR is usually compared by means of classification metrics. These metrics are mostly designed for binary classification problems, but they can be adapted also to measure multi-class classification performance. Amongst the many alternatives, TSR mostly relies on accuracy [77,78], which measures the overall correct and incorrect classifications. Correct classifications reside in the diagonal of the confusion matrix, whereas any other item of the confusion matrix is counted as a misclassification.

It should be noticed that this is a quite conservative metric for TSR as it considers all misclassifications at the same level. Instead, we may not be too worried about misclassifying an informative sign (e.g., Category 8 in Table 2) with a stop sign, whereas the opposite represents a very dangerous event. That being said, for ease of comparison with existing studies, we calculate accuracy according to its traditional formulation, thus considering each misclassification as equally harmful.

4.5. Traditional Supervised Classifiers and Hyper-Parameters

Traditional Supervised classifiers process features extracted from images. Amongst the many alternatives, we summarize below those algorithms that frequently appear in most studies about TSR.

- **K Nearest Neighbors (K-NN)** algorithm [13] classifies a data point based on the class of its neighbors, or rather other data points that have a small Euclidean Distance with respect to the novel data point. The size k of the neighborhood has a major impact on classification, and therefore, needs careful tuning, which is mostly achieved through grid or random searches.
- **Support Vector Machines (SVMs)** [14], instead, separate the input space through hyperplanes, whose shape is defined by a kernel. This allows performing either linear or non-linear (e.g., radial basis function RBF kernel) classification. When SVM is used

for multi-class classification, the problem is divided into multiple binary classification problems [79].

- **Decision Tree** provides a branching classification of data and is widely used to approximate discrete functions [36]. The split of internal nodes is usually driven by the discriminative power of features, measured either with Gini or Entropy Gain. Training of decision trees employs a given number of iterations and a final pruning step to limit overfitting.
- **Boosting (AdaBoostM2)** [39] ensembles combine multiple (weak) learners to build a strong learner by weighting the results of individual weak learners. Those are created iteratively by building specialized decision stumps that focus on "hard" areas of input space.
- **Linear Discriminant Analysis (LDA)** is used to find out the linear combination of features that efficiently separates different classes by distributing samples into the same type of category [38]. This process uses a derivation of Fisher discriminant to fit multi-class problems.
- **Random Forests** [37] build ensembles of Decision Trees, each of them trained with a subset of the training set extracted by random sampling with replacement of examples.

Each supervised algorithm has its own set of hyper-parameters. To such an extent, we identified the following parameter values to exercise grid searches.

- K-NN with different values of k, i.e., different odd values of k from 1 to 25. Additionally, we observe that DITS contains nine categories of traffic signs: therefore, we disregard the usage of k = 9 to further avoid ties.
- SVM: we used three different kernels: Linear, RBF and Polynomial (quadratic), leaving other parameters (e.g., nu) as default.
- Decision Tree: we used the default configuration of MATLAB which assigns MaxNumSplits = training sample size 1, with no depth limits on decision trees.
- Boosting: we created boosting ensembles with AdaBoostM2 by building 25, 50, 75, and 100 trees (decision stumps) independently.
- Random Forest: we build forests of 25, 50, 75, or 100 decision trees.
- LDA: we Trained LDA using different discriminants, namely: pseudo-linear, diag-linear, diag-quadratic, and pseudo-quadratic.

4.6. Deep Learners and Hyper-Parameters

Deep learners may be either built from scratch or more likely—by adapting existing models to a given problem through transfer learning (i.e., knowledge transfer). Through transfer learning, we fine tune the fully connected layers of the deep model, letting all convolutional layers remain unchanged. Commonly used deep learners for the classification of images and object recognition are below.

- **AlexNet** [34] is composed of eight layers, i.e., five convolutional layers and three fully connected layers that were previously trained on the ImageNet database [80], which contains images of 227 × 227 pixels with RGB channels. The output of the last fully connected layer is provided to the SoftMax function, which provides the distribution of overall categories of images.
- **InceptionV3** is a deep convolutional neural network built by 48 layers that were trained using the ImageNet database [80], which includes images (299 × 299 with RGB channels) belonging to 1000 categories. InceptionV3 builds on (i) the basic convolutional block, (ii) the Inception module and finally (iii) the classifier. A 1x1 convolutional kernel is used in the Inceptionv3 model to accelerate the training process by decreasing the number of feature channels; further speedup is achieved by partitioning large convolutions into small convolutions [40].
- **MobileNet-v2** [41] embeds 53 layers trained on ImageNet database [80]. Differently from others, it can be considered a lightweight and efficient deep convolutional neural network with fewer parameters to tune for mobile and embedded computer vision

applications. MobileNet-v2 embeds two types of blocks: the residual block and a downsizing block, with three layers each.

Those deep learners can be tailored to TSR through transfer learning. Fully connected layers are trained on defined categories of traffic signs with different learning rates (LR) to fine-tune the models which are already trained on the ImageNet database of 1000 categories. Additionally, we employ data augmentation to avoid model overfitting; this was conducted through X and Y translation with a random value between $[-30, 30]$ and scale range within a range $[0.7, 1]$.

The hyper-parameter learning rate controls how fast weights are updated in response to the estimated errors, and therefore, controls both the time and the resources needed to train a neural network. Choosing the optimal learning rate is usually a tricky and time-consuming task: learning rates that are too big may result in fast but unstable training, while small learning rates usually trigger a heavier training phase which may even get stuck without completing correctly. In our experiments, we varied learning rate as follows: {0.05, 0.01, 0.005, 0.001, 0.0005, 0.0001} for Inceptionv3 and MobileNet-v2, and {0.0001, 0.0005, 0.00001, 0.00005, 0.000005, 0.000001} for AlexNet, which resulted in very low accuracy when using the same learning rates of the Inceptionv3 and MobileNet-v2. Noticeably, training a deep classifier with the highest learning rate in the interval reduce the training time with respect to using the smallest value in the interval (e.g., training Inceptionv3 with a learning rate of 0.05 instead of using 0.0001).

We set a minimum batch size of 32, with 10 train epochs and stochastic gradient descent with momentum (sgdm) optimizer for all the experiments on each dataset to fine-tune the models for TSR. Furthermore, we used the loss function 'crossentropyex' at the classification layer and the fully connected weights and biases were updated with a learning factor (different from learning rate) of 10. We had the weights vector size associated with the last fully connected layers [Num_cat × 4096], [Num_cat × 1280], and [Num_cat × 2048] for Alexnet, MobileNet-v2 and Inceptionv3 models, respectively, where Num_cat represented the number of traffic sign categories in each dataset.

4.7. Stacking Meta-Level Learners

Stacking meta-learners orchestrate a set of base-learners, which provide meta-data to the meta-level learner. In our study, we foresee the usage of different meta-level learners as listed below.

- **Majority Voting** [42] commits the final decision based on the class the majority of base-learners agree upon. This technique is not very sophisticated, albeit it was and is widely used to manage redundancy in complex systems [81] and to build robust machine learners [82].
- **Discrete Hidden Markov Model (DHMM)** [43]. For each class, a separate Discrete HMM returns the probability of an image belonging to that class. The classification result of the frames within the sliding window is given as input to all three DHMMs. Each DHMM returns the likelihood of the sequence to a specific class. The higher the likelihood to a specific class is decided as a final label for that specific sequence.
- **Supervised Classifiers in Section 4.5.** These classifiers can be employed as meta-level learners as meta-data resembles a set of features coming from base-learning episodes.

The parameters we used to execute grid searches and train meta-level learners above are as follows.

- Majority Voting: no parameter is needed.
- Each DHMM model was trained with 500 iterations.
- Supervised Classifiers: we used the same parameter values we already presented in Section 4.5.

4.8. Long-Short Term Memory (LSTM) Networks

LSTM networks are artificial recurrent neural networks, which efficiently process sequences of images, and therefore, suit the classification of sequences of traffic signs. LSTM networks are trained on all 12 feature sets in Section 4.3 independently considering three different training functions or optimizers, i.e., 'adam', 'sgdm', and 'rmsprop' with a learning rate of 0.001.

5. Results and Discussion

This section reports and discusses the results of our experimental campaign. We split the results into two sub-sections: Section 5.1 describes the experimental results of single-frame classifiers, while Section 5.2 reports on the results achieved by classifiers that consider sliding windows of frames.

5.1. TSR Based on Single Frame

First, we elaborate on the classification performance of TSR systems that process frames individually.

5.1.1. Highest Accuracy for Each Dataset

Figure 4 depicts a bar chart diagram reporting the highest accuracy achieved by classifiers in each of the three datasets. It is clear from the blue solid bars in Figure 4 that almost all classifiers give better performance on the GTSRB dataset compared to the other two datasets, i.e., BelgiumTSC and DITS. All classifiers in the figure but Decision Tree and LDA achieve perfect accuracy on the GTSRB dataset. The reason behind the high accuracy may be the higher number of training samples and better image quality of the GTSRB dataset compared to the other two datasets. Instead, SVM provides the highest accuracy of 95.94% in DITS, with LDA that comes close at 95.85%. Consequently, the highest accuracy in each dataset is not always achieved by the same algorithm, despite K-NN, SVM and LDA performing better overall compared to other supervised classifiers.

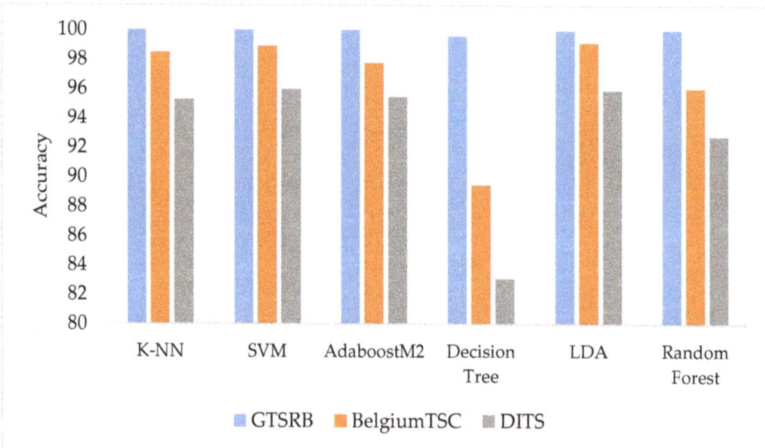

Figure 4. Highest accuracy achieved by traditional supervised classifiers on each dataset.

5.1.2. Impact of Feature Descriptors

Table 3 further elaborates on the impact of features on accuracy scores achieved by supervised classifiers on each dataset. Supervised classifiers achieve perfect accuracy with all feature descriptors on GTSRB. Instead, the combination of AFeat and RFeat builds a feature descriptor that allows algorithms to achieve the highest accuracy of 95.94% for DITS and 99.12% for BelgiumTSC. Additionally, AFeat and RFeat descriptors provide features that allow algorithms to reach higher accuracy. By using just a single feature descriptor AFeat always achieves the highest accuracy on all three datasets, while the second highest accuracy is achieved by RFeat. Instead, using only LBP, HOG or their combination generates accuracy scores that are lower than potential alternatives. Moreover, it is worth noticing how combining feature descriptors provides features that increase the classification performance of supervised classifiers, such as: from 95.51% to 95.94% in DITS, and from 98.84% to 99.12% in BelgiumTSC.

Table 3. Highest accuracy achieved using different feature descriptors on each dataset (bold highlighted values represent the highest achieved accuracy across each dataset).

Feature Descriptor (s)	GTSRB	DITS	BelgiumTSC
AFeat	**100.00**	95.51	98.84
RFeat	**100.00**	94.13	97.76
LBP	**100.00**	79.98	93.49
HOG	**100.00**	87.92	96.24
HOG ∪ LBP	**100.00**	88.26	96.56
AFeat ∪ RFeat	**100.00**	**95.94**	**99.12**
AFeat ∪ HOG	**100.00**	95.68	98.96
AFeat ∪ LBP	**100.00**	95.85	98.96
RFeat ∪ HOG	**100.00**	95.51	98.72
RFeat ∪ LBP	**100.00**	95.85	98.80
AFeat ∪ HOG ∪ LBP	**100.00**	95.42	98.88
RFeat ∪ HOG ∪ LBP	**100.00**	95.34	98.84

5.1.3. Results of Deep Classifiers

We explore the results of the deep classifiers considered in this study with the aid of Table 4, which shows accuracy scores achieved by those classifiers for different learning rates.

MobileNet-v2 achieves the highest accuracy out of the three deep learners for the GTSRB dataset with a learning rate of 0.001, whereas a learning rate of 0.00005 maximizes the accuracy scores of AlexNet on the BelgiumTSC dataset. Instead, the learning rate of 0.0001 allows InceptionV3 to reach the maximum accuracy of 96.03% for the DITS dataset, outperforming MobileNet-v2 and AlexNet, which instead achieves the highest accuracy in the BelgiumTSC dataset with a learning rate of 0.0005. Interestingly, whereas accuracy scores for GTSRB do not vary a lot when using different learning rates, the choice of the learning rate becomes of paramount importance when classifying DITS and BelgiumTSC datasets. Particularly, the bottom of Table 4, the third column, shows a 14.97% accuracy on the BelgiumTSC dataset using learning rates of 0.05 and 0.005, which is a very poor achievement. For these learning rates, the training process was unstable, with weights that were updated too fast and ended up with a classifier that has semi-random classification performance. Unfortunately, we could not identify a single deep classifier that outperforms others in all three datasets.

5.2. *TSR Based on Sliding Windows*

This section elaborates on the classification performance of TSR systems that process a sliding window of multiple frames.

5.2.1. Meta Learning with Traditional Base Classifiers

Table 5 reports scores achieved by stacking meta-learners built using (i) the three traditional supervised classifiers that performed better in Section 5.1.1 as base learners, and (ii) different meta-level learners, such as K-NN, SVM, LDA, Decision Tree, Majority Voting,

Boosting, Random Forest and DHMM. The GTSRB dataset does not appear in Table 5 since single-frame traditional classifiers alone already achieved perfect classification. The table reports the highest accuracy scores achieved by each stacking meta-level classifier by using different combinations of base-learners (K-NN, SVM, LDA) and window sizes of two and three items.

Table 4. Accuracy achieved by deep learners for each of the three datasets with varying learning rates (bold highlighted values represent the highest achieved accuracy across each dataset by deep classifiers).

	InceptionV3		MobileNet-v2		AlexNet	
	LR	**Acc**	**LR**	**Acc**	**LR**	**Acc**
GTSRB	0.01	96.62	0.01	96.11	0.0001	95.98
	0.05	93.56	0.05	93.38	0.0005	94.92
	0.001	96.95	0.001	**99.35**	0.00001	94.86
	0.005	97.06	0.005	96.42	0.00005	95.64
	0.0001	96.81	0.0001	96.83	0.000001	95.83
	0.0005	**98.03**	0.0005	96.65	0.000005	**96.07**
DITS	0.01	80.06	0.01	93.52	0.0001	87.40
	0.05	80.67	0.05	85.93	0.0005	86.45
	0.001	88.17	0.001	94.99	0.00001	**95.51**
	0.005	84.98	0.005	88.78	0.00005	92.06
	0.0001	**96.03**	0.0001	95.77	0.000001	92.23
	0.0005	91.88	0.0005	**95.94**	0.000005	95.16
BelgiumTSC	0.01	89.58	0.01	97.16	0.0001	99.24
	0.05	14.97	0.05	94.49	0.0005	92.57
	0.001	98.12	0.001	98.72	0.00001	99.52
	0.005	14.97	0.005	94.73	0.00005	**99.72**
	0.0001	99.64	0.0001	**99.24**	0.000001	97.92
	0.0005	**99.68**	0.0005	98.96	0.000005	99.24

Table 5. Results achieved using different meta-learners considering traditional classifiers as base classifiers varying window size (WS). We bolded the highest achieved accuracy using different combinations across each dataset and low accuracy achieved through AdaBoostM2 and Random Forest are italicized in the 10th and 11th columns.

Dataset	Base-Level Classifier	Single Frame Accuracy	WS	Stacking Meta-Level Classifier							
				Majority Voting	K-NN	SVM	LDA	Decision Tree	AdaBoostM2	Random Forest	DHMM
BelgiumTSC	KNN	98.44	2	99.40	99.04	98.8	98.80	98.68	98.80	*51.86*	98.56
	SVM	98.88		99.52	99.64	**99.76**	98.68	99.64	99.28	98.80	99.04
	LDA	99.12		99.64	99.40	99.28	99.28	98.32	98.92	97.84	99.40
	KNN	98.44	3	99.40	99.40	99.04	98.92	98.44	98.32	63.47	98.20
	SVM	98.88		99.52	99.52	**99.64**	99.40	98.56	*14.97*	99.28	98.80
	LDA	99.12		**99.64**	**99.64**	99.40	98.2	99.28	97.96	98.32	98.80
DITS	KNN	95.25	2	97.56	97.56	97.56	96.75	97.56	97.56	82.93	96.75
	SVM	95.94		96.75	97.56	**98.37**	97.56	95.12	*31.71*	95.12	95.93
	LDA	95.85		**98.37**	**98.37**	97.56	97.56	**98.37**	95.93	96.75	96.75
	KNN	95.25	3	99.00	99.00	99.00	99.00	99.00	99.00	85.00	98.00
	SVM	95.94		99.00	**100.00**	99.00	99.00	97.00	*36.00*	97.00	96.00
	LDA	95.85		99.00	**100.00**	98.00	**100.00**	99.00	98.00	99.00	98.00

Overall, LDA as a base-level classifier with a K-NN meta-level classifier is the preferred choice (bolded values in Table 5) on DITS and on BelgiumTSC with a sliding window of three items. Instead, using ensembles of Decision Trees as AdaBoost and Random Forests sparingly gives very low accuracy scores (see italicized numbers in the 10th and 11th columns of Table 5), showing that those two classifiers do not always adequately play the role of a meta-level classifier for a stacker.

Results for DITS in Table 5 show that using a sliding window of three items generally improves accuracy with respect to using a sliding window of only two items. A sliding window of three items allowed stacking meta-learners, which used K-NN or LDA as meta-level classifiers to reach perfect accuracy (100%) on the DITS dataset using either LDA or SVM as base-learners. This result was largely expected: the more information is available (i.e., wider sliding window), the fewer misclassifications we expect from a given classifier.

Instead, we obtained maximum accuracy for the BelgiumTSC by using a sliding window of two items, whereas using three items often degrades classification performance. At a first glance, this result is counter-intuitive with respect to previous discussions. However, the reader should note that the BelgiumTSC dataset reports on a set of images of the same traffic signs which are captured with multiple input cameras without any temporal order. Consequently, the sliding window for the BelgiumTSC contains images of the traffic sign which are taken from different angles and may lead the meta-learner to lean towards misclassifications rather than improving accuracy. In fact, for this dataset, there is no direct relation between the size of the window and accuracy values, which instead turned out to be evident for the other datasets.

5.2.2. Meta Learning with Base-Level Deep Classifiers

Table 6 has a structure similar to Table 5 but employs base-level deep classifiers to build the stacking meta-learner, and also reports on all datasets as deep classifiers based on a single frame but did not achieve perfect accuracy on any of the three datasets. Deep base-level classifiers in conjunction with K-NN as a meta-level classifier achieved perfect classification on all three datasets, as shown by bold values in Table 6. GTSRB turns out to be the dataset that provides the higher average accuracy by using a different base and meta-level classifiers. The highest achieved accuracies are highlighted in Table 6 with bold typeset. It is very interesting to discuss that all three deep learning models (base-level classifiers) with meta-level classifiers K-NN, LDA, Boosting and Random Forest give 100% accuracy, while MobileNet-v2 achieves 100% accuracy with all meta-level classifiers for a sliding window of size 2 or 3 on the GTSRB dataset. Inceptionv3 and MobileNetv2 with meta-level classifiers (K-NN, AdaboostM2) achieve 100% accuracy on the DITS dataset for sliding windows of size 2 and 3, respectively, While AlexNet base-level classifier with Majority voting and K-NN as the meta-level classifier achieves 100% accuracy for both sliding windows of size 2 & 3 on BelgiumTSC dataset.

Similarly, to Table 5, we observe that AdaboostM2 does not show up as a reliable meta-level classifier as it provides very low accuracy for the BelgiumTSC with a sliding window of three frames. All meta-level classifiers with base-level classifier Mobilenet-v2 achieve 100% accuracy on the GTSRB dataset, whose sequences contain 30 images of the same traffic sign, and therefore, provide much information for the stacking classifier to classify traffic signs as the window slides.

5.2.3. Results of LSTM Networks

Table 7 reports accuracy scores of LSTM networks on the BelgiumTSC and DITS datasets with a sliding window of size 2 or 3. Similarly to Section 5.2.1, we omit the GTSRB dataset since it is perfectly classified by single-frame traditional classifiers. We independently trained the LSTM by using each of the 12 feature sets in Section 4.3, with different window sizes (WS) and by using three different optimizers: adam, sgdm and rmsprop.

Table 7 reports the highest accuracy achieved by LSTM by using a given WS and optimizer function. It is evident how the adam optimizer always allows achieving the highest accuracy scores in both datasets and with different WS. Additionally, accuracy is always higher when using a window of size 3 with respect to a window containing only two items: this was expected for DITS, whose images are time-ordered, but it is also verified for the BelgiumTSC, which does not have such ordering. Overall, the results of the

LSTM are slightly lower than stacking meta-learners using traditional base-level classifiers and clearly worse than stacking using deep base-level classifiers, which achieves perfect accuracy on all datasets.

Table 6. Results achieved using different meta learners with deep learners as base classifiers with varying window size (WS). We bolded the perfect classification (100% accuracy).

Dataset	Base-Level Classifier	Single Frame Accuracy	WS	Majority Voting	K-NN	SVM	LDA	Decision Tree	AdaBoostM2	Random Forest	DHMM
BelgiumTSC	AlexNet	99.72	2	99.88	**100.00**	99.64	99.88	99.88	99.40	99.16	99.76
	InceptionV3	99.68		99.88	99.88	99.64	99.88	99.52	99.88	99.64	99.64
	MobileNetv2	99.24		99.52	99.88	99.64	99.04	99.64	99.52	98.68	99.64
	AlexNet	99.72	3	**100.00**	**100.00**	99.28	99.88	99.88	99.76	99.76	99.28
	InceptionV3	99.68		99.88	99.88	99.40	99.52	99.40	*14.97*	99.76	99.52
	MobileNetv2	99.24		99.88	99.88	98.80	99.16	99.40	*14.97*	99.76	99.52
DITS	AlexNet	95.51	2	96.75	97.56	97.56	97.56	96.74	96.74	96.74	98.37
	InceptionV3	96.03		98.37	**100.00**	97.56	98.37	98.37	96.74	95.93	99.19
	MobileNetv2	95.94		97.56	99.18	99.18	99.18	99.18	**100.00**	98.37	99.19
	AlexNet	95.51	3	97.00	99.00	99.00	99.00	99.00	99.00	99.00	99.00
	InceptionV3	96.03		98.00	**100.00**	99.00	98.00	99.00	**100.00**	99.00	98.00
	MobileNetv2	95.94		98.00	**100.00**	99.00	**100.00**	99.00	**100.00**	**100.00**	**100.00**
GTSRB	AlexNet	96.07	2	97.37	**100.00**	99.76	**100.00**	98.09	**100.00**	**100.00**	98.09
	InceptionV3	98.03		**100.00**	**100.00**	**100.00**	**100.00**	**100.00**	**100.00**	**100.00**	99.76
	MobileNetv2	99.35		**100.00**	**100.00**	**100.00**	**100.00**	**100.00**	**100.00**	**100.00**	**100.00**
	AlexNet	96.07	3	0.9737	**100.00**	99.76	**100.00**	98.09	**100.00**	**100.00**	98.09
	InceptionV3	98.03		**100.00**	**100.00**	**100.00**	**100.00**	**100.00**	**100.00**	**100.00**	99.76
	MobileNetv2	99.35		**100.00**	**100.00**	**100.00**	**100.00**	**100.00**	**100.00**	**100.00**	**100.00**

Table 7. Accuracy of LSTM with window sizes 2 and 3.

Dataset	WS	Optimizer		
		adam	sgdm	rmsprop
DITS	2	**97.56**	**97.56**	96.74
DITS	3	**99.00**	**99.00**	98.00
BelgiumTSC	2	**99.40**	99.16	99.16
BelgiumTSC	3	**99.64**	99.28	99.40

5.3. Comparing Sliding Windows and Single-Frame Classifiers

Independent analyses and discussions of results in Sections 5.1 and 5.2 provided interesting findings concerning both traditional supervised and deep base-level classifiers and the usage of sliding windows to improve the classification performance through meta-learning.

Traditional supervised classifiers, such as K-NN, SVM, AdaboostM2, and Random Forests achieved a perfect classification of each image contained in the GTRSB dataset. Moreover, we observed how combining deep features descriptor {AFeat ∪ RFeat} allowed traditional classifiers to reach the highest accuracy in any of the three datasets, achieving 100%, 95.94%, 99.12% on the GTSRB, DITS and BelgiumTSC datasets, respectively. On the other hand, deep classifiers outperform traditional classifiers on the DITS and BelgiumTSC datasets but still cannot reach a perfect classification accuracy.

Noticeably, stacking meta-learners that take advantage of sliding windows achieve perfect classification accuracy on all three datasets when using deep base-level classifiers and K-NN as meta-level classifiers. These results show that orchestrating sliding windows critically increases the classification performance compared to single frame classifiers. Differently, LSTM networks achieve 97.56% and 99% of accuracy on the DITS dataset for a sliding window of size 2 or 3, respectively, which is better than single frame classifier performance, but still inferior with respect to stacking meta-learners.

Figure 5 compares the accuracy achieved by stacking meta-learners and LSTM networks by means of a bar chart. Base-level traditional supervised classifiers with stacking meta learners achieved 98.37% and 100% accuracy on the DITS dataset considering a sliding window of two and three inputs, respectively, which is slightly higher than the LSTM scores. A similar trend can be observed for the BelgiumTSC, while the GTSRB scores are not reported in the chart as it does not require sliding windows to achieve perfect accuracy.

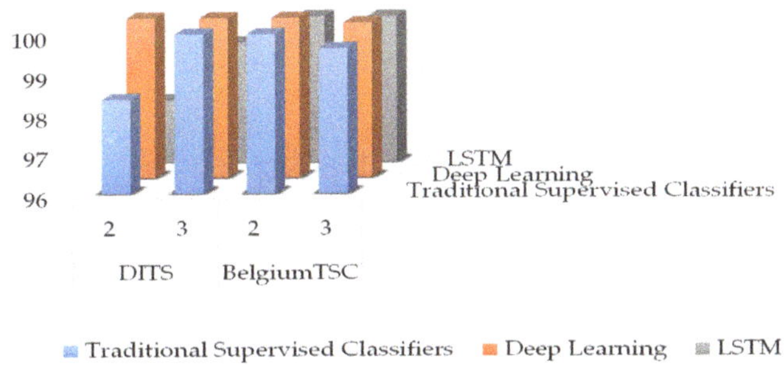

Figure 5. Highest accuracy achieved by LSTM, stacker with supervised base-level, and stacker with deep base-level on BelgiumTSC and DITS.

5.4. In-Depth View of BelgiumTSC

Similarly, to the GTSRB and DITS, we observed perfect classification by using a stacker with deep base-level classifiers also with the BelgiumTSC dataset, which contains unordered sets of images rather than sequences. Consequently, our meta-learning strategy proves to be beneficial even if images in the sliding window are not time-ordered.

However, Table 7 showed that a sliding window of three items performs poorly with respect to using only two items, which may seem counterintuitive. Figure 6 shows one of those cases in which using a window of two items is beneficial with respect to using three items. The upper part of Figure 6 represents the process adopted for the classification of a Diamond traffic sign (Category 7) when using a window of three images. All the three images taken from different viewpoints are individually classified by the base-level classifier AlexNet, which returns the probabilities PTS of belonging to all classes (see Base-Classifier output in the figure). These three probability vectors (which match the PTS_i in Section 3.2) are fed to the meta-level classifier to commit the final decision. We observe that PTS_1 and PTS_2 give almost a certain probability of belonging to class 7 (0.999), while PTS_3 gives a higher probability for class 1 (i.e., stop traffic sign). With those results, the SVM meta-learner decides that the traffic sign is a stop sign, ending up with a misclassification. Clearly, the third image is taken from a different angle, has some blurring and makes the meta-learner lean towards a misclassification rather than helping.

Instead, Figure 7 shows the process to classify the same inputs using a window of two items. When $\{PTS_1, PTS_2\}$ are provided as meta features to a meta-level classifier, the final output shows a high likelihood of being a category 7 which is indeed a correct classification. Meanwhile, providing $\{PTS_2, PTS_3\}$ or $\{PTS_1, PTS_3\}$ as meta features lead the stacker to misclassify the set of images as a stop sign (category 1): the predicted final output is class 6 which is a wrong prediction. This enforces the conjecture that in this case using the third image constitutes noise that causes misclassification.

5.5. Timing Analysis

This section expands on the time required for classification using the different setups in this paper. Table 8 reports the average and standard deviation of time required for (i) feature extraction, (ii) single-frame classification, and (iii) stacking meta-learning across test images of three datasets.

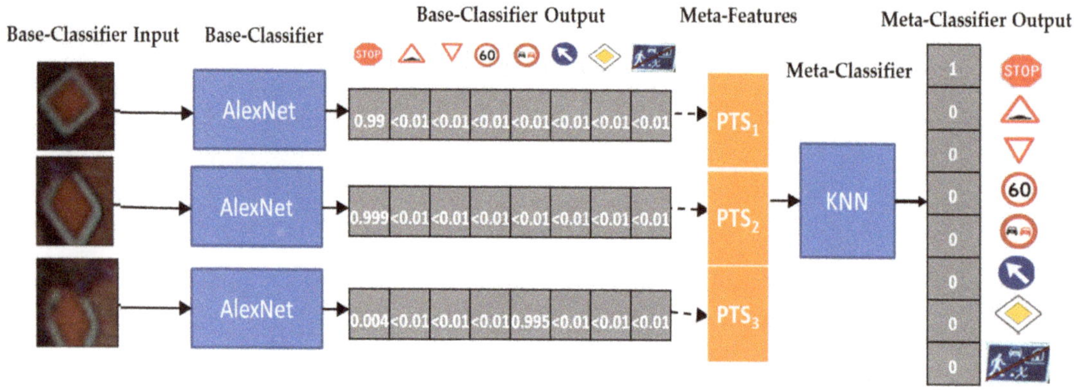

Figure 6. Instantiation of the stacking-meta learner with AlexNet base-learner and SVM meta-level learner, managing a sliding window of size 3 for BelgiumTSC. The three frames we use as input describe a Diamond sign (Category 7) which is misclassified using all three frames.

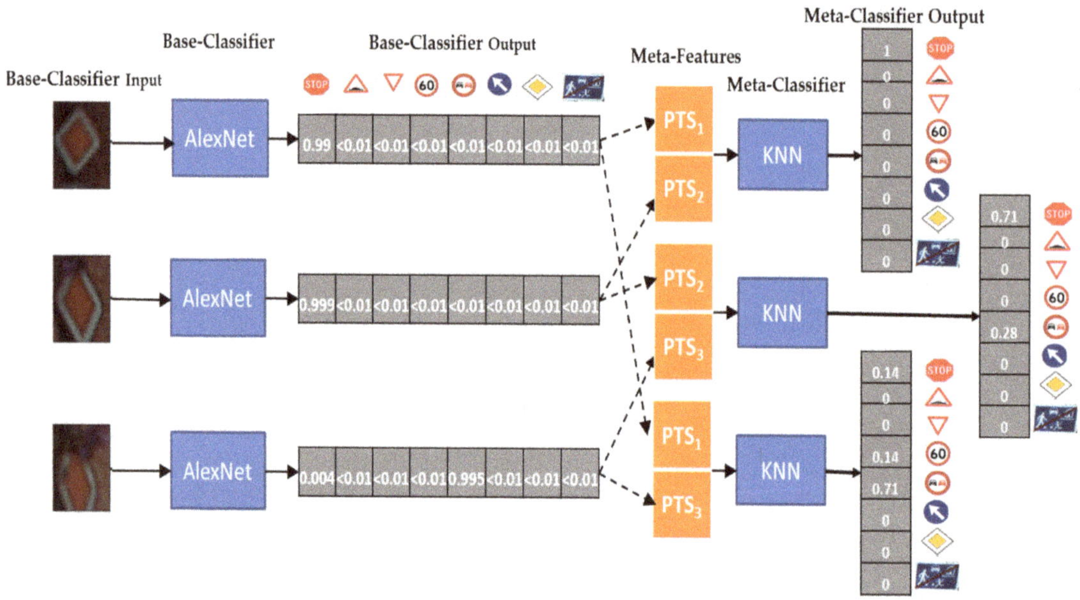

Figure 7. Instantiation of the Stacking-Meta learner with AlexNet Base-learner and SVM meta-level learner, managing a sliding window of size 2 for the BelgiumTSC. The three frames we use as input describe a Diamond sign (Category 7) which is misclassified using all three frames (Figure 6) but may be classified correctly by using a shorter window.

Starting from feature extraction on the left of the table, it turns out that the extraction of handcrafted features takes slightly less time compared to deep features. However, even extracting deep features through ResNet-18 from a single image does not require on average more than 0.04 s (roughly 40 ms). Instead, the time required for exercising single-frame TSR classifiers varies a lot: traditional supervised classifiers need at most 200 ms to classify a given input set, whereas deep classifiers need more than half a second to classify an image with our hardware setup, depending on the number of layers of deep models. Indeed, the reader should note that whereas deep classifiers embed feature extraction through convolutional layers, traditional classifiers have the prerequisite of feature extraction. In fact, on the right of Table 8, we show that a TSR system that relies on AFeat ∪ RFeat features (i.e., most useful ones according to Table 3) provided to an SVM classifier takes on average 0.1974 s to classify an image: this includes feature extraction and classification itself. A perfect parallelization of the feature extractors cuts down this timing to 0.1756 and will be easily achievable on basic multi-core systems.

Table 8. Time required for (left) feature extraction, (middle) exercising individual classifiers, and (right) different TSR strategies, either sequential or parallel execution.

Feature Extractor	Time in Seconds Avg ± St. Dev	Individual Classifier	Time in Seconds Avg ± St. Dev	TSR Strategy	Average Time in Seconds	
					(Sequential)	(Parallel)
HOG	0.0204 ± 0.0024	SVM	0.1344 ± 0.0364	Single-frame (AFeat ∪ RFeat + SVM)	0.1974	0.1756
LBP	0.0196 ± 0.0023	KNN	0.1205 ± 0.0302	Single-frame (InceptionV3)	1.4205	1.4205
AFeat	0.0218 ± 0.0023	LDA	0.1034 ± 0.0256	Stacking with WS = 2 (AFeat ∪ RFeat + SVM + KNN)	0.4036	0.3636
RFeat	0.0412 ±0.0034	InceptionV3	1.4205 ± 0.6613			
		MobileNetV2	0.6391 ± 0.2180	Stacking with WS = 2 (AlexNet + KNN)	0.5621	0.5621
		AlexNet	0.3749 ± 0.1407	Stacking with WS = 2 (InceptionV3 + KNN)	1.6085	1.6085

Table 8 also shows the time needed to perform other TSR strategies we discussed in this paper. Particularly, the third to sixth line on the right of the table show the time needed to classify an image using a sliding window of two or three items with different base-levels and meta-level learners. The time required for base-level learning equals single-frame classification: only the most recent frame in the window is processed, whereas probabilities assigned by classifiers to older frames are stored, and therefore, do not need to be re-computed again. The table reports on different base-learners but always uses K-NN as the meta-level learner, as this was the classifier that allowed reaching high scores in Section 5.2. K-NN takes on average 0.188 to classify a sliding window of two items, (i.e., two PTS vectors of 8/9 numbers each), and only slightly more time to process a sliding window of three items.

Overall, we can observe how most TSR systems that embed sliding windows are able to classify a new image in less than a second, whereas heavier deep learners make classification time lean towards two seconds. We believe that such timing performance albeit slower than using single-frame classifiers is still efficient enough to be installed on a vehicle, which only rarely samples more than a frame per second for TSR tasks. Nevertheless, using more efficient hardware, especially GPUs, could help in reducing, even more, the time required for classification.

5.6. Comparison to the State of the Art TSR

Ultimately, we recap the accuracy scores achieved by studies we already referred to as related works in Sections 2.2 and 2.3, to compare their scores with ours. Therefore, Table 9 summarizes those studies, the datasets they used, and the accuracy they achieved. At a first glance, those studies conclude that their single-frame classifiers are often far from perfect classification. In fact, even in this study, we observed that single-frame TSR in the BelgiumTSC and DITS datasets cannot reach perfect accuracy (i.e., second-last row in the table). Unfortunately, promising studies [64,65], which describe multi-frame classifiers, do not rely on our datasets, and therefore, we cannot directly compare them.

To summarize, our experiment ended up achieving perfect classification on all datasets thanks to sliding windows (see last row of Table 9), dramatically improving existing studies on those datasets, for which perfect accuracy was hardly achieved by existing studies.

Table 9. Comparison with state of the art approaches.

Studies	Sequences of Frames	Achieved Accuracy (%)		
		GTSRB	BelgiumTSC	DITS
Stallkamp et al. [31]	No	98.98		
Atif et al. [29]	No	100.00	99.80	99.31
Agrawal et al. [45]	No	* 77.43		
Youssef et al. [33]	No	95.00		98.20
Mathias et al. [1]	No		97.83	
Huang et al. [44]	No	* 95.56		
Lin et al. [51]	No		* 99.18	
Li et al. [28]	No	97.40	98.10	
Li and Wang [3]	No	99.66		
Zeng et al. [83]	No		95.40	
Our Approach	**No**	**100.00**	**99.72**	**96.03**
	Yes	**100.00**	**100.00**	**100.00**

Note: Accuracy values with * represent average accuracy across different classes of traffic signs no balanced accuracy was provided.

6. Concluding Remarks

To conclude the paper, we report in this section the limitations to the validity of our study, we summarize the findings and lessons learned in this paper and ultimately discuss future works.

6.1. Limitations to Validity

We report here possible limitations to the validity and the applicability of our study. These are not to be intended as showstoppers when considering the conclusions of this paper. Instead, they should be interpreted as boundaries or possible future implications which may impact the validity of this study.

6.1.1. Usage of Public Data

The usage of public image datasets and public tools to run algorithms was a prerequisite of our analysis, to allow reproducibility and to rely on proven-in-use data. However, the heterogeneity of data sources and their potential lack of documentation may limit the understandability of data. In addition, such datasets are not under our control; therefore, possible actions, such as changing the way data is generated, are out of consideration. For example, creating longer sequences of traffic signs or creating a time-sequenced version of the BelgiumTSC is not possible at all.

6.1.2. Parameters of Classifiers

Each classifier relies on its own parameters. Finding the optimal values of parameters is a substantial process that requires sensitive analyses and is directly linked with the scenario in which the classifier is going to be exercised. When applying classifiers to different datasets it is not always possible to precisely tune these parameters: instead, in this study, we perform grid searches, which run a classifier with different parameter values and choose the parameter that maximizes accuracy. This does not guarantee finding the absolute optimum value of a parameter for a given classifier on a given dataset but constitutes a good approximation [84].

6.2. Lessons Learned

This section highlights the main findings and lessons learned from this study.

- We observed that classifying images in the DITS dataset is harder than classifying the BelgiumTSC and GTSRB datasets as both base-level traditional supervised and deep classifiers' performances are low comparatively. This is mostly due to the amount of training images and their quality, which is higher in the GTSRB compared to the other two datasets.
- Combining feature descriptors allows for improving classification performance. Particularly, we found that the {AFeat ∪ RFeat} descriptor allows traditional classifiers to maximize accuracy.
- Single-frame traditional supervised classifiers achieved perfect classification on the GTSRB dataset, while on the BelgiumTSC and DITS they show a non-zero amount of misclassifications. To the best of our knowledge, this result is due to the number of training samples, which is higher in the GTSRB with respect to the BelgiumTSC and DITS, and image quality, which again is better for the GTSRB. On the other hand, we achieved 100% accuracy by adopting a sliding windows based TSR strategy on all three considered datasets.
- There is no clear benefit in adopting deep classifiers over traditional classifiers for single-frame classification as they show similar accuracy scores. Additionally, both are outperformed, when using sliding windows for TSR.
- LSTM networks often, but not always, outperform single-frame classifiers but show lower accuracy than stacking meta-learners in orchestrating sliding windows.
- A stacking meta-learner with deep base-level classifiers and K-NN as meta-level classifier can perfectly classify traffic signs on all three datasets with any window size $WS \geq 2$.
- For datasets that contain sequences (time-series) of images, enlarging the sliding window never decreases accuracy and, in most cases, raises the number of correct classifications.
- Deep learning models require more time compared to traditional supervised classifiers, especially because there are many layers, e.g., InceptionV3.
- Sliding windows based classification takes more time compared to single-frame classifiers but has remarkably higher classification performance across all three datasets.
- Overall, adopting classifiers that use a sliding window rather than a single-frame classifier allows reducing misclassifications, consequently raising accuracy.

6.3. Current and Future Works

Our study showed how the adoption of a stacking meta-learner in conjunction with sliding windows allows achieving perfect classification on the public GTSRB, BelgiumTSC and DITS datasets. Those datasets contain images taken in different parts of the world and mostly taken in semi-ideal lighting and environmental conditions. Therefore, they may not completely represent what a real TSR system installed on a vehicle will face during its life. As a result, we plan to explore the robustness of classifiers used in this study by injecting different types of faults/perturbations in the captured images [85], tracking the likely growth of misclassifications of individual classifiers. After this test, we plan to re-train (either from scratch or through transfer learning) classifiers using both original images from datasets and those faulty images. Furthermore, we plan to inject adversarial attacks to traffic sign images and using them both (i) as a test set, to observe the degradation of accuracy (if any) when processing corrupted frames, and (ii) during training, to learn a more reliable model. We believe that this process will allow us to build robust classifiers with very high accuracy, even when classifying faulty, adversarial, or corrupted images.

Author Contributions: Conceptualization, M.A., T.Z., M.G. and A.B.; data curation, M.A.; methodology, M.A., T.Z., M.G. and A.B.; experiments-design, evaluation and investigation, M.A.; writing-original draft preparation, M.A.; writing-review and editing, T.Z., M.G. and A.B.; Funding acquisition, A.B. All authors have read and agreed to the published version of the manuscript.

Funding: This work has been partially supported by the H2020 programme under the Marie Sklodowska-Curie grant agreement 823788 (ADVANCE) project and by the Regione Toscana under the POR FESR 2014-2020 SPaCe project.

Conflicts of Interest: The authors declare no conflict of interest.

References

1. Mathias, M.; Timofte, R.; Benenson, R.; Van Gool, L. Traffic Sign Recognition—How Far Are We from the Solution? In Proceedings of the 2013 International Joint Conference on Neural Networks (IJCNN), Dallas, TX, USA, 4–9 August 2013; pp. 1–8.
2. Mammeri, A.; Boukerche, A.; Almulla, M. Design of Traffic Sign Detection, Recognition, and Transmission Systems for Smart Vehicles. *IEEE Wirel. Commun.* **2013**, *20*, 36–43. [CrossRef]
3. Li, J.; Wang, Z. Real-Time Traffic Sign Recognition Based on Efficient CNNs in the Wild. *IEEE Trans. Intell. Transp. Syst.* **2019**, *20*, 975–984. [CrossRef]
4. Avizienis, A.; Laprie, J.C.; Randell, B.; Landwehr, C. Basic Concepts and Taxonomy of Dependable and Secure Computing. *IEEE Trans. Dependable Secur. Comput.* **2004**, *1*, 11–33. [CrossRef]
5. Carvalho Barbosa, R.; Shoaib Ayub, M.; Lopes Rosa, R.; Zegarra Rodríguez, D.; Wuttisittikulkij, L. Lightweight PVIDNet: A Priority Vehicles Detection Network Model Based on Deep Learning for Intelligent Traffic Lights. *Sensors* **2020**, *20*, 6218. [CrossRef]
6. Application of Machine Learning Algorithms in Lane-Changing Model for Intelligent Vehicles Exiting to off-Ramp: Transportmetrica A: Transport Science: Volume 17, No 1. Available online: https://www.tandfonline.com/doi/abs/10.1080/23249935.2020.1746861 (accessed on 18 March 2022).
7. Sasikala, G.; Ramesh Kumar, V. Development of Advanced Driver Assistance System Using Intelligent Surveillance. In *International Conference on Computer Networks and Communication Technologies*; Springer: Singapore, 2019; pp. 991–1003. [CrossRef]
8. Jose, A.; Thodupunoori, H.; Nair, B.B. A Novel Traffic Sign Recognition System Combining Viola–Jones Framework and Deep Learning. In *Soft Computing and Signal Processing*; Springer: Singapore, 2019; pp. 507–517. [CrossRef]
9. Kuo, C.Y.; Lu, Y.R.; Yang, S.M. On the Image Sensor Processing for Lane Detection and Control in Vehicle Lane Keeping Systems. *Sensors* **2019**, *19*, 1665. [CrossRef]
10. McDonald, A.; Carney, C.; McGehee, D.V. *Vehicle Owners' Experiences with and Reactions to Advanced Driver Assistance Systems (Technical Report)*; AAA Foundation for Traffic Safety: Washington, DC, USA, 2018.
11. Hurtado, S.; Chiasson, S. An Eye-Tracking Evaluation of Driver Distraction and Unfamiliar Road Signs. In Proceedings of the 8th International Conference on Automotive User Interfaces and Interactive Vehicular Applications, (AutomotiveUI '16), Ann Arbor, MI, USA, 24–26 October 2016; pp. 153–160.
12. Wali, S. Comparative Survey on Traffic Sign Detection and Recognition: A Review. *Prz. Elektrotech.* **2015**, *1*, 40–44. [CrossRef]
13. Guo, G.; Wang, H.; Bell, D.; Bi, Y.; Greer, K. KNN Model-Based Approach in Classification. In *On The Move to Meaningful Internet Systems 2003: CoopIS, DOA, and ODBASE*; Springer: Berlin/Heidelberg, Germany, 2003; pp. 986–996.
14. Hsu, C.W.; Lin, C.J. A Comparison of Methods for Multiclass Support Vector Machines. *IEEE Trans. Neural Netw.* **2002**, *13*, 415–425. [CrossRef]
15. Wahyono; Jo, K.H. A Comparative Study of Classification Methods for Traffic Signs Recognition. In Proceedings of the 2014 IEEE International Conference on Industrial Technology (ICIT), Busan, Korea, 26 February 2014; pp. 614–619.
16. Ojala, T.; Pietikainen, M.; Maenpaa, T. Multiresolution Gray-Scale and Rotation Invariant Texture Classification with Local Binary Patterns. *IEEE Trans. Pattern Anal. Mach. Intell.* **2002**, *24*, 971–987. [CrossRef]
17. Dalal, N.; Triggs, B. Histograms of Oriented Gradients for Human Detection. In Proceedings of the 2005 IEEE Computer Society Conference on Computer Vision and Pattern Recognition (CVPR'05), San Diego, CA, USA, 20–25 June 2005; Volume 1, pp. 886–893.
18. Stallkamp, J.; Schlipsing, M.; Salmen, J.; Igel, C. Man vs. Computer: Benchmarking Machine Learning Algorithms for Traffic Sign Recognition. *Neural Netw.* **2012**, *32*, 323–332. [CrossRef]
19. Yang, X.; Qu, Y.; Fang, S. Color Fused Multiple Features for Traffic Sign Recognition. In Proceedings of the 4th International Conference on Internet Multimedia Computing and Service, Wuhan, China, 9–11 September 2012; pp. 84–87.
20. Szegedy, C.; Liu, W.; Jia, Y.; Sermanet, P.; Reed, S.; Anguelov, D.; Erhan, D.; Vanhoucke, V.; Rabinovich, A. Going Deeper with Convolutions. In Proceedings of the IEEE Conference on Computer Vision and Pattern Recognition, Boston, MA, USA, 7–12 June 2015; pp. 1–9.
21. Manisha, U.K.D.N.; Liyanage, S.R. An Online Traffic Sign Recognition System for Intelligent Driver Assistance. In Proceedings of the 2017 Seventeenth International Conference on Advances in ICT for Emerging Regions (ICTer), Colombo, Sri Lanka, 6–9 September 2017; pp. 1–6.

22. Matoš, I.; Krpić, Z.; Romić, K. The Speed Limit Road Signs Recognition Using Hough Transformation and Multi-Class Svm. In Proceedings of the 2019 International Conference on Systems, Signals and Image Processing (IWSSIP), Osijek, Croatia, 5–7 June 2019; pp. 89–94.
23. Liu, C.; Li, S.; Chang, F.; Wang, Y. Machine Vision Based Traffic Sign Detection Methods: Review, Analyses and Perspectives. *IEEE Access* **2019**, *7*, 86578–86596. [CrossRef]
24. Soni, D.; Chaurasiya, R.K.; Agrawal, S. *Improving the Classification Accuracy of Accurate Traffic Sign Detection and Recognition System Using HOG and LBP Features and PCA-Based Dimension Reduction*; Social Science Research Network: Rochester, NY, USA, 2019.
25. Hasan, N.; Anzum, T.; Jahan, N. Traffic Sign Recognition System (TSRS): SVM and Convolutional Neural Network. In *Inventive Communication and Computational Technologies*; Springer: Singapore, 2021; pp. 69–79. [CrossRef]
26. Rahmad, C.; Rahmah, I.F.; Asmara, R.A.; Adhisuwignjo, S. Indonesian Traffic Sign Detection and Recognition Using Color and Texture Feature Extraction and SVM Classifier. In Proceedings of the 2018 International Conference on Information and Communications Technology (ICOIACT), Yogyakarta, Indonesia, 6–7 March 2018; pp. 50–55.
27. Cao, J.; Song, C.; Peng, S.; Xiao, F.; Song, S. Improved Traffic Sign Detection and Recognition Algorithm for Intelligent Vehicles. *Sensors* **2019**, *19*, 4021. [CrossRef] [PubMed]
28. Li, W.; Li, D.; Zeng, S. Traffic Sign Recognition with a Small Convolutional Neural Network. *IOP Conf. Ser. Mater. Sci. Eng.* **2019**, *688*, 044034. [CrossRef]
29. Atif, M.; Zoppi, T.; Gharib, M.; Bondavalli, A. Quantitative Comparison of Supervised Algorithms and Feature Sets for Traffic Sign Recognition. In Proceedings of the 36th Annual ACM Symposium on Applied Computing, Virtual, 22–26 March 2021; Association for Computing Machinery: Gwangju, Korea, 2021; pp. 174–177. [CrossRef]
30. Vilalta, R.; Drissi, Y. A Perspective View and Survey of Meta-Learning. *Artif. Intell. Rev.* **2002**, *18*, 77–95. [CrossRef]
31. Stallkamp, J.; Schlipsing, M.; Salmen, J.; Igel, C. The German Traffic Sign Recognition Benchmark: A Multi-Class Classification Competition. In Proceedings of the 2011 International Joint Conference on Neural Networks, San Jose, CA, USA, 31 July–5 August 2011; pp. 1453–1460.
32. Timofte, R.; Zimmermann, K.; Van Gool, L. Multi-View Traffic Sign Detection, Recognition, and 3D Localisation. *Mach. Vis. Appl.* **2014**, *25*, 633–647. [CrossRef]
33. Youssef, A.; Albani, D.; Nardi, D.; Bloisi, D.D. Fast Traffic Sign Recognition Using Color Segmentation and Deep Convolutional Networks. In *Advanced Concepts for Intelligent Vision Systems*; Springer: Lecce, Italy, 2016; pp. 205–216.
34. Krizhevsky, A.; Sutskever, I.; Hinton, G.E. ImageNet Classification with Deep Convolutional Neural Networks. In Proceedings of the Advances in Neural Information Processing Systems 25: 26th Annual Conference on Neural Information Processing Systems 2012, Lake Tahoe, NV, USA, 3–6 December 2012.
35. He, K.; Zhang, X.; Ren, S.; Sun, J. Deep Residual Learning for Image Recognition. In Proceedings of the IEEE Conference on Computer Vision and Pattern Recognition, Las Vegas, NV, USA, 27–30 June 2016; pp. 770–778.
36. Carbonell, J.G.; Michalski, R.S.; Mitchell, T.M. 1—An Overview of Machine Learning. In *Machine Learning*; Michalski, R.S., Carbonell, J.G., Mitchell, T.M., Eds.; Morgan Kaufmann: San Francisco, CA, USA, 1983; pp. 3–23. ISBN 978-0-08-051054-5.
37. Breiman, L. Random Forests. *Mach. Learn.* **2001**, *45*, 5–32. [CrossRef]
38. Fisher, R.A. The Use of Multiple Measurements in Taxonomic Problems. *Ann. Eugen.* **1936**, *7*, 179–188. [CrossRef]
39. Freund, Y. A More Robust Boosting Algorithm. *arXiv* **2009**, arXiv:0905.2138.
40. Szegedy, C.; Vanhoucke, V.; Ioffe, S.; Shlens, J.; Wojna, Z. Rethinking the Inception Architecture for Computer Vision. In Proceedings of the IEEE Conference on Computer Vision and Pattern Recognition, Las Vegas, NV, USA, 27–30 June 2016; pp. 2818–2826.
41. Sandler, M.; Howard, A.; Zhu, M.; Zhmoginov, A.; Chen, L.C. MobileNetV2: Inverted Residuals and Linear Bottlenecks. In Proceedings of the IEEE Conference on Computer Vision and Pattern Recognition, Salt Lake City, UT, USA, 18–23 June 2018; pp. 4510–4520.
42. Lam, L.; Suen, S.Y. Application of Majority Voting to Pattern Recognition: An Analysis of Its Behavior and Performance. *IEEE Trans. Syst. Man Cybern.-Part A Syst. Hum.* **1997**, *27*, 553–568. [CrossRef]
43. Yasuda, H.; Takahashi, K.; Matsumoto, T. A Discrete Hmm for Online Handwriting Recognition. *Int. J. Pattern Recognit. Artif. Intell.* **2000**, *14*, 675–688. [CrossRef]
44. Huang, Z.; Yu, Y.; Gu, J. A Novel Method for Traffic Sign Recognition Based on Extreme Learning Machine. In Proceedings of the 11th World Congress on Intelligent Control and Automation, Shenyang, China, 29 June–4 July 2014; pp. 1451–1456.
45. Agrawal, S.; Chaurasiya, R.K. Ensemble of SVM for Accurate Traffic Sign Detection and Recognition. In Proceedings of the International Conference on Graphics and Signal Processing, Singapore, 24–27 June 2017; pp. 10–15.
46. Myint, T.; Thida, L. Real-Time Myanmar Traffic Sign Recognition System Using HOG and SVM. *Int. J. Trend Sci. Res. Dev.* **2019**, *3*, 2367–2371. [CrossRef]
47. Abedin, M.Z.; Dhar, P.; Deb, K. Traffic Sign Recognition Using SURF: Speeded up Robust Feature Descriptor and Artificial Neural Network Classifier. In Proceedings of the 2016 9th International Conference on Electrical and Computer Engineering (ICECE), Dhaka, Bangladesh, 20–22 December 2016; pp. 198–201.
48. Farag, W. Traffic Signs Classification by Deep Learning for Advanced Driving Assistance Systems. *Intell. Decis. Technol.* **2019**, *13*, 305–314. [CrossRef]

49. Kamal, U.; Tonmoy, T.I.; Das, S.; Hasan, M.K. Automatic Traffic Sign Detection and Recognition Using SegU-Net and a Modified Tversky Loss Function With L1-Constraint. *IEEE Trans. Intell. Transp. Syst.* **2020**, *21*, 1467–1479. [CrossRef]
50. Liu, C.; Chang, F.; Chen, Z.; Liu, D. Fast Traffic Sign Recognition via High-Contrast Region Extraction and Extended Sparse Representation. *IEEE Trans. Intell. Transp. Syst.* **2016**, *17*, 79–92. [CrossRef]
51. Lin, C.; Li, L.; Luo, W.; Wang, K.C.P.; Guo, J. Transfer Learning Based Traffic Sign Recognition Using Inception-v3 Model. *Period. Polytech. Transp. Eng.* **2019**, *47*, 242–250. [CrossRef]
52. Zaki, P.S.; William, M.M.; Soliman, B.K.; Alexsan, K.G.; Khalil, K.; El-Moursy, M. Traffic Signs Detection and Recognition System Using Deep Learning. *arXiv* **2020**, arXiv:2003.03256.
53. Abdel-Salam, R.; Mostafa, R.; Abdel-Gawad, A.H. RIECNN: Real-Time Image Enhanced CNN for Traffic Sign Recognition. *Neural Comput. Appl.* **2022**, *34*, 6085–6096. [CrossRef]
54. Mangshor, N.N.A.; Paudzi, N.P.A.M.; Ibrahim, S.; Sabri, N. A Real-Time Malaysian Traffic Sign Recognition Using YOLO Algorithm. In *12th National Technical Seminar on Unmanned System Technology 2020*; Lecture Notes in Electrical Engineering; Springer: Singapore, 2022; pp. 283–293.
55. Naim, S.; Moumkine, N. LiteNet: A Novel Approach for Traffic Sign Classification Using a Light Architecture. In *WITS 2020, Proceedings of the 6th International Conference on Wireless Technologies, Embedded, and Intelligent Systems*; Springer: Singapore, 2022; pp. 37–47.
56. Nartey, O.T.; Yang, G.; Asare, S.K.; Wu, J.; Frempong, L.N. Robust Semi-Supervised Traffic Sign Recognition via Self-Training and Weakly-Supervised Learning. *Sensors* **2020**, *20*, 2684. [CrossRef]
57. Vennelakanti, A.; Shreya, S.; Rajendran, R.; Sarkar, D.; Muddegowda, D.; Hanagal, P. Traffic Sign Detection and Recognition Using a CNN Ensemble. In Proceedings of the 2019 IEEE International Conference on Consumer Electronics (ICCE), Las Vegas, NV, USA, 11–13 January 2019; pp. 1–4.
58. Lu, X.; Wang, Y.; Zhou, X.; Zhang, Z.; Ling, Z. Traffic Sign Recognition via Multi-Modal Tree-Structure Embedded Multi-Task Learning. *IEEE Trans. Intell. Transp. Syst.* **2017**, *18*, 960–972. [CrossRef]
59. Bi, Z.; Yu, L.; Gao, H.; Zhou, P.; Yao, H. Improved VGG Model-Based Efficient Traffic Sign Recognition for Safe Driving in 5G Scenarios. *Int. J. Mach. Learn. Cybern.* **2021**, *12*, 3069–3080. [CrossRef]
60. Bayoudh, K.; Hamdaoui, F.; Mtibaa, A. Transfer Learning Based Hybrid 2D–3D CNN for Traffic Sign Recognition and Semantic Road Detection Applied in Advanced Driver Assistance Systems. *Appl. Intell.* **2021**, *51*, 124–142. [CrossRef]
61. Sermanet, P.; LeCun, Y. Traffic Sign Recognition with Multi-Scale Convolutional Networks. In Proceedings of the 2011 International Joint Conference on Neural Networks, San Jose, CA, USA, 31 July–5 August 2011; pp. 2809–2813.
62. Yang, Y.; Luo, H.; Xu, H.; Wu, F. Towards Real-Time Traffic Sign Detection and Classification. *IEEE Trans. Intell. Transp. Syst.* **2016**, *17*, 2022–2031. [CrossRef]
63. Islam, K.T.; Raj, R.G. Real-Time (Vision-Based) Road Sign Recognition Using an Artificial Neural Network. *Sensors* **2017**, *17*, 853. [CrossRef] [PubMed]
64. Luo, H.; Yang, Y.; Tong, B.; Wu, F.; Fan, B. Traffic Sign Recognition Using a Multi-Task Convolutional Neural Network. *IEEE Trans. Intell. Transp. Syst.* **2018**, *19*, 1100–1111. [CrossRef]
65. Yuan, Y.; Xiong, Z.; Wang, Q. An Incremental Framework for Video-Based Traffic Sign Detection, Tracking, and Recognition. *IEEE Trans. Intell. Transp. Syst.* **2017**, *18*, 1918–1929. [CrossRef]
66. Park, J.; Lee, K.; Kim, H.Y. *Recognition Assistant Framework Based on Deep Learning for Autonomous Driving: Restoring Damaged Road Sign Information*; Social Science Research Network: Rochester, NY, USA, 2022.
67. Zakir Hussain, K.M.; Kattigenahally, K.N.; Nikitha, S.; Jena, P.P.; Harshalatha, Y. Traffic Symbol Detection and Recognition System. In *Emerging Research in Computing, Information, Communication and Applications*; Springer: Singapore, 2022; pp. 885–897.
68. Lahmyed, R.; Ansari, M.E.; Kerkaou, Z. Automatic Road Sign Detection and Recognition Based on Neural Network. *Soft Comput.* **2022**, *26*, 1743–1764. [CrossRef]
69. Gautam, S.; Kumar, A. Automatic Traffic Light Detection for Self-Driving Cars Using Transfer Learning. In *Intelligent Sustainable Systems*; Springer: Singapore, 2022; pp. 597–606.
70. Schuszter, I.C. A Comparative Study of Machine Learning Methods for Traffic Sign Recognition. In Proceedings of the 2017 19th International Symposium on Symbolic and Numeric Algorithms for Scientific Computing (SYNASC), Timisoara, Romania, 21–24 September 2017; pp. 389–392.
71. Brazdil, P.; Carrier, C.G.; Soares, C.; Vilalta, R. *Metalearning: Applications to Data Mining*; Springer Science & Business Media: Berlin/Heidelberg, Germany, 2008; ISBN 978-3-540-73262-4.
72. Vanschoren, J. Understanding Machine Learning Performance with Experiment Databases. Ph.D. Thesis, Katholieke Universiteit Leuven—Faculty of Engineering Address, Leuven, Belgium, May 2010.
73. Wolpert, D.H. Stacked Generalization. *Neural Netw.* **1992**, *5*, 241–259. [CrossRef]
74. Breiman, L. Bagging Predictors. *Mach. Learn.* **1996**, *24*, 123–140. [CrossRef]
75. Hochreiter, S.; Schmidhuber, J. Long Short-Term Memory. *Neural Comput.* **1997**, *9*, 1735–1780. [CrossRef]
76. Li, Y.; Zhu, Z.; Kong, D.; Han, H.; Zhao, Y. EA-LSTM: Evolutionary Attention-Based LSTM for Time Series Prediction. *Knowl.-Based Syst.* **2019**, *181*, 104785. [CrossRef]
77. Sokolova, M.; Lapalme, G. A Systematic Analysis of Performance Measures for Classification Tasks. *Inf. Process. Manag.* **2009**, *45*, 427–437. [CrossRef]

78. Mortaz, E. Imbalance Accuracy Metric for Model Selection in Multi-Class Imbalance Classification Problems. *Knowl.-Based Syst.* **2020**, *210*, 106490. [CrossRef]
79. Chamasemani, F.F.; Singh, Y.P. Multi-Class Support Vector Machine (SVM) Classifiers—An Application in Hypothyroid Detection and Classification. In Proceedings of the 2011 Sixth International Conference on Bio-Inspired Computing: Theories and Applications, Penang, Malaysia, 27–29 September 2011; pp. 351–356.
80. Deng, J.; Dong, W.; Socher, R.; Li, L.J.; Li, K.; Fei-Fei, L. ImageNet: A Large-Scale Hierarchical Image Database. In Proceedings of the 2009 IEEE Conference on Computer Vision and Pattern Recognition, Miami, FL, USA, 20–25 June 2009; pp. 248–255.
81. Popov, G.; Raynova, K. Diversity in Nature and Technology—Tool for Increase the Reliability of Systems. In Proceedings of the 2017 15th International Conference on Electrical Machines, Drives and Power Systems (ELMA), Sofia, Bulgaria, 1–3 June 2017; pp. 464–466.
82. Shang, R.; Xu, K.; Shang, F.; Jiao, L. Sparse and Low-Redundant Subspace Learning-Based Dual-Graph Regularized Robust Feature Selection. *Knowl.-Based Syst.* **2020**, *187*, 104830. [CrossRef]
83. Zeng, Y.; Xu, X.; Shen, D.; Fang, Y.; Xiao, Z. Traffic Sign Recognition Using Kernel Extreme Learning Machines With Deep Perceptual Features. *IEEE Trans. Intell. Transp. Syst.* **2017**, *18*, 1647–1653. [CrossRef]
84. Jiménez, Á.B.; Lázaro, J.L.; Dorronsoro, J.R. Finding Optimal Model Parameters by Discrete Grid Search. In *Innovations in Hybrid Intelligent Systems*; Springer: Berlin/Heidelberg, Germany, 2007; pp. 120–127. [CrossRef]
85. Secci, F.; Ceccarelli, A. On Failures of RGB Cameras and Their Effects in Autonomous Driving Applications. In Proceedings of the 2020 IEEE 31st International Symposium on Software Reliability Engineering (ISSRE), Coimbra, Portugal, 12–15 October 2020; pp. 13–24.

sensors

MDPI

Article

Enhancing Detection Quality Rate with a Combined HOG and CNN for Real-Time Multiple Object Tracking across Non-Overlapping Multiple Cameras

Lesolė Kalake [1,*], Yanqiu Dong [1], Wanggen Wan [1] and Li Hou [2]

1 School of Communications and Information Engineering, Institute of Smart City, Shanghai University, Shanghai 200444, China; yanqiu_dong@shu.edu.cn (Y.D.); wanwg@staff.shu.edu.cn (W.W.)
2 School of Information Engineering, Huangshan University, Huangshan 245041, China; houli@shu.edu.cn
* Correspondence: tumelok1@shu.edu.cn; Tel.: +86-198-2121-4680

Abstract: Multi-object tracking in video surveillance is subjected to illumination variation, blurring, motion, and similarity variations during the identification process in real-world practice. The previously proposed applications have difficulties in learning the appearances and differentiating the objects from sundry detections. They mostly rely heavily on local features and tend to lose vital global structured features such as contour features. This contributes to their inability to accurately detect, classify or distinguish the fooling images. In this paper, we propose a paradigm aimed at eliminating these tracking difficulties by enhancing the detection quality rate through the combination of a convolutional neural network (CNN) and a histogram of oriented gradient (HOG) descriptor. We trained the algorithm with an input of 120 × 32 images size and cleaned and converted them into binary for reducing the numbers of false positives. In testing, we eliminated the background on frames size and applied morphological operations and Laplacian of Gaussian model (LOG) mixture after blobs. The images further underwent feature extraction and computation with the HOG descriptor to simplify the structural information of the objects in the captured video images. We stored the appearance features in an array and passed them into the network (CNN) for further processing. We have applied and evaluated our algorithm for real-time multiple object tracking on various city streets using EPFL multi-camera pedestrian datasets. The experimental results illustrate that our proposed technique improves the detection rate and data associations. Our algorithm outperformed the online state-of-the-art approach by recording the highest in precisions and specificity rates.

Keywords: convolutional neural network; histogram oriented graphic; multi-camera multi-object tracking; detection quality

Citation: Kalake, L.; Dong, Y.; Wan, W.; Hou, L. Enhancing Detection Quality Rate with a Combined HOG and CNN for Real-Time Multiple Object Tracking across Non-Overlapping Multiple Cameras. *Sensors* **2022**, *22*, 2123. https://doi.org/10.3390/s22062123

Academic Editors: Euntai Kim, Kang Ryoung Park and Sangyoun Lee

Received: 30 December 2021
Accepted: 5 March 2022
Published: 9 March 2022

1. Introduction

The visualization and tracking of multiple objects in surveillance applications are enormously dominating topics in computer vision's security field. In recent years, there has been a drastic change in point of focus for enhancing the handling of security issues on these applications [1]. Many researchers are attracted, and several techniques and algorithms emerged are applied continuously on various smart city projects to ensure residence safety. However, most rely on the traditional convolutional neural network (CNN) to improve the detection quality rate and object classification [2]. The CNN provides an effective and quick solution to extract high-level contour features and record a significant state-of-the-art performance on real-time multiple-object-tacking (MOT) [3]. It is considered to be more effective compared to HOG descriptor algorithms which mainly focus on global features process handling [4].

Despite the state-of-the-art achievement, the traditional CNN proposed algorithms tend to ignore the global features [5]. Their detectors are mainly based on the local features extraction for the application to understand the image information [6]. Therefore, they continue to suffer from identifying the shape and boundary characteristics from the captured images [7]. Thus, this contributes to their incapability for handling the detection accuracy on light, appearance distortion, deformation, and motion-blurred images. Furthermore, it results in poor detection quality and high false positives, hence, its failure in representing human-like application systems [8]. Other studies tried to eliminate this grey area by exploiting the HOG descriptor technique and recorded satisfactory results but suffered from the speed and classification of huge samples during the training phase [9].

Therefore, to ensure both contour and global features are effectively incorporated into the neural network to represent a human-like system. In this paper, we propose to build a new model by combining the HOG descriptors and a traditional CNN to form an HCNN algorithm for tracking multi-object across non-overlapping cameras. We further propose to improve the detection quality rate by removing the background information and ensuring that the appearance and motion variations are well maintained throughout the tracking process. This paper is arranged into five sections: Section 1 introduces the background, Section 2 details the related work, Section 3 describes details of our approach, Section 4 presents experimental results, Section 5 discusses an interpretation of the results and comparison with state-of-the-art algorithms, and finally Section 6 concludes the paper.

2. Related Works

The techniques that implement multiple view angles provide additional information that enables the computer vision applications to acquire more knowledge and understanding of the object's characteristics. This has proven its effectiveness in enriching the target-related shape, features, and location in sequential video frames [3]. It further resulted in the emergence of various multiple view object tracking approaches to solve the persisting challenges such as partial inclusion, shape deformation, illumination variations, and background cluttering. The approaches are online or offline depending on the criteria, such as handcrafted features or deep features handlings. The handcrafted feature-based trackers are manually defined, whereas the deep features trackers use neural networks [10]. However, both categories tend to ignore the preprocessing of input images to reduce interferences. Therefore, integration has emerged to achieve fast and accurate human-like detection application systems [11]. Thus, in this section, we summarize these previously proposed state-of-the-art tracking methods by classifying them into two themes: (i) histogram of oriented gradient (HOG) and (ii) convolutional neural network (CNN) learning-based methods.

The histogram of oriented gradient (HOG) descriptor is one of the most popular approaches in computer vision used to extract significant features from images. It discards the futile information by relying heavily on the extracted features to compute accurate objects detections and classifications [12].

Zhang et al. [11] were inspired by these capabilities and proposed a combined local and global feature handling algorithm to simulate a human-like application. They trained both features (local and global) with traditional CNN and set the number of hidden layer nodes to 3000 to distinguish the fool images. However, the technique is most efficient in offline mode and recorded few false alarms compared to CNN solely based paradigms. It further illustrated the incapability of learning features recursively and resulted in slow detection performance, decreased accuracy, and posed challenges to implement online. To eliminate these challenges, Zhang et al. [13] introduced the model detection and classification of moving objects in video and used HOG to remove the noisy background. This strengthened the approach in detecting the moving objects accurately in food and agricultural traceability analysis. However, it failed to obtain adequate features from the selection and resulted in a poor detection rate and data association. Najva et al. [14] proposed improving the detection rate by combining tensor features with scale invariant feature transform (SIFT) features. The technique merged the handcrafted features with a deep convolutional neural network (DCNN) and served as the concrete foundation to expand in the computer vision field. Then

Lipetski et al. [5] took advantage of the laid foundation and combined the HOG descriptor with CNN to form the HCNN model for improving the pedestrian detection quality rate. They extracted HOG features and fed them into the CNN as input to increase classification and detection rates. This reduced the processing time of the overall detector and proved that the concept enhances the capabilities of the overlapping window to handle real-time object tracking processes. The development gained the attention of Rui et al. [15], who proposed an algorithm that takes various features maps from the first CNN layer as input to HOG and extracts the HOG features. However, the performance results illustrated that a single feature map was not comprehensive enough to reflect all the necessary information on the original image. Thus, the technique performed worse than the original HOG paradigm but proved that pedestrian detection with HOG-based multi-convolutional features could obtain a high detection accuracy and stabilized network performance. Then Sujanaa et al. [16] proposed to eliminate pedestrian detection and classification issues by introducing the combined pyramid histogram of oriented gradient (PHOG) and CNN algorithm for real-time object tracking. They used the PHOG descriptor to create pyramid histograms over the entire image and attach them into a single vector, whereas the CNN is used as the classifier for the PHOG features extracted from the window's raw image data. The first layer of the CNN moved adequately over the input image window thus that the second layer could transfer functions to the input image window. Lastly, the hidden layer unit is used to connect to each input through a separate weight. This reduced computational cost, adaptable parameters during training, and proved the technique compatible for real-time object tracking. However, it suffered from low performance with a high misdetection rate under heavy light variations.

Qi et al. [17] proposed an internet of things (IoT) based on a key frame extraction algorithm to enhance detection quality rate in videos. They modeled and trained the CNN to generate a predicted score to indicate the quality of faces in the frame. The selected key frames fed into the neural network to enhance face detection accuracy. This enhanced the extraction of feature vectors and increased face recognitions and detections on poor-quality captured images. Angeline et al. [1] capitalized on the progress and proposed to enhance efficiency on face recognition applications in real-time object tracking. They used HOG descriptor detections to enhance accuracy and train CNN with a linear support vector machine (SVM) to handle blurred motions, occlusions, and pose variation. However, the algorithm used a small dataset and struggled with misfeeding. Thus, Yudin et al. [18] used video streams of specified IP cameras to access more data through the server module. They augmented the IoT application with the HOG descriptor and masked R-CNN architecture for accurate detection of a human head on low-quality and light variations images. This enabled the application to carry out client requests from various computers connected to the network. However, the updating of people counting results performed once per minute hindered overall speed performance. This contributed to the misdetection rate where objects' motion changes.

Madan et al. [6] proposed a hybrid model based on a combination of HOG-speeded-up robust features (SURF) features and CNN. They used extracted HOG features as an input into the network (CNN) and reduced the dimensions. The application embedding from the first layer and second layer of the CNN passes through the fully connected layer. Therefore, this reduced the model parameter's computational cost by filtering out the fool images at an early stage. It further improved the detection and classification accuracy rate. Bao et al. [7] showed appreciation of these developments when proposing the merging of both HOG feature space and traditional CNN to ease the plant species identification and classification from a leaf pattern in botany. They extracted HOG features through 8×8 dimension cells and 2×2 cells per block for the input image. These attributes are passed into the network for further processing and classification. However, the algorithm is an offline mode and recorded a noticeable improvement in the overall performance.

3. Proposed HCNN for Real-Time MOT

The main task is to track and re-identify the target across these multiple cameras [19–21]. We, therefore, designed our algorithm to detect, track and re-identify the object of interest across several non-overlapping cameras using the multi-object tracking process. We implemented the proposed algorithm using the dataset that contains different poses of persons [22] and different illumination conditions. The algorithm is divided into two modules, namely, detection and tracking. The detection module buttressed [23,24] by the inclusion of HOG descriptors which have been proven to cater to both texture and contour features [8,21,22]. We train the model on the EPFL dataset with multiple pedestrians' videos using the HOG detector. However, the HOG descriptor is slowing down overall algorithm performance. Therefore, we combined the HOG detector module with CNN to create an HCNN to enhance classification and identify the association in tracking multiple people. According to our best knowledge, there is no similar proposed algorithm for real-time object tracking across multiple non-overlapping cameras.

The algorithm's process of determining an object's background is split into several separated steps to eliminate backgrounds that might otherwise be classified [25–27]. This is embraced by subtracting the objects' background and computing a foreground mask on colored video frames [28] and gray images captured from multiple surveillance cameras. The proposed algorithm takes an input of the 120 × 32 images, cleans and converts them into binary format, and then smoothens the pixels on binary images by applying morphological operations that are followed by the implementation of a Laplacian of Gaussian model (LOG) mixture after blobs. The images undergo further feature extraction and computation with the HOG descriptor. We stored these features into a 2-dimensional array and passed them to the fully connected multi-layer neural network for further classifications and matching computation, as shown in Figure 1. The CNN flattens the given 2D array into a single feature vector that is used to determine the object of interest's class. Then an output from the HOG descriptor compared them with the object of interest on the input frame based on the connected components, image region properties, and window binary mask. The sliding window tactics were applied on input frames to reduce the data size, processing time, and to improve the object locating during tracking in one step. The normalized cross-function is used to obtain the object centroids on these images. Finally, we considered the use of the Kalman filter to track the object of interest, based on the computed centroids.

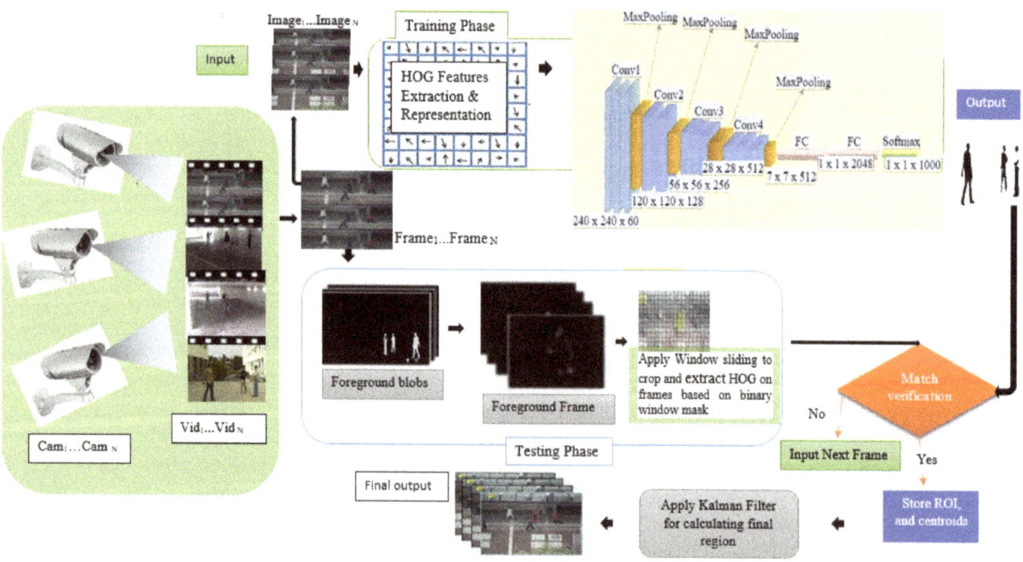

Figure 1. Proposed HCNN method implementation architecture overview.

3.1. Background Segmenting Modeling

Background motion has always been a throwback for many conventional methods to achieve the desired accuracy [23,25]. However, we applied the background subtraction model to ensure that our algorithm overcomes these challenges. In this modeling, we set the threshold pixel value to 0.5 to ensure the detection of every blob for all objects shapes. We further applied the splitting of the image into foreground and background for our algorithm to efficiently classify the pixels [29]. However, the recent history of each pixel value is observed with a mixture of Gaussian distributions, and the new pixel values are considered as the major components to update the model. These new pixel values at a given time (pV_t) are further checked against generated Gaussian distribution until matches are obtained [30]. The pixels with similar velocity at given x and y directions are considered as a point of interest of the same object representing its velocity. These matches are then defined with a standard deviation (σ) of the distribution. This improved the foreground masks, connectivity between neighboring pixels, speed mapping of the moving object, and the capability to distinguish the non-stationary detections from the foreground blobs.

However, when there are no matches found in the T generated distribution, the probability of the distribution of the previous action is replaced with the current mean (μ) value, highest variance (σ^2) and the lowest weight (w) of the object. Thus, we observe the probability of the pixel values as follows.

$$P(X_t) = \sum_{i=1}^{T} W_{i,t} \times \Psi(X_t,\ \mu_{i,t}, \sum_{i,t}^{T}) \tag{1}$$

where $\{X_1, X_2 \dots X_t\}$ represent recent pixels history and $1 \leq i \leq t$; T denotes the number of the distributions, whereas $W_{i,t}$ represents an estimated weight of the ith The Gaussian mixture at given time t, $\mu_{i,t}$ and $\sum_{i,t}^{T}$ respectively denotes the mean and covariance matrix. Then Ψ denotes the Gaussian probability density function and is computed as follows.

$$\Psi(X_t, \mu, \Sigma) = 1/((2\Pi)^{n/2}\left|\Sigma\right|_{1/2})e^{1/2(X_t-\mu_t)^T\Sigma^{-1}(X_t-\mu_t)} \tag{2}$$

Then the weight of the T distribution at a given time is updated as follows.

$$W_{T,t} = (1-\alpha) \times W_{T,(t-1)} + \alpha \times (\Xi_{T,t}) \tag{3}$$

where α denote the learning rate, T is equivalent to the available memory and computation power usage, $\Xi_{T,t}\ \epsilon(1,0)$ where one denotes model matching is true, zero represents that model as unmatched. The advantage of this background technique we applied is that our background model is updated without destroying the existing model. This is achieved by ensuring that after the weights normalizations, the mean and the variance corresponds with the conditions of the distribution and are updated only when conditions change by using the following equations, respectively.

$$\mu_t = (1-p)\mu_{(t-1)} + pX_t \tag{4}$$

and

$$\sigma^2{}_t = (1-p)\sigma^2{}_{(t-1)} + p(X_t - \mu_t) \tag{5}$$

We further ensured that the learning factor p adapts to the current distributions by computing it as:

$$p = \alpha \times \Psi(X_t|\mu_t, \sigma_t) \tag{6}$$

3.2. Foreground Blobs Windowing Modeling

In this modeling, we applied a sliding window approach on both images and foreground frames. This helped our algorithm to avoid detection of non-moving background and shadows of the objects in motion. Therefore, the binary foreground image is used to fulfill the desired window output that is extracted and expressed with the following equation.

$$\zeta_{xy}^{w} = \left\{ \delta_{xy}^{w} \ \Sigma_{x,y=1}^{w_x w_y} \delta_{xy}^{wb} \geq \kappa_p \times 50\%; \ \ w_x \epsilon \text{hieght}, \ w_y \epsilon \text{width} \right. \tag{7}$$

where ζ_{xy}^{w} denotes desired output window for the given window input image δ_{xy}^{w} which is extracted through a sliding window on the binary foreground image δ_{xy}^{wb} with a sum of the total number of pixels $\Sigma_{x,y=1}^{w_x w_y} \delta_{xy}^{wb}$ on binary window κ_p. In the next section, we discuss the HoG descriptor implemented in this paper in detail.

3.3. HOG Descriptor's Features Extraction

Hog is a feature extraction technique that extracts features from every position of the image by constructing logic histograms of the object from the images [7]. In this paper, the images are first passed through the HOG descriptor for data size reduction and searching for an object to detect. Thereafter, the histograms are created and computed over the whole images that are retrieved from several video frames. These histograms are then appended into a single feature vector using the exponential equation 2^{ℓ}, representing the grid level (ℓ) for all cells along the dimensions. However, the correspondence on the whole input images between the vectors and histograms bins is ensured by limiting the level (ℓ) to ≤ 3 and computed using the following equation.

$$\nu = \mathcal{K} \sum_{i=1}^{\ell} 4^{\ell}; \ i \leq 3 \tag{8}$$

where ν, denotes vector dimensions, $\mathcal{K}$ denotes bins, ℓ defines grid level. This equation ensures that all images that are extremely large and rich in texture are weighted the same as low texture images within the set parameters. It is also used to guard and control our algorithm against overfitting.

In our detection module, a two-dimensional (2D) array of the detected object is constructed. It is passed to the CNN, wherein the process of targeted object recognition is flattened into a single vector using two fully connected layers. The CNN is also used to classify that the person detected by the HOG descriptor is either associated with the assigned ID (e.g., ID1 or other IDs) [31].

3.4. Structure of the Convolutional Neural Network

The structure of the CNN incorporated into our algorithm is shown in Figure 2. We considered extracting the appropriate features first from the window's raw data. There are four convolutional layers with three max-pooling layers, two fully convolutional layers, and a softmax activation function. The first layer is used to map various small features that are cited as local receptive fields (LRF) that move satisfactorily over the input image window on the grid. The second layer contains one or many fully connected output neurons that are applied to transfer functions to the inputs during the training phase. Therefore, the hidden layer of the multiple layer perception is used to connect each input with a separate weight.

The LRF was applied to all image portions using the same weights, and this contributed to the reduction of adaptable parameters. However, when the network has biased weights, the output weights becomes the element of the transferred functions, which are applied to the first and second layer, respectively. The object is then recognized from the foreground frame's sliding window, and its parameters such as x and y coordinates for the starting position, height, width, and centroids are calculated. This avoided network overfitting and provided the current location of the object being detected [24,25]. Finally, the Kalman filter was applied to track the object of interest based on the computed centroids and assigned unique identities throughout the frames.

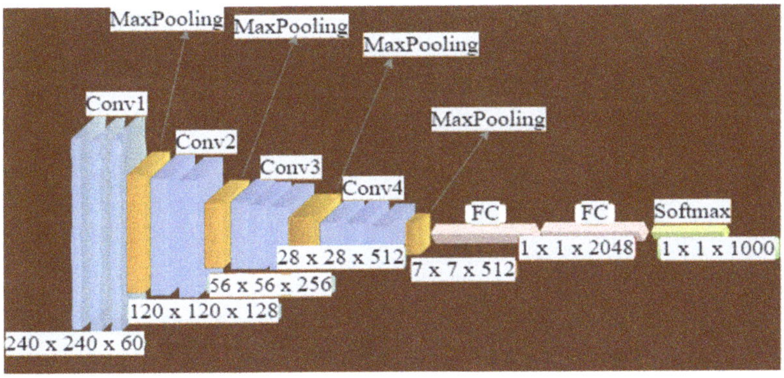

Figure 2. Overview of the CNN structure incorporated into HOG.

3.5. Designing Kalman Filter for Our HCNN Algorithm

In most cases, computer vision algorithms' frequent task is based on object detection and localization [26,32]. Therefore, in this paper, we considered the design and the incorporation of a simple and robust procedure to engage complex scenes with minimum resources [27]. We integrated the computed object centroids into the Kalman filter's object motion and measurement noise [29]. This strengthened the processing of noises and the estimation of the object's next position in the next frame at a given speed and time [12]. However, it also made our algorithm entitled to efficiently re-detect the moving object during occlusions, scaling, illuminations, appearance changes, and rapid motion on both training and validation phases [33]. Therefore, to solve these challenges, we enabled the Kalman filer to model and associate the target ID that is assigned based on the computed centroids. This improved the observations, predictions, measurements, corrections, and updating of the object's whereabouts and directions.

Thus, observations are effectively used to locate the object and provide a direction at a given velocity and measurement using the following equation.

$$Z = X + \mathcal{E}_r \ ; \tag{9}$$

where Z denotes measurements, X represents the location of the object being tracked, and $\mathcal{E}_r$ is distributed normally ($\mathcal{E}_r \sim N\ (0, \sigma^2)$) and denotes noisy measurements due to uncertainty of the current object location. Although this guarantees that our algorithm can handle the noises, we prognosticate that our detector might be imperfect due to the combination of $\mathcal{E}_r$ and velocity (v) variations that will affect the tracker to locate and track the object of interest effectively. Thus, to handle these uncertainties, we estimated the trajectories of the moving object from the initial state to the final state of direction by incorporating the $\mathcal{E}_r$ into the converted matrix formulae of motion measurement as follows.

$$\overline{X}_t = \begin{bmatrix} X_t \\ V_t \end{bmatrix}; \tag{10}$$

denoting location X, and speed V of an object at a particular time

$$\overline{Z_t} = [Z_t]; \tag{11}$$

denoting the distance measurement of an object at a particular time

Thus, the Equations (10) and (11) are combined and expanded to express the location of an object being tracked as follows:

$$Z_t = X_t + \mathcal{E}_r \tag{12}$$

which is further converted into a matrix equation and used to handle both noisy measurements and speed variation.

$$\overline{Z_{t+1}} = \begin{bmatrix} 1 & 0 \end{bmatrix} \overline{X_t} + \overline{\mathcal{E}_r}; \tag{13}$$

$\begin{bmatrix} 1 & 0 \end{bmatrix}$ denote H state control matrix at time $t + 1$.

In short, Equation (13) is expressed as $\overline{Z_{t+1}} = H\overline{X_t} + \overline{\mathcal{E}_r}$.

However, the Equation (13) estimations do not adapt to the speed changes. Therefore, to incorporate speed variations and locate the position of the object correctly in the next frame, we calculated the algorithm evaluation through time (t) at acceleration (a) and changes in time (Δt) using the equation below.

$$X_{t+1} = X_t + V_t \times \Delta t + \frac{1}{2}at^2 \tag{14}$$

where X_{t+1} denotes our prediction corrections, X_t denotes the location of the object at a given time (t), V_t denote the speed of the object at a given time (t), and $\Delta t + \frac{1}{2}at^2$ represent speed integration at a given time (t). However, the speed is not constant for the object in motion. Hence, we accommodated its changes through different frames scenes by adapting velocity variations using the equation below.

$$V_{t+1} = V_t + a\Delta t \tag{15}$$

We further expanded Equation (14) for time evolution handling and to ensure that the motion and object feature representation on both foreground frames and binary images are correctly captured and predicted. Hence, the newly desired formulae are expressed as follows:

$$\overline{X_{t+1}} = \begin{bmatrix} 1 & \Delta t \\ 0 & 1 \end{bmatrix} + \overline{X_t}\{\text{Previous state}\} + \begin{bmatrix} \frac{1}{2}\Delta t^2 \\ \Delta_t \end{bmatrix} a; \tag{16}$$

where $\begin{bmatrix} 1 & \Delta t \\ 0 & 1 \end{bmatrix}$ denotes state transition matrix function (F), a denote object's acceleration and is distributed normally with mean 0 and variance of the noise measurements, $a \sim N(0, \sigma_r^2)$. Therefore, Equation (16) is further expressed in short, as $\overline{X_{t+1}} = F\overline{X_t} + GV_t$ where G represents a vector $\begin{bmatrix} \frac{1}{2}\Delta t^2 \\ \Delta_t \end{bmatrix}$, which is the object's uncertainty in time changes. Finally, we used these equations into the Kalman filter to predict and correct the object velocity based on the pixels found in the x and y directions. We predicted the steps and propagated the state as follows:

$$\overline{X_t} \Rightarrow \overline{X_{t+1}}, \tag{17}$$

$$if(\overline{X_t} \sim N(\hat{\overline{X}}_t, \acute{P}_t)) \tag{17a}$$

where X_t is a random variable of a normal distribution with a mean $\hat{\overline{X}}_t$ and covariance $\acute{P}_t$.

$$then\ \hat{\overline{X}}_{t+1} = F \cdot \hat{\overline{X}}_t \tag{17b}$$

where F represents the previous state with a certain speed at a particular time. Therefore, we expanded the covariance equation to estimate and update time as follows.

$$P_{t+1} = FP_tF^T + G\sigma_a^2G^T \tag{17c}$$

where P_{t+1} defines the estimated error covariance matrix in the next frame. Thus, knowledge of the measurement (Z_t) steps are now incorporated into the moving object's estimate state vector $(\overline{X_t})$ and the (a) Measuring residual error, (b) Residual covariance, and (c) Kalman gain are computed as respectively as follows.

$$\overline{Y} = \overline{Z}_t - H \cdot \hat{\overline{X}}_t\ ;\ \hat{\overline{X}}_t = \mu \tag{18a}$$

$$S_t = HP_tH^T + R; \text{ where } R \text{ denote } \sigma_r^2 \tag{18b}$$

$$K = P_tH^TS_k^{-1} \tag{18c}$$

Therefore, after this measurement steps incorporation, we can finally update the variable position estimates in the next frame by updating the mean and covariance based on the Kalman gain using the equations below.

$$\hat{\overline{X}}_{|z} = \hat{\overline{X}}_t + K \cdot \overline{Y}; \text{ where } \hat{\overline{X}}_t \text{ denote the previous mean} \tag{18d}$$

$$P_{|z} = (I - K \cdot H)P_t\,; \tag{18e}$$

and I is a 4×4 identity matrix: $\begin{bmatrix} 1 & 0 & 0 & 0 \\ 0 & 1 & 0 & 0 \\ 0 & 0 & 1 & 0 \\ 0 & 0 & 0 & 1 \end{bmatrix}$.

4. Experiments

Experimental Setup

We performed experiments on the EPFL datasets based on campus passengers and subway scenes that contain lots of poses and illumination variations. The algorithm is implemented on Dell, G15 Corei7 11800H Processor, NVidia GeForce RT 350Ti GPU, 4 GB GDDR6, 16 GB RAM with Python 3 (Dell, Pretoria, South Africa).

Datasets and Evaluation Metrics: The EPFL dataset is used and contains campus and passageway scenes that are both outdoor sequences. The campus scene consists of 6 videos, while the passageway has 4 videos. The videos are split into training and validation sets, where we selected 4 campus scenes videos, 3 passageway videos and split them into frames, and retrieved 40,000 images for training. The remaining videos are used for validation in the testing phase.

The algorithm training is conducted with 30,000 multi-view angle positive images and 10,000 negative images of size 120×32. These images are subsets of the frames of the video. We show their instances, labels associations, and correlations in Figure 3. The algorithm is trained with the use of the HOG descriptor, which resized images and activated the object detection module. The HOG descriptor is integrated with the structured CNN illustrated in Figure 2 that is applied as an additional processing mechanism and also a classification mechanism. The training of this proposed system was conducted with 3000 iterations at a learning rate of 0.001.

We evaluated our algorithm's performance with CLEAR MOT metrics that include the precisions(P), recall(R), identity F1 score(IDF1), mean average precisions(mAP), multiple object tracking accuracy(MOTA), multiple object tracking precisions(MOTP), mostly tracked(ML), mostly lost(ML) and ID switches(IDs). The P is the ratio of the correct positive predictions out of all the positive predictions made, whereas R is the ratio of the number of correct positive predictions made out of all positive predictions that could have been made. The mAP was used to evaluate our detection model by comparing the ground truth-bounding box with the detected box. However, the MT and ML account for the ground-truth trajectories that are the ratio of 80% and 20% correctly identified detections over the mAP returned scores respectively [28]. These metrics are defined as follows:

$$\text{Precision} = \frac{\text{TruePositives}}{(\text{TruePositives} + \text{FalsePositives})} \tag{19}$$

$$\text{Recall} = \frac{\text{TruePositives}}{(\text{TruePositives} + \text{FalseNegatives})} \tag{20}$$

$$\text{IDF1 scores} = 2[\frac{P \times R}{(P + R)}] \tag{21}$$

where P and R denote precision and recall respectively.

$$mAP = \frac{1}{2}\sum_{k=1}^{k=n} AP_k\,;\; AP = \sum_{k=0}^{k=n-1} [R_k - R_{k+1}] \times P_k \tag{22}$$

where $R_n = 0$, $P_n = 1$ and n denotes the number of thresholds. The k represents the number of classes.

$$\text{MOTA} = 1 - \left[\frac{\sum_t^N (fn_t + fp_t + IDs_t)}{\sum_t^N G_t}\right] \tag{23}$$

where fn_t, fp_t and IDs_t denote the number of false-negative or missed detections, the false positive, and the miss-match errors in frame t. The G_t represent the ground truth.

$$\text{MOTP} = 1 - [\frac{\sum_{i,t}^N d_t^i}{\sum_t^N c_t}] \tag{24}$$

where d_t^i denotes the distance between the localization of objects in the ith ground truth and the detection output in frame t. The c_t is the total matches made between ground truth and the detection output in frame t.

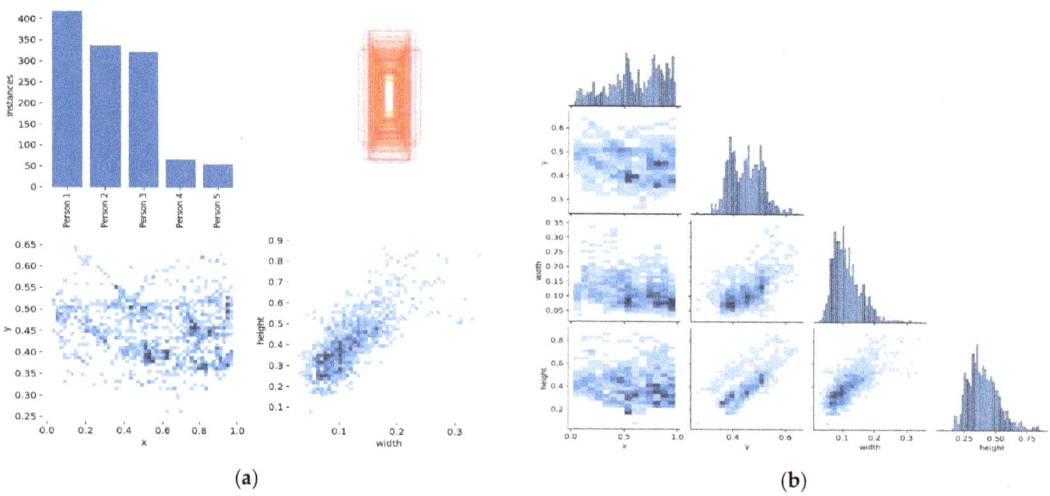

Figure 3. The left (**a**) shows the scatter plot of each instance and label associates, and (**b**) illustrates the correlation relations on the campus scenes images dataset. At most, 90% of the instances are correctly associated with the labels throughout the scenes.

Parameter Settings. Our Algorithm reacted to a new entry object by initiating a Kalman filter for object tracking [29]. The tracker continues to track and check if the new object falls within the acceptance region of the trajectories by using the Kalman filter predicting equations [12]. The error between the actual observation and the predicted observation is normalized by the computation of a covariance matrix from the Kalman filter update equations [32]. Thus, the determination of whether the new object observation is associated with an existing track is performed by the threshold value test on the residual error (covariance matrix values) [12]. This defines the acceptance relations for each object being tracked and updates the state where the threshold test satisfies. All trajectories that are shorter than 80 milliseconds are deleted. However, when an object observation does not fall within any acceptable trajectory region, the tracker establishes a new track. This endorsed the auto-labeling correlations showed in Figures 3a,b and 4a,b. Therefore, the instances are only associated with a single label, and this has increased the label correlations, precisions, and recall in our experimented dataset [30,31]. It also led to the highest MOTA and MOTP, as shown in Tables 1–3. The metrics results and analysis are discussed in the next section, Results Analysis.

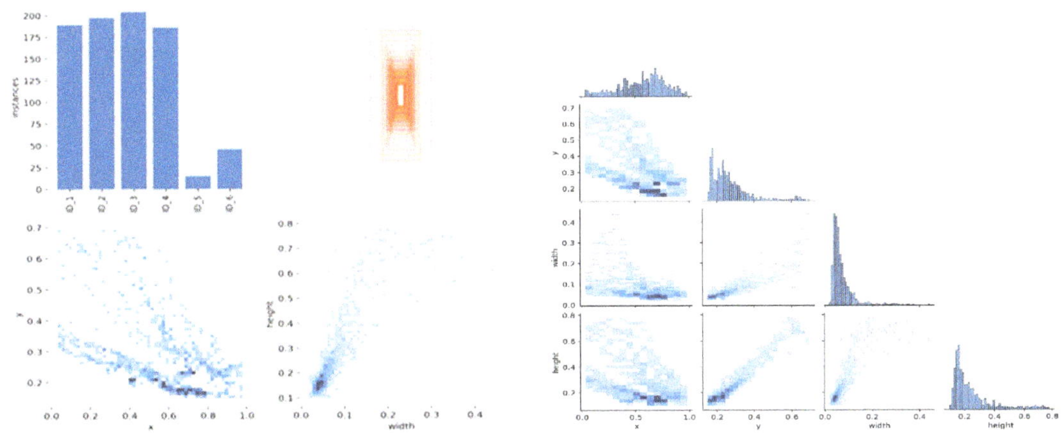

(a) (b)

Figure 4. The left (**a**) shows the scatter plot of each instance and label associates, and (**b**) illustrates the correlation relations on the passageway scenes images dataset. Mostly, 92% of the instances are correctly associated with the labels throughout the scenes.

Table 1. Comparison with state-of-the-art methods based on MOT Classification Accuracy.

Methods	Precision ↑	Causality
Improved HOG [4]	86.70%	Online
HOG + 1DCNN [16]	90.23%	Offline
HOG + DCNN Net [32]	96.74	Offline
HOG + CNN [33]	94.14%	Offline
Ours	91.00%	Online

Table 2. Performance evaluation metrics on EPFL dataset campus sequence.

Sequences	Precision ↑	Recall ↑	IDF Score ↑	MOTA ↑	MOTP ↑	IDS ↓	ML ↓	MT ↑	FM ↓
CAM#4_scene0	99.4%	96.0%	95.9%	94.0%	91.9%	1	1%	96.0%	2
CAM#4_scene1	98.0%	97.0%	98.0%	93.0%	92.0%	1	1%	94.0%	1
CAM#4_scene2	98.0%	94.0%	96.0%	93.0%	89.0%	2	2%	92.0%	3
CAM#7_scene0	68.0%	80.2%	76.4%	63.3%	75.0%	4	3%	82.0%	5
CAM#7_scene1	88.9%	87.6%	88.2%	83.9%	82.5%	3	2%	88%	2
CAM#7_scene2	95.0%	96.8%	96.3%	90.0%	91.8%	1	1%	92%	1
Overall performance	91.22%	91.93%	91.80%	86.20%	87.03%	2	1.67%	90.67%	3

Table 3. Performance evaluation metrics on EPFL dataset passageway sequence.

Sequences	Precision ↑	Recall ↑	IDF Score ↑	MOTA ↑	MOTP ↑	IDS ↓	ML ↓	MT ↑	FM ↓
CAM#1_scene0	94.0%	92.0%	93.0%	89.4%	87.0%	2	2.0%	88.0%	3
CAM#2_scene1	83.0%	82.0%	82.0%	78.0%	76.8%	4	3.0%	86.0%	5
CAM#3_scene2	97.0%	90.8%	93.8%	92.3%	85.8%	2	1.0%	93.0%	2
CAM#4_scene3	76.0%	71.2%	73.5%	71.0%	66.3%	4	4.0%	81.0%	8
Overall performance	87.50%	84.00%	85.58%	82.68%	78.98%	3	2.50%	87.00%	5

5. Results Analysis

In this section, we analyze our HCNN algorithm's results obtained from the experimented dataset. We trained and evaluated our detector to classify with a coupled HOG descriptor and CNN using the EPFL dataset with the selected scenes (campus and passage) for real-time multi-object tracking. The objects are observed and tracked by use of Kalman Filter, as shown in Figure 1. Figures 5–8 illustrate the overall performance and effectiveness of our algorithm's detector and classify for both training and validation phases.

The algorithm has proven to be effective with high performance in precision and recall, accompanied by the high confidence values on the campus scene dataset. It achieved a greater balance between precision and recall, with a mean average precision of 95.1% at a 0.5 threshold for all classes. This demonstrated in Figure 9 that the algorithm could be trusted for accurately detecting and correctly classifying the objects of interest. However, through this process, the algorithm at the beginning of training and the testing phases had challenges of the unrepresentative data but gradually converged well with more training epochs. This is shown in Figures 6 and 8, with the ups and downs of the jumping of the stats values in either training or validation phase graphs. Thus, it led to the high numbers of false-positive classification and miss matching as clearly advocated in Figures 5 and 7, and Tables 1–3. It is emphasized in Figures 6 and 8, where the algorithm training losses and gains on 200 and 100 epochs are projecting the performance well on both the campus and passageway sequences scenes, respectively.

However, the algorithm demonstrated better performance on passageway scenes, which had more difficult challenges such as illumination variations, and different poses compared to the outdoor environment (campus scenes). This is well illustrated in Figures 6 and 8 performance comparisons, where our algorithm recorded the highest performance in precision, recall, and IDF1 scores on the passageway scenes dataset than on the campus scenes dataset. It recorded an absolute 100% for all those metrics with satisfactory confidence values. It is illustrated in Figure A1a,b that our algorithm has mostly identified all the objects of interest under various heavy conditions [32]. This proves that the algorithm is robust against various heavy illuminations and different poses or skewed view angles. However, Figure 9 shows that though the algorithm performed better, it had similar challenges of the unrepresentative data, mostly in the middle of training and testing phases. However, it quickly converged better compared to campus scenes. This proved that our algorithm in the training phase had been fitted with enough data, although at the beginning of our training on the campus scenes, it could be seen struggling or not receiving enough data. The up and downs jumping [33] could be due to data fit because we can see that when we trained the algorithm with more epochs, we obtained better and more stable results for both passageway and campus scenes datasets.

To demonstrate our algorithm's classification accuracy (CA) and specificity, we compared our precision results with state-of-the-art paradigms. The results are summarized in Table 1. Our approach achieved better results compared to the online approach and short just 5.74% to the current state-of-the-art paradigm.

Thus, for real-time tracking, we evaluated our algorithm with several video frames taken from two different sequences of the EPFL dataset, as shown in Tables 2 and 3. The CLEAR MOT is used for evaluations, where ↑ denotes high performance and ↓ represents lower performance. In both sequences, our approach recorded an average overall performance above 80% with very few fragmentations and ID switches in all metrics. Further training and testing were conducted on our algorithm without Kalman filter using the 8000 frames from the real-time overlapping multiple cameras dataset (EPFL-RCL multi-cameras). In the comparison exercise, we found that the model's MOTA, MOTP, precision, and recall performance were very low compared to the one with the Kalman filter in Tables 2–4. It had a low detection ratio and a high ID switches ratio that adversely affected the overall tracking results. This is displayed in Table 5 and illustrated well in Figure 9e,f, where the Kalman Filter and segmentation technique are removed from our proposed HCNN algorithm. However, the proposed HCNN with Kalman filter performed very closely to the Yolo5Deep model in Table 4. This proves that the proposed model provides a better data affinity of a close equivalent to the Yolo5Deep model in real-time multiple object tracking.

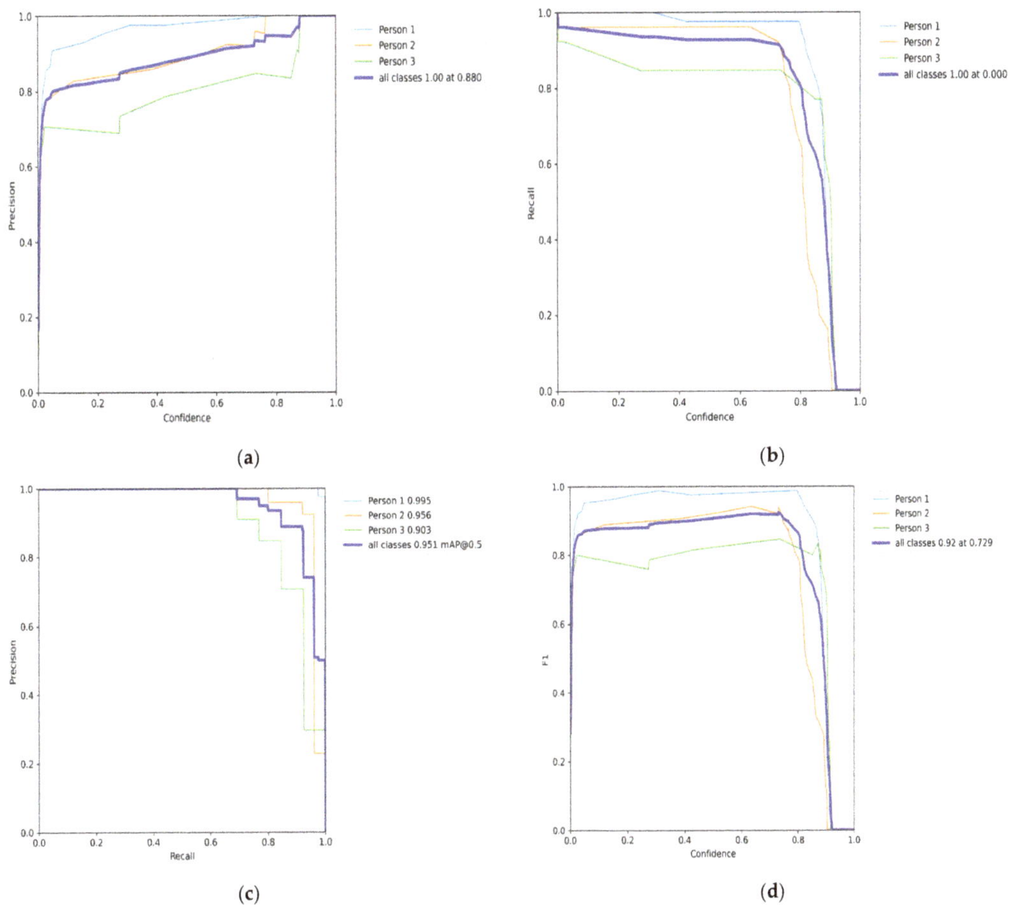

Figure 5. The label (**a**) shows the precision (P) versus confidence (C) graph, (**b**) the recall (R) versus confidence (C), (**c**) is the mean average precision based on comparing the truth bounding box and detection box, and (**d**) the IDF1 score at 92% with confidence of 0.729, advocates the balancing between the P and R based on Campus scenes images dataset. The mAP for all classes is high and accurately modeling detections at 95.1% with a threshold of 0.5. The P and R are high at 88.0%, and 87.5%, respectively, and more confidence at 0.8 and 0.78, respectively, for all classes.

Table 4. Performance evaluation analysis of fine-tuned Yolo5Deep on EPFL dataset (campus and passageway).

Sequences	Precision ↑	Recall ↑	IDF Score ↑	MOTA ↑	MOTP ↑	IDS ↓	ML ↓	MT ↑	FM ↓
Campus scenes	96.0%	90.6%	91.5%	92.0%	85.0%	2	1.0%	93.0%	2
Passageway scenes	94.0%	92.0%	93.0%	89.4%	87.0%	2	2.0%	88.0%	3

Table 5. Performance evaluation analysis of the proposed algorithm without Kalman filter on EPFL-RCL overlapping multi-cameras.

Sequences	Precision ↑	Recall ↑	IDF Score ↑	MOTA ↑	MOTP ↑	IDS ↓	ML ↓	MT ↑	FM ↓
Overall performance	65.0%	56.2%	58.5%	52.0%	46.3%	24	34.0%	54.0%	14

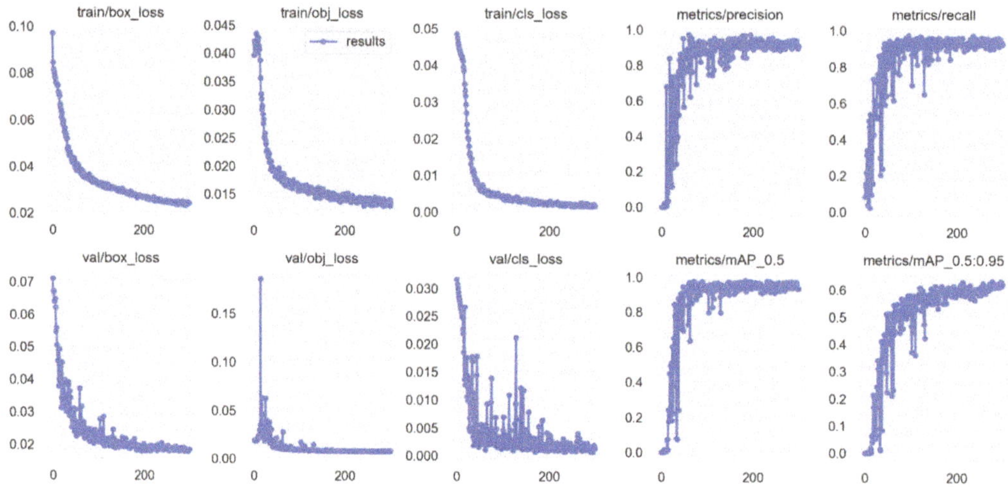

Figure 6. Shows both training and validations losses of the HCNN algorithm's object detector and classification on 200 epochs for campus scenes dataset. The precision and recall metrics in the training and validation phase converge at the highest of 95.7% accuracy, whereas the mAP converges at 95% with a 0.5 threshold.

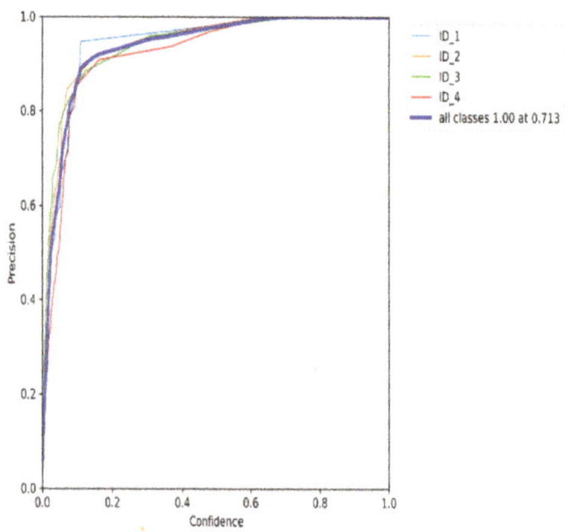

(**a**)

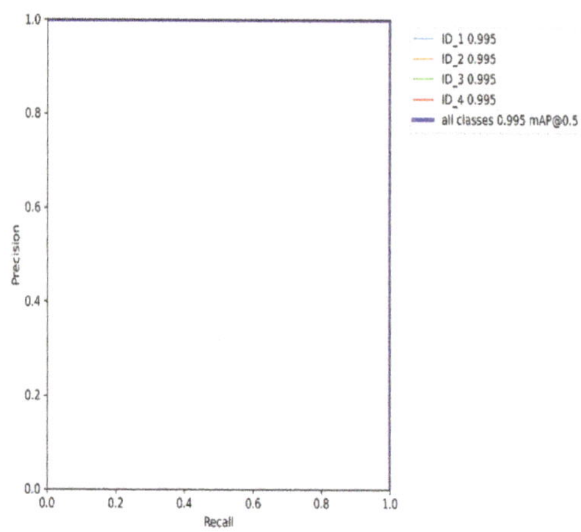

(**b**)

Figure 7. *Cont.*

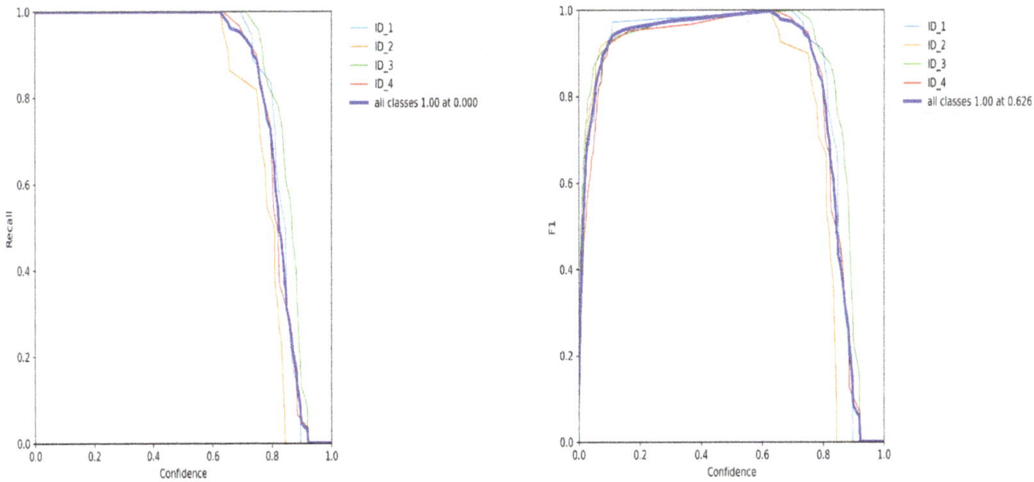

(c) (d)

Figure 7. The label (**a**) shows the precision(P) versus confidence(C) graph, (**b**) the recall(R) versus confidence(C), (**c**) is the mean average precision(mAP) based on comparing the truth bounding box and detection box, and (**d**) the IDF1 score at 100% with confidence of 0.626, which advocates the balance between P and R based on passageway scenes dataset. The mAP for all classes is high and accurately modeling detections at 95.1% with a threshold of 0.5. The P and R are high at 100% and 100%, respectively, and more confidence at 0.713 and 0.0 respectively for all classes.

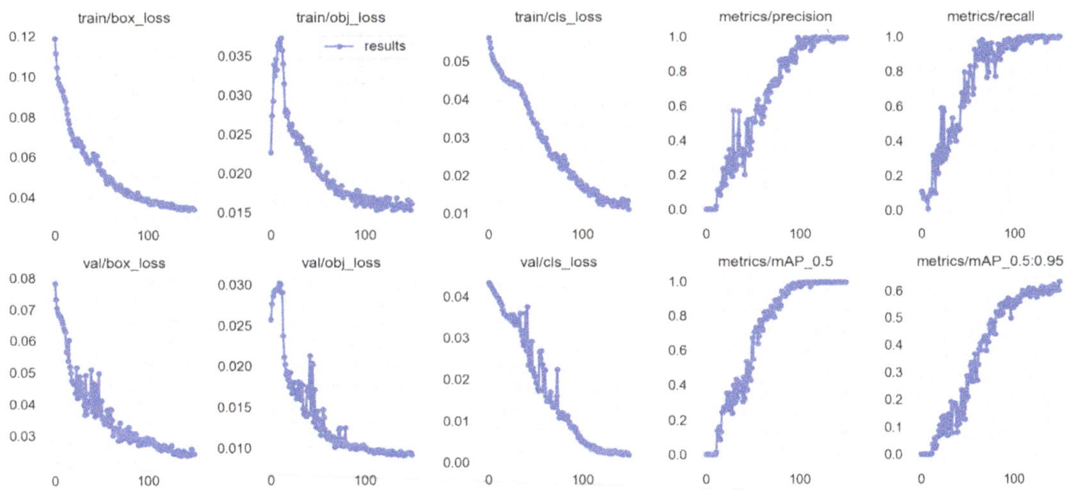

Figure 8. Shows both training and validations of the HCNN algorithm's object detector and classification loss converging on 100 epochs for passageway scenes dataset. The precision and recall metrics in the training and validation phase converge at the highest of 95.7% accuracy, whereas the mAP converges at 95% with a 0.5 threshold.

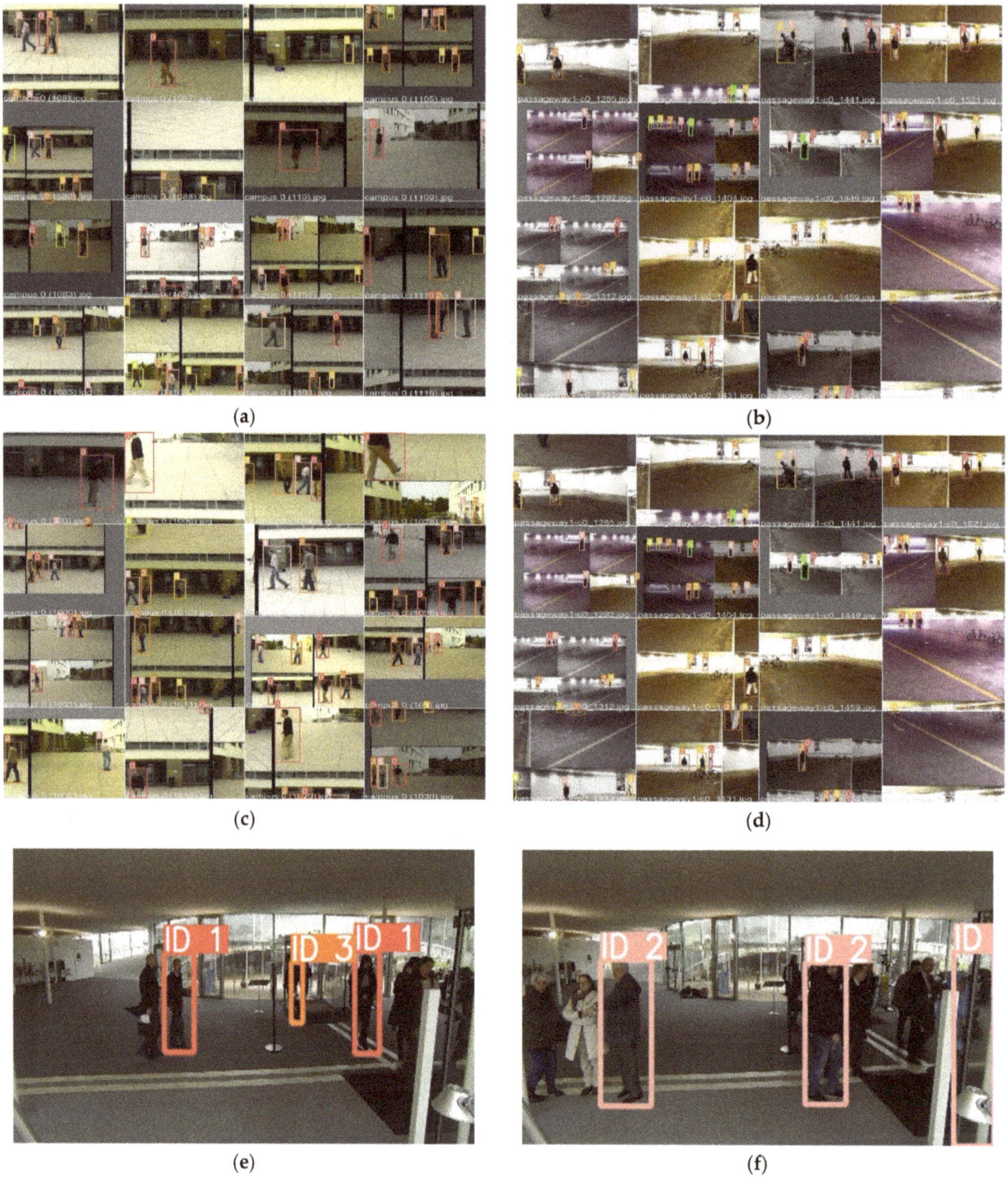

Figure 9. The first row shows visualize (**a**,**b**), the tracking results on validations of both sequences (Campus and Passageway, respectively) with proposed the HCNN algorithm's tracker. While (**c**,**d**) shows the tracking results of the fine-tuned Yolov5 + Deepsort, (Yolo5Deep) model integrated with HOG and Kalman Filter. (**e**,**f**) shows the EPFL-RCL Multi-cameras frame results for the proposed HCNN without a Kalman Filter and segmentation technique. Compared to our detector and tracker with Yolo5Deep, our proposed algorithm increased positive detections and improved the precision of detection boxes. Moreover, the method is robust for occlusion, illumination, and re-appearance variations.

Benchmark Evaluation Results

Results on EPFL multi-camera pedestrian datasets: In Table 6, we summarized the results of the EPFL multi-camera pedestrians tracking testing set. We compared our algorithm to several state-of-the-art methods. However, some of these approaches could only be analyzed offline.

Table 6. Comparison with state-of-the-art methods on testing the subset of EPF multi-cameras pedestrian dataset.

Method	MOTA ↑	MOTP ↑	Causality
NCA-Net [32]	64.5%	78.2%	Offline
CNN + HOG Template Matching [11]	94.0%	80.9%	Offline
Yolo + Deepsort [33]	86.1%	88.6%	Online
MCMOT HDM [34]	62.4%	78.2%	Offline
Ours	68.2%	65.0%	Online

For the offline mode, our approach performs poorly. Interestingly, we found that in real-time tracking settings, our approach recorded results that were close to the best state-of-the-art approach. However, in ablation studies, as shown in Figure 9e,f, our approach suffered from overlapping detection boxes and resulted in high misdetection and object re-identification.

6. Conclusions

Our study presents an efficient algorithm for multi-view pedestrian detection, identification, and tracking based on combined HOG descriptors and CNN. The background subtraction technique was used to eliminate noise from video frames taken from the EPFL dataset. Extensive experiments were conducted on selected sequences (campus and passageway) of the outdoor environments, where the Kalman filter was used to track the multiple objects and to test the robustness of the proposed system under difficult tracking conditions. Our algorithm demonstrated that contour and global features handling enhances real-time multi-object tracking performance. The results showed that the proposed technique produces better detection rates and data associations. Therefore, our feature work will involve the implementation of the algorithm for tracking multiple fast-moving objects on a huge dataset with more objects such as vehicles.

Author Contributions: Conceptualization, L.K.; methodology, L.K.; software, L.K.; validation, L.K. and Y.D.; formal analysis, L.K. and L.H.; investigation, Y.D.; data curation, L.K.; writing, L.K.; supervision, W.W.; funding acquisition, W.W. and L.H. All authors have read and agreed to the published version of the manuscript.

Funding: This work is supported by the Shanghai Science and Technology Committee International Cooperation Project No. 18510760300, and China Postdoctoral Science Foundation No. 2020M681264.

Informed Consent Statement: The public dataset used to conduct the study. Hence, informed consent was obtained from all subjects involved in the study.

Data Availability Statement: The dataset that resulted from this study can be found at the link: https://www.epfl.ch/labs/cvlab/data/data-pom-index-php/ (last access date: 26 December 2021).

Acknowledgments: Authors would like to thank Hou Li and the School of Information Engineering at Huangshan University for their tremendous support.

Conflicts of Interest: Authors declare no conflict of interest.

Appendix A

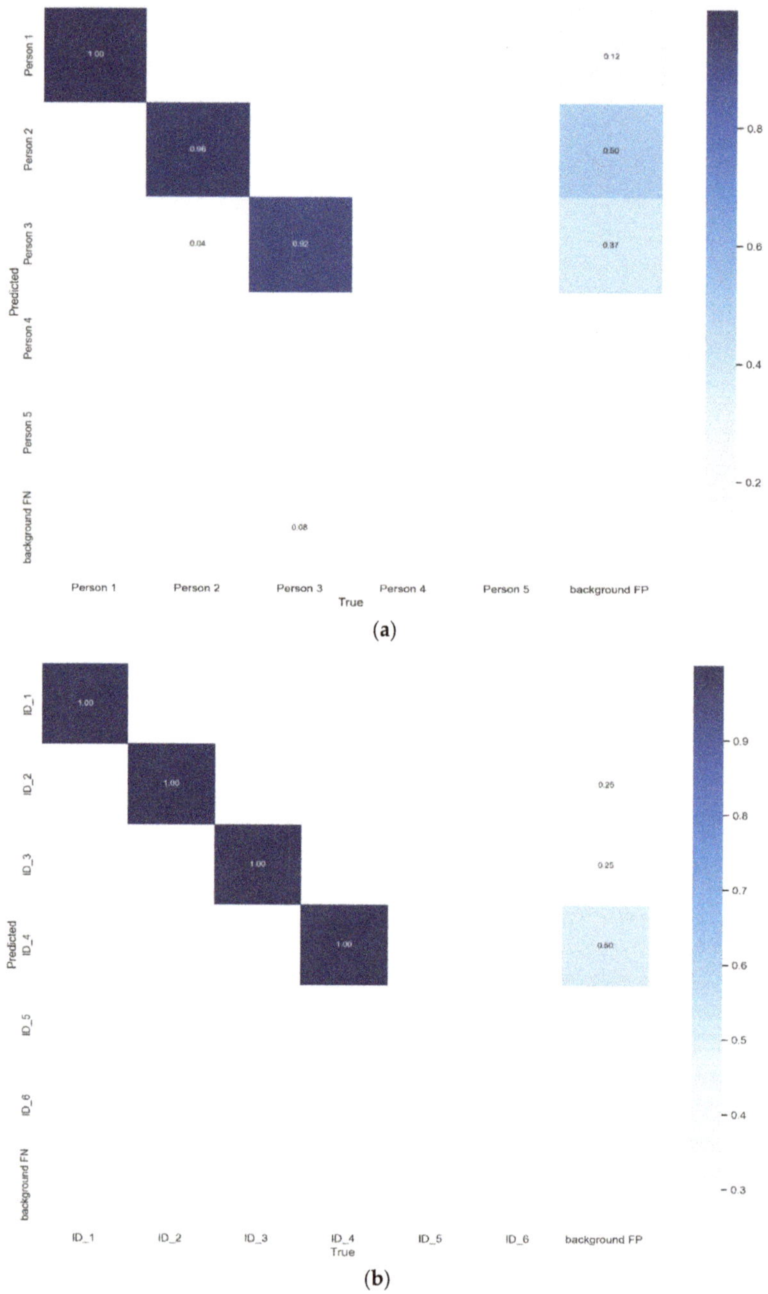

Figure A1. Illustrate the confusion matrices on both (**a**) passageway sequence, and (**b**) campus sequence, respectively.

References

1. Angeline, R.; Kavithvajen, K.; Balaji, T.; Saji, M.; Sushmitha, S.R. CNN integrated with HOG for efficient face recognition. *Int. J. Recent Technol. Eng.* **2019**, *7*, 1657–1661.
2. Zhang, C.; Patras, P.; Haddadi, H. Deep Learning in Mobile and Wireless Networking: A Survey. *IEEE Commun. Surv. Tutor.* **2019**, *21*, 2224–2287. [CrossRef]
3. Sanchez-Matilla, R.; Poiesi, F.; Cavallaro, A. Online multi-target tracking with strong and weak detections. In *Lecture Notes in Computer Science (Including Subseries Lecture Notes in Artificial Intelligence and Lecture Notes in Bioinformatics)*; Springer: Cham, Switzerland, 2016; Volume 9914, pp. 84–99. [CrossRef]
4. Bai, Y.X.; Zhang, S.-H.; Fan, Z.; Liu, X.-Y.; Zhao, X.; Feng, X.-Z.; Sun, M.-Z. Automatic multiple zebrafish tracking based on improved HOG features. *Sci. Rep.* **2018**, *8*, 10884. [CrossRef] [PubMed]
5. Lipetski, Y.; Sidla, O. A combined HOG and deep convolution network cascade for pedestrian detection. *IS T Int. Symp. Electron. Imaging Sci. Technol.* **2017**, *2017*, 11–17. [CrossRef]
6. Madan, R.; Agrawal, D.; Kowshik, S.; Maheshwari, H.; Agarwal, S.; Chakravarty, D. Traffic sign classification using hybrid HOG-SURF features and convolutional neural networks. In Proceedings of the 8th International Conference on Pattern Recognition Applications and Methods (ICPRAM 2019), Prague, Czech Republic, 19–21 February 2019; pp. 613–620. [CrossRef]
7. Bao, T.Q.; Kiet, N.T.T.; Dinh, T.Q.; Hiep, H.X. Plant species identification from leaf patterns using histogram of oriented gradients feature space and convolution neural networks. *J. Inf. Telecommun.* **2020**, *4*, 140–150. [CrossRef]
8. Bahri, H.; Chouchene, M.; Sayadi, F.E.; Atri, M. Real-time moving human detection using HOG and Fourier descriptor based on CUDA implementation. *J. Real-Time Image Process.* **2020**, *17*, 1841–1856. [CrossRef]
9. Kalake, L.; Wan, W.; Hou, L. Analysis Based on Recent Deep Learning Approaches Applied in Real-Time Multi-Object Tracking: A Review. *IEEE Access* **2021**, *9*, 32650–32671. [CrossRef]
10. Kumar, K.C.A.; Jacques, L.; De Vleeschouwer, C. Discriminative and Efficient Label Propagation on Complementary Graphs for Multi-Object Tracking. *IEEE Trans. Pattern Anal. Mach. Intell.* **2017**, *39*, 61–64. [CrossRef]
11. Zhang, T.; Zeng, Y.; Xu, B. HCNN: A neural network model for combining local and global features towards human-like classification. *Int. J. Pattern Recognit. Artif. Intell.* **2016**, *30*, 1655004. [CrossRef]
12. Aslan, M.F.; Durdu, A.; Sabanci, K.; Mutluer, M.A. CNN and HOG based comparison study for complete occlusion handling in human tracking. *Meas. J. Int. Meas. Confed.* **2020**, *158*, 107704. [CrossRef]
13. Zhang, J.; Cao, J.; Mao, B. Moving Object Detection Based on Non-parametric Methods and Frame Difference for Traceability Video Analysis. *Procedia Comput. Sci.* **2016**, *91*, 995–1000. [CrossRef]
14. Najva, N.; Bijoy, K.E. SIFT and Tensor Based Object Detection and Classification in Videos Using Deep Neural Networks. *Procedia Comput. Sci.* **2016**, *93*, 351–358. [CrossRef]
15. Rui, T.; Zou, J.; Zhou, Y.; Fang, H.; Gao, Q. Pedestrian detection based on multi-convolutional features by feature maps pruning. *Multimed. Tools Appl.* **2017**, *76*, 25079–25089. [CrossRef]
16. Sujanaa, J.; Palanivel, S. HOG-based emotion recognition using one-dimensional convolutional neural network. *ICTACT J. Image Video Process.* **2020**, *11*, 2310–2315. [CrossRef]
17. Qi, X.; Liu, C.; Schuckers, S. IoT edge device based key frame extraction for face in video recognition. In Proceedings of the 18th IEEE/ACM International Symposium on Cluster, Cloud and Grid Computing, CCGRID 2018, Washington, DC, USA, 1–4 May 2018; pp. 641–644. [CrossRef]
18. Yudin, D.; Ivanov, A.; Shchendrygin, M. Detection of a human head on a low-quality image and its software implementation. *Int. Arch. Photogramm. Remote Sens. Spat. Inf. Sci.-ISPRS Arch.* **2019**, *42*, 237–241. [CrossRef]
19. Girdhar, R.; Gkioxari, G.; Torresani, L.; Paluri, M.; Tran, D. Detect-and-Track: Efficient Pose Estimation in Videos. *Proc. IEEE Comput. Soc. Conf. Comput. Vis. Pattern Recognit.* **2018**, *1*, 350–359. [CrossRef]
20. Perwaiz, N.; Fraz, M.M.; Shahzad, M. Stochastic attentions and context learning for person re-identification. *PeerJ Comput. Sci.* **2021**, *7*, e447. [CrossRef]
21. Mewada, H.; Al-Asad, J.F.; Patel, A.; Chaudhari, J.; Mahant, K.; Vala, A. A Fast Region-based Active Contour for Non-rigid Object Tracking and its Shape Retrieval. *PeerJ Comput. Sci.* **2021**, *7*, e373. [CrossRef]
22. Fiaz, M.; Mahmood, A.; Jung, S.K. Tracking Noisy Targets: A Review of Recent Object Tracking Approaches. *arXiv* **2018**, arXiv:1802.03098. Available online: http://arxiv.org/abs/1802.03098 (accessed on 10 October 2021).
23. Patel, D.M.; Jaliya, U.K.; Vasava, H.D. Multiple Object Detection and Tracking: A Survey. *Int. J. Res. Appl. Sci. Eng. Technol.* **2018**, *6*, 809–813.
24. Abdelhafiz, D.; Yang, C.; Ammar, R.; Nabavi, S. Deep convolutional neural networks for mammography: Advances, challenges and applications. *BMC Bioinform.* **2019**, *20* (Suppl. 11), 281. [CrossRef] [PubMed]
25. Liu, P.; Li, X.; Liu, H.; Fu, Z. Online learned siamese network with auto-encoding constraints for robust multi-object tracking. *Electronics* **2019**, *8*, 595. [CrossRef]
26. Stojnić, V.; Risojević, V.; Muštra, M.; Jovanović, V.; Filipi, J.; Kezić, N.; Babić, Z. A method for detection of small moving objects in UAV videos. *Remote Sens.* **2021**, *13*, 653. [CrossRef]
27. Ahmad, M.; Ahmed, I.; Khan, F.A.; Qayum, F.; Aljuaid, H. Convolutional neural network–based person tracking using overhead views. *Int. J. Distrib. Sens. Netw.* **2020**, *16*, 1–12. [CrossRef]

28. Zhao, D.; Fu, H.; Xiao, L.; Wu, T.; Dai, B. Multi-object tracking with correlation filter for autonomous vehicle. *Sensors* **2018**, *18*, 2004. [CrossRef] [PubMed]
29. Bhuvana, V.P.; Schranz, M.; Regazzoni, C.S.; Rinner, B.; Tonello, A.M.; Huemer, M. Multi-camera object tracking using surprisal observations in visual sensor networks. *Eurasip J. Adv. Signal Process.* **2018**, *2016*, 50. [CrossRef]
30. Hu, C.; Huang, H.; Chen, M.; Yang, S.; Chen, H. Video object detection from one single image through opto-electronic neural network. *APL Photonics* **2021**, *6*, 046104. [CrossRef]
31. Milan, A.; Leal-Taixe, L.; Reid, I.; Roth, S.; Schindler, K. MOT16: A Benchmark for Multi-Object Tracking. *arXiv* **2016**, arXiv:1603.00831. Available online: http://arxiv.org/abs/1603.00831 (accessed on 18 September 2021).
32. Rahman, M.M.; Nooruddin, S.; Hasan, K.M.A.; Dey, N.K. HOG + CNN Net: Diagnosing COVID-19 and Pneumonia by Deep Neural Network from Chest X-Ray Images. *SN Comput. Sci.* **2021**, *2*, 371–386. [CrossRef]
33. Ghosh, S.K.; Islam, M.R. Bird Species Detection and Classification Based on HOG Feature Using Convolutional Neural Network. *Commun. Comput. Inf. Sci.* **2019**, *1035*, 363–373. [CrossRef]
34. Lee, B.; Erdenee, E.; Jin, S.; Nam, M.Y.; Jung, Y.G.; Rhee, P.K. Multi-class multi-object tracking using changing point detection. In *Lecture Notes in Computer Science (Including Subseries Lecture Notes in Artificial Intelligence and Lecture Notes in Bioinformatics)*; Springer: Cham, Switzerland, 2016; Volume 9914, pp. 68–83. [CrossRef]

sensors

MDPI

Article

An Efficient Approach Using Knowledge Distillation Methods to Stabilize Performance in a Lightweight Top-Down Posture Estimation Network

Changhyun Park [1], Hean Sung Lee [1], Woo Jin Kim [1], Han Byeol Bae [2], Jaeho Lee [1] and Sangyoun Lee [1,*]

[1] Department of Electrical and Electronic Engineering, Yonsei University, Seoul 03722, Korea; qkrckd2002@yonsei.ac.kr (C.P.); hslee2860@yonsei.ac.kr (H.S.L.); woojinkim0207@yonsei.ac.kr (W.J.K.); jhlee430@yonsei.ac.kr (J.L.)

[2] Department of Artificial Intelligence Convergence, Kwangju Women's University, Gwangju 62396, Korea; kwu_BHB@kwu.ac.kr

* Correspondence: syleee@yonsei.ac.kr; Tel.: +82-2-2123-5768

Abstract: Multi-person pose estimation has been gaining considerable interest due to its use in several real-world applications, such as activity recognition, motion capture, and augmented reality. Although the improvement of the accuracy and speed of multi-person pose estimation techniques has been recently studied, limitations still exist in balancing these two aspects. In this paper, a novel knowledge distilled lightweight top-down pose network (KDLPN) is proposed that balances computational complexity and accuracy. For the first time in multi-person pose estimation, a network that reduces computational complexity by applying a "Pelee" structure and shuffles pixels in the dense upsampling convolution layer to reduce the number of channels is presented. Furthermore, to prevent performance degradation because of the reduced computational complexity, knowledge distillation is applied to establish the pose estimation network as a teacher network. The method performance is evaluated on the MSCOCO dataset. Experimental results demonstrate that our KDLPN network significantly reduces 95% of the parameters required by state-of-the-art methods with minimal performance degradation. Moreover, our method is compared with other pose estimation methods to substantiate the importance of computational complexity reduction and its effectiveness.

Keywords: pose estimation; convolutional neural network; lightweight; knowledge distillation

Citation: Park, C.; Lee, H.S.; Kim, W.J.; Bae, H.B.; Lee, J.; Lee, S. An Efficient Approach Using Knowledge Distillation Methods to Stabilize Performance in a Lightweight Top-Down Posture Estimation Network. *Sensors* **2021**, *21*, 7640. https://doi.org/10.3390/s21227640

Academic Editor: Cosimo Distante

Received: 8 September 2021
Accepted: 12 November 2021
Published: 17 November 2021

1. Introduction

The demand for human pose estimation has increased over time as it is essential for detecting human behaviors and for numerous applications such as human-computer interaction [1], human action recognition [2], and human performance analysis [3]. Previously, human pose estimation has been studied as a close-up technique requiring a balance between accuracy and low computational complexity. Traditional approaches such as histogram of oriented gradient (HOG) [4] and Edgelet [5] extract discriminative features from images and assign a class to the feature vector. However, they cannot adequately determine the accurate location of body parts in a human figure [6].

Recent advances in convolutional neural networks (CNNs) that enable robust feature extraction have afforded significant improvements in pose estimation. Therefore, owing to the feature extraction capabilities of CNNs, the research paradigm of human pose estimation shifted from classic approaches to deep learning [7–9]. Two main approaches, i.e., bottom-up and top-down approaches, of deep-learning-based methods, have been employed to overcome the limitations of handcrafting-based methods during the transition.

Bottom-up approaches [10–16] first detect human body poses and then group them using clustering algorithms. Compared to top-down approaches, they are faster in testing and thus require lower computational complexities during model building. However, the bottom-up approaches are unable to amplify the details of each person, and subsequently, they yield lower accuracies than top-down approaches.

In contrast, the keypoint prediction process in top-down approaches is a two-step operation. Generally, top-down approaches [17–23] first detect all the people in an image and crop the person region and then input the cropped image into a single-person pose estimation model. Due to the two-step operation, they yield better results than bottom-up approaches. To accurately estimate the keypoints of people in an image, top-down approaches construct network layers deeper than bottom-up approaches. However, top-down approaches are unable to solve the speed degradation issue that arises when deeply constructing network layers for estimating keypoints.

Most previous multi-person estimation methods require high computational complexity to accurately estimate the keypoints of people in an image. Additionally, to guarantee accuracy, the network layers need to be deeply designed, which decreases the estimation speed. Due to these limitations, the accuracy and speed need to be balanced in multi-person estimations.

In this paper, we present a lightweight top-down human pose estimation network that uses a knowledge distillation method to overcome the balance limitations. Inspired by PeleeNet [24], we propose knowledge distilled lightweight top-down pose network (KDLPN), a network that minimizes computational complexity while shuffling the pixels in the decoder previously introduced in the dense upsampling convolution (DUC) layer [25] to reduce the number of channels.

To effectively and efficiently resolve the performance degradation occurring in the lightweight networks, the knowledge distillation approach [26] is applied in our model. To satisfy the complexity and performance requirements of deploying state-of-the-art deep neural models in our proposed system, compact and fast models are trained by transferring knowledge from extremely deep and powerful teacher models. Additionally, experiments are conducted on the MSCOCO dataset [27], a widely-used human pose estimation dataset to verify the model effectiveness, and the balance between the estimation accuracy and the speed of our approach is demonstrated.

The remainder of this paper is organized as follows. Section 2 presents a review of the related work. In Section 3, the proposed two-dimensional (2D) human pose estimation model is presented and the features of each part are formulated. Sections 3.1 and 3.2 discuss the study motivations. Section 3.3 describes the building of a lightweight 2D human pose estimation network and replacement of the decoder to reduce complexity. In Section 3.4, to prevent performance degradation, the proposed method is adopted for knowledge distillation in training. Section 4 presents the simulation results and discussion on the MSCOCO dataset. Finally, Section 4.5 presents the conclusions.

2. Related Work

2.1. Multi-Person Pose Estimation

Recently, multi-person pose estimation has drawn increasing attention because of its applicability in real-life applications, such as postural correction [28], action recognition [29], and health care [30]. Multi-person pose estimation using neural networks can be determined via two main approaches. The first is the bottom-up approaches that obtain all the pose keypoints in input images and assemble them as distinct people using methods such as part affinity field (PAF) [11]. The other is the top-down approaches that employ human detection to obtain bounding boxes and input the cropped image batch to the pose estimation neural network.

Bottom-up approach: Bottom-up pose estimation methods [10–16] detect the identity-free human joints of all people within an image and assemble the joints using an algorithm. While traditional multi-person pose estimation models focus on human structural characteristics, contemporary models focus more on measuring the body itself by adopting strong CNN models.

To accurately estimate multi-person poses, DeepCut [12] proposed the derivation of a joint detection and pose estimation formulation that was casted as an integer linear program problem and a new formulation that was casted as a joint subset partitioning and labeling problem. Further, DeeperCut [13], an improved DeepCut employs a residual network (ResNet) [31] to extract more robust body parts representations. Moreover, it adopts image-conditioned pairwise terms to achieve better performance. Next, Associative Embedding [14] detects heatmaps with pixelwise embedding and groups the keypoint candidates by comparing the distance between the embeddings to generate the result. To improve the estimation accuracy, Openpose [11] employed PAFs to learn to associate body parts with individuals in an image. Openpose obtains a heatmap around the grouped joints of people using a multi-stage CNN (initialized by the first ten layers of VGG-19 [32]) and fine-tunes and then yields multi-person poses using PAFs (Figure 1a).

Current bottom-up pose estimation approaches have high estimation speeds, and can be implemented in mobile devices without additional human detection networks. However, their performances are significantly affected by complex backgrounds and human outer walls.

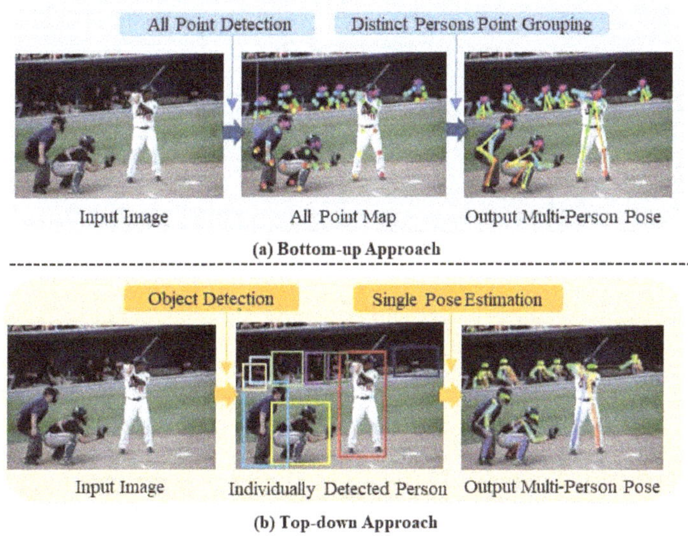

Figure 1. Human Pose Estimation algorithm. Column (**a**): The example of a bottom-up approach. Column (**b**): The example of a top-down approach.

Top-down approach: Top-down approaches [17–23] employ a two-step process to estimate pose keypoints. They first detect all the people within an image using object detection. Then, each cropped image is processed into a single-person pose estimation network model. A Cascaded Pyramid Network (CPN) [20] has been proposed to robustly detect "hard" keypoints and divide keypoints into simple levels. CPN comprises a pyramid architecture as the backbone network, including GlobalNet and RefineNet. In RefineNet, CPN selects the hard keypoints online based on the training L2 loss. George et al. [19] predicted heatmaps and offsets using a fully convolutional ResNet and a faster RCNN detector to detect bounding boxes, and they then predicted the final location output using the heatmaps, offsets, and keypoint-based non maximum suppression. Regional multi-person pose estimation (RMPE) [17] comprises a human detection model and a skeleton registration model [33–37] for estimating the multi-person poses in the image. The detected single human bounding boxes in batches from the detection model are input into the skeleton registration model to detect the skeleton keypoints (Figure 1b).

To utilize the characteristics of the superior performance of top-down approaches and overcome its shortcomings, we propose a lightweight top-down pose estimation approach to improve computational efficiency while improving the performance.

2.2. Lightweight Neural Network

The effectiveness of neural networks has significantly improved the performance of applications that use various memory locations and operations. However, computing power and memory have not kept up with the development of neural networks. Consequently, lightweight networks with low computational complexity have been proposed to meet the demand for mobile devices.

MobileNets [38–40] proposed the construction of a lightweight model that can run on mobile devices by minimizing the number of network parameters. It minimizes the overall computation using a depth-wise convolution to convert each channel into its respective kernel and by applying a 1×1 convolution to change the output channel to pointwise convolution. MobileNetsV3 [40] proposed the platform-aware network architecture search method, which automatically optimizes each network block. It utilizes a module based on squeeze and excitation in the bottleneck structure to minimize the network parameters while improving the performance. PeleeNet [24] is a network model that performs various tunings based on DenseNet [41] for mobile devices. It utilizes the architecture of DenseNet that concatenates the feature map of the layers. Additionally, it uses stemblock and two-way dense layers to reduce the computational cost and adopts a structure with varying number of layers on each stage. The computational complexity of PeleeNet is significantly low, which allows its operation in mobile devices. It affords better accuracy and over 1.8 times faster speed than MobileNet and MobileNetV2 on the ImageNet ILSVRC 2012 dataset [42].

2.3. Knowledge Distillation

Deep learning models are basically wide and deep; thus, feature extension works efficiently if the number of parameters and operations are high. Subsequently, the object classification or detection performance, which is the purpose of the model, is improved. However, deep learning cannot be configured using large and deep networks owing to device limitations, such as computing resources (CPU and GPU) and memory. Therefore, considering these device environments, a deep learning model with a small size and improved performance is required. This demand has led to the development of various algorithms that can afford similar performance to large networks, and among them, knowledge distillation is attracting immense attention [26,43].

Knowledge distillation is the information transfer between different neural networks with distinct capacities. Bucilua et al. [44] were the first to propose model compression to use the information from a large model for the training of a small model without a substantial drop in accuracy. This is mainly based on the idea that student models reflect teacher models and afford similar performances. Hinton et al. [43] employed a well-trained large and complex network to help train a small network. Yim et al. [45] compared an original network and a network trained using the original network, as a teacher network. They determined that the student network that learned the distilled knowledge is optimized much quicker than the original model, and it outperforms the original network.

This is because the teacher model provides extra supervision in the form of class probabilities, feature representations [46,47], or an inter-layer flow. Recently, this principle has also been applied to accelerate the model training process of large-scale distributed neural networks and transfer knowledge between multiple layers [48] or between multiple training states [49]. In addition to the conventional two-stage training-based offline distillation, one-stage online knowledge distillation has been attempted, and advantageously, it provides more efficient optimization and learning. Furthermore, knowledge distillation has been used to distil easy-to-train large networks into harder-to-train small networks. Alashkar et al. [50] presented a makeup recommendation and synthesis system wherein the makeup art domain knowledge and makeup expert experience are both incorporated into a neural network to boost the performance of the makeup recommendation. Although knowledge distillation in deep neural networks has been successfully applied to solve the problems of visual relationship detection, sentence sentiment analysis and name entity recognition, its application in the fashion domain has been limited.

In this work, we adopt a knowledge distillation method to advantageously employ the complex teacher network knowledge to guide lightweight neural models.

3. Proposed Method

3.1. Overview

In this section, we propose a lightweight multi-person pose estimation network using a top-down-based approach. The top-down method basically comprises a detector, which detects people, and a single-person pose estimation (SPPE), which predicts a single pose from the detected person. Although the speed reduces based on the number of people in the top-down approach compared to the bottom-up approach, the top-down approach affords better performance. Moreover, the speed problem can be alleviated by minimizing the number of network parameters. As several fast and accurate approaches [33–37] exist for human detection, we mainly focus on making the SPPE of the pose estimation model lightweight.

SPPE comprises an encoder model, which extracts features from the detected person as input, and a decoder model, which acquires the heatmap to the keypoints of that person by upsampling from the extracted features. As shown in Figure 2, we changed the encoder model to the proposed optimal lightweight model. Concurrently, we reduced the number of parameters by applying a new structure to the upsampling layer of the decoder model. To avoid the performance degradation when reducing the number of parameters, we employed knowledge distillation using a teacher network with high performance.

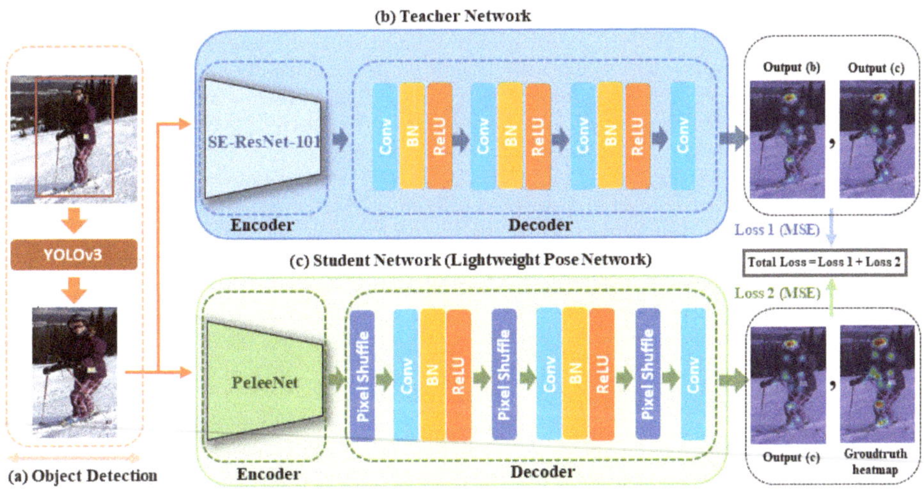

Figure 2. Overall lightweight human pose estimation network.

In the next section, we present the overview of our method. Then, we illustrate the lightweight network corresponding to the top-down-based SPPE in Section 3.2 and the decoder of the lightweight network in Section 3.3. Finally, we present the knowledge distillation method that can minimize the performance reduction associated with lightweightedness in Section 3.4.

3.2. *Preliminary Processing*

Human pose estimation aims to localize the body joints of all the detected people in a given image. In the top-down mode, the detector first yields the bounding box of detection information about people in images.

We use YOLOV3 [37] to quickly and efficiently detect people. The detected images are passed through a spatial transformer network [51], which is a parametric network that automatically selects areas of interest and appears prior to the SPPE input, and the detected information about the human region is converted into high quality information of the same size. Then, using the converted detection information, the SPPE extracts the heatmap, which represents the location information of the human body joints. The original resolution and size of the extracted heatmap (H) is determined by the inverse conversion of the spatial de-transformer network. Finally, we estimate the posture of every person in the image by connecting the body joints based on the heat maps extracted from each person.

3.3. *Network Architecture*

3.3.1. Lightweight Network Encoder

Top-down methods, which detect people from images and estimate poses from within bounding boxes, are more accurate than bottom-up methods, which estimate all the key-points in an image and correlate them. However, disadvantageously, in top-down methods, the detected bounding boxes need to be cropped and the estimation speed reduces if multiple people are present in the images.

Although many studies have been conducted on top-down methods [17–23], the limitations of heavy and slow models have not yet been overcome. As a representative example, Alpha-pose based on RMPE [17,18] utilizes a very heavy encoder structure with SE-ResNet. Therefore, after conducting multiple experiments to determine a suitable encoder structure that lightens the multi-person pose estimation network, we selected PeleeNet as the optimal encoder structure. PeleeNet is a lightweight model of DenseNet [41] and has been widely used as a feature extractor that reduces the size of the input image by four times the width and length, which makes the entire architecture cost-effective. Additionally, it can increase the feature expression ability with a small amount of computation. Moreover, to obtain the receptive fields at various scales, PeleeNet utilizes a two-way dense layer, where DenseNet only comprises a combination of 1×1 convolution and a 3×3 convolutions in the bottleneck layer. Instead of a depth-wise convolution layer, it utilizes a simple convolution layer to improve its implementation efficiency. Owing to its efficient methods and small number of calculations, its speed and performance are superior to those of typical methods, such as MobileNetV1 [38], V2 [39], and ShuffleNet [52]. Furthermore, because of its simple convolution, the use of additional techniques could likely afford a much more efficient detector. Various types of network decoders can be added via simple convolutions of the encoder while applying various training methods.

3.3.2. Lightweight Network Decoder

To speed up the computation in the decoder, we designed a novel network structure using the DUC proposed in Figure 3. Table 1 summarizes the structure of the entire decoder comprising the proposed DUC layer. The DUC layer contains pixel shuffle operations, which increase the resolution and reduce the number of channels, and 3×3 convolution operations. When the input feature map is set to $(H) \times$ width $(W) \times$ channel (C), pixel shuffle reduces the number of channels to C/d^2 and increases the resolution to $dH \times dW$ as shown in Figure 3. Here, d denotes the upsampling coefficient and is set as 2, i.e., the same as that in the standard deconvolution-based upsampling method. This helps substantially reduce the number of parameters to C/d^2 during upsampling. The feature that reduces the channel to C/d^2 size using the pixel shuffle layer again expands the number of channels to C/d through the convolution layer. This minimizes performance degradation by embedding the same amount of information into the feature as that before the reduction of the number of input channels. The entire decoder structure includes three DUC layers and outputs heatmaps showing the positions of each keypoint in the last layer. The proposed decoder network substantially reduces the number of parameters and speeds up the computation compared to the standard deconvolution-based decoder.

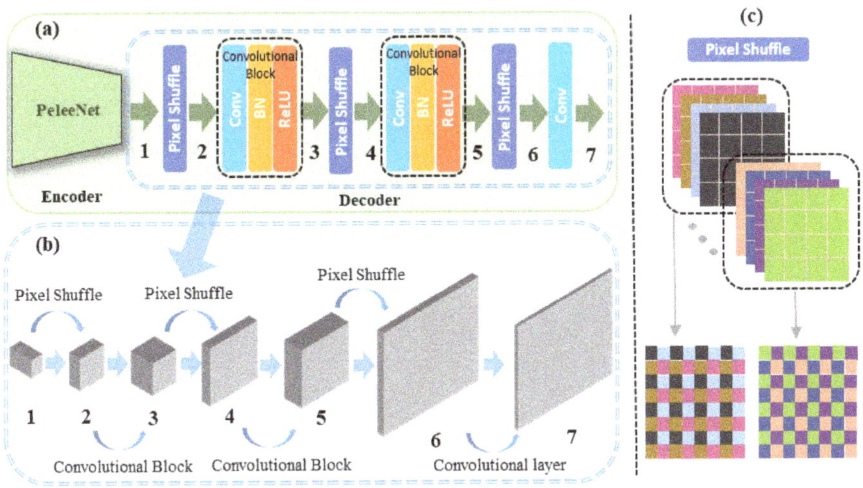

Figure 3. Specifications of the decoder of our proposed algorithm. (**a**): Block diagram of proposed algorithm. (**b**): The process of decoding. (**c**): The example operation of PixelShuffle.

Table 1. Decoder architecture.

Stage		Layer	Output Shape
	Input		$12 \times 8 \times 704$
DUC Stage 0	PixelShuffle	PixelShuffle	$24 \times 16 \times 176$
	Convolutional Block	conv2d 3×3 BatchNorm2d ReLU	$24 \times 16 \times 352$
DUC Stage 1	PixelShuffle	PixelShuffle	$48 \times 32 \times 88$
	Convolutional Block	conv2d 3×3 BatchNorm2d ReLU	$48 \times 32 \times 176$
DUC Stage 2	PixelShuffle	PixelShuffle	$96 \times 64 \times 44$
	Convolutional layer	conv2d 3×3	$96 \times 64 \times 17$

3.4. Knowledge Distillation Method

Accuracy and speed must both be considered in multi-person pose estimation. However, most existing methods only focus on accuracy and thus consume considerable computing resources and memory. However, lightweight networks exhibit performance degradation because of the reduced computing resources.

To overcome these shortcomings, we applied knowledge distillation to alleviate the performance degradation of the lightweight multi-person pose estimation network.

(1). We trained a large pose model of the teacher network. Then, we selected SE-ResNet-101 as the teacher network because it utilizes squeeze-and-excitation blocks to perform channel-wise feature extraction.

(2). Thereafter, we trained a target student model using the knowledge learned by the teacher model. The training model is capable of handling wrong pose joint annotations, e.g., when the pretrained teacher predicts more accurate joints than the manually assigned wrong and missing labels.

As stated by [26], knowledge distillation mainly aims to designs an appropriate mimicry loss function that can effectively extract a teachers' knowledge and transfer it to student model training. The previous distillation functions were designed for single-label based softmax cross-entropy loss in the context of object categorization and are thus unsuitable for transferring the structured pose knowledge in a 2D image space.

To address this problem, we employed a joint confidence map dedicated pose distillation loss function, as given below:

$$\mathcal{L}_{total} = \alpha_{KD}(\frac{1}{N}\sum_{n=0}^{N}({m_n}^S - {m_n}^{GT})^2) + (1-\alpha_{KD})(\frac{1}{N}\sum_{n=0}^{N}({m_n}^S - {m_n}^T)^2) \quad (1)$$

Here, α_{KD} is the knowledge distillation balancing parameter. Additionally, N denotes the number of joints, and ${m_n}^S$, ${m_n}^T$, and ${m_n}^{GT}$ denote the heatmaps for the n-th joint predicted by the in-training student target model, pretrained teacher model, and corresponding ground truth of the prediction, respectively. Then, to maximize the comparability with the pose supervised learning loss, we set the mean squared error as the distillation quantity to measure the divergence between the estimation and its label. By employing these knowledge distillation techniques, learning was performed using superior and complex networks rather than the existing ground truth alone, which boosted the performance of lightweight networks to match that of the superior networks.

4. Experiments and Results

4.1. Dataset and Evaluation Matrix

We used the MSCOCO dataset [27] to train and evaluate our method. The dataset comprises more than 200 k images including 250 k person instances with 17 keypoints per instance. We trained our method on the training set of the MSCOCO dataset, comprising 56 k images including 150 k person instances. We used the official evaluation metric of the MSCOCO keypoints challenge dataset, i.e., average precision (AP) based on object keypoint similarity (OKS). OKS is a measure of how close a predicted keypoint is to the ground truth, and is defined as follows:

$$OKS = \frac{\sum_i exp(-d_i^2/2s^2k_i^2)\delta(v_i > 0)}{\sum_i \delta(v_i > 0)}, \quad (2)$$

where d_i is the Euclidean distance between a detected keypoint and its corresponding ground truth, v_i is the visibility flag of the ground truth, s is the object scale, and k_i is a per-keypoint constant that controls fall off. We report the standard AP and recall scores from MSCOCO dataset: AP^{50} (AP at OKS = 0.50), AP^{75}, AP (the mean of AP scores at OKS = 0.50, 0.55, …, 0.90, and 0.95), AP^M for medium objects, AP^L for large objects, and AR (the mean of recalls at OKS = 0.50, 0.55, …, 0.90 and 0.95).

4.2. Training Details

A YOLOV3 detector that was pretrained on the MSCOCO dataset was utilized to detect humans in images. Each detected image was resized to 384 × 256 and was randomly flipped horizontally to augment the data. PeleeNet was used as the encoder of the proposed method, and it was pretrained using ImageNet. The model was trained for 120 epochs, and the initial learning rate was set to 0.0001, which decreased by 10% at both the 60th and 90th epochs. Then, the model was optimized using an Adam optimizer [53] and the batch size was set as 8. An SE-ResNet-based RMPE that was pretrained on the MSCOCO dataset was used as the teacher network, and the knowledge distillation parameter alpha was set as 0.8.

4.3. Ablation Study

4.3.1. Lightweight Network Structure

Additional experiments were conducted to evaluate the effectiveness of employing PeleeNet as the backbone for human pose estimation. The number of parameters (M) and FLOPS (G) and AP were measured using various encode models and compared PeleeNet, the encoder of our model, was compared with popular lightweight networks such as MobileNetV1, V2, and V3 [38–40], ShuffleNetV2 [54], MnasNet [55], and Hourglass [56] with 1, 2, and 4 stacks. The knowledge distillation parameter α_{KD} was set as 0.8. The results are summarized in Table 2.

Table 2. Results for the MSCOCO validation sets in a lightweight network with $\alpha_{KD} = 0.8$.

Encoder	AP	AP^{50}	AP^{75}	AP^{M}	AP^{L}	Param (M)	FLOPS (G)
Hourglass (4-stack)	64.8	82.1	71.3	60.6	71.6	26.0	46.6
Hourglass (2-stack)	62.6	81.1	69.0	58.2	69.4	13.5	23.3
Hourglass (1-stack)	55.4	78.8	60.9	51.0	62.4	7.17	11.7
ShufflenetV2 [54]	52.5	76.9	57.5	48.2	59.1	2.73	1.26
MobileNetV3 [40]	60.8	81.1	67.9	56.2	68.0	3.94	1.36
MobileNetV2 [39]	56.1	79.0	62.0	52.1	63.0	4.54	2.12
MobileNetV1 [38]	54.8	77.9	59.9	50.1	61.7	4.69	2.11
MnasNet [55]	57.7	79.4	63.8	53.9	64.5	5.42	2.14
PeleeNet	61.9	82.0	68.5	57.6	68.7	2.80	1.49

As shown in Table 2, from the perspective of AP, PeleeNet affords better performance than the other encoders. Moreover, PeleeNet achieves significantly better accuracy and lower complexity than MobileNetV1, V2, and V3 and MnasNet. Compared to ShuffleNetV2, PeleeNet exhibits better AP by 7.1. Although models with Hourglass with 2 stacks and Hourglass with 4 stacks exhibited better accuracy than our KDLPN, the number of their network parameters was significantly higher. The table also shows that when PeleeNet is used as the encoder, stable performance can be obtained even with a small number of parameters. Figure 4 shows a schematic diagram of Table 2.

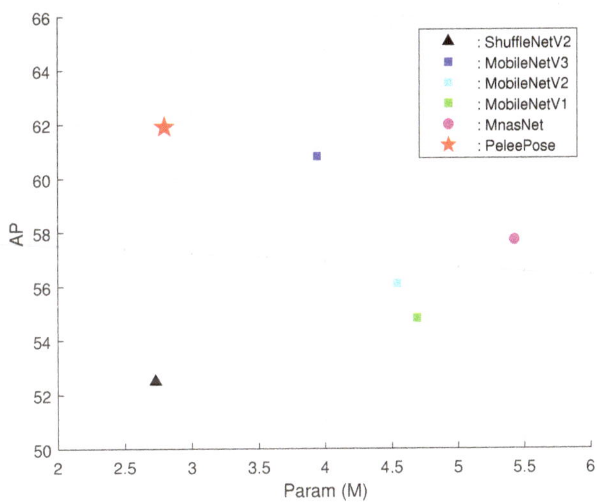

Figure 4. Comparison of the parameters and accuracies of lightweight networks for MSCOCO validation Sets.

4.3.2. Decoder Structure

To select the best decoder for KDLPN, we performed experiments to evaluate the performance of decoder approaches using the knowledge distillation method with overall loss alpha $\alpha_{KD} = 0.8$. In comparison, we also attempted to implement a three-step deconvolution layer decoder. For the three-step deconvolution layer decoder, experiments were performed by changing the number of channels from 352 to 44 for each decoder layer condition. To evaluate the accuracy and efficiency, Table 3 shows comparison of the number of parameters in the networks.

In Table 3, the parameters in the proposed decoder are reduced as compared to the parameters of the deconvolution decoder. From the computational complexity perspective, KDLPN with DUC exhibits the best performance. It uses only 38% of the parameters but affords a competitive performance to the deconvolution decoder with an AP difference of 1.5. The parameter and FLOPS of this decoder model were reduced to nearly seven and ten times that of deconvolution decoder, respectively, with competitive performance.

Table 3. Comparison of the network parameters with $\alpha_{KD} = 0.8$.

Encoder	Decoder	Decoder Param (M)	Decoder FLOPS (G)	AP
PeleeNet	Deconv (512 512 512)	14.11	34.49	62.8
	Deconv (256 256 256)	4.98	9.19	63.5
	Deconv (128 128 128)	1.98	2.58	62.8
	Deconv (64 64 64)	0.86	0.79	43.3
	Half-step deconv	5.21	4.58	63.4
	Ours	0.71	0.47	61.9

To demonstrate the effectiveness of our method, we performed various experiments such as simplifying the number of channels, and simply reducing the parameters of the proposed DUC model, and measured the corresponding performance and complexity. First, we constructed a baseline model by combining the encoder of the PeleeNet and a decoder comprising three deconvolution layers. Then the number of deconvolution channels in the lightweight model was modified from (256, 256, 256) (baseline model) to (512, 512, 512), (128, 128, 128), (64, 64, 64) and half-step channel. The half-step channel has the same output channel size as the DUC decoder model proposed as (176, 88, 44). The resultant performance, memory size, and FLOPS obtained by lightweighting the model in the aforementioned manner are presented in Table 3. In the experiment on reducing the number of channels, highest performance was afforded when the output channel size was reduced and modified to (256, 256, 256). Models with reduced output channel size of (128, 128, 128) and (64, 64, 64) exhibited performance degradation of 1.1% and 31.8%, respectively. The performance of the proposed DUC layer reduced by 2% on average compared to the existing model; however, the FLOPS and memory size considerably reduced to 85.4% and 60.5%, respectively, compared to those of the baseline model with the output channel size of (256, 256, 256). Moreover, compared to the model with the smallest output channel size of (64, 64, 64), FLOPS and memory size decreased further to 41.0% and 17.2%, respectively. Moreover, in comparison with the half-step deconv model with the same channel standard as the DUC decoder, FLOPS and memory size decreased to 86.4% and 89.7%, respectively. This indicates that the proposed DUC method is more efficient in lightweighting than the simple reduction of the number of channels. Considering the computational cost and performance of these methods presented in Table 3, KDLPN with DUC is the optimal model that can balance accuracy and efficient performance.

4.3.3. Knowledge Distillation Method

To demonstrate and optimize the effect of the knowledge distillation (Section 3.4) on the proposed network, experiments were performed on the proposed model with respect to α_{KD}. Table 4 shows the results of the experiments with varying α_{KD} using the teacher network. The table also shows the APs for each overall function α_{KD} in the same backbone network and DUC decoder. The α_{KD} values were varied from 0.3 to 1.0 for each dataset. The knowledge distillation method afforded better performances across all intervals than the PeleeNet network with DUC (57.4 AP). Furthermore, $\alpha_{KD} = 0.8$ afforded the best performance in this experiment; thus, we selected $\alpha_{KD} = 0.8$ for model training.

Table 4. Comparison of experiments on knowledge distillation.

Encoder	Decoder	α_{KD}	AP
PeleeNet	DUC	0.3	59.6
		0.4	59.6
		0.5	60.4
		0.6	60.6
		0.7	61.5
		0.8	61.9
		0.9	61.6
		1.0	60.9

During training through knowledge distillation, the knowledge of a teacher network can be advantageously learned, which is relatively accessible compared to the ground truth, which is difficult to learn. Accordingly, we first prepared a large and deep pretrained network using the teacher network and then trained a student network to apply knowledge distillation using the teacher network. If the teacher and student networks are simultaneously trained, the performance decreases since the teacher network is not converged. Similarly, when training a teacher network that has already converged, the test performance of the student network deteriorates as the optimally trained teacher network is overfitted to the training set. Regarding the above case, we conducted additional experiments, and the graph below displays the performance comparison between the model where the teacher and student network are simultaneously trained and the original training scheme. As shown in Figure 5, the performance of the simultaneously trained model, indicated in orange color, is decreased than that of the existing model, indicated in blue color. The reason why the performance difference between the two experiments is small is that both models used the same pretrained teacher model. However, since the teacher model is already pretrained, it can be overfitted to the training set during simultaneous learning, and the performance may degrade due to the probability of deviating from the optimal point. For this reason, the proposed original training scheme shows higher performance.

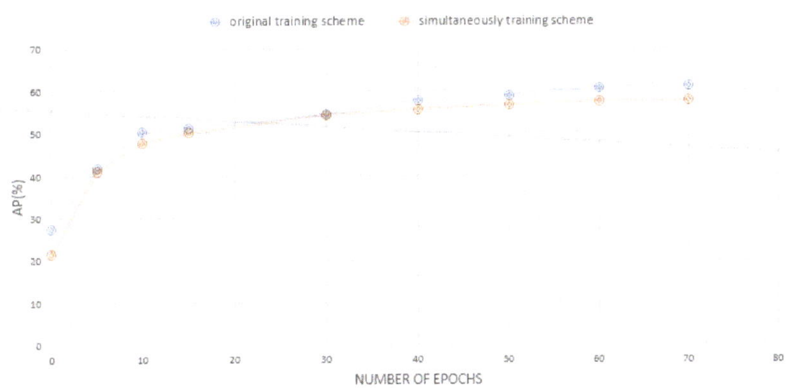

Figure 5. A graph comparing performance according to epoch of simultaneous training method and existing training method on MSCOCO validation dataset.

4.4. Results and Analysis

4.4.1. Overall Results

We compared our methods to other current state-of-the-art top-down-based human pose estimation methods such as RMPE, Mask-RCNN [57], and G-RMI [19]. For fair comparison, we used the same human detector for the top-down approach, to evaluate the pose estimation network performance of these methods based on a uniform criterion.

To further clarify the effectiveness of our scheme, we conducted additional experiments and modified only for the top-down algorithms using the same approach as the proposed method and fairly and accurately compared the amount of parameters. Table 5 below illustrates the validation results comparison of AP values, total parameters used, and FLOPS values. Our proposed model exhibits similar performance as the existing top-down-approach-based pose estimation networks and requires very few parameters in comparison as shown in Figure 6. We achieved an AP of 61.9 with only 2.80 M parameters and 1.49 FLOPS. Particularly, the amount of parameter used can be reduced by 90% compared to G-RMI with significantly lower computational complexity.

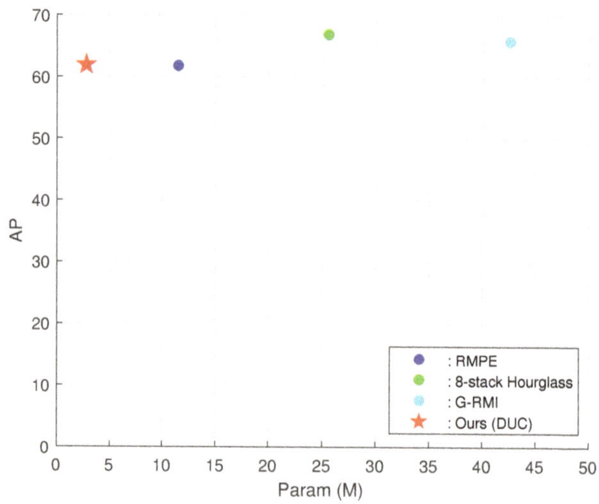

Figure 6. Parameter and accuracy comparison of top-down pose networks.

Table 5. Validation results comparison of AP values, total parameters used, and FLOPS values on MSCOCO dataset Params and FLOPS are calculated for the pose estimation network, and those for human detection and keypoint grouping are not included.

Method	Encoder	Decoder	AP	Param (M)	FLOPS (G)
RMPE	4-stack hourglass	Deconv	62.3	14.8	-
8-Stage Hourglass	Hourglass	(dev)	66.9	25.6	26.2
G-RMI	ResNet-101	(dev)	65.8	42.6	57.0
Ours	PeleeNet	DUC	61.9	2.80	1.49

We further conducted experiments on the MPII dataset [58] to demonstrate the generalization of our model. The MPII dataset is a popular open dataset on human pose that contains 25 k images with over 40 k people with annotated pose points acquired from YouTube. We conducted knowledge distillation learning on the same teacher network and validated the performance for 16 keypoints that are different from the MSCOCO dataset.

Table 6 describes the performance results of the experiment conducted on the MPII dataset. As illustrated in the table, the proposed method affords a mean of 86.9 (mean@0.5) via PCKh evaluation. We achieved a PCKh of 86.9 with 2.80 M parameters and 1.48 FLOPS. Compared to DU-Net (8), our model performed better 2.2 PCKh, and memory size decreased further to 64.6%, respectively. Moreover, in comparison with a state-of-the-art algorithm, our model achieved a balance between complexity and accuracy. Thus, the proposed pose network affords better performance than other methods.

Table 6. Comparison results on the MPII validation dataset (PCKh@0.5).

Method	PCKh(@0.5)	Param (M)	FLOPS (G)
DU-Net (8) [59]	84.7	7.90	-
Tang et al. [60]	87.5	15.5	33.6
Yang et al. [61]	88.5	7.30	16.2
Bulat et al. [61]	89.5	8.50	9.9
EfficientPoseIV [62]	89.75	6.56	72.9
Ours	86.9	2.8	1.48

Figure 7 illustrates the effectiveness of applying knowledge distillation to our method. Figure 7b–d display the pose heat map visualizations of the input image (a). (1) to (4) of Figure 7, the proposed algorithm is close to the ground-truth according to the teaching network for the moving and standing sequences. (6) of Figure 7 shows the example results when a person is occluded by an object. In fact, the ground-truth heatmap has missing keypoint labels due to occlusion object, but the teacher model identifies the missing ground-truth keypoint labels. Accordingly, the teacher model labels extra poses that assists the student network to learn as shown in 5, 6 (b) of Figure 7. Figure 8 visualizes the example image results from the validation set of the MSCOCO dataset. Therefore, our method achieved robust and stabilized pose estimation, even for difficult cases when joints are occluded by objects.

4.4.2. Discussions

We focused on introducing an approach to address the imbalance between performance and computational complexity, which is a fundamental problem of pose estimation. We introduced a method that reduces complexity using a fast lightweight network with few parameters and that compensates for the insufficient performance using the knowledge distillation method. Our proposed approach does not require the designing of a new network to ensure performance or speed. Furthermore, our approach is not limited to the network capacity of the network, as well as it advantageously complements performance by combining networks in various ways.

The proposed model has a limitation: its performance is relatively lower than that of existing deep and heavy networks such as [17–23]. However, our lightweighting scheme has demonstrated that its performance is high even when the resources are limited, such as low memory and computing power. Furthermore, a highly efficient learning method using the DUC layer is expected for pose estimation in mobile devices or embedded devices requiring low memory size with real time.

Figure 7. Pose estimation results of our model on the MSCOCO validation set. Column (**a**): The input images. Column (**b**): the keypoint heatmaps of Propoesd algorithm. Column (**c**): the keypoint heatmaps by the teacher model. Column (**d**): the ground-truth keypoint heatmaps. Row (1,2): Moving sequence on MSCOCO dataset. Row (3,4): Standing sequence on the MSCOCO dataset. Row (5,6): Body sequence hidden by objects in the MSCOCO dataset.

4.5. Conclusions

In this paper, we propose a new lightweight top-down multi-person pose estimation approach. The main challenge of a top-down approach is the achievement of a balance between the complexity and accuracy. Traditional top-down pose estimation approaches afford high performances, but the high complexity and high computing load make them time consuming. To resolve this dilemma, we introduced KDLPN, which operates efficiently and has low computational complexity. Moreover, the pixel shuffling operation in the decoder allows the reduction of the number of parameters. We applied the knowledge distillation method to prevent performance degradation and improve accuracy. Overall, our proposed algorithm achieved a balance between complexity and accuracy, as demonstrated by the qualitative and quantitative evaluation on the MSCOCO and MPII datasets.

Figure 8. Qualitative results of our model in MSCOCO. Column (**a**): Images that contain single person. Column (**b**): Images that contain two people. Column (**c**): Images that contain three & four people. Column (**d**): Images of group of people.

Author Contributions: Conceptualization, C.P. and H.S.L.; methodology, C.P.; software, C.P.; validation, C.P. and H.S.L.; formal analysis, C.P. and H.S.L.; investigation, C.P. and W.J.K.; resources, S.L.; data curation, C.P., H.S.L. and W.J.K.; writing: original draft preparation, C.P.; writing: review and editing, W.J.K., H.B.B., J.L. and S.L; visualization, C.P. and H.B.B.; supervision, S.L.; project administration, S.L.; funding acquisition, S.L. All authors have read and agreed to the published version of the manuscript.

Funding: This work was supported by Hanwha Techwin.

Institutional Review Board Statement: Ethical review and approval were waived for this study, due to the use of public databases, which were conducted according to the guidelines of the Declaration of Helsinki and approved by the relevant review boards. We complied with the terms of use of the databases regarding the publication of data.

Informed Consent Statement: According to the documentation of the used public databases, informed consent was obtained from all subjects involved.

Data Availability Statement: Not applicable.

Acknowledgments: We would like to express our gratitude to Hanwha Techwin and Yonsei University Office of Research Affairs.

Conflicts of Interest: The authors declare no conflicts of interest.

References

1. Zhang, Z. Microsoft Kinect sensor and its effect. *IEEE Multimed.* **2012**, *19*, 4–10. [CrossRef]
2. Fan, Z.; Zhao, X.; Lin, T.; Su, H. Attention-Based Multiview Re-Observation Fusion Network for Skeletal Action Recognition. *IEEE Trans. Multimed.* **2019**, *21*, 363–374. [CrossRef]
3. Torres, C.; Fried, J.C.; Rose, K.; Manjunath, B.S. A multiview multimodal system for monitoring patient sleep. *IEEE Trans. Multimed.* **2018**, *20*, 3057–3068. [CrossRef]
4. Dalal, N.; Triggs, B.; Dalal, N.; Triggs, B. Histograms of oriented gradients for human detection to cite this version: Histograms of oriented gradients for human detection. In Proceedings of the 2005 IEEE Computer Society Conference on Computer Vision and Pattern Recognition (CVPR'05), San Diego, CA, USA, 20–25 June 2005; pp. 886–893.
5. Wu, B.; Nevatia, R. Detection of Multiple, Partially Occluded Humans in a Single Image by Bayesian Combination of Edgelet Part Detectors. In Proceedings of the Tenth IEEE International Conference on Computer Vision (ICCV'05), Beijing, China, 17–21 October 2005.
6. Yang, Y.; Ramanan, D. Articulated human detection with flexible mixtures of parts. *IEEE Trans. Pattern Anal. Mach. Intell.* **2012**, *35*, 2878–2890. [CrossRef] [PubMed]
7. Toshev, A.; Szegedy, C. Deeppose: Human Pose Estimation via Deep Neural Networks. In Proceedings of the IEEE Conference on Computer Vision and Pattern Recognition, Columbus, OH, USA, 23–28 June 2014; pp. 1653–1660.
8. Zheng, C.; Wu, W.; Yang, T.; Zhu, S.; Chen, C.; Liu, R.; Shen, J.; Kehtarnavaz, N.; Shah, M. Deep learning-based human pose estimation: A survey. *arXiv* **2020**, arXiv:2012.13392.
9. Gong, W.; Zhang, X.; Gonzàlez, J.; Sobral, A.; Bouwmans, T.; Tu, C.; Zahzah, E.-H. Human Pose Estimation from Monocular Images: A Comprehensive Survey. *Sensors* **2016**, *16*, 1966. [CrossRef]
10. Cao, Z.; Simon, T.; Wei, S.E.; Sheikh, Y. Realtime Multi-Person 2D Pose Estimation using Part Affinity Fields. In Proceedings of the 2017 IEEE Conference on Computer Vision and Pattern Recognition (CVPR), Honolulu, HI, USA, 21–26 July 2017; pp. 7291–7299.
11. Cao, Z.; Hidalgo Martinez, G.; Simon, T.; Wei, S.; Sheikh, Y.A. OpenPose: Realtime Multi-Person 2D Pose Estimation using Part Affinity Fields. *IEEE Trans. Pattern Anal. Mach. Intell.* **2019**, *43*, 172–186. [CrossRef]
12. Pishchulin, L.; Insafutdinov, E.; Tang, S.; Andres, B.; Andriluka, M.; Gehler, P.; Schiele, B. DeepCut: Joint Subset Partition and Labeling for Multi Person Pose Estimation. In Proceedings of the 2016 IEEE Conference on Computer Vision and Pattern Recognition (CVPR), Las Vegas, NV, USA, 27–30 June 2016; pp. 4929–4937.
13. Insafutdinov, E.; Pishchulin, L.; Andres, B.; Andriluka, M.; Schiele, B. Deepercut: A deeper, stronger, and faster multi-person pose estimation model. In *Proceedings of the European Conference on Computer Vision*; Springer: Cham, Switzerland, 2016; pp. 34–50.
14. Newell, A.; Huang, Z.; Deng, J. Associative embedding: End-to-end learning for joint detection and grouping. In Proceedings of the 31st International Conference on Neural Information Processing Systems, Long Beach, CA, USA, 4–9 December 2017; pp. 1–11.
15. Kocabas, M.; Karagoz, S.; Akbas, E. Multiposenet: Fast multi-person pose estimation using pose residual network. In Proceedings of the European Conference on Computer Vision (ECCV), Munich, Germany, 8–14 September 2018; Springer International Publishing: Berlin/Heidelberg, Germany, 2018; pp. 417–433.
16. Cheng, B.; Xiao, B.; Wang, J.; Shi, H.; Huang, T.S.; Zhang, L. Higherhrnet: Scale-aware representation learning for bot-tom-up human pose estimation. In Proceedings of the International Conference on Computer Vision and Pattern Recogni-tion (CVPR), Seattle, WA, USA, 16–28 June 2020; pp. 5386–5395.
17. Fang, H.-S.; Xie, S.; Tai, Y.-W.; Lu, C. RMPE: Regional Multi-person Pose Estimation. In Proceedings of the 2017 IEEE International Conference on Computer Vision (ICCV), Venice, Italy, 22–29 October 2017; pp. 2353–2362.
18. Machine Vision and Intelligence Group. *AlphaPose*. Available online: https://github.com/MVIG-SJTU/AlphaPose (accessed on 5 February 2018).
19. Papandreou, G.; Zhu, T.; Kanazawa, N.; Toshev, A.; Tompson, J.; Bregler, C.; Murphy, K. Towards accurate multi-person pose estimation in the wild. In Proceedings of the IEEE Conference on Computer Vision and Pattern Recognition, Honolulu, HI, USA, 21–26 July 2017; pp. 4903–4911.
20. Chen, Y.; Wang, Z.; Peng, Y.; Zhang, Z.; Yu, G.; Sun, J. Cascaded Pyramid Network for Multi-Person Pose Estimation. In Proceedings of the IEEE Computer Society Conference on Computer Vision and Pattern Recognition (CVPR), Salt Lake City, UT, USA, 18–23 June 2018; pp. 7103–7112.
21. Xiao, B.; Wu, H.; Wei, Y. Simple baselines for human pose estimation and tracking. In Proceedings of the European Conference on Computer Vision (ECCV), Munich, Germany, 8–14 September 2018; pp. 466–481.
22. Ning, G.; Zhang, Z.; He, Z. Knowledge-guided deep fractal neural networks for human pose estimation. *IEEE Trans. Multimed.* **2017**, *20*, 1246–1259. [CrossRef]
23. Lin, T.Y.; Dollár, P.; Girshick, R.; He, K.; Hariharan, B.; Belongie, S. Feature pyramid networks for object detection. In Proceedings of the IEEE Conference on Computer Vision and Pattern Recognition (CVPR), Honolulu, HI, USA, 21–26 July 2017; pp. 2117–2125.
24. Wang, R.J.; Li, X.; Ling, C.X. Pelee: A real-time object detection system on mobile devices. In Proceedings of the IEEE Conference on Computer Vision and Pattern Recognition (CVPR), Salt Lake City, UT, USA, 18–23 June 2018; pp. 1963–1972.
25. Wang, P.; Chen, P.; Yuan, Y.; Liu, D.; Huang, Z.; Hou, X.; Cottrell, G. Understanding Convolution for Semantic Segmentation. In Proceedings of the 2018 IEEE Winter Conference on Applications of Computer Vision (WACV), Lake Tahoe, NV, USA, 12–15 March 2018; pp. 1451–1460.

26. Bissacco, A.; Yang, M.H.; Soatto, S. Fast human pose estimation using appearance and motion via multi-dimensional boosting regression. In Proceedings of the 2007 IEEE Conference on Computer Vision and Pattern Recognition, Minneapolis, MN, USA, 17–22 June 2007; pp. 1–8.
27. Lin, T.Y.; Maire, M.; Belongie, S.; Hays, J.; Perona, P.; Ramanan, D.; Dollár, P.; Zitnick, C.L. Microsoft COCO: Common objects in context. In Proceedings of the European Conference on Computer Vision, Zurich, Switzerland, 6–12 September 2014; Springer: Berlin/Heidelberg, Germany, 2014; pp. 740–755.
28. Chen, S.; Saiki, S.; Nakamura, M. Nonintrusive Fine-Grained Home Care Monitoring: Characterizing Quality of In-Home Postural Changes Using Bone-Based Human Sensing. *Sensors* **2020**, *20*, 5894. [CrossRef] [PubMed]
29. Lin, F.-C.; Ngo, H.-H.; Dow, C.-R.; Lam, K.-H.; Le, H.L. Student Behavior Recognition System for the Classroom Environment Based on Skeleton Pose Estimation and Person Detection. *Sensors* **2021**, *21*, 5314. [CrossRef] [PubMed]
30. Sadeghi-Niaraki, A.; Choi, S.-M. A Survey of Marker-Less Tracking and Registration Techniques for Health & Environmental Applications to Augmented Reality and Ubiquitous Geospatial Information Systems. *Sensors* **2020**, *20*, 2997.
31. He, K.; Zhang, X.; Ren, S.; Sun, J. Deep residual learning for image recognition. In Proceedings of the 2016 IEEE Conference on Computer Vision and Pattern Recognition (CVPR), Las Vegas, NV, USA, 27–30 June 2016; pp. 770–778.
32. Simonyan, K.; Zisserman, A. Very Deep Convolutional Networks for Large-Scale Image Recognition. 2015. Available online: http://www.robots.ox.ac.uk/ (accessed on 8 February 2021).
33. Cheng, B.; Wei, Y.; Feris, R.; Xiong, J.; Hwu, W.M.; Huang, T.; Shi, H. Decoupled classification refinement: Hard false positive suppression for object detection. *arXiv* **2018**, arXiv:1810.04002.
34. Ren, S.; He, K.; Girshick, R.; Sun, J. Faster R-CNN: Towards Real-Time Object Detection with Region Proposal Networks. *IEEE Trans. Pattern Anal. Mach. Intell.* **2017**, *39*, 1137–1149. [CrossRef]
35. Cheng, B.; Wei, Y.; Shi, H.; Feris, R.; Xiong, J.; Huang, T. Revisiting RCNN: On awakening the classication power of Faster RCNN. In Proceedings of the European Conference on Computer Vision, Munich, Germany, 8–14 September 2018.
36. Li, X.; Lai, T.; Wang, S.; Chen, Q.; Yang, C.; Chen, R.; Lin, J.; Zheng, F. Weighted feature pyramid networks for object detection. In Proceedings of the 2019 IEEE International Conference on Parallel & Distributed Processing with Applications, Big Data & Cloud Computing, Sustainable Computing & Communications, Social Computing & Networking (ISPA/BDCloud/SocialCom/SustainCom), Xiamen, China, 16–18 December 2019; pp. 1500–1504.
37. Redmon, J.; Farhadi, A. Yolov3: An incremental improvement. *arXiv* **2018**, arXiv:1804.02767.
38. Howard, A.G.; Zhu, M.; Chen, B.; Kalenichenko, D.; Wang, W.; Weyand, T.; Andreetto, M.; Adam, H. Mobilenets: Efficient convolutional neural networks for mobile vision applications. *arXiv* **2017**, arXiv:1704.04861.
39. Sandler, M.; Howard, A.; Zhu, M.; Zhmoginov, A.; Chen, L.C. Mobilenetv2: Inverted residuals and linear bottlenecks. In Proceedings of the IEEE Conference on Computer Vision and Pattern Recognition (CVPR), Salt Lake City, UT, USA, 18–23 June 2018; pp. 4510–4520.
40. Howard, A.; Sandler, M.; Chu, G.; Chen, L.C.; Chen, B.; Tan, M.; Wang, W.; Zhu, Y.; Pang, R.; Vasudevan, V.; et al. Searching for mobilenetv3. In Proceedings of the IEEE/CVF International Conference on Computer Vision (ICCV), Seoul, Korea, 27–28 October 2019; pp. 1314–1324.
41. Iandola, F.; Moskewicz, M.; Karayev, S.; Girshick, R.; Darrell, T.; Keutzer, K. Densenet: Implementing Efficient Convnet Descriptor Pyramids. *arXiv* **2014**, arXiv:1404.1869.
42. ImageNet. Large Scale Visual Recognition Challenge (ILSVRC): Competition. 2012. Available online: http://www.image-net.org/challenges/LSVRC/ (accessed on 27 December 2016).
43. Hinton, G.; Vinyals, O.; Dean, J. Distilling the knowledge in a neural network. *arXiv* **2015**, arXiv:1503.02531.
44. Buciluǎ, C.; Caruana, R.; Niculescu-Mizil, A. Model compression. In Proceedings of the 12th ACM SIGKDD International Conference on Knowledge Discovery And Data Mining, Philadelphia, PA, USA, 20–23 August 2006; pp. 535–541.
45. Yim, J.; Joo, D.; Bae, J.; Kim, J. A gift from knowledge distillation: Fast optimization, network minimization and transfer learning. In Proceedings of the IEEE Conference on Computer Vision and Pattern Recognition, Honolulu, HI, USA, 21–26 July 2017; pp. 4133–4141.
46. Ba, J.; Caruana, R. Do Deep Nets Really Need to Be Deep? *Adv. Neural Inf. Process. Syst.* **2014**, *27*, 2654–2662.
47. Romero, A.; Ballas, N.; Kahou, S.E.; Chassang, A.; Gatta, C.; Bengio, Y. Fitnets: Hints for thin deep nets. *arXiv* **2014**, arXiv:1412.6550.
48. Lan, X.; Zhu, X.; Gong, S. Person search by multi-scale matching. In Proceedings of the European Conference on Computer Vision (ECCV), Munich, Germany, 8–14 September 2018; pp. 536–552.
49. Lan, X.; Zhu, X.; Gong, S. Self-referenced deep learning. In *Proceedings of the Asian Conference on Computer Vision;* Springer: Cham, Switzerland, 2018; pp. 284–300.
50. Alashkar, T.; Jiang, S.; Wang, S.; Fu, Y. Examples-Rules Guided Deep Neural Network for Makeup Recommendation. In Proceedings of the AAAI Conference on Artificial Intelligence, San Francisco, CA, USA, 4–9 February 2017; pp. 941–947.
51. Jaderberg, M.; Simonyan, K.; Zisserman, A.; Kavukcuoglu, K. Spatial Transformer Networks. In Proceedings of the International Conference on Neural Information Processing Systems, Montreal, QC, Canada, 7–12 December 2015; MIT Press: Montreal, QC, Canada, 2015; Volume 28.
52. Zhang, X.; Zhou, X.; Lin, M.; Sun, J. Shufflenet: An extremely efficient convolutional neural network for mobile devices. In Proceedings of the IEEE Conference on Computer Vision and Pattern Recognition, Salt Lake City, UT, USA, 18–23 June 2018; pp. 6848–6856.

53. Kingma, D.; Ba, J. Adam: A method for stochastic optimization. *arXiv* **2014**, arXiv:1412.6980.
54. Ma, N.; Zhang, X.; Zheng, H.T.; Sun, J. Shufflenet V2: Practical guidelines for efficient cnn architecture design. In Proceedings of the European Conference on Computer Vision (ECCV), Munich, Germany, 8–14 September 2018; pp. 116–131.
55. Tan, M.; Chen, B.; Pang, R.; Vasudevan, V.; Sandler, M.; Howard, A.; Le, Q.V. MnasNet: Platform-Aware Neural Architecture Search for Mobile. In Proceedings of the 2019 Computer Vision and Pattern Recognition, Long Beach, CA, USA, 16–20 June 2019. Available online: https://arxiv.org/abs/1807.11626 (accessed on 29 May 2019).
56. Newell, A.; Yang, K.; Deng, J. Stacked Hourglass Networks for Human Pose Estimation. In Proceedings of the European Conference on Computer Vision, Amsterdam, The Netherlands, 11–14 October 2016; pp. 483–499
57. He, K.; Gkioxari, G.; Dollár, P.; Girshick, R. Mask r-cnn. In Proceedings of the IEEE International Conference on Computer Vision, Santiago, Chile, 13–16 December 2015; pp. 2961–2969.
58. Andriluka, M.; Pishchulin, L.; Gehler, P.; Schiele, B. 2D Human Pose Estimation: New Benchmark and State of the Art Analysis. In Proceedings of the 2014 IEEE Conference on Computer Vision and Pattern Recognition, Columbus, OH, USA, 23–28 June 2014; pp. 3686–3693.
59. Tang, Z.; Peng, X.; Geng, S.; Wu, L.; Zhang, S.; Metaxas, D. Quantized Densely Connected U-Nets for Efficient Landmark Localization. In Proceedings of the 2018 IEEE Conference on Computer Vision and Pattern Recognition (CVPR), Salt Lake City, UT, USA, 18–22 June 2018.
60. Tang, W.; Yu, P.; Wu, Y. Deeply learned compositional models for human pose estimation. In Proceedings of the European Conference on Computer Vision (ECCV), Munich, Germany, 8–14 September 2018; pp. 190–206.
61. Bulat, A.; Kossaifi, J.; Tzimiropoulos, G.; Pantic, M. Toward fast and accurate human pose estimation via soft-gated skip connections. In Proceedings of the 2020 15th IEEE International Conference on Automatic Face and Gesture Recognition (FG 2020), Buenos Aires, Argentina, 16–20 November 2020; pp. 8–15.
62. Groos, D.; Ramampiaro, H.; Ihlen, E.A. EfficientPose: Scalable single-person pose estimation. *Appl. Intell.* **2020**, *51*, 2518–2533. [CrossRef]

sensors

MDPI

Article

Deep-Learning-Based Stress Recognition with Spatial-Temporal Facial Information

Taejae Jeon [1], Han Byeol Bae [2], Yongju Lee [1], Sungjun Jang [1] and Sangyoun Lee [1,*]

1 Department of Electrical and Electronic Engineering, Yonsei University, Seoul 03722, Korea; jtj7587@yonsei.ac.kr (T.J.); paulyongju@yonsei.ac.kr (Y.L.); jeu2250@yonsei.ac.kr (S.J.)
2 Department of Artificial Intelligence Convergence, Kwangju Women's University, 45 Yeodae-gil, Gwangsan-gu, Gwangju 62396, Korea; kwu_BHB@kwu.ac.kr
* Correspondence: syleee@yonsei.ac.kr; Tel.: +82-2-2123-5768

Abstract: In recent times, as interest in stress control has increased, many studies on stress recognition have been conducted. Several studies have been based on physiological signals, but the disadvantage of this strategy is that it requires physiological-signal-acquisition devices. Another strategy employs facial-image-based stress-recognition methods, which do not require devices, but predominantly use handcrafted features. However, such features have low discriminating power. We propose a deep-learning-based stress-recognition method using facial images to address these challenges. Given that deep-learning methods require extensive data, we constructed a large-capacity image database for stress recognition. Furthermore, we used temporal attention, which assigns a high weight to frames that are highly related to stress, as well as spatial attention, which assigns a high weight to regions that are highly related to stress. By adding a network that inputs the facial landmark information closely related to stress, we supplemented the network that receives only facial images as the input. Experimental results on our newly constructed database indicated that the proposed method outperforms contemporary deep-learning-based recognition methods.

Keywords: deep learning; stress recognition; stress database; spatial attention; temporal attention; facial landmark

Citation: Jeon, T.; Bae, H.B.; Lee, Y.; Jang, S.; Lee, S. Deep-Learning-Based Stress Recognition with Spatial-Temporal Facial Information. *Sensors* **2021**, *21*, 7498. https://doi.org/10.3390/s21227498

Academic Editor: Sheryl Berlin Brahnam

Received: 27 August 2021
Accepted: 9 November 2021
Published: 11 November 2021

1. Introduction

People in contemporary society are under immense stress due to various factors [1]. As stress is a cause of various diseases and affects longevity, it is vital to keep it under control [2–4]. A system that detects a user's stress level in real time and provides feedback about how to lower stress is the need of the hour [5–7]. To develop such a system, high-accuracy stress recognition technology is required. In response to this need, research on stress recognition technology has been actively conducted. Reliable stress recognition technology will be useful in various fields, such as driver stress monitoring [8,9] and online psychological counseling.

Most stress-recognition studies have been conducted using a two-class classification, which divides subjects into stressed or relaxed, or using three classes, i.e., low, medium, and high stress [10]. Several stress recognition studies have been conducted on physiological signals acquired through wearable devices [8,11–17]. Physiological-signal-based approaches effectively recognize human stress because they use signals that immediately reveal a person's condition, such as respiration rate, heart rate, skin conductivity, and body temperature. However, this method involves additional costs because a special wearable device is required to acquire physiological signals, which users may find too expensive or feel reluctant to wear.

Other studies have identified and classified stress using life-log data such as mobile app usage records obtained from smartphones [18–21]. As smartphones are always attached to their users, it is possible to ascertain the user's status by accumulating data over a certain period. This approach is suitable for recognizing stress over a specific period, but fails to recognize an instantaneous stress state. By contrast, images, such as thermal images showing blood flow and respiratory rate and visual images portraying body movements and pupil size, can be used for stress recognition [22–24]. Some stress-recognition studies use only visual images, especially facial images, which have the advantage of only requiring a camera; the subjects need not wear additional equipment [25,26]. However, in many of these methods, handcrafted features continue to be used. In some recent studies, a neural network with handcrafted features is used in the feature extraction process [27–29].

Some recent studies have recognized stress using only deep learning. Zhang et al. [30] proposed a deep-learning-based method that detects the presence or absence of stress using the video footage of a person watching a video clip that induces or does not induce stress. In this method, when the face-level representation was first learned, an emotion recognition network was used to learn the emotion change between the two frames with the largest emotion difference. Furthermore, the action-level representation was learned by using motion information and an attention module that passes the entire feature through one fully connected layer. The resolution of both the facial image and upper body image was 64×64, which rendered the detection of small facial changes difficult.

By contrast, our study focuses on a more difficult task: subdividing stressful situations into low-stress and high-stress situations. Furthermore, the attention used was subdivided into spatial and temporal attention, and since it had a precise structure, it could be advantageously used for learning attention for each purpose. Additionally, the face-level representation was learned using all frame information, and the resolution of the facial image used was 112×112, which was more advantageous for detecting small facial changes. Moreover, since the proposed method does not use motion information, it can show higher performance in situations where only a face is visible or for people without bodily motion. The experimental results in Section 5.4 show that the proposed method could detect overall spatial and temporal changes in the face related to stress and that it is superior to the method presented in the previous work [30].

In a previous study [31], we constructed a database and performed deep-learning-based stress recognition using facial images. In this database, data were acquired in both the speaking and nonspeaking stages. However, this resulted in a challenge: the learning proceeds in such a way that the network classifies speaking and nonspeaking states. Moreover, the amount of data was insufficient for detecting minute changes in the face because images were stored at a rate of about five images per second. Furthermore, the stress recognition network was not designed in detail to find minute changes in facial expressions, but was instead designed as a combination of a convolutional neural network (CNN) and a deep neural network (DNN) with a simple structure.

Therefore, in this study, the database construction and network design were improved so as to alleviate the aforementioned concerns. High-quality data were acquired by designing a more sophisticated scenario, and the recognition model also had a more sophisticated design. We acquired additional data because a large-capacity image database is required to use deep learning, but there is no existing database that can be used for stress recognition. Therefore, we built a large image database by conducting a stress-inducing experiment and released the database publicly. We propose a deep-learning-based stress-recognition method using facial images from this stress recognition database.

In the proposed method, we used time-related information, which is unavailable in still images. Given that our database contains images captured from video data, we use a temporal attention module that assigns a high weight to frames related to stress when viewed from the time axis. Furthermore, we used a spatial attention module that assigns a high weight to the stress-related areas in the image to improve the performance further. One study [32] found that peoples' eye, mouth, and head movements differ when under stress. Therefore, to accurately capture these movements, a network that receives facial landmark information was added. Accordingly, we supplemented the network, which receives only facial images as the input. In addition, designing a proper loss function when using the deep-learning method is crucial. Therefore, we designed a loss function that is suitable for our database and trained the proposed method end-to-end.

Our contributions are as follows:

1. We built and released a large-capacity stress recognition image database that can be used for deep learning;
2. We applied a multi-attention structure to the deep learning network, and the proposed method was trained end-to-end;
3. We trained a feature with stronger discriminating power by adding a network that uses facial landmarks.

The remainder of this paper is organized as follows. In Section 2, previous studies related to stress recognition and deep learning are described. In Section 3, we introduce the construction process and contents of our database. In Section 4, the proposed method is presented in detail. In Section 5, the experimental settings are described and the experimental results are analyzed. Finally, Section 6 concludes this study.

2. Related Work

2.1. Facial-Action-Unit-Based Stress Recognition Methods

Many studies have attempted to recognize stress using facial action unit information that defines the movements of the eyes, nose, mouth, and head [25,32–34]. There are several types of facial action units, and among them, units that are highly related to stress, such as inner brow raise, nose wrinkle, and jaw drop, are used often. In previous studies, the movement of each facial action unit was used as a feature, and classical classifiers such as random forest and support vector machine (SVM) were used for classification. Some studies recognized stress primarily using pupil size [24,35]. The pupil diameter and pupil dilation acceleration were used as features, and the SVM and decision tree were used as classifiers. Pampouchidou et al. [36] recognized stress using mouth size as a primary characteristic. Stress was recognized using normalized openings per minute and the average openness intensity obtained from mouth openness. In another study, stress was recognized by observing breathing patterns through changes in the nostril area [27]. After discovering breathing patterns through temperature changes near the nostrils, two-dimensional respiration variability spectrogram sequences were constructed using these data and were used to recognize stress. Giannakakis et al. [37] recognized stress based on facial action unit information obtained from nonrigid 3D facial landmarks, the histogram of oriented gradients (HOG), and the SVM. The limitations of the aforementioned methods are that they cannot utilize the changes in the facial colors and the full facial image because the entire image information is not used.

2.2. Facial-Image-Based Stress Recognition Methods

In one popular method of recognizing stress using facial images, unlike the facial action unit, a comprehensive feature is extracted from the entire image. In some studies, the HOG features were extracted from the eye, nose, and mouth regions in RGB images and used as features [26,29]. In these methods, a CNN and a method combining the SVM and slant binary tree algorithm were used as classifiers. Some studies used features extracted from thermal images or nearinfrared (NIR) images [9,22,38]. In the methods using thermal images, stress was recognized based on the tissue oxygen saturation value extracted from the thermal image or by applying a CNN to the thermal image itself. In the method using NIR images, stress recognition was performed using an SVM after extracting scale-invariant feature transform (SIFT) descriptors around facial landmarks. In other studies, stress was recognized by fusing RGB and thermal images [28,39,40]. In these methods, stress was recognized using the features extracted from super-pixels and local binary patterns on the three orthogonal plane (LBP-TOP) descriptor. All the methods introduced above used handcrafted features, but there was also a method using deep learning. This method recognizes stress by fusing facial images and motion information such as hand movements [30]. In this method, optical flow images were used to obtain motion information, and stress was recognized by applying attention to facial features and motion features. Most of the facial-image-based stress recognition studies have used handcrafted features. Many image recognition studies have shown great performance improvement through deep learning. If deep learning is used, the stress recognition performance can be further improved because stress-related high-dimensional features can be learned from images. Recently, a study [30] that recognized stress using deep learning came out, and we also tried to recognize stress using deep learning for better performance.

2.3. Facial-Image-Based Emotion Recognition Methods

Many studies on facial-image-based emotion recognition are being conducted, and there are similarities between emotion recognition and stress recognition studies since emotion and stress are related. Among studies on emotion recognition methods, many studies using facial landmark information are underway [41–43]. As changes in facial expressions are highly correlated with changes in facial landmarks, these studies input the coordinates of facial landmarks directly into a network or images created from facial landmarks. Palestra et al. [44] classified emotions using a random forest classifier after extracting geometrical features from facial landmark information. Studies on recognizing emotions in videos are also being actively conducted. For such emotion recognition, various methods for using time-related information are being studied. These include a method that uses a 3D-CNN [41,45] and a method that combines a 2D-CNN and a recurrent neural network (RNN) [42,43]. Furthermore, many recent deep-learning-based studies have improved the recognition performance by using simple modules such as the attention module [46,47]. The attention module creates attention maps that are multiplied by the input feature maps and then refines those feature maps to improve recognition performance. For example, Zhu et al. [48] proposed a hybrid attention module comprising a self-attention module and a spatial attention module to detect regions with large differences in facial expressions. Meng et al. [49] proposed a frame attention module that assigns higher weights to frames with higher importance among multiple frames when video data are input. The difference between our method and the above methods is that the former were designed to detect overall spatial and temporal changes in the face. First, attention was divided into spatial attention and temporal attention to emphasize spatially and temporally important parts, respectively. We then designed a network that could effectively detect facial changes by using preprocessed facial landmark images. We showed that the proposed method is superior to other methods through various ablation studies and performance evaluation experiments.

3. Database Construction

Several databases [50,51] containing data for stress recognition are available, but most contain physiological signal data; few have image-related information. As far as we know, there is only one database, i.e., the SWELL-KW database [51], that includes facial image information. This database provides four types of information: computer interactions, facial expressions, body postures, and physiology. It provides four pieces of information related to facial expressions. First, the orientation of the head in three dimensions is provided. Second, ten pieces of information related to facial movements, such as gaze direction and whether the mouth is closed, are provided. Third, 19 pieces of information related to facial action units such as inner brow raise, nose wrinkle, and chin raise are provided. Finally, probability values are provided for eight emotions such as neutral, happy, and sad. However, this database does not provide images, but only the above high-level information obtained from images. Therefore, this database cannot be used for deep-learning-based stress-recognition methods that take images as the input.

Therefore, a new database is required to recognize stress using deep learning, so we built a large image database. The database we built consists of the subject's facial images and information on whether the subject's stress level belongs to one of three levels (neutral, low stress, or high stress). As this study involved human participants, our database was built with the approval of the Institutional Review Board of Yonsei University, and the study was conducted upon it. We created this database by designing an experimental scenario that included stress-inducing situations. The designed stress-inducing experiment scenario is depicted in Figure 1.

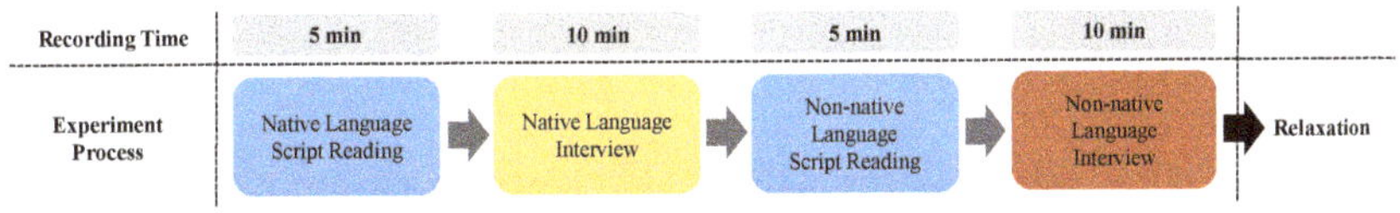

Figure 1. Progress of the designed stress-inducing experimental scenario, including the recording time for each stage.

As research results indicated that an interview induces stress in the subject [52,53] and that the subject is stressed when asked to use a non-native language [54,55], the experimental scenario was designed in accordance with these studies' results. Therefore, the stress-inducing situation comprised interviews in native and non-native languages. The former was established as a situation that induces low stress and the latter as a situation that induces high stress. We recruited subjects near our school. As most of the population is Korean, Koreans were selected as test subjects, and accordingly, Korean was used as the native language. English was selected as the non-native language because it is the most popular non-native language used by Koreans.

Situations in which the test subject reads scripts written in the native or non-native languages were used as the comparison group. These were considered situations that did not cause stress (i.e., neutral). If the nonspeaking situations were set as a comparison group, the network can learn to classify speaking and nonspeaking situations. Thus, the comparison group was limited to situations in which subjects read scripts. The experiment time for each stage was 5 min for the native and non-native language script reading and 10 min for the native and non-native language interviews. We set the experiment time for each script-reading stage to 5 min because we designed both script-reading stages to be stress-free so that the sum of the experiment time of the two stages would be the same (10 min) as the other stress-inducing stages in the experiment. We shot a single video at each experimental stage for each subject. As there were four experimental steps, the number of videos for each subject was four.

We collected data by recruiting 50 men and women in their 20 s and 30 s. We chose this age group because the experimental stages included reading scripts and interviewing in a non-native language. We believed that this task would be difficult for older people. In addition, the population in their 20s and 30s in the subject recruitment area was large. During the experiment involving situations that do and do not induce stress, the subject's appearance was photographed using a Kinect v2 camera.

The data acquisition environment was as follows. The data were acquired in a windowless location so that the lighting could be kept constant. The camera was set so that only a white wall appeared behind the subject, eliminating any potential interference from a complex background. The camera was positioned in front of the subject so that the subject's frontal face could be photographed. To ensure that the subject's face would always be visible, hair or accessories other than glasses were not allowed to cover the subject's face. The reason for this constraint is that if hair or accessories cover the face, they interfere with the observation of the subject's facial changes. We enforced these constraints because the purpose of this study is to detect overall spatial and temporal changes in the face related to stress. The resolution of the recorded video is 1920×1080. When the data were acquired, about 24 images were saved per second, and the entire database comprises 2,020,556 images. The summary information about the database construction settings and database contents is depicted in Appendix A.

As presented in Table 1, this database comprises a large number of images for deep learning, which is considered highly useful, and was released as the Yonsei Stress Image Database on IEEE DataPort (https://dx.doi.org/10.21227/17r7-db23 (accessed on 8 November 2021)). It is publicly available for stress recognition research. We measured the stress recognition accuracy after labeling the acquired data according to the scenario we designed. We labeled the data acquired during the native language interview as low stress, the data acquired during the non-native language interview as high stress, and the data acquired while reading the script produced in the native language or non-native language as neutral.

Table 1. Number of images acquired at each stage of the database construction.

Designed State	Experimental Stage	Total Images
Neutral	Native Language Script Reading	366,121
	Non-native Language Script Reading	368,991
Low Stress	Native Language Interview	656,624
High Stress	Non-native Language Interview	628,820

We annotated the data in this manner because many stress recognition studies still use this method [10]. The reason why this labeling method continues to be popular is that it is difficult to annotate stress data in real time. In the case of an emotion database, an annotator can examine the facial expression of a subject and label the subject's emotions as positive or negative in real time. This is possible because in the case of facial expressions, the emotion is visually apparent, and therefore, other people can judge to some extent whether it is positive or negative. However, in the case of stress, it is difficult to judge it solely from facial expressions. For example, while a subject may actually be stressed, it may not be evident from his/her facial expressions, or he/she may fake a smile. Therefore, many studies have created a stress-inducing situation, and all data obtained from that situation were labeled as corresponding to a stress state. We trained and tested how accurately the proposed method and other methods classified data into these three labels, and the performance of each method was compared using the test accuracy. The ablation studies and comparative experiments conducted using the established database are described in Section 5.

4. Proposed Methodology

In this section, we describe the structure of the proposed method for recognizing stress using facial information and multiple attention. We look at the proposed method's overall structure and then look at the spatial attention module, facial landmark feature module, temporal attention module, and loss function, in that order.

4.1. Overall Structure

The proposed method predicts a person's stress level from video data based on facial information. A flowchart for the proposed method is depicted in Figure 2, and the details are described below.

First, one clip was entered as the input for the proposed method. This clip was created by dividing all 5 or 10 min videos acquired in the database construction experiment into 2 s clips. As the data acquisition rate was 24 frames per second (fps), one clip consisted of 48 frames, and we used all 48 frames as the input. The size of the original image was 1920×1080, but when training and testing, the face area was detected, cropped, and resized to 112×112. A multitask cascaded convolutional network [56] was used to detect and localize the facial area. When the facial image passes through the ResNet-18 residual network [57], feature maps are generated. Furthermore, as these feature maps pass through the spatial attention module and global average pooling (GAP) [58], a facial image feature is generated.

In the spatial attention module, a high weight was assigned to the positionally important parts of the feature maps, and a lower weight was assigned to the positionally unimportant parts of the feature maps. The details of the spatial attention module are described later in Section 4.2. When a facial image passed through the facial landmark detector, 68 facial landmarks were obtained. After creating a facial landmark image by marking 68 facial landmark points as white dots on a black image, the facial landmark feature network and GAP were applied to obtain a facial landmark feature. The details of the facial landmark feature module are described later in Section 4.3. The resulting 48 facial image features and 48 facial landmark features were concatenated for each frame and then passed through the temporal attention module to obtain a final feature.

In the temporal attention module, a high weight was assigned to frame features that were highly related to stress, while a low weight was assigned to frame features that were less related to stress. The details of the temporal attention module are described later in Section 4.4. When the final obtained feature passed through the fully connected layer, a stress prediction result was finally produced. We divided the stress state into neutral, low, and high stress. Therefore, the stress prediction result would be one of these three states. While the learning was in progress, the part that was actually learned is marked with a red box in Figure 2.

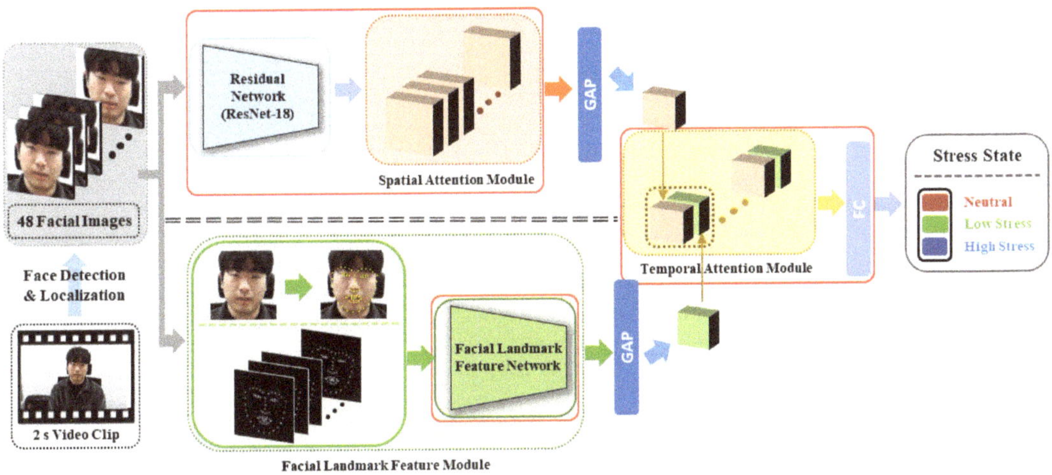

Figure 2. Flowchart of the proposed method. The residual network ResNet-18 extracts feature maps from facial images. GAP: global average pooling; FC: fully connected layer.

4.2. Spatial Attention Module

Chen et al. [59] used a spatial attention module to pinpoint to the network the relevant parts of the feature map that should be viewed more closely. Since then, the spatial attention module's structure has continued to develop. As the module proposed by Woo et al. [47] demonstrated both light and high performance, we used it to obtain the spatial attention weight. The spatial attention module's overall structure is depicted in Figure 3, and the details are described below.

First, the feature maps were extracted by inputting the facial image into ResNet-18. This network is light and has high performance, so it is widely used in various recognition fields. We did not use a pretrained network; only the structure of ResNet-18 was used and trained from the beginning after initializing the weights. After obtaining the feature maps, average pooling and max pooling were performed on the channel axis. The two results were concatenated along the channel axis. Chen et al. [59] demonstrated that performing the pooling operation on the channel axis emphasizes locational importance. The average pooling operation used by Zhou et al. [60] is frequently used because it is effective for aggregating information.

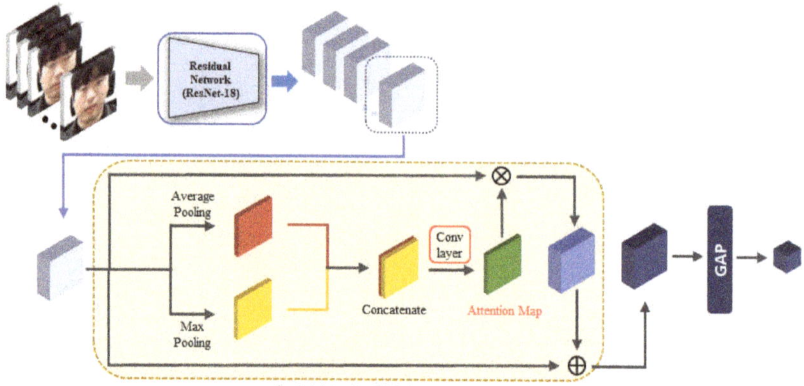

Figure 3. Structure of the spatial attention module. For efficient learning, the multiplication result from the original feature maps and the spatial attention map is added to the original feature maps. GAP: global average pooling.

Furthermore, Woo et al. [47] found that the max pooling operation reveals important information that differs from that revealed by the average pooling operation. Therefore, if the results obtained by performing both the average pooling and max pooling operations on the feature maps are concatenated and a convolutional operation is performed, it is possible to obtain an attention map that highlights stress-relevant regions by considering multiple perspectives. In our design, the sigmoid function was used to obtain the final spatial attention map. By multiplying the obtained spatial attention map by the original feature maps, feature maps with applied spatial attention can be obtained.

In the next step, the final feature maps were obtained by adding attention-applied feature maps to the original feature maps. This addition to the previous layer's result is called identity mapping. This structure reduces the amount of information that the layer must learn so that learning can be performed more effectively [57]. Finally, the facial image feature was obtained by applying GAP to the final feature maps. While the learning was in progress, the part that was actually learned is marked with a red box in Figure 3. The facial image feature was obtained using the following equation:

$$M_{sa} = \sigma(conv^{7\times7}([AvgPool(F); MaxPool(F)])), \tag{1}$$

$$f_{facial\ image} = GAP(F + M_{sa} \circ F), \tag{2}$$

where σ is the sigmoid function, $conv^{7\times7}$ denotes the convolutional operation with a 7×7 filter, F denotes the feature maps extracted from ResNet-18, the symbol ; denotes the concatenation operation, GAP indicates the GAP operation, and $\circ$ is the product of the attention weight and feature value for each position in the feature map. The residual network's structure and spatial attention module are depicted in Table 2. As can be seen from Table 2, the size of the feature space of the facial image feature was $4 \times 4 \times 512$. This module was automatically trained through an end-to-end learning process. The importance of the spatial attention module is evaluated in Section 5.3.2.

Table 2. Network structure of the residual network and spatial attention module.

	Unit	Layer	Filter/Stride	Output Size
Input	0			112 × 112 × 3
Residual Network	1	Conv-BN-ReLU	7 × 7, 64/2	56 × 56 × 64
		Max Pooling	3 × 3/2	28 × 28 × 64
	2	Conv-BN-ReLU	3 × 3, 64/1	28 × 28 × 64
		Conv-BN	3 × 3, 64/1	28 × 28 × 64
	3	Conv-BN	1 × 1, 128/2	14 × 14 × 128
		Conv-BN-ReLU	3 × 3, 128/1	14 × 14 × 128
		Conv-BN	3 × 3, 128/1	14 × 14 × 128
	4	Conv-BN	1 × 1, 256/2	7 × 7 × 256
		Conv-BN-ReLU	3 × 3, 256/1	7 × 7 × 256
		Conv-BN	3 × 3, 256/1	7 × 7 × 256
	5	Conv-BN	1 × 1, 512/2	4 × 4 × 512
		Conv-BN-ReLU	3 × 3, 512/1	4 × 4 × 512
		Conv-BN	3 × 3, 512/1	4 × 4 × 512
Spatial Attention Module	6	AvgPool		4 × 4 × 1
		MaxPool		4 × 4 × 1
		AvgPool+MaxPool		4 × 4 × 2
		Conv-Sigmoid	7 × 7, 1/1	4 × 4 × 1
	7	Product (5 ∘ 6)		4 × 4 × 512
Output	8	GlobalAvgPool		512

BN: batch normalization. In Unit 7, the outputs of Units 5 and 6 are multiplied for each position in the feature maps.

4.3. Facial Landmark Feature Module

Giannakakis et al. [32] indicated that peoples' eye, mouth, and head movements during stressful situations differ from those during nonstressful situations. To accurately capture these movements, we designed a network that receives facial landmark points representing the eye, mouth, and head positions as the input. The feature extracted from this network is used along with the facial image feature to complement its discriminating power. The process of extracting the facial landmark feature is depicted in Figure 4, and the details are described below.

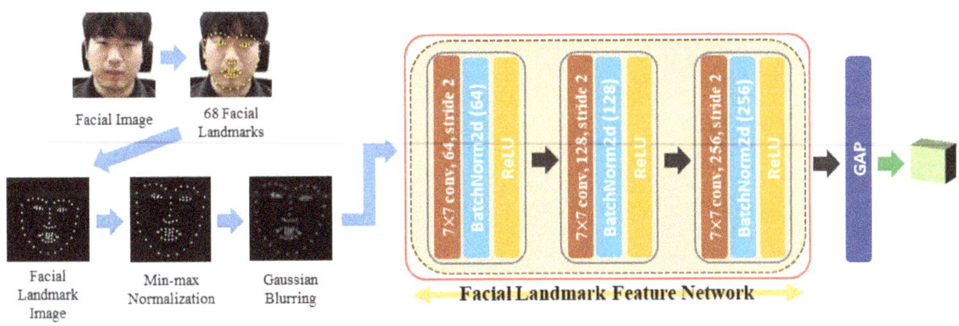

Figure 4. Facial landmark feature extraction process. A simple network with three convolutional layers is used to extract the facial landmark feature. GAP: global average pooling.

A facial image was first input into the facial landmark detector to extract the facial landmark feature, where the detector was an ensemble of the regression tree algorithm [61]. Passing through the facial landmark detector, 68 facial landmarks were obtained and displayed as white dots on a black image to create a facial landmark image. The facial landmark image was used because it better captures the movement of the facial landmarks when input into the CNN, which uses spatial information, rather than simply entering the facial landmark coordinate values into the fully connected neural network. In the method proposed by Wu et al. [41], the facial landmark image was used to utilize the facial location, and it was shown that fine movements could be captured well. Therefore, we also tried to capture the minute movements of the face by proposing a method to utilize a facial landmark image by paying attention to this aspect.

Furthermore, two preprocessing steps were performed on the facial landmark image; one is min–max normalization, and the other is Gaussian blurring. Min–max normalization was used because the position of the area where the human face is detected in each frame of the video jitters slightly, so the face is stationary, but appears to be moving. If the location of the face area moves slightly, the location of the facial landmark detected in the facial area also moves slightly. Consequently, the head is stationary, but it may appear to move, which may adversely affect stress recognition. By performing min–max normalization, this phenomenon can be prevented because the positions of the facial landmarks are evenly aligned in all frames. In the face detection stage, we roughly aligned the positions of the eyes, nose, and mouth through alignment, but these positions were not always precisely fixed. Therefore, min–max normalization was additionally applied to reduce this phenomenon as much as possible.

After min–max normalization, Gaussian blurring was performed because jittering also occurred in the facial landmark detector result, and the effects that arise from these phenomena can be reduced when blurring is performed by spreading the data around a point rather than merely displaying that point. After performing these two preprocessing steps, the image was passed through the CNN. The structure of this network comprises three convolutional layers. The content of the facial landmark image is simple. Useful information can be extracted even by a simple network, so we chose a simple network to avoid unnecessary complexity. Finally, the facial landmark feature was obtained by performing a GAP operation on the feature maps that passed through the CNN. While the learning was in progress, the part that was actually learned is marked with a red box in Figure 4. The facial landmark feature module network structure is depicted in Table 3. As can be seen from Table 3, the size of the feature space of the facial landmark feature was $9 \times 9 \times 256$.

Table 3. Network structure of the facial landmark feature module.

	Unit	Layer	Filter/Stride	Output Size
Input	0			$112 \times 112 \times 1$
Facial Landmark Feature Network	1	Conv-BN-ReLU Conv-BN-ReLU Conv-BN-ReLU	$7 \times 7, 64/2$ $7 \times 7, 128/2$ $7 \times 7, 256/2$	$53 \times 53 \times 64$ $24 \times 24 \times 128$ $9 \times 9 \times 256$
Output	2	GlobalAvgPool		256

BN: batch normalization. The stride is 2, but the feature map size is reduced by more than 0.5-times because padding is not performed during convolution.

4.4. Temporal Attention Module

Meng et al. [49] used a temporal attention module to observe the information in all frames to determine on which frame to focus. As the structure is simple and demonstrated high performance in facial expression recognition, we modified this module and used it to obtain the temporal attention weight. The temporal attention module's overall structure is depicted in Figure 5, and the details are described below.

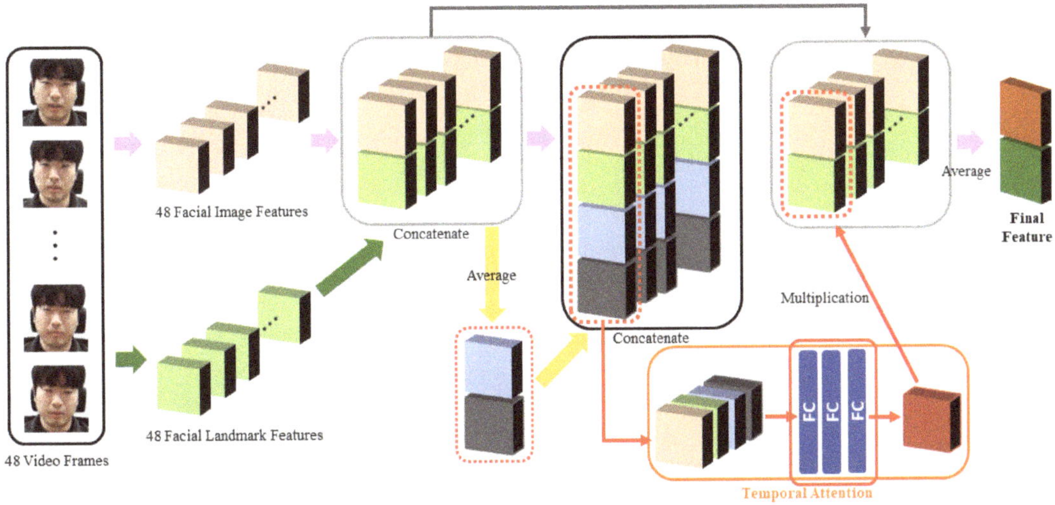

Figure 5. Structure of the temporal attention module. The attention weight increases when the frame is highly related to stress, considering the average feature representing 48 frames and the feature of a specific frame. FC: fully connected layer.

First, 48 video frames passed through the ResNet-18 network and spatial attention module, and 48 facial image features were extracted. Then, these frames passed through the facial landmark detector and facial landmark feature network, and 48 facial landmark features were extracted. When the 48 extracted facial image features and 48 extracted facial landmark features entered the temporal attention module, they were first concatenated frame-by-frame to create 48 concatenated features. Thus, frames highly related to stress were found by considering the facial image features, as well as the facial landmark features.

The 48 concatenated features were averaged to obtain the average feature, and the average feature was concatenated into 48 concatenated features to generate 48 final concatenated features. The average feature can be regarded as containing all information for all frames. When the temporal attention weight is calculated using these final concatenated features, it becomes possible to obtain each frame's temporal attention weight by comprehensively viewing the information of the entire frame, as well as the information of individual frames. Therefore, each final concatenated feature was passed through three fully connected layers to obtain each frame's temporal attention weight. It is possible to attach the 49th slice and calculate the weight at once, but the weight of the target individual feature and the total feature decreases, so the desired weight value cannot be obtained. Therefore, we did not proceed in this manner.

When the obtained temporal attention weight for each frame is multiplied by the concatenated feature from the corresponding frame's facial image feature and facial landmark feature, the concatenated feature reflects the importance of the corresponding frame. Accordingly, after obtaining the concatenated features that reflect the importance of all 48 frames, the final feature was obtained by applying the average operation. By applying a fully connected layer to this feature, the stress recognition result was output. While the learning was in progress, the part that was actually learned is marked with a red box in Figure 5. The final feature was obtained using the following equations:

$$f^i_{concat} = [f^i_{facial\ image}; f^i_{facial\ landmark}], \quad (3)$$

$$f^i_{total\ concat} = [f^i_{concat}; Avg(\boldsymbol{f_{concat}})], \quad (4)$$

$$W^i_{ta} = fc^1(fc^{1536}(fc^{1536}(f^i_{total\ concat}))), \quad (5)$$

$$f_{final} = Avg(\mathbf{W_{ta}} \cdot \mathbf{f_{concat}}), \quad (6)$$

where f^i is the feature of the ith frame, Avg denotes the averaging operation on the time axis, W^i is the weight of the ith frame, fc^n represents a fully connected layer with n output nodes, and the symbol · denotes the multiplication operation for each frame. The bold notation indicates a vector of features or weights for all frames. The network structure of the temporal attention module is depicted in Table 4.

Table 4. Network structure of the temporal attention module.

	Unit	Layer	Output Size
Input	0	Facial Image Feature	512 × 48 (frames)
	1	Facial Landmark Feature	256 × 48 (frames)
Temporal Attention Module	2	Concatenate (0 + 1)	768 × 48 (frames)
	3	Average (48 frames)	768
	4	Concatenate (2 + 3)	1536 × 48 (frames)
	5	Fully Connected Fully Connected Fully Connected	1536 × 48 (frames) 1536 × 48 (frames) 1 × 48 (frames)
	6	Multiplication (2 · 5)	768 × 48 (frames)
	7	Average (48 frames)	768
Output	8	Fully Connected	3 or 4

In Unit 4, the outputs of Units 2 and 3 are concatenated for each frame. In Unit 6, the outputs of Units 2 and 5 are multiplied for each frame.

4.5. Loss Function

We trained and tested the proposed method using the constructed database. Given the database's characteristics, the choice of the loss function influenced the training result considerably. For the constructed database, the difference in facial changes observed by the same person in different stress states is minute, so the difference between classes within the same subject's data is not large.

In contrast, even in the same stress state, each person has a unique face, and a difference in the pattern of facial changes occurs. Accordingly, the difference between subjects within data from the same class is large. Therefore, if the distance between features for data from different classes within the data for the same subject is increased and the distance between features for data from different subjects within the data for the same class is decreased, it is possible to prevent ineffective learning caused by database characteristics.

Previous studies have proposed several loss functions to prevent this phenomenon, such as the widely used contrastive loss [62] and triplet loss [63] functions. For contrastive loss, only one positive data point and one negative data point are used in the loss function, but this may result in less efficiency than using both. For triplet loss, one formula handles both, reducing the distance between data for the same class and increasing the distance between data for different classes. However, this approach can reduce the learning ability when compared with methods that handle these tasks separately and then combine the results. Therefore, considering this information, we propose a new loss function by combining the two loss functions.

The first component of the proposed loss function reduces the Euclidean distance between the features extracted from the anchor data and the positive data to zero. The second component changes the Euclidean distance between the features extracted from the anchor data and the negative data to a value called the margin. The final loss function was completed by adding three cross-entropy losses to the proposed loss function. The three cross-entropy losses were obtained from the prediction scores of the anchor, positive, and negative data and the ground truth for each data point. The final loss function was obtained using the following equations:

$$L_{CE} = -\sum_{c=1}^{C} t_c \log(s_c), \tag{7}$$

where C is the number of classes, t_c indicates the ground truth of class c, and s_c is the prediction score of class c.

$$L_{MSE}(f_1, f_2) = \frac{1}{N}\sum_{i=1}^{N}(f_1^i - f_2^i)^2, \tag{8}$$

$$\begin{aligned} L_{final} &= L_{CE-anchor} + L_{CE-pos} + L_{CE-neg} \\ &+ L_{MSE}(f_{anchor}, f_{pos}) \\ &+ \max(0, m - L_{MSE}(f_{anchor}, f_{neg})), \end{aligned} \tag{9}$$

where N denotes the feature dimension, f^i is the ith element of the feature, and m represents the margin. Furthermore, t_c is one when the ground truth of a data point is class c and zero for the rest, and L_{CE-x} is the cross-entropy loss of x data.

Positive and negative data input into the final loss function were selected considering the characteristics of the constructed database. The positive data were selected to have the same class as the anchor data, with the selected subject being different from the anchor data. The negative data were selected to be a different class from the anchor data, with the selected subject being the same as the anchor data. The proposed method was learned end-to-end using this newly proposed loss function.

5. Experimental Results

This section explains the experiment we conducted. First, the experimental setting and dataset are described. Second, the results of the ablation study experiment performed to design the proposed method are presented. Finally, the results of the performance comparison experiment between the proposed method and other methods are explained and analyzed.

5.1. Experimental Setting

PyTorch, a deep-learning library, was used to implement the proposed method. We divided the training set and the testing set using a five-fold cross-validation method to evaluate the performance. When training, the parameters were set as follows. First, in the final loss function (9), the margin was set to 2, and for the optimizer, a stochastic gradient descent optimizer was used. The momentum was set to 0.9, and the weight decay was set to 0.0001. The training epoch was set to 45, and the initial value for the learning rate was set to 0.001 and decreased by 0.1 every 15 epochs. The batch size was set to maximize the GPU memory and set to 6 in the proposed method. We divided the data into a training set, a validation set, and a testing set in a ratio of 3:1:1, and the best hyperparameter set was determined by conducting experiments with various hyperparameter combinations for the validation set. During the division of the data, it was ensured that a subject's data belonged to only one set, since if the same subject's image were to be included in both the training and test sets, the subject's appearance could be learned and the performance could hence be abnormally high.

In the experiments, the performance comparison between the methods used accuracy values obtained by dividing the number of correctly predicted clips in the testing set by the number of all clips in the testing set. As we used the five-fold cross-validation method, we used the average of five accuracy values from five testing sets.

5.2. Dataset

We used the Yonsei Stress Image Database previously described in Section 3 to evaluate the stress recognition performance. A total of 42,023 clips were created by dividing 2,020,556 images of 50 subjects into 48 consecutive frames, and the clips were used as the input for training and testing. The reason for defining a clip as 48 consecutive frames, i.e., two seconds in length, is as follows. In university labs, GPUs with 11GB of memory are often used. When learning the proposed method using this GPU, if 48 frames are input to the GPU, the maximum batch size is 6. If the batch size is too small, the performance deteriorates, so we could use up to 48 frames at once. Additionally, we conducted an ablation study (described in Section 5.3.4) to investigate the variation of the performance with the clip length. To match the experimental conditions as much as possible, 48 frames were randomly selected and used for clips longer than 2 s. In the experimental results, the 2 s clip showed the highest performance, so we used the 2 s clip as a training and test unit.

The facial images were cropped from the original images and input into the network. Examples of the facial images are depicted in Figure 6. For four randomly selected subjects, the various facial expressions displayed by them are presented for each situation.

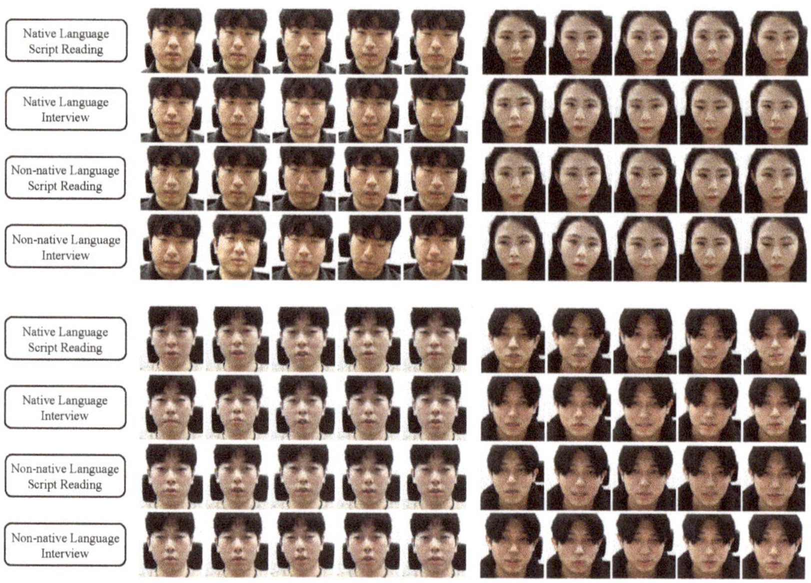

Figure 6. Samples of cropped facial images from the constructed database.

5.3. Ablation Study

In this subsection, we describe the settings and results of the experiments conducted to select the structure of the proposed method. We also present the results of the experiments and examine the effect of the clip settings.

5.3.1. Loss Function

First, an experiment was conducted to determine the loss function that most effectively improved the learning. The proposed loss function was designed with reference to the contrastive loss [62] and triplet loss [63] to ensure effective learning considering the characteristics of these databases. The performance was compared with these functions to determine whether the proposed loss function was effective. The results are listed in Table 5.

Table 5. Comparison of different loss functions.

Method	Accuracy (%)
ResNet-18 + Cross-Entropy Loss	60.0895
ResNet-18 + Cross-Entropy Loss + Contrastive Loss	63.1357
ResNet-18 + Cross-Entropy Loss + Triplet Loss	62.8771
ResNet-18 + Cross-Entropy Loss + Proposed Loss	64.1865

As depicted in the experimental results, the best performance was achieved when the cross-entropy loss and proposed loss were used together. When learning using the proposed loss function, the distance between the data from the same class was reduced, and the distance between the data from different classes was increased when compared with using other loss functions.

5.3.2. Attention Module

With several types of attention modules available, we experimented to determine the best combination by fusing several attention modules. The attention modules used in the experiment are common: the spatial attention module, channel attention module, and temporal attention module. The spatial and channel attention modules were proposed by Woo et al. [47], and the temporal attention module was a modified version of that proposed by Meng et al. [49]. Table 6 presents the experimental results for various combinations of attention modules.

Table 6. Comparison of various combinations of attention modules.

Method	Accuracy (%)
ResNet-18	64.1865
ResNet-18 + Spatial Att	64.7608
ResNet-18 + Channel Att	65.1097
ResNet-18 + Temporal Att	65.2569
ResNet-18 + Channel Att + Temporal Att	64.3173
ResNet-18 + Spatial Att + Temporal Att	65.3396
ResNet-18 + Spatial Att + Channel Att	64.4165
ResNet-18 + Spatial Att + Channel Att + Temporal Att	64.8969

The experimental results demonstrated that the highest performance occurred when the spatial attention and temporal attention modules were both used. Accordingly, finding a channel with a high correlation to stress on the feature maps did not significantly affect the performance, whereas finding a location and frame with a high correlation to stress significantly affected the performance.

5.3.3. Facial Landmark Feature Module

Furthermore, 68 facial landmarks were imaged and entered into the network to extract facial landmark features, and an experiment was conducted to determine the best method for processing and inputting these facial landmark images. As the results of the face detector and facial landmark detector illustrated a jittering pattern, we examined the extent to which the stress recognition performance was affected when this phenomenon was prevented by applying min–max normalization and Gaussian blurring to the facial landmark images. The experimental results are listed in Table 7.

Table 7. Comparison of facial landmark feature extraction methods.

Method	Accuracy (%)
ResNet-18 + Att	65.3396
ResNet-18 + Att + Landmark Image	63.0085
ResNet-18 + Att + Landmark Image + Norm	64.0012
ResNet-18 + Att + Landmark Image + Blur	66.1854
ResNet-18 + Att + Landmark Image + Norm + Blur	66.8409

Att: spatial and temporal attention modules, Norm: min–max normalization, Blur: Gaussian blurring.

The experimental results demonstrated that the performance decreased when only the landmark image was used or only min–max normalization was applied. However, when min–max normalization and Gaussian blurring were both applied to the landmark images, the performance increased. Thus, when both min–max normalization and Gaussian blurring were used, the jittering phenomenon was prevented.

5.3.4. Clip Length and Number of Frames

Finally, we analyzed the impact of the proposed method on the performance by varying the clip length and number of frames. First, we experimented by changing the clip length, which is a unit used in training and testing, to 1 s, 2 s, 5 s, 10 s, and 30 s; the results are listed in Table 8. To match the experimental conditions as much as possible, we used 24 frames for 1 s, and 48 frames were used in the remaining experiments.

Table 8. Effect of clip length on the performance.

Method	Clip Length (s)	Accuracy (%)
Ours	1	65.9470
	2	66.8409
	5	65.6555
	10	65.8282
	30	65.6207

The experimental results demonstrated that the best performance occurred when the clip length was 2 s. It was possible to identify the cues that indicated stress in 2 s clips, and the temporal change was learned well using 48 consecutive frames. In contrast, we randomly selected 48 frames for clips longer than 2 s and used them for training and testing; hence, the discontinuity between frames could have an adverse effect on learning the temporal change. Next, we experimented by changing the number of frames constituting one clip to 8, 16, 32, 48, and 64, and the results were the same as in Table 9. To match the experimental conditions as much as possible, we used 2.7 s clips for 64 frames, while the other experiments used 2 s clips.

Table 9. Effect on the performance of the number of frames.

Method	Number of Frames	Accuracy (%)
Ours	8	65.0138
	16	64.8687
	32	66.1900
	48	66.8409
	64	64.4527

The experimental results demonstrated that the highest performance was achieved when 48 frames were used. This setting exhibited the highest performance when all 48 frames of the 2 s clips were used because it is necessary to find the overall spatial and temporal facial changes when recognizing stress. In contrast, when the clip length exceeded

2 s, recognition was hampered by the increased amount of unnecessary information, as in the above experiment.

5.4. Comparison with Other Methods

We evaluated the stress recognition performance of the proposed method, as well as various other methods. We compared the proposed method with widely used deep-learning networks that have demonstrated high performance [46,47,57,64–67]. The HOG–SVM method, which combines the widely used handcrafted features, HOG [68], and the classical classifier SVM [69], was used for comparison. In addition, current deep-learning-based recognition methods [41–43,45] using spatial–temporal facial information were also used for performance comparison. These methods were used because an emotion recognition network could be considered similar to a stress recognition network.

The experimental results of the proposed method and other methods are listed in Table 10, along with each method's feature dimension. In general, a higher feature dimension indicates a higher discriminating power, but because the computational complexity increases, lower feature dimensions that exhibit high performance are preferable.

Table 10. Stress recognition accuracy, sensitivity, and specificity on the constructed database.

Method	Feature Dimension	Accuracy (%)	Sensitivity (%)	Specificity (%)
HOG-SVM [68,69]	1764	50.9153	50.4360	64.3488
VGG-16 [65]	2048	56.9125	56.4093	71.0178
CBAM-ResNet-18 [47]	512	58.8559	58.1435	72.4161
ResNet-50 [57]	2048	60.0093	59.4649	74.2789
ResNet-18 [57]	512	60.0895	59.4573	74.4877
Inception v3 [66]	2048	63.4185	62.8578	77.6015
AlexNet [64]	4096	64.1588	63.4871	78.3340
DenseNet-121 [67]	1024	64.9408	64.4349	78.2179
SE-ResNet-18 [46]	512	65.7013	65.1206	79.2945
2D-CNN + LSTM + Facial Landmark [42]	768	58.3432	57.7521	72.5148
3D-CNN + Facial Landmark Image [41]	4096	62.5361	62.1877	76.1710
2D-CNN + GRU + Multimodel [43]	512	65.8770	65.3907	79.3543
3D-CNN + Hyperparameter Optimize [45]	4096	65.9372	65.4369	79.7895
Zhang et al. [30]	47104	64.6481	64.0199	78.3209
Ours (w/o Facial Landmark Feature)	512	65.3396	64.5928	78.9639
Ours	768	66.8409	66.1292	80.0959

As depicted in the experimental results, the proposed method had the highest accuracy, 66.8409%, even though features with a relatively low dimensional number of 768 were used. Even when the facial landmark feature was not used, it exhibited an accuracy of 65.3396% with a small 512-dimensional feature. SE-ResNet-18 had the highest performance, at 65.7013%, among the widely used deep-learning networks. This network uses attention modules, which seems to have a positive effect on the stress recognition performance.

By contrast, VGG-16 and ResNet-50 exhibited low performance despite using a relatively high number of feature dimensions, i.e., 2048. This result demonstrates that these methods have a network structure that is unsuitable for stress recognition. The HOG–SVM method used a relatively high number of feature dimensions, i.e., 1764, but exhibited the lowest performance, i.e., 50.9153%. Thus, it was demonstrated that the discriminating power of the handcrafted features was lower than that of the deep-learning networks.

Examining the results of methods using spatial–temporal facial information, the method using the 2D-CNN, LSTM, and facial landmarks demonstrated low performance, i.e., 58.3432%. This result indicates that the facial landmark information was not utilized satisfactorily because the coordinates of the facial landmarks were simply input into the network. Furthermore, the method using the 3D-CNN with hyperparameter optimization exhibited high performance at 65.9372%. Thus, even a simple network can exhibit high performance through appropriate hyperparameter optimization.

We also compared the performance with the method using a physiological signal database [13]. It can be seen that the performance of that method was higher than ours at 74.1%. However, unlike our method, which classified three stress states, this method classified two stress states. In addition, since this method uses physiological signal data, a direct comparison with our method is not possible. Therefore, our approach, which showed the highest performance when there were three stress states, was quite competitive as it offered finer distinctions. Furthermore, as mentioned before, our method does not require biosensors and has the advantage of being able to be used for more diverse applications using images.

Furthermore, we compared the performance with the previous video-based stress-recognition method [30]. The performance of the method was high at 64.6481%, but the performance was lower than that of our proposed method. Therefore, the experimental results in Table 10 show that the performance of the proposed method was higher than that of the other methods. These results indicate that the proposed method is superior to other methods in detecting the overall spatial and temporal changes of the face.

We present the sensitivity and specificity rates along with classification accuracy in Table 10. It can be confirmed that the proposed method showed the best performance in both sensitivity and specificity, as well as accuracy.

We also output the feature maps and attention map obtained from the spatial attention module, and the results are shown in Figure 7. In the case of the attention map, it can be seen that a higher weight was assigned to the lower part of the face. However, in the case of the feature map, it can be seen that it is difficult to identify which features have been learned because the resolution was as low as 4×4. Therefore, we drew a picture of the Grad-cam [70], which shows which part of the image was mainly viewed and determined the prediction. We drew the Grad-cam results for the facial image, as well as the facial landmark image, and the results are shown in Figure 7. As can be seen from Figure 7, the network predicted the stress level primarily by considering the areas around the eyes and mouth.

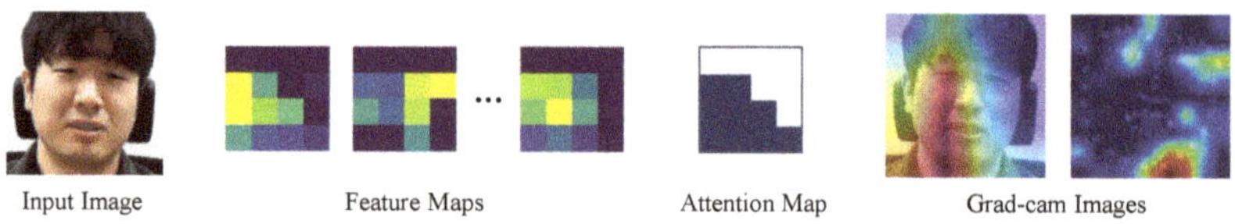

Figure 7. Feature maps, attention map, and Grad-cam images output from an example facial image.

In addition, temporal attention weights were visualized to check whether temporal attention was well applied, and the result is as shown in Figure 8. In the neutral state, the change rate of the weight was not large; however, in the stressed state, the change rate was large. It can be seen that the weight was higher for images in which the change in facial expression was large. This showed that the temporal attention module was working properly.

The classification accuracy for each of the proposed method's classes is listed in Table 11. When the facial landmark feature was used, the proposed method demonstrated higher performance for all three classes than when it was not used. This result implies that the facial landmark feature effectively complements the facial image feature. However, even if the facial landmark feature is used in the proposed method, its classification of the neutral state was superior to its classification of the stress states. Thus, it is challenging to find overall spatial and temporal facial changes that appear when people are under stress. Especially under low stress, the changes are smaller, so they are more difficult to pinpoint. We also output the confusion matrix of the proposed method without and with the facial landmark feature, and the results are shown in Figure 9. Figure 9 shows that the overall performance improved when the facial landmark feature was used compared with not using it.

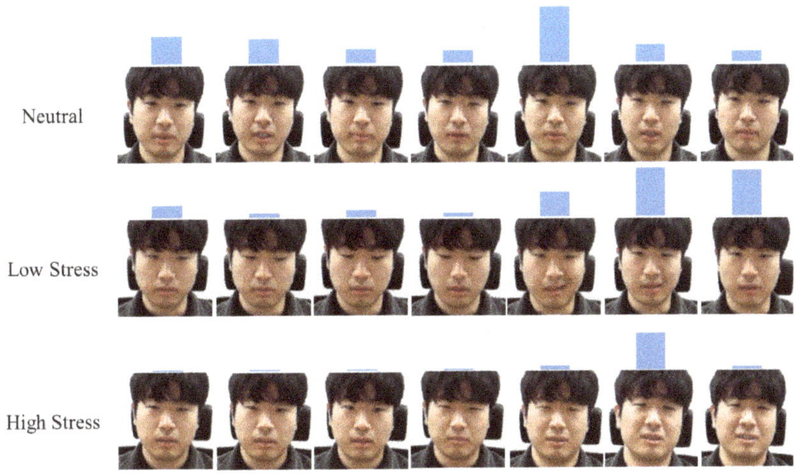

Figure 8. Visualization of the temporal attention weight in three stress states. The higher the height of the bar on the image, the greater the weight is.

Table 11. The proposed method's classification accuracy for each stress state with and without the facial landmark feature.

Stress State	Accuracy (%)	
	Ours (w/o Facial Landmark Feature)	Ours
Neutral	79.9396	80.5567
Low Stress	49.4030	51.5811
High Stress	64.4358	66.2499

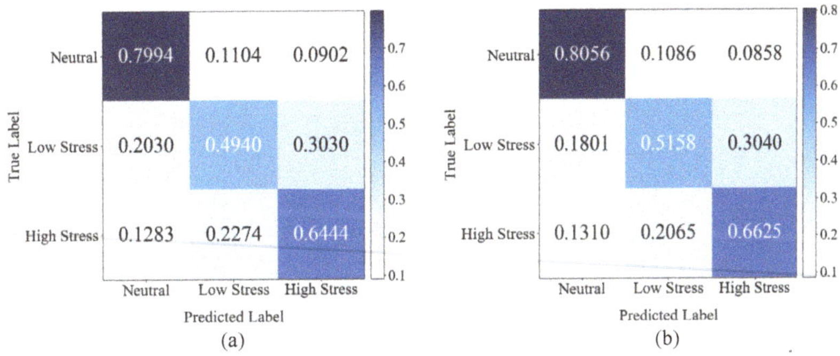

Figure 9. Confusion matrix of the proposed method (**a**) without and (**b**) with the facial landmark feature.

We plotted a histogram of the accuracy of each subject in the proposed method, as shown in Figure 10. The histogram shows how the average performance of the three classes is distributed for all subjects. More specifically, five subjects with an accuracy of 30%∼40% means that the number of subjects with an average performance of three classes between 30% and 40% is five. The interval with the largest number of subjects was between 60% and 70%, and the average performance of the three classes in our method from Table 11 also involved this interval.

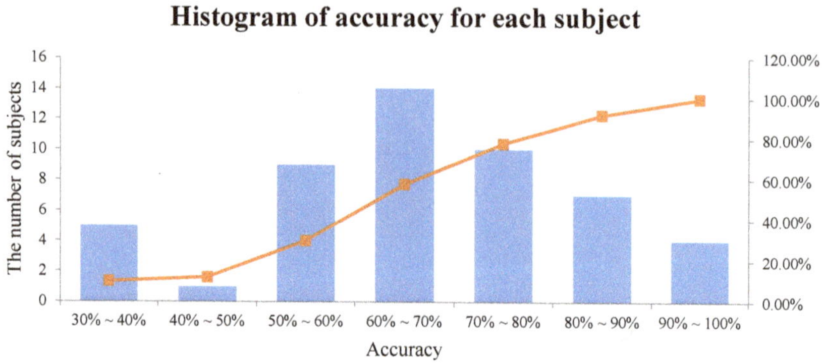

Figure 10. Histogram of the accuracy for each subject in the proposed method.

To evaluate the performance of the video unit, we performed classification by dividing all 5 min and 10 min videos into 2 s clips. For each subject, there were two 5 min videos for the neutral class and one 10 min video for the low- and high-stress classes. If the ratio of correctly classified clips was greater than the threshold, the video was counted as correctly classified and the accuracy was measured. The video unit performance of the proposed method is shown in Table 12, and it is possible to grasp the trend of the performance change according to the threshold change. Since the accuracy was calculated using the results of Table 10 learned by the cross-validation method, the cross-validation method was also applied to these results. If the threshold was set to 50%, the video unit performance was better than the 2 s clip unit performance. For the three classes, the threshold value of 50% can be seen as a reasonable value.

Table 12. Video-based stress recognition accuracy in the proposed method obtained by changing the threshold.

	Accuracy (%)		
Threshold	**40%**	**50%**	**60%**
Neutral	84.0000	79.0000	77.0000
Low Stress	58.0000	56.0000	52.0000
High Stress	79.5918	73.4694	67.3469
Total	73.8639	69.4898	65.4490

6. Conclusions

In this paper, a stress-recognition method using spatial–temporal facial information was proposed using deep learning. To use deep learning technology, we built and released a large image database for stress recognition. In the proposed method, we used a spatial attention module that assigns a high weight to the stress-related regions of the facial image. Using a temporal attention module that assigns a high weight to frames that are highly related to stress from among several frames in the video, we improved the feature's discriminating power. Furthermore, using features extracted from the facial landmark information, we supplemented the discriminating power of the feature extracted from the facial image.

We designed the loss function so that the network learning proceeds effectively, considering the characteristics of the constructed database. We evaluated the proposed method on our constructed database, and it exhibited higher performance than existing deep-learning-based recognition methods. However, our approach has a limitation in that it would find it difficult to recognize stress in people who do not display much change in their facial expressions. In the future, to mitigate this limitation, a study on stress recognition based on multimodal data will be conducted using voice data, which is closely related to stress, along with the images. In addition, research in more difficult environments such as occlusion on the face will be conducted as future work.

Author Contributions: T.J. developed the methodology, led the entire research including the evaluations, and wrote and revised the manuscript. H.B.B., Y.L. and S.J. designed the experiments and analyzed the results. S.L. guided the research direction and verified the research results. All authors have read and agreed to the published version of the manuscript.

Funding: This work was supported by Institute of Information & communications Technology Planning & Evaluation (IITP) grant funded by the Korea government, Ministry of Science and ICT (MSIT) (No.2016-0-00197, Development of the high-precision natural 3D view generation technology using smart-car multi sensors and deep learning).

Institutional Review Board Statement: The study was conducted according to the guidelines of the Declaration of Helsinki and approved by the Institutional Review Board of Yonsei University (date of approval: 26 September 2019).

Informed Consent Statement: Informed consent was obtained from all subjects involved in the study.

Data Availability Statement: The data presented in this study are openly available on IEEE DataPort at https://dx.doi.org/10.21227/17r7-db23 (accessed on 8 November 2021).

Acknowledgments: We would like to express our gratitude to the Korea government, Ministry of Science and ICT (MSIT) and Yonsei University Office of Research Affairs.

Conflicts of Interest: The authors declare no conflict of interest.

Appendix A

Table A1. Summary of the database construction settings and database contents.

Item	Description
Number of Subjects	50
Age of Subjects	20–39 y
Gender Ratio of Subjects	1:1
Nationality of Subjects	Korea
Number of Experimental Stages	4
Number of Stress States	3
Number of Videos per Subject	4
Camera Used for Recording	Kinect v2
Image Resolution	1920 × 1080
Data Acquisition Rate	24 frames/s
Total Number of Images	2,020,556
Total Length of Recorded Videos	1403 min
Illumination	Keep the lights constantly bright
Background	Only clean, white walls
Head Orientation	Almost straight ahead
Occlusion	Hair or accessories do not cover the face (excluding glasses)

References

1. Wainwright, D.; Calnan, M. *Work Stress: The Making of a Modern Epidemic;* McGraw-Hill Education (UK): London, UK, 2002.
2. Selye, H. *The Stress of Life;*Mc Gran-Hill Book Company Inc.: New York, NY, USA, 1956.
3. McEwen, B.S.; Stellar, E. Stress and the individual: Mechanisms leading to disease. *Arch. Intern. Med.* **1993**, *153*, 2093–2101. [CrossRef]
4. Segerstrom, S.C.; Miller, G.E. Psychological stress and the human immune system: A meta-analytic study of 30 years of inquiry. *Psychol. Bull.* **2004**, *130*, 601. [CrossRef] [PubMed]
5. Costa, J.; Adams, A.T.; Jung, M.F.; Guimbretière, F.; Choudhury, T. EmotionCheck: Leveraging bodily signals and false feedback to regulate our emotions. In Proceedings of the 2016 ACM International Joint Conference on Pervasive and Ubiquitous Computing, Heidelberg, Germany, 12–16 September 2016; pp. 758–769.

6. Akmandor, A.O.; Jha, N.K. Keep the stress away with SoDA: Stress detection and alleviation system. *IEEE Trans. Multi-Scale Comput. Syst.* **2017**, *3*, 269–282. [CrossRef]
7. Hollis, V.; Konrad, A.; Springer, A.; Antoun, M.; Antoun, C.; Martin, R.; Whittaker, S. What does all this data mean for my future mood? Actionable analytics and targeted reflection for emotional well-being. *Hum. Comput. Interact.* **2017**, *32*, 208–267. [CrossRef]
8. Chui, K.T.; Lytras, M.D.; Liu, R.W. A generic design of driver drowsiness and stress recognition using MOGA optimized deep MKL-SVM. *Sensors* **2020**, *20*, 1474. [CrossRef] [PubMed]
9. Gao, H.; Yüce, A.; Thiran, J.P. Detecting emotional stress from facial expressions for driving safety. In Proceedings of the 2014 IEEE International Conference on Image Processing (ICIP), Paris, France, 27–30 October 2014; pp. 5961–5965.
10. Can, Y.S.; Arnrich, B.; Ersoy, C. Stress detection in daily life scenarios using smart phones and wearable sensors: A survey. *J. Biomed. Inform.* **2019**, *92*, 103139. [CrossRef] [PubMed]
11. Cho, H.M.; Park, H.; Dong, S.Y.; Youn, I. Ambulatory and laboratory stress detection based on raw electrocardiogram signals using a convolutional neural network. *Sensors* **2019**, *19*, 4408. [CrossRef]
12. Akbar, F.; Mark, G.; Pavlidis, I.; Gutierrez-Osuna, R. An empirical study comparing unobtrusive physiological sensors for stress detection in computer work. *Sensors* **2019**, *19*, 3766. [CrossRef]
13. Siirtola, P.; Röning, J. Comparison of regression and classification models for user-independent and personal stress detection. *Sensors* **2020**, *20*, 4402. [CrossRef]
14. Can, Y.S.; Chalabianloo, N.; Ekiz, D.; Ersoy, C. Continuous stress detection using wearable sensors in real life: Algorithmic programming contest case study. *Sensors* **2019**, *19*, 1849. [CrossRef]
15. Chen, J.; Abbod, M.; Shieh, J.S. Pain and stress detection using wearable sensors and devices—A review. *Sensors* **2021**, *21*, 1030. [CrossRef] [PubMed]
16. Affanni, A. Wireless sensors system for stress detection by means of ECG and EDA acquisition. *Sensors* **2020**, *20*, 2026. [CrossRef] [PubMed]
17. Zhang, B.; Morère, Y.; Sieler, L.; Langlet, C.; Bolmont, B.; Bourhis, G. Reaction time and physiological signals for stress recognition. *Biomed. Signal Process. Control* **2017**, *38*, 100–107. [CrossRef]
18. Peternel, K.; Pogačnik, M.; Tavčar, R.; Kos, A. A presence-based context-aware chronic stress recognition system. *Sensors* **2012**, *12*, 15888–15906. [CrossRef] [PubMed]
19. Vildjiounaite, E.; Kallio, J.; Kyllönen, V.; Nieminen, M.; Määttänen, I.; Lindholm, M.; Mäntyjärvi, J.; Gimel'farb, G. Unobtrusive stress detection on the basis of smartphone usage data. *Pers. Ubiquitous Comput.* **2018**, *22*, 671–688. [CrossRef]
20. Fukazawa, Y.; Ito, T.; Okimura, T.; Yamashita, Y.; Maeda, T.; Ota, J. Predicting anxiety state using smartphone-based passive sensing. *J. Biomed. Inform.* **2019**, *93*, 103151. [CrossRef]
21. Sysoev, M.; Kos, A.; Pogačnik, M. Noninvasive stress recognition considering the current activity. *Pers. Ubiquitous Comput.* **2015**, *19*, 1045–1052. [CrossRef]
22. Chen, T.; Yuen, P.; Richardson, M.; Liu, G.; She, Z. Detection of psychological stress using a hyperspectral imaging technique. *IEEE Trans. Affect. Comput.* **2014**, *5*, 391–405. [CrossRef]
23. Aigrain, J.; Spodenkiewicz, M.; Dubuiss, S.; Detyniecki, M.; Cohen, D.; Chetouani, M. Multimodal stress detection from multiple assessments. *IEEE Trans. Affect. Comput.* **2016**, *9*, 491–506. [CrossRef]
24. Baltacı, S.; Gökçay, D. Role of pupil dilation and facial temperature features in stress detection. In Proceedings of the 2014 22nd Signal Processing and Communications Applications Conference (SIU), Trabzon, Turkey, 23–25 April 2014; pp. 1259–1262.
25. Viegas, C.; Lau, S.H.; Maxion, R.; Hauptmann, A. Towards independent stress detection: A dependent model using facial action units. In Proceedings of the 2018 International Conference on Content-Based Multimedia Indexing (CBMI), La Rochelle, France, 4–6 September 2018; pp. 1–6.
26. Prasetio, B.H.; Tamura, H.; Tanno, K. Support Vector Slant Binary Tree Architecture for Facial Stress Recognition Based on Gabor and HOG Feature. In Proceedings of the 2018 International Workshop on Big Data and Information Security (IWBIS), Jakarta, Indonesia, 12–13 May 2018; pp. 63–68.
27. Cho, Y.; Bianchi-Berthouze, N.; Julier, S.J. DeepBreath: Deep learning of breathing patterns for automatic stress recognition using low-cost thermal imaging in unconstrained settings. In Proceedings of the 2017 Seventh International Conference on Affective Computing and Intelligent Interaction (ACII), San Antonio, TX, USA, 23–26 October 2017; pp. 456–463.
28. Feng, S. Dynamic Facial Stress Recognition in Temporal Convolutional Network. In Proceedings of the 26th International Conference on Neural Information Processing (ICONIP), Sydney, NSW, Australia, 12–15 December 2019; pp. 698–706.
29. Prasetio, B.H.; Tamura, H.; Tanno, K. The facial stress recognition based on multi-histogram features and convolutional neural network. In Proceedings of the 2018 IEEE International Conference on Systems, Man, and Cybernetics (SMC), Miyazaki, Japan, 7–10 October 2018; pp. 881–887.
30. Zhang, H.; Feng, L.; Li, N.; Jin, Z.; Cao, L. Video-based stress detection through deep learning. *Sensors* **2020**, *20*, 5552. [CrossRef]
31. Jeon, T.; Bae, H.; Lee, Y.; Jang, S.; Lee, S. Stress Recognition using Face Images and Facial Landmarks. In Proceedings of the 2020 International Conference on Electronics, Information, and Communication (ICEIC), Barcelona, Spain, 19–22 January 2020; pp. 1–3.
32. Giannakakis, G.; Pediaditis, M.; Manousos, D.; Kazantzaki, E.; Chiarugi, F.; Simos, P.G.; Marias, K.; Tsiknakis, M. Stress and anxiety detection using facial cues from videos. *Biomed. Signal Process. Control* **2017**, *31*, 89–101. [CrossRef]

33. Gavrilescu, M.; Vizireanu, N. Predicting depression, anxiety, and stress levels from videos using the facial action coding system. *Sensors* **2019**, *19*, 3693. [CrossRef] [PubMed]
34. Pediaditis, M.; Giannakakis, G.; Chiarugi, F.; Manousos, D.; Pampouchidou, A.; Christinaki, E.; Iatraki, G.; Kazantzaki, E.; Simos, P.G.; Marias, K.; et al. Extraction of facial features as indicators of stress and anxiety. In Proceedings of the 2015 37th Annual International Conference of the IEEE Engineering in Medicine and Biology Society (EMBC), Milano, Italy, 25–29 August 2015; pp. 3711–3714.
35. Mokhayeri, F.; Akbarzadeh-T, M. Mental stress detection based on soft computing techniques. In Proceedings of the 2011 IEEE International Conference on Bioinformatics and Biomedicine, Atlanta, GA, USA, 12–15 November 2011; pp. 430–433.
36. Pampouchidou, A.; Pediaditis, M.; Chiarugi, F.; Marias, K.; Simos, P.; Yang, F.; Meriaudeau, F.; Tsiknakis, M. Automated characterization of mouth activity for stress and anxiety assessment. In Proceedings of the 2016 IEEE International Conference on Imaging Systems and Techniques (IST), Chania, Crete Island, Greece, 4–6 October 2016; pp. 356–361.
37. Giannakakis, G.; Koujan, M.R.; Roussos, A.; Marias, K. Automatic stress detection evaluating models of facial action units. In Proceedings of the 2020 15th IEEE International Conference on Automatic Face and Gesture Recognition (FG 2020),Buenos Aires, Argentina, 16–20 November 2020; pp. 728–733.
38. Yuen, P.; Hong, K.; Chen, T.; Tsitiridis, A.; Kam, F.; Jackman, J.; James, D.; Richardson, M.; Williams, L.; Oxford, W.; et al. Emotional & physical stress detection and classification using thermal imaging technique. In Proceedings of the 3rd International Conference on Imaging for Crime Detection and Prevention (ICDP), London, UK, 3 December 2009; pp. 1–6.
39. Sharma, N.; Dhall, A.; Gedeon, T.; Goecke, R. Thermal spatio-temporal data for stress recognition. *EURASIP J. Image Video Process.* **2014**, *2014*, 28. [CrossRef]
40. Irani, R.; Nasrollahi, K.; Dhall, A.; Moeslund, T.B.; Gedeon, T. Thermal super-pixels for bimodal stress recognition. In Proceedings of the 2016 Sixth International Conference on Image Processing Theory, Tools and Applications (IPTA), Oulu, Finland, 12–15 December 2016; pp. 1–6.
41. Wu, H.; Lu, Z.; Zhang, J.; Li, X.; Zhao, M.; Ding, X. Facial Expression Recognition Based on Multi-Features Cooperative Deep Convolutional Network. *Appl. Sci.* **2021**, *11*, 1428. [CrossRef]
42. Huang, K.; Li, J.; Cheng, S.; Yu, J.; Tian, W.; Zhao, L.; Hu, J.; Chang, C.C. An efficient algorithm of facial expression recognition by tsg-rnn network. In Proceedings of the 26th International Conference on Multimedia Modeling (MMM), Daejeon, South Korea, 5–8 January 2020; pp. 161–174.
43. Kollias, D.; Zafeiriou, S.P. Exploiting multi-cnn features in cnn-rnn based dimensional emotion recognition on the omg in-the-wild dataset. *IEEE Trans. Affect. Comput.* **2020**, *12*, 595–606. [CrossRef]
44. Palestra, G.; Pettinicchio, A.; Del Coco, M.; Carcagnì, P.; Leo, M.; Distante, C. Improved performance in facial expression recognition using 32 geometric features. In Proceedings of the 18th International Conference on Image Analysis and Processing (ICIAP), Genova, Italy, 7–11 September 2015; pp. 518–528.
45. Haddad, J.; Lézoray, O.; Hamel, P. 3D-CNN for Facial Emotion Recognition in Videos. In Proceedings of the 15th International Symposium on Visual Computing (ISVC), San Diego, CA, USA, 5–7 October 2020; pp. 298–309.
46. Hu, J.; Shen, L.; Sun, G. Squeeze-and-excitation networks. In Proceedings of the IEEE Conference on Computer Vision and Pattern Recognition, Salt Lake City, UT, USA, 18–22 June 2018; pp. 7132–7141.
47. Woo, S.; Park, J.; Lee, J.Y.; So Kweon, I. Cbam: Convolutional block attention module. In Proceedings of the European Conference on Computer Vision (ECCV), Munich, Germany, 8–14 September 2018; pp. 3–19.
48. Zhu, X.; Ye, S.; Zhao, L.; Dai, Z. Hybrid attention cascade network for facial expression recognition. *Sensors* **2021**, *21*, 2003. [CrossRef] [PubMed]
49. Meng, D.; Peng, X.; Wang, K.; Qiao, Y. Frame attention networks for facial expression recognition in videos. In Proceedings of the 2019 IEEE International Conference on Image Processing (ICIP), Taipei, Taiwan, 22–25 September 2019; pp. 3866–3870.
50. Schmidt, P.; Reiss, A.; Duerichen, R.; Marberger, C.; Van Laerhoven, K. Introducing wesad, a multimodal dataset for wearable stress and affect detection. In Proceedings of the 20th ACM International Conference on Multimodal Interaction, Boulder, CO, USA, 16–20 October 2018; pp. 400–408.
51. Koldijk, S.; Sappelli, M.; Verberne, S.; Neerincx, M.A.; Kraaij, W. The swell knowledge work dataset for stress and user modeling research. In Proceedings of the 16th International Conference on Multimodal Interaction, Istanbul, Turkey, 12–16 November 2014; pp. 291–298.
52. Dimsdale, J.E.; Stern, M.J.; Dillon, E. The stress interview as a tool for examining physiological reactivity. *Psychosomatic Med.* **1988**, *50*, 64–71. [CrossRef]
53. Johnson, D.T. Effects of interview stress on measure of state and trait anxiety. *J. Abnorm. Psychol.* **1968**, *73*, 245. [CrossRef]
54. Horwitz, E.K. Preliminary evidence for the reliability and validity of a foreign language anxiety scale. *Tesol Q.* **1986**, *20*, 559–562. [CrossRef]
55. Woodrow, L. Anxiety and speaking English as a second language. *RELC J.* **2006**, *37*, 308–328. [CrossRef]
56. Zhang, K.; Zhang, Z.; Li, Z.; Qiao, Y. Joint face detection and alignment using multitask cascaded convolutional networks. *IEEE Signal Process. Lett.* **2016**, *23*, 1499–1503. [CrossRef]
57. He, K.; Zhang, X.; Ren, S.; Sun, J. Deep residual learning for image recognition. In Proceedings of the IEEE Conference on Computer Vision and Pattern Recognition, Las Vegas, NV, USA, 26 June–1 July 2016; pp. 770–778.
58. Lin, M.; Chen, Q.; Yan, S. Network in network. *arXiv* **2013**, arXiv:1312.4400.

59. Chen, L.; Zhang, H.; Xiao, J.; Nie, L.; Shao, J.; Liu, W.; Chua, T.S. Sca-cnn: Spatial and channel-wise attention in convolutional networks for image captioning. In Proceedings of the IEEE Conference on Computer Vision and Pattern Recognition, Honolulu, HI, USA, 21–26 July 2017; pp. 5659–5667.
60. Zhou, B.; Khosla, A.; Lapedriza, A.; Oliva, A.; Torralba, A. Learning deep features for discriminative localization. In Proceedings of the IEEE Conference on Computer Vision and Pattern Recognition, Las Vegas, NV, USA, 26 June–1 July 2016; pp. 2921–2929.
61. Kazemi, V.; Sullivan, J. One millisecond face alignment with an ensemble of regression trees. In Proceedings of the IEEE Conference on Computer Vision and Pattern Recognition, Columbus, OH, USA, 23–28 June 2014; pp. 1867–1874.
62. Hadsell, R.; Chopra, S.; LeCun, Y. Dimensionality reduction by learning an invariant mapping. In Proceedings of the 2006 IEEE Computer Society Conference on Computer Vision and Pattern Recognition (CVPR'06), New York, NY, USA, 17–22 June 2006; Volume 2, pp. 1735–1742.
63. Schroff, F.; Kalenichenko, D.; Philbin, J. Facenet: A unified embedding for face recognition and clustering. In Proceedings of the IEEE Conference on Computer Vision and Pattern Recognition, Boston, MA, USA, 7–12 June 2015; pp. 815–823.
64. Krizhevsky, A.; Sutskever, I.; Hinton, G.E. Imagenet classification with deep convolutional neural networks. *Commun. ACM* **2017**, *60*, 84–90. [CrossRef]
65. Simonyan, K.; Zisserman, A. Very deep convolutional networks for large-scale image recognition. *arXiv* **2014**, arXiv:1409.1556.
66. Szegedy, C.; Vanhoucke, V.; Ioffe, S.; Shlens, J.; Wojna, Z. Rethinking the inception architecture for computer vision. In Proceedings of the IEEE Conference on Computer Vision and Pattern Recognition, Las Vegas, NV, USA, 26 June–1 July 2016; pp. 2818–2826.
67. Huang, G.; Liu, Z.; Van Der Maaten, L.; Weinberger, K.Q. Densely connected convolutional networks. In Proceedings of the IEEE Conference on Computer Vision and Pattern Recognition, Honolulu, HI, USA, 21–26 July 2017; pp. 4700–4708.
68. Dalal, N.; Triggs, B. Histograms of oriented gradients for human detection. In Proceedings of the 2005 IEEE Computer Society Conference on Computer Vision and Pattern Recognition (CVPR'05), San Diego, CA, USA, 20–26 June 2005; Volume 1, pp. 886–893.
69. Suykens, J.A.; Vandewalle, J. Least squares support vector machine classifiers. *Neural Process. Lett.* **1999**, *9*, 293–300. [CrossRef]
70. Selvaraju, R.R.; Cogswell, M.; Das, A.; Vedantam, R.; Parikh, D.; Batra, D. Grad-cam: Visual explanations from deep networks via gradient-based localization. In Proceedings of the IEEE International Conference on Computer Vision, Venice, Italy, 22–29 October 2017; pp. 618–626.

sensors

MDPI

Article

Restoration of Motion Blurred Image by Modified DeblurGAN for Enhancing the Accuracies of Finger-Vein Recognition

Jiho Choi, Jin Seong Hong, Muhammad Owais, Seung Gu Kim and Kang Ryoung Park *

Division of Electronics and Electrical Engineering, Dongguk University, 30 Pildong-ro 1-gil, Jung-gu, Seoul 04620, Korea; choijh1027@dongguk.edu (J.C.); turtle1990@dgu.ac.kr (J.S.H.); owais2018@dongguk.edu (M.O.); ismysg104@dgu.ac.kr (S.G.K.)
* Correspondence: parkgr@dongguk.edu

Abstract: Among many available biometrics identification methods, finger-vein recognition has an advantage that is difficult to counterfeit, as finger veins are located under the skin, and high user convenience as a non-invasive image capturing device is used for recognition. However, blurring can occur when acquiring finger-vein images, and such blur can be mainly categorized into three types. First, skin scattering blur due to light scattering in the skin layer; second, optical blur occurs due to lens focus mismatching; and third, motion blur exists due to finger movements. Blurred images generated in these kinds of blur can significantly reduce finger-vein recognition performance. Therefore, restoration of blurred finger-vein images is necessary. Most of the previous studies have addressed the restoration method of skin scattering blurred images and some of the studies have addressed the restoration method of optically blurred images. However, there has been no research on restoration methods of motion blurred finger-vein images that can occur in actual environments. To address this problem, this study proposes a new method for improving the finger-vein recognition performance by restoring motion blurred finger-vein images using a modified deblur generative adversarial network (modified DeblurGAN). Based on an experiment conducted using two open databases, the Shandong University homologous multi-modal traits (SDUMLA-HMT) finger-vein database and Hong Kong Polytechnic University finger-image database version 1, the proposed method demonstrates outstanding performance that is better than those obtained using state-of-the-art methods.

Keywords: Finger-vein recognition; motion blur image restoration; modified DeblurGAN; CNN

Citation: Choi, J.; Hong, J.S.; Owais, M.; Kim, S.G.; Park, K.R. Restoration of Motion Blurred Image by Modified DeblurGAN for Enhancing the Accuracies of Finger-Vein Recognition. *Sensors* **2021**, *21*, 4635. https://doi.org/10.3390/s21144635

Academic Editor: Loris Nanni

Received: 10 May 2021
Accepted: 3 July 2021
Published: 6 July 2021

1. Introduction

There are several types of measurable human biometrics, including those of voice, face, iris, fingerprint, palm print, and finger-vein recognition. Among these, finger-vein recognition has the following advantages, (1) finger-vein patterns are hidden under the skin. Therefore, they are generally invisible, making them difficult to forge or steal. (2) Non-invasive image capture ensures both convenience and cleanliness and is more suitable for a user. (3) As most people have ten fingers, if an unexpected accident occurs with one finger, the other finger can be used for authentication [1]. However, due to various factors such as light scattering in the skin layer caused by near-infrared (NIR) light, focus mismatch of a camera lens, differences in finger thickness, differences in depth between the surface of the skin and vein, and finger movements, blurring may occur when capturing finger-vein images. Blurred images generated in these kinds of blur can significantly reduce finger-vein recognition performance. Therefore, image restoration through a deblurring method is necessary. Extensive research has been conducted for restoring skin scattering blur that occurs frequently [2–9], and several studies have been conducted on optical blur caused by the difference in the distance from a camera lens to the finger vein and finger thickness [10,11]. Motion blur can occur frequently, due to finger movement. However, no study has been conducted for motion blurred finger-vein image restoration.

Although, during finger-vein image capture, a finger is fixed to the image capturing device to some extent; however, Parkinson's disease, physiologic tremors, dystonia, and excessive stress may cause hand tremors. Due to these reasons, motion blur can occur. Furthermore, with the recent expansion of non-contact devices due to COVID-19, motion blur occurs more in the input image, and the resulting motion blurred image causes severe performance degradation during finger-vein recognition. To solve this problem, the restoration of a motion blurred finger-vein image is essential.

Conventional image restoration methods can be categorized into blind and non-blind deblurring [12]. Early non-blind deblurring methods perform deblurring, assuming that blur kernels are known. Blur kernel is deduced from knowledge of the image formation process (e.g., amount of motion or defocus blur and camera sensor optics), calculated from the test image, or measured through point spread function (PSF) [13]. Using these methods, the original sharp image can be obtained through deconvolution by estimating the blur kernel. However, when a non-blind restoration method is applied, the recognition performance can be reduced if images are acquired from various devices and show difference blurring characteristics in the spatial domain. Moreover, there are limitations to applying non-blinded methods to each case because various types of distortions occur when capturing an image in actual environments. Also, most blur kernels are unknown in actual environments, and it is time-consuming to estimate blur kernels.

Contrary to the non-blind deblurring method, the blind deblurring methods proceed with deblurring, assuming that blur kernels are unknown [12,14,15]. A generative adversarial network (GAN) that combines the blind deblurring method and the training-based method has also been studied to solve the problems arising from non-blinded deblurring [12,14]. GAN is a network that generates an image by finding an optimal filter using weights trained from the training data. Therefore, using GAN has the advantage of being robust even if images have various distortions. Also, there is no need to estimate the blur kernel directly, and restoration can be performed through training. Considering these reasons, we propose a method of performing motion blurred finger-vein image restoration using the newly proposed modified DeblurGAN and a method of performing restored finger-vein image recognition using deep CNN. The main contributions of our paper are as follows:

- This is the first study on motion blur finger-vein image restoration that can occur in actual environments.
- For restoration of motion blur finger-vein image, we propose a modified DeblurGAN. The proposed modified DeblurGAN has differences in comparison with the original DeblurGAN, (1) dropout layer removal, (2) number of trainable parameters reduction by modifying the number of the residual block structure, (3) and uses feature-based perceptual loss in the first residual block.
- Training is conducted by separating the modified DeblurGAN and the deep CNN, therefore, reducing training complexity while improving convergence.
- The modified DeblurGAN, a deep CNN, and a non-uniform motion blurred image database are published in [16] to allow other researchers to perform fair performance evaluations.

This paper is organized as follows: Section 2 provides an overview of the previous studies, and the proposed method is explained in Section 3. In Section 4, comparative experiments and experimental results with analysis are described. Finally, in Section 5, the conclusions of this paper are explained.

2. Related Works

Previous studies on blurred finger-vein image restoration have been conducted on the restoration of skin scattering or optical blur, and studies related to motion blur restoration have not been conducted. Therefore, previous studies were analyzed in terms of finger-vein recognition without blur restoration, with skin scattering blur restoration, and with optical blur restoration. Such methods can be further categorized into handcrafted feature-based and deep-feature-based finger-vein recognition for analysis.

2.1. Finger-Vein Recognition without Blur Restoration

For the handcrafted feature-based finger-vein recognition without blur restoration method, Lee et al. [17] proposed a method for finger-vein recognition by aligning the image using minutia points extracted from the finger-vein region, extracting finger-vein features using a local binary pattern (LBP), and calculating the Hamming distance using the extracted features. Peng et al. [18] applied Gabor filters having eight orientations to the original finger-vein image and extracted the finger-vein pattern by the fusion of the image with the vein pattern highlighted. They proposed a scale-invariant feature transform (SIFT) feature matching method based on the extracted finger-vein patterns. The method proposed in the study of [18] has the advantage that recognition performance is improved when an optimal filter is accurately modeled. However, this method can cause performance degradation when the filter is applied to finger-vein images having multiple characteristics, and since this experiment was conducted in a constraint environment, it is not robust to image variants, such as illumination or misalignment. Moreover, they did not consider the blur that could occur when capturing a finger-vein image.

Deep feature-based methods have been studied to overcome the drawbacks of these handcrafted feature-based methods. Although a deep-learning-based method was not used, Wu et al. [19] performed dimension reduction and feature extraction of a finger-vein image using a principal component analysis (PCA) and a linear discriminant analysis (LDA). They proposed a finger-vein pattern identification method based on a support vector machine (SVM), which used the PCA- and LDA-extracted features. Hong et al. [20] and Kim et al. [21] proposed finger-vein verification methods to distinguish genuine (authentic) matching (matching images of the same class), and imposter matching (matching images of different classes) using the difference image of enrolled and input images as input to a CNN. Qin et al. [22] created vein-pattern maps, calculated the finger-vein feature probability for each pixel, and labeled veins and backgrounds. Subsequently, training was conducted by dividing the original image into an N×N size, and the probability that the final input image was the vein pattern was calculated. Song et al. [23] and Noh et al. [24,25] proposed a shift-matching finger-vein recognition method using a composite image. Qin et al. [26] proposed a finger-vein verification method that combined a CNN and long short-term memory (LSTM). They assigned labels through handcrafted finger-vein image segmentation techniques and extracted finger-vein features using stacked convolutional neural networks and long short-term memory (SCNN-LSTM). Genuine and imposter matching were verified using feature matching between supervised feature encoding and enrollment databases using extracted features. These studies on deep feature-based finger-vein recognition have a limitation that an intensive training process is required, and there is a disadvantage that they did not consider blur that can occur when capturing finger-vein images.

2.2. Finger-Vein Recognition with Skin Scattering Blur Restoration

Lee et al. [2] proposed a method for restoring skin scattering blur by measuring a PSF of a skin scattering blur and using a constrained least squares (CLS) filter. Yang et al. [3,4] performed scattering-removal by calculating light-scattering components of a biological optical model (BOM). Yang et al. [5] performed scattering effects removal from finger-vein images by considering an anisotropic diffusion, and gamma correction (ADAGC), weighted biological optical model (WBOM), Gabor wavelet, non-scattered transmission map (NSTM), and inter-scale multiplication operation. Shi et al. [6] used haze-removal techniques based on Koschmieder's law to remove scattering effects in finger-vein images. Yang et al. [7] used multilayered PSF and BOM to restore blurred images. Furthermore, Yang et al. [8] proposed a scattering-effect removal method using a BOM-based algorithm that measured the scattering component with the transmission map. You et al. [9] designed a bilayer diffusion model to simulate light scattering and measured the parameters of a bilayer diffusion model through blur-Steins unbiased risk estimate (blur-SURE). Image restoration methods were also proposed based on these parameters with the multi-Wiener linear expansion thresholds (SURE-LET). However, these studies have the disadvantages that scattering blur parameters must be accurately estimated, and parameters must be re-estimated when the domain between the image used for estimation and the test image is different.

2.3. Finger-Vein Recognition with Optical Blur Restoration

Lee et al. [10] proposed a blurred finger-vein image restoration method that considers both optical and scattering blur using PSF and CLS filters. They restored blurred finger-vein images by considering both optical blur components and scattering blur components and improved recognition performance. However, this method requires that parameters should be accurately predicted when measuring two PSFs to improve performance, causing extensive processing time. Choi et al. [11] proposed a finger-vein recognition method by restoring the optical blur included in the original finger-vein image based on modified conditional GAN. This method has the advantage that it can be applied to images acquired from various environments but has the disadvantage that it does not consider more complex motion blur that can occur during image acquisition.

As such, most of the previous studies did not focus on motion blur that can occur from the movement of fingers in finger-vein recognition and did not consider the image restoration associated with the motion blur. Therefore, we propose a new method of restoring a motion blurred finger-vein image using the modified DeblurGAN and recognizing the restored image using a deep CNN.

Point spread functions (PSFs) for skin scattering and optically blurred images are completely different from that for motion blurred images [2–11,27,28]. Therefore, the methods developed for skin scattering or optically blurred images cannot be used directly to solve the motion blurring issue. In the case of the handcrafted feature-based method of Table 1, the PSFs for skin scattering or optically blurred images should be replaced by the PSF for motion blurred images with optimal parameters of PSF. In case of deep feature-based method of Table 1, the CNN and GAN models for skin scattering or optically blurred images should be retrained with motion blurred images in addition to the modification of layers or filters of CNN and GAN models.

Table 1. Comparisons of the previous and proposed finger-vein image restoration methods.

Category		Methods	Advantages	Disadvantages
Without considering blur restoration	Handcrafted feature-based	LBP-based feature extraction + Hamming distance [17] Gabor filter + SIFT feature matching [18]	Recognition performance is improved when an optimal filter is accurately modeled	- Performance degradation when the modeled optimal filter is applied to images having different characteristics - Not robust to image variants, such as illumination or misalignment, because the research was conducted in a constrained environment - A blur that may occur when capturing finger-vein images is not considered
	Deep feature-based	PCA + LDA + SVM [19] Difference image + CNN [20,21] Vein-pattern maps + CNN [22] Composite image + shift matching + CNN [23–25] SCNN-LSTM [26]	- No need to directly model an optimal filter - Robust to image variation as various image features are trained	- Requires intensive training process - Not consider a blur that may occur during image capturing
Skin scattering blur restoration	Handcrafted feature-based	PSF + CLS filter [2] BOM [3,4] WBOM + ADAGC + NSTM + Gabor wavelets [5] Haze removal techniques [6] Multilayered PSF + BOM [7] Optical model-based scattering removal [8] Bilayer diffusion model + blur-SURE + multi-Wiener SURE-LET [9]	Performance is significantly improved if scattering blur parameters are accurately estimated	- Scattering blur parameters must be accurately estimated - Parameters must be re-estimated when the domain between the image used for estimation and the test image is different

Table 1. *Cont.*

Category		Methods	Advantages	Disadvantages
Optical-blur restoration	Handcrafted feature-based	PSF for optical blur + PSF for scattering blur + CLS filter [10]	Image restoration considering both optical and skin scattering blur	- Performance can be improved only when the parameters of two PSFs are accurately predicted from the perspective of skin structure and camera optics - Processing time is long because optical blur restoration and skin scattering blur restoration are processed simultaneously
	Deep feature-based	Conditional GAN + CNN [11]	Applicable to images captured from various environments	Did not consider the motion blur
Motion blur restoration	Deep feature-based	Modified DeblurGAN-based method + CNN (Proposed method)	Recognition performance improved after restoration considering a motion blur that may occur when capturing finger-vein images	Networks for restoration and recognition require large data and take a long time to train.

Table 1 presents a comparison of the advantages and disadvantages of the proposed method and the previous studies.

3. Proposed Method

3.1. Overview of the Proposed Method

Figure 1 shows the overall flowchart of the proposed method. After acquiring finger images (step (1)), the finger region of interest (ROI) is detected using preprocessing method (step (2)). Then, the motion blurred finger-vein image is restored using the proposed modified DeblurGAN (step (3)). One difference image is then generated from the restored enrolled and recognized images (step (4)). Lastly, based on the output score obtained by inputting the difference image in the deep CNN, finger-vein recognition is performed to distinguish genuine (authentic) or imposter matching (step (5)).

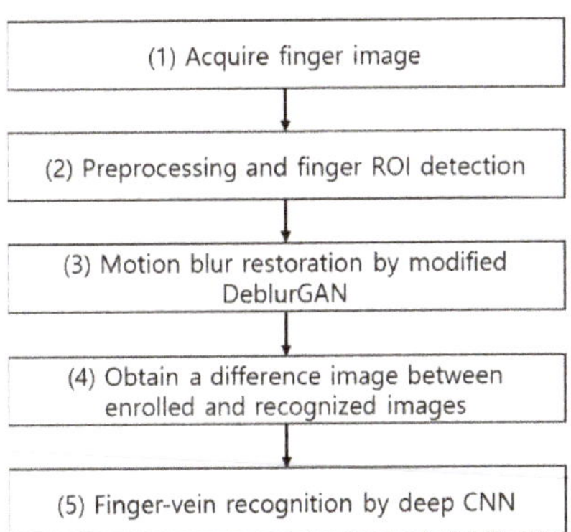

Figure 1. Flowchart of the proposed method.

3.2. Preprocessing the Finger-Vein Image

The first part of preprocessing removes unnecessary background regions and finds the finger-vein ROI. The captured image is then binarized to obtain the image shown in Figure 2b. However, even if binarization is performed, the background is not completely removed, so an edge map is created using a Sobel filter. A difference image is then generated using the created edge map and the binarized image. By applying the area threshold method [29] to the generated difference image, an image with the background removed as shown in Figure 2c is obtained. Then, in order to correct misalignment caused by in-plane rotation of the finger image, which degrades recognition performance, second-order moments of the binarized mask R (Figure 2c), are calculated using Equation (1).

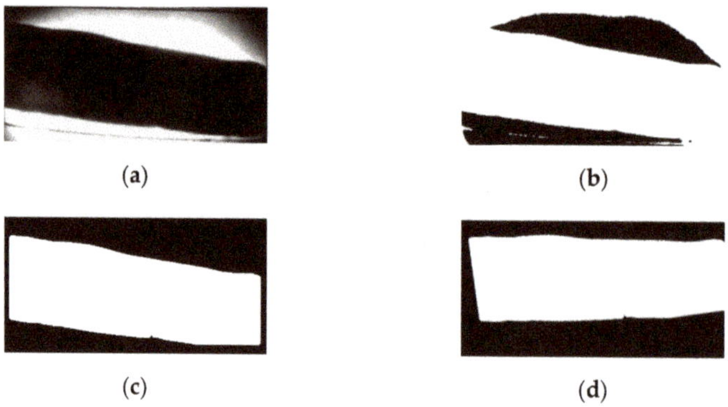

Figure 2. Example of background removal and in-plane rotation compensation: (**a**) original image; (**b**) binarized image; (**c**) background removed image; (**d**) in-plane rotation compensation.

Note that $f(x,y)$ and $(m_x,\ m_y)$ represent image pixel values and central coordinates, respectively. Based on these values, the rotation angle θ, in Equation (2) is calculated to compensate for the in-plane rotation [30]. The compensated image, shown in Figure 2d, is obtained from this process.

$$\begin{aligned} a_{11} &= \frac{\sum_{(x,y)\in R}\left(y-m_y\right)^2\cdot f(x,y)}{\sum_{(x,y)\in R} I(x,\ y)} \\ a_{12} &= \frac{\sum_{(x,y)\in M}\left(x-m_x\right)\left(y-m_y\right)\cdot f(x,y)}{\sum_{(x,y)\in R} I(x,\ y)} \\ a_{22} &= \frac{\sum_{(x,y)\in R}\left(x-m_x\right)^2\cdot f(x,y)}{\sum_{(x,y)\in R} I(x,\ y)} \end{aligned} \tag{1}$$

$$\theta = \begin{cases} tan^{-1}\left\{\dfrac{a_{11}-a_{22}+\sqrt{\left(a_{11}-a_{22}\right)^2+4a_{12}^2}}{-2a_{12}}\right\} if\ a_{11} > a_{22} \\ tan^{-1}\left\{\dfrac{-2a_{12}}{a_{22}-a_{11}+\sqrt{\left(a_{22}-a_{11}\right)^2+4a_{12}^2}}\right\} if\ a_{11} \le a_{22} \end{cases} \tag{2}$$

As shown in Figure 3a, the left and right ends of the finger are the regions of the thick area or region with a fingernail where NIR lighting is not well-transmitted. Thus, these regions are inappropriate for recognition because vein patterns are not likely to be captured accurately. Therefore, the image, shown in Figure 3c, is obtained by removing the left and right sides by a predetermined size to which in-plane rotation compensation is applied. By performing erosion operation, component labeling process, and dilation operation [27], the unnecessary region for finger-vein recognition such as the upper right corner of Figure 3c, is removed. As a result of this process, an image as shown in Figure 3d is created. Since the vein pattern is not acquired by bright illumination, the black area of the finger area is not required for recognition. An ROI mask is obtained by using a 4×20 mask to fill the black area with the average pixel values around it (Figure 3e). In details, as shown in the red-dashed circles of the lower boundary of finger in Figure 3a, there exists bright pixels inside of finger caused by excessive illumination, which causes the error of binarization of lower boundary as shown in Figure 3b–d. Therefore, we applied 4×20 mask to the binarized image of Figure 3d. At each convolution position of mask, the average pixel value within 4×20 area (except for the black pixels of Figure 3d) is assigned to the binarized image of Figure 3d. That is, if the majority pixels within 4×20 area is white (255), white pixel is assigned. Then, the inaccurate black pixels of the red-dashed circles of Figure 3d are replaced by the white pixels of finger region as shown in the lower boundary of Figure 3e.

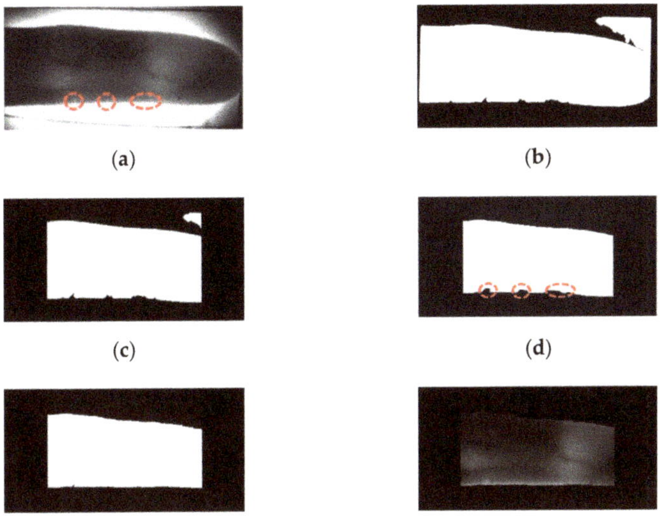

(a) (b) (c) (d) (e) (f)

Figure 3. Extracting finger-vein ROI: (**a**) original finger image, (**b**) in-plane rotation compensation, (**c**) left and right areas removal, (**d**) component labeling, (**e**) ROI mask after filling black area of finger region, and (**f**) obtained ROI image.

The mask size (4 × 20) generalizes on the images of different resolutions. To confirm this, we used two open databases of the Shandong University homologous multi-modal traits (SDUMLA-HMT) finger-vein database [31] and the Hong Kong Polytechnic University finger-image database version 1 [29] in our research.

3.3. Modified DeblurGAN-Based Finger-Vein Image Restoration

The principal objective of enhancement is to process the image so that the result is more suitable than the original image for a specific application [27]. Therefore, although image enhancement is mostly a subjective process, while image restoration is a generally objective process. Because image restoration is an attempt to reconstruct a degraded image using prior knowledge of degradation, the restoration method must focus on applying degradation modeling to restore the original image and the inverse process. The blur model based on the above process can be expressed as follows [28]:

$$g(x,\ y) = h(x,\ y) * f(x,\ y) + \eta(x,\ y) \tag{3}$$

Here, $g(x,\ y)$ is a degraded (blurred) image, $h(x,\ y)$ is a spatial representation of a degradation function (H), $*$ is a convolution operation, $f(x,\ y)$ is an input image, and $\eta(x,\ y)$ is an additive noise. If the above conditions are given, the goal of restoration is to obtain $\hat{f}(x,\ y)$, which is the estimation of an original image. The more accurately $h(x,\ y)$ and $\eta(x,\ y)$ are estimated, $\hat{f}(x,\ y)$ and $f(x,\ y)$ become closer [28]. However, from $g(x,\ y)$, which is the image obtained from various environments, it is extremely difficult to estimate $h(x,\ y)$ and $\eta(x,\ y)$ accurately. Furthermore, when images having different characteristics than those used for estimation are input, the estimated $h(x,\ y)$ and $\eta(x,\ y)$ may sometimes not be applicable. Considering these facts, this study proposes a training-based restoration model, the modified DeblurGAN, and we aim to ensure the restored finger-vein image F^{res}, becomes similar to the original finger-vein image F^{ori}, through training without separately estimating $h(x, y)$ and $\eta(x, y)$ when a motion blurred finger-vein image G^{blur}, is given.

A deblurring task can be generally divided into blind and non-blind deblurring. For the non-blind deblurring method, deblurring is performed assuming that the blur kernel ($h(x, y)$) is known, whereas, for the blind deblurring method, deblurring is performed assuming that the blur kernel is not known [12]. In a general environment, a blind kernel is not known, and it is time-consuming to directly estimate it. In this study, we assume that the blur kernel is unknown, similar to the general environment. Also, it proposes a restoration method applicable for motion blurred finger-vein images obtained from various environments, so this study can be considered a blind deblurring task. Because the original DeblurGAN exhibits good performance in a blind motion-deblurring task [12], we determined that it would be effective in this study as well. Therefore, we propose a modified DeblurGAN. The generator of the modified DeblurGAN used in this study is shown in Figure 4 and Table 2, and the discriminator is shown in Figure 5 and Table 3. A more detailed explanation is provided in the next subsection.

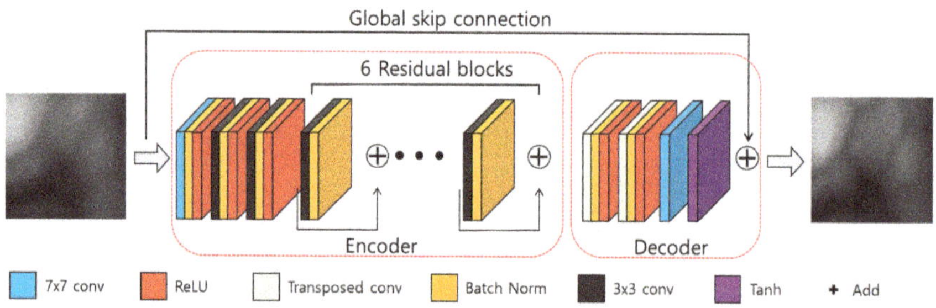

Figure 4. Generator of the modified DeblurGAN.

Table 2. Descriptions of generator in modified DeblurGAN.

	Layer	Number of Filters	Size of Feature Map (Height × Width × Channel)	Size of Kernel (Height × Width × Channel)	Number of Strides (Height × Width)	Number of Paddings (Height × Width)
Encoder	Image input layer		256 × 256 × 3			
	1st convolutional layer Batch normalization ReLU	64	256 × 256 × 64	7 × 7 × 3	1 × 1	3 × 3
	2nd convolutional layer Batch normalization ReLU	128	128 × 128 × 128	3 × 3 × 64	2 × 2	1 × 1
	3rd convolutional layer Batch normalization ReLU	256	64 × 64 × 256	3 × 3 × 128	2 × 2	1 × 1
	Residual Blocks × 6 [3 × 3 conv, Batch normalization]	256	64 × 64 × 256	3 × 3 × 256	1 × 1	1 × 1
Decoder	1st transposed layer Batch normalization ReLU	128	128 × 128 × 128	3 × 3 × 256	2 × 2	
	2nd transposed layer Batch normalization ReLU	64	256 × 256 × 64	3 × 3 × 128	2 × 2	
	4th convolutional layer Batch normalization ReLU	3	256 × 256 × 3	7 × 7 × 64	1 × 1	3 × 3
	Output (input + 4th convolutional layer)		256 × 256 × 3			

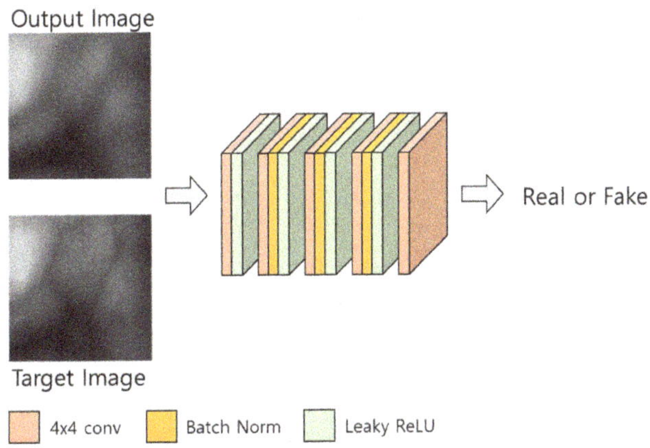

Figure 5. Discriminator of the modified DeblurGAN.

Table 3. Descriptions of the discriminator in modified DeblurGAN (* means the output image or target image of Figure 5).

Layer	Number of Filters	Size of Feature Map (Height × Width × Channel)	Size of Kernel (Height × Width × Channel)	Number of Strides (Height × Width)	Number of Paddings (Height × Width)
* Image input layer		256 × 256 × 3			
1st convolutional layer Leaky ReLU	64	129 × 129 × 64	4 × 4 × 3	2 × 2	2 × 2
2nd convolutional layer Batch normalization Leaky ReLU	128	65 × 65 × 128	4 × 4 × 64	2 × 2	2 × 2
3rd convolutional layer Batch normalization Leaky ReLU	256	33 × 33 × 256	4 × 4 × 128	2 × 2	2 × 2
4th convolutional layer Batch normalization Leaky ReLU	512	34 × 34 × 512	4 × 4 × 256	1 × 1	2 × 2
5th convolutional layer	1	35 × 35 × 1	4 × 4 × 512	1 × 1	2 × 2

3.3.1. Generator

A GAN generally comprises generator and discriminator models in which the adversarial training between the two gradually improves the performance of both. The generator of the original DeblurGAN has one convolution block, two strided convolution blocks with strides of 1/2, nine residual blocks (ResBlocks) [32], and two transposed convolution blocks [12]. Each ResBlock consists of a convolution layer, an instance normalization layer, and a rectified linear unit (ReLU) for activation [33]. Compared with the original DeblurGAN, the following two aspects were modified for this study.

First, a dropout [34] is removed. In the original DeblurGAN, a dropout ratio of 0.5 is applied to each residual block of the generator, and the same ratio is applied for inference. Generally, a dropout is effective as a regularization method for avoiding overfitting, but it can cause the modification of a vein pattern in the restored output image, due to the randomness of a dropout when applied to a restoration task. The modified vein pattern then has different features from the original finger-vein image, which results in degraded performance. Rather than creating a variety of outputs in which the vein pattern is deformed, the generated pattern information needs a deterministic output that is similar to the original as possible, therefore, dropout.

Second, the number of parameters is reduced by modifying the residual blocks. Large parameters can increase the inference time when applied to an actual environment, and increased inference time can cause the inefficiency of the system. The original DeblurGAN used the GoPro [35] and Kohler datasets [36] and applied nine residual blocks to the generator. In this study, the existing nine residual blocks were reduced to six to shorten the inference time by reducing the number of parameters. Also, by modifying the residual blocks as shown in Figure 6, feature information is maintained in the layer prior to the next convolution layer, and the number of parameters is reduced. The width and height of feature map are reduced by passing through convolution layer, which usually causes the reduction of important feature information [32]. Therefore, by comparing Figure 6a,b, the second 3×3, 256 Conv layer is removed in our modified residual block, which can maintain feature information in the layer prior to the next convolution layer. In addition, the number of parameters is reduced by removing the second 3×3, 256 Conv layer in the modified residual block. Consequently, the number of parameters of the generator is reduced from 6.0 to 4.2 million.

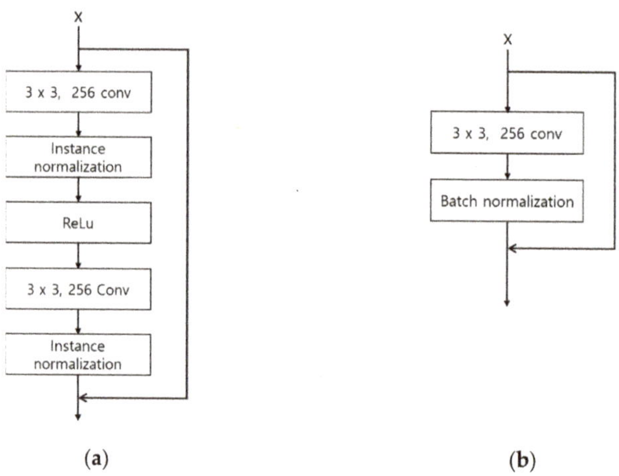

Figure 6. Architectures of original and modified residual blocks: (**a**) residual block in original DeblurGAN; (**b**) a modified residual block in the modified DeblurGAN.

3.3.2. Discriminator

The structure of the discriminator is shown in Figure 5 and Table 3. The discriminator of the modified DeblurGAN proposed in this study has the same structure as the discriminator of the original DeblurGAN, which used the Wasserstein WGAN gradient penalty (GP) [37]. For a GAN, the Nash equilibrium in a non-convex system must be found using continuous and high-dimensional parameters for smooth training, however, the existing GAN [38] cannot solve this problem, therefore, it fails to converge [39]. In the case of DeblurGAN, WGAN-GP is used as a critic function using Wasserstein distance and the gradient penalty methods proposed in [37]. Thus, a structure that is robust to generator structure selection and at the same time enables stable training is proposed. In this study, these advantages of the discriminator of the original DeblurGAN are adopted.

3.3.3. Loss

In the case of the original DeblurGAN, a perceptual loss is applied to perceptually hard to distinguish between the generated image and the real sharp image and to restore finer texture detail [12]. A perceptual loss refers to the difference in feature maps between the generated and target images, which can produce better results than the loss that generates blurry results by calculating the pixel-wise average difference such as L1 or L2 loss. The perceptual loss function used in this study, based on the previous study [12], can be defined as follows:

$$\mathcal{L}_X = \frac{1}{W_\varnothing H_\varnothing} \sum_{x=1}^{W_\varnothing} \sum_{y=1}^{H_\varnothing} \left(\varnothing\left(F^{ori}\right)_{x,y} - \varnothing\left(G_{\theta_G}\left(F^{blur}\right)\right)_{x,y} \right)^2 \quad (4)$$

where $\varnothing$ is the feature maps extracted from the ImageNet pretrained network. For the original DeblurGAN, the feature maps extracted from the third convolution layer before the third max-pooling layer in the visual geometry group (VGG)-19 [40] are used for perceptual loss. $W_\varnothing$ and $H_\varnothing$ are the width and height of feature maps, respectively. In a classification network, such as that of a VGG, abstracted features extracted from a higher layer preserve the overall spatial structure, whereas low-level features, such as color, corner, edge, and texture, cannot be preserved [41,42]. In terms of finger-vein images, it is important to restore the high-level features of restored output image similar to those of the original image, however, restoring low-level features is important as well, because vein patterns and texture are slightly different for each class, and performance can be varied due to differences in low-level features during recognition. Because of these reasons, unlike the original DeblurGAN that applied perceptual loss by extracting feature maps from the middle layer of the ImageNet pretrained VGG-19, in this study, feature sets are extracted from the generated image and target image in the first residual block (conv2_x) using the ImageNet pretrained ResNet-34 [32] model, respectively, and the difference between the two feature sets is applied as a perceptual loss. In a typical neural network, vanishing gradient and explosion occur as the layer gets deeper, eventually resulting in performance degradation. In ResNet, however, this problem is solved by applying a residual learning method. The residual block applying the residual learning method is trained so that identity mapping $F(x) + x$ that is mapping between output $F(x)$ of the weight layer and output x of the layer just before the weight layer, and plain layer output $H(x)$ are the same ($H(x) = F(x) + x$). From the characteristics of the residual block that identity mapping the output information of the previous layer to the next layer, we inferred that low-level features such as color, corner, edge, and texture of the finger-vein can be preserved during restoration training. For this reason, a perceptual loss is applied from the conv2_x layer of ResNet-34 instead of the original VGG-19.

3.3.4. Summarized Differences between Original DeblurGAN and Proposed Modified DeblurGAN

The differences between the original DeblurGAN and the proposed modified DeblurGAN are as follows.

- A dropout is applied to the generator of the original DeblurGAN, whereas a dropout is not applied to the generator of the modified DeblurGAN because the vein patterns of the restored image can be modified. The dropout layer usually helps avoiding overfitting. However, the dropout layer can also bring about the excessive sparsity of activation and features with coarser features compared to the case without the dropout layer [34,43], which can cause the consequent modification of a vein pattern in the restored output image. Therefore, we do not use the dropout layer in the generator of proposed modified DeblurGAN.
- In the original DeblurGAN, nine residual blocks (convolutional layer—normalization layer—activation layer—convolutional layer—normalization layer) were used for the generator. In the modified DeblurGAN, to reduce the inference time, the number of parameters was reduced by reducing the structure of the residual block (convolutional layer-normalization layer) and reducing the total number of residual blocks to six.
- In the original DeblurGAN, high-level feature maps extracted from the third convolution layer prior to the third max-pooling layer of the ImageNet-pretrained VGG-19 were applied to a perceptual loss. However, it is equally important to restore the information of low-level features, such as color, corner, edge, and texture, during

finger-vein restoration. Hence, a perceptual loss was applied to the first residual block (conv2_x) using the ImageNet-pretrained ResNet-34 in the modified DeblurGAN.

3.4. Finger-Vein Recognition by Deep CNN

In this study, the difference image between registered (enrolled) and recognized images was used as an input for CNN-based finger-vein recognition. An image differencing method determines the changes in images where the differences are determined by calculating the pixel differences, and a new image is then created based on the calculation results [44]. Thus, an image differencing method reacts sensitively to the changes in images. For the finger-vein datasets used in this study, if the same class images are used, the pixel difference between the two images is small. So, in general, a pixel value with a low difference image, that is, an image with many black areas is an output. Whereas in the case of other classes, since the pixel difference between the two images is large, the difference image has generally a high pixel value, that is, an image with many bright areas is output. An image differencing method has the advantage of expressing the characteristics of genuine and imposter matching with one output image. Here, genuine matching refers to matching when the input image and the enrolled image are the same class, and imposter matching refers to matching when the input image and the enrolled image are the different class. The finger-vein datasets used in this study have a high similarity of vein patterns between intra-class, but a low similarity between inter-class. Therefore, the finger-vein recognition performance can be verified in the difference image. The examples of finger-vein difference images generated from the dataset used in this study are shown in Figure 7c,f.

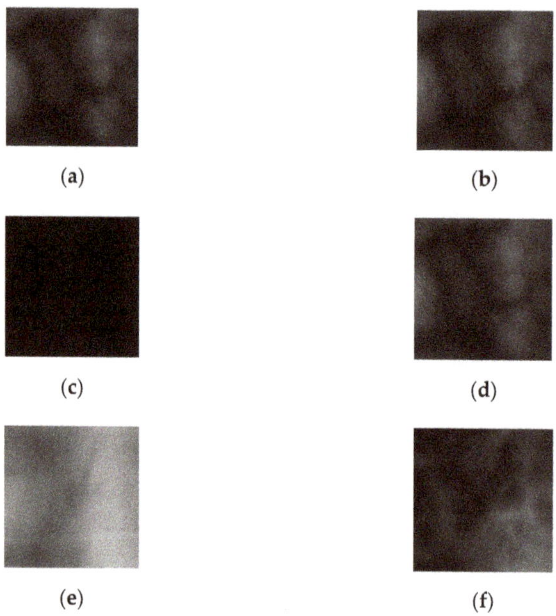

Figure 7. Difference images of registered and input images. (**a**) Registered image, (**b**) input image of same class as registered image, (**c**) difference image of (**a**,**b**), (**d**) registered image, (**e**) input image of different class as registered image, and (**f**) difference image of (**d**,**e**).

The generated difference image is then used as an input to deep CNN. DenseNet-161 [45] is used to recognition of finger-vein images. DenseNet adopts dense connectivity in which the feature maps of a previous layer are concatenated in the current layer.

$$x_l = H_l([x_0, x_1, \ldots, x_{l-1}]) \tag{5}$$

Equation (5) represents dense connectivity, where $[x_0, x_1, \ldots, x_{l-1}]$ means the feature map concatenation from layers 0 to $l-1$. A dense block performs feature map concatenation of the previous and the current layer and transfers the concatenated feature maps to the following layer. H_l is a composite function and is composed of batch normalization [46], ReLU [33], and a convolution layer. Generally, as the network becomes deeper, the number of channels of feature maps caused by dense connectivity increases, resulting in an increased number of network parameters. To mitigate the increasing parameters, a bottleneck layer is added to the dense block of DenseNet. As a result, utilizing the bottleneck structure reduces computational costs. However, the output of a dense block concatenates all layers within the block. As the layer gets deeper or the number of layers in the dense block increases, the size and depths of the feature map increase enormously. To solve this problem, a transition layer was added between the dense blocks to reduce the size and depths of the feature maps. The transition layer cuts the number of feature map depths by half through 1×1 convolutional computation and reduces width and height by half using 2×2 average pooling. In addition, by specifying a growth rate, DenseNet controls the number of output feature map channels. Dense block outputs the feature map at the size of the designated growth rate. In this research, the growth rate is set to 48.

In this study, for finger-vein recognition, the DenseNet-161 pretrained with the ImageNet database [47] is fine-tuned with the finger-vein training data. Difference images are used for the training and testing process, and these images are created using the output restored images by the proposed modified DeblurGAN. The number of output classes of DenseNet-161 is set to 2, genuine matching and imposter matching. The criterion for this is based on the output score obtained from the last layer of the DenseNet. With respect to the threshold of the equal error rate (EER) of genuine and imposter matching distributions of the CNN output score obtained from the training data, it is determined as genuine matching if the CNN output score of the testing data is below the threshold. And imposter matching is determined if the output score is greater than the threshold. The EER is the rate of error at the point where the false rejection rate (FRR) which is the error rate of falsely rejecting genuine matching as an imposter matching and the false acceptance rate (FAR) which is the error rate of falsely accepting imposter matching as genuine matching are equal.

4. Experimental Results

4.1. Two Open Databases for Experiments

In this study, experiments were conducted using two types of open finger-vein databases, SDUMLA-HMT finger-vein database [31] and session 1 images from the Hong Kong Polytechnic University finger-image database version 1 [29]. In SDUMLA-HMT finger-vein database, 6 images from the ring, middle, and index finger from both hands were obtained respectively, from 106 individuals, a total of 3816 images were obtained (2 hands $\times$ 3 fingers $\times$ 6 images from 106 individuals). In session 1 from the Hong Kong Polytechnic University finger-image database version 1, 6 images from the middle and index finger images were obtained respectively, from 156 individuals, a total of 1872 images were obtained (2 fingers $\times$ 6 images from 156 individuals). In this study, the finger-vein database of the SDUMLA-HMT is referred to as SDU-DB, and the session 1 finger-image database version 1 of the Hong Kong Polytechnic University is referred to as PolyU-DB. Figure 8 shows examples from the same finger for PolyU-DB and SDU-DB. The image resolution of SDUMLA-HMT is 320×240 pixels, and that of the Hong Kong Polytechnic University finger-image database is 513×256 pixels.

SDU-DB consists of 636 classes, whereas PolyU-DB consists of 312 classes. All experiments adopted two-fold cross-validation. Through the two-fold cross-validation method, data of the same class were not used for training and testing (open-world setting). The average accuracy measured through two-fold cross-validation was adopted as the final recognition accuracy.

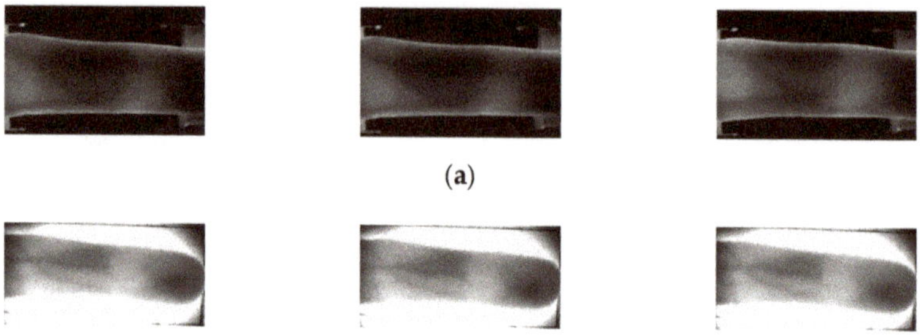

(a)

(b)

Figure 8. Images obtained from the same finger. (**a**) SDU-DB and (**b**) PolyU-DB.

4.2. Motion Blur Datasets for Finger-Vein Image Restoration

In the case of PolyU-DB and SDU-DB, which are open databases used in this study, motion blurred finger-vein datasets were not constructed. Therefore, to proceed with this study, a motion blurred finger-vein database was constructed by applying motion blurring kernels to the two open databases. When constructing the database, non-uniform (random) motion blurring kernels were applied instead of uniform motion blurring kernels to closely resemble the actual environment. For the random motion blurring kernels, the method proposed by Kupyn et al. [12] was used. Figures 9 and 10 show original and generated motion blurred images of SDU-DB and PolyU-DB.

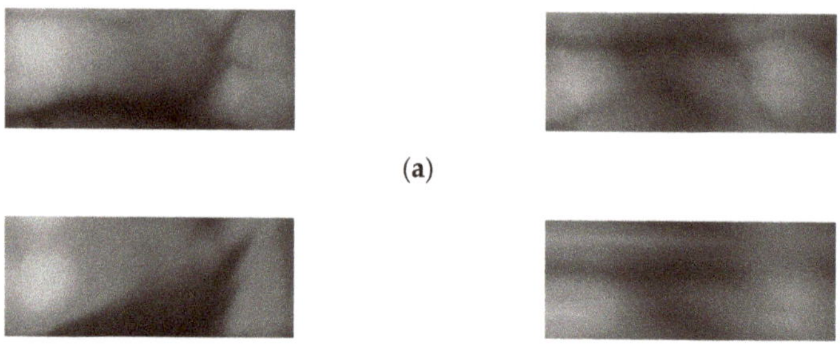

(a)

(b)

Figure 9. Examples of original images and motion blurred images of SDU-DB. (**a**) Original images; (**b**) motion blurred images.

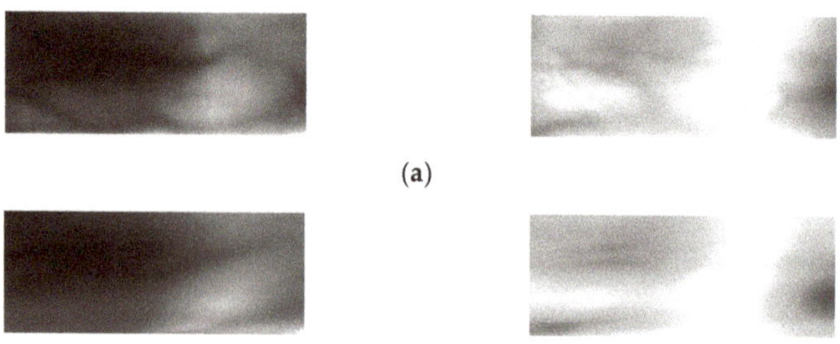

(a)

(b)

Figure 10. Examples of original images and motion blurred images of PolyU-DB. (**a**) Original images; (**b**) motion blurred images.

4.3. Data Augmentation and Experimental Setup

The datasets used in this study do not contain enough images to train a deep CNN, which would result in overfitting. To solve this problem, a data augmentation method was applied to increase the number of training data. For this method, 5 pixel shifting was applied for each image based on 8 directions in a combination of the top, bottom, left, and right. Therefore, each image was increased to 9 times including the original image. Table 4 presents the descriptions of original and augmented data from PolyU-DB and SDU-DB datasets. From the data augmentation, 54 images were generated that increased 9 times from 6 images per class. When training DenseNet-161 for finger-vein recognition, only 1 image among 54 augmented images was selected as an enrolled image, and the other images were used as input images. A difference image was generated using the enrolled image and input image to determine genuine and imposter matching. In the case of SDU-DB, the number of imposter matching was 317 times that of genuine matching, and it was 155 times that of genuine matching for the PolyU-DB. When training this data as it is, a bias on the majority class occurs due to data imbalance. In order to solve this problem, when genuine matching data is augmented with the same number as imposter matching data, training time is increased due to a large number of data, and an overfitting problem for genuine matching data can occur. Therefore, in this study, we applied a random selection method for the imposter matching data. Augmentation and random selection methods were applied to both SDU-DB and PolyU-DB in the same manner, but only to the training data. The original images that were not augmented were used as the testing data.

Table 4. Descriptions of experimental databases by data augmentation.

			SDU-DB	PolyU-DB
Original images		# of images	3816	1872
		# of people	106	156
		# of hands	2	1
		# of fingers	3 (index, middle, and ring fingers)	2 (index and middle fingers)
		# of classes (# of images per class)	636 (6)	312 (6)
Training for 1st or 2nd fold cross validation	Training of modified DeblurGAN	# of images (original + augmented data)	17,172 (6 images × 9 times × 318 classes)	8424 (6 images × 9 times × 156 classes)
	Training of CNN for finger-vein recognition	# of images for genuine matching	16,854 ((6 images × 9 times − 1) × 318 classes)	8268 ((6 images × 9 times − 1) × 156 classes)
		# of images for imposter matching	16,854 (Random selection)	8268 (Random selection)

The training and testing were performed on a desktop computer equipped with NVIDIA GeForce GTX 1070 graphics processing unit (GPU) [48] and Intel® Core™ i7-9700F CPU with 16 GB RAM.

4.4. Training of Modified DeblurGAN Model for Motion Blur Restoration

For the training parameter of modified DeblurGAN, the max epoch was set to 100, the mini-batch size was set to 4, and the learning rate was set to 0.0005. Adaptive moment estimation (Adam) optimization [49] was used for the generator and discriminator to train the modified DeblurGAN. Figures 11a,b and 12a,b show the graphs of training loss of the proposed modified DeblurGAN according to the epoch for SDU-DB and PolyU-DB, respectively. The loss values converged as the training progresses, confirming that the proposed modified DeblurGAN was trained sufficiently, as shown in the figures. The trained model with excessive larger number of epochs usually causes the model overfitting. Therefore, we used 10% of training data as validation set which was not used as training. With the trained model of each epoch, the accuracies of validation set was measured, and the model which showed the best validation accuracy was selected for measuring testing accuracy with testing data. We included the validation performances with validation set in Figures 11c and 12c, which confirms that our model was not overfitted with training data.

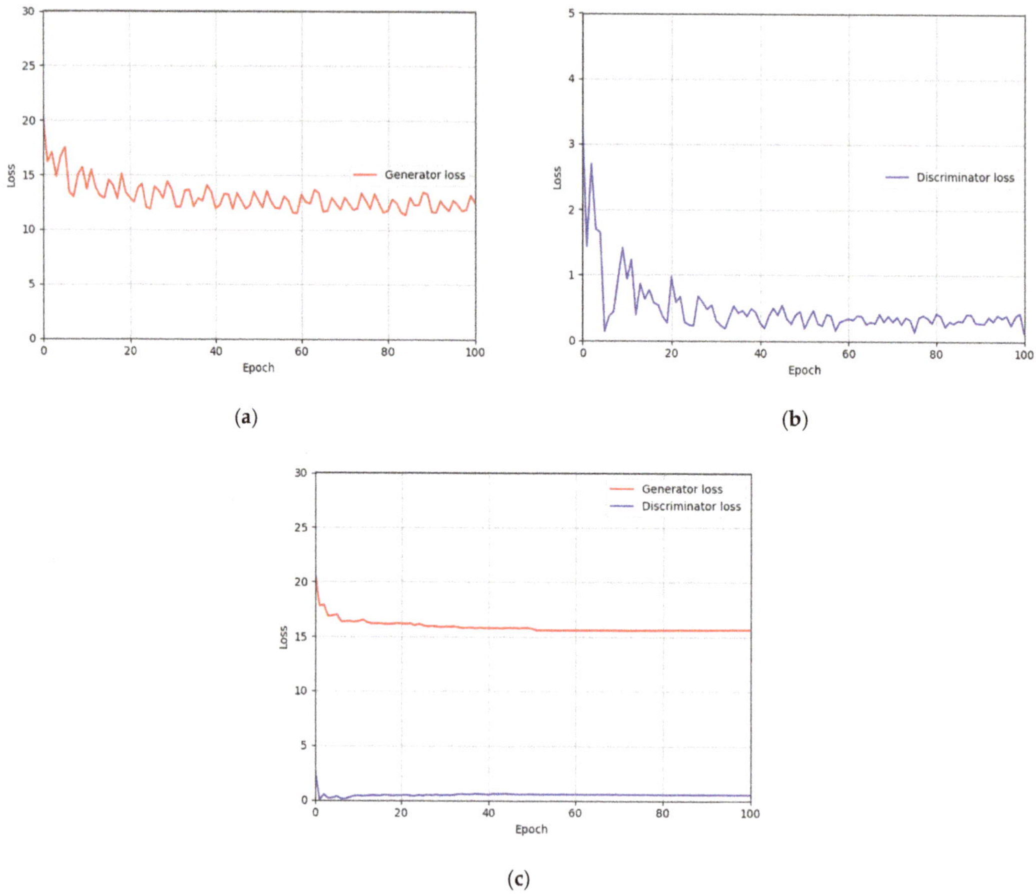

Figure 11. Training and validation loss graphs of the modified DeblurGAN (SDU-DB): training loss graphs of (**a**) generator and (**b**) discriminator. (**c**) Validation loss graphs of generator and discriminator.

4.5. Training of DenseNet-161 for Finger-Vein Recognition

A stochastic gradient descent (SGD) optimization method [50] was used to train the CNN model for finger-vein recognition. This method involves multiplying a gamma value by the learning rate for every step size at a mini-batch unit to reduce the learning rate, thereby rapidly converging training accuracy and loss. As explained in Section 3.4, DenseNet-161 was used in this study for training and testing. The number of output classes was set to two (authentic and imposter-matching), the number of max epochs was set to 30. The mini-batch size was set to 4, the learning rate was set to 0.001, the step size was set to 16 epochs, the momentum was set to 0.9, and the gamma value was set to 0.1. All the hyperparameters were determined with training data. In detail, the optimal hyperparameters (with which the highest accuracies of finger-vein recognition were obtained with training data) were selected. The search spaces for the number of max epochs, mini-batch size, and learning rate were 10~50, 1~10, and 0.0001~0.01, respectively.

The search spaces for the step size, momentum, and gamma value are 5~25 epochs, 0.1~1, and 0.1~1, respectively.

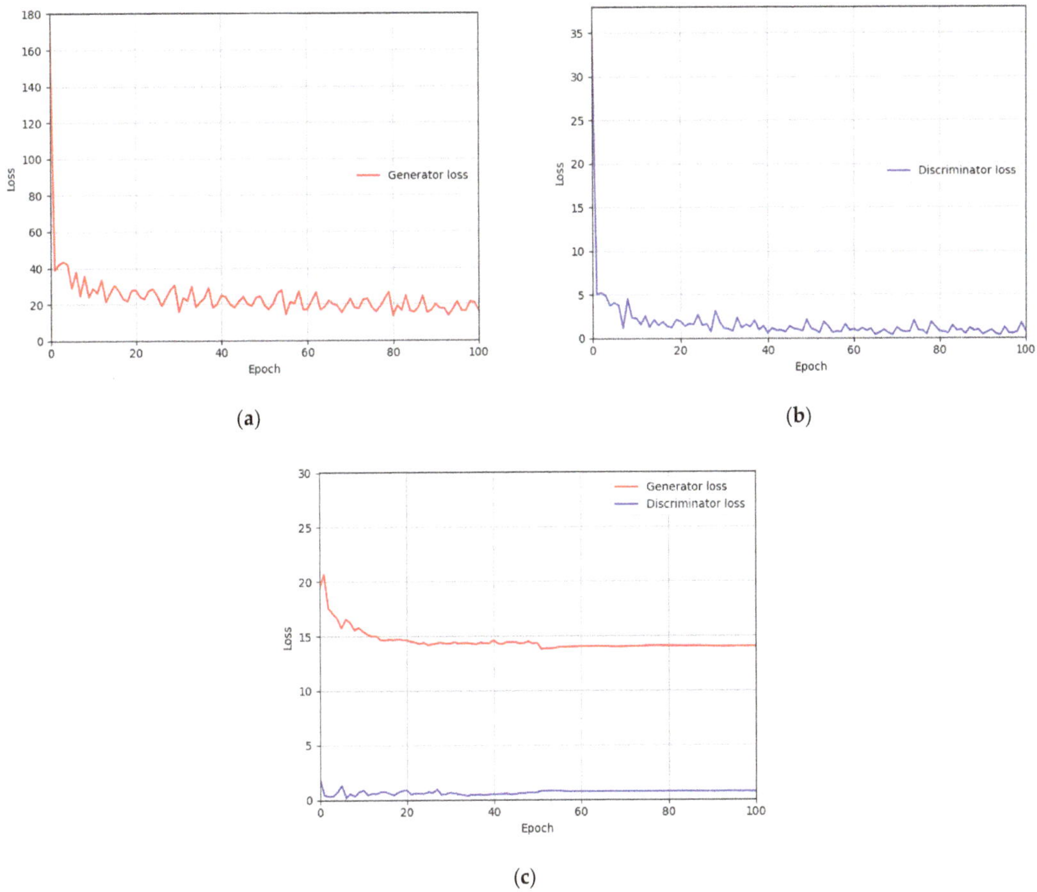

Figure 12. Training and validation loss graphs of the modified DeblurGAN (PolyU-DB): training loss graphs of (**a**) generator and (**b**) discriminator. (**c**) Validation loss graphs of generator and discriminator.

Figures 13 and 14 show the training loss and accuracy graphs of DenseNet-161, which used a difference image restored by the modified DeblurGAN as input. As shown in the training graphs, training loss converged to nearly zero, whereas accuracy converged to nearly 100, indicating that the CNN model for finger-vein recognition was sufficiently trained.

4.6. Testing Results of Proposed Method

4.6.1. Ablation Studies

As ablation studies, experiments were conducted according to with or without motion blur is applied, and the methods can be largely divided into 4 schemes. Scheme 1 means that DenseNet-161 trained with the original training data without blurring was used to perform finger-vein recognition with the original testing data to measure the EER. Scheme 2 means that DenseNet-161 trained with the original training data was used to perform finger-vein recognition with the motion blurred testing data to measure the EER. Scheme 3 represents that DenseNet-161 trained with the motion blurred training data was used to perform finger-vein recognition with the motion blurred testing data to measure the EER. Lastly, scheme 4 represents that DenseNet-161 trained with the training data restored with the modified DeblurGAN proposed in this study was used to perform finger-vein recognition with testing data restored using the modified DeblurGAN to measure the EER. As shown in schemes 2 and 3 in Tables 5 and 6, the vein-pattern region and other regions were difficult to distinguish, due to motion blur, resulting in degradation of recognition performance. Also, in all cases, compared with schemes 2 and 3, when training was performed with the training data restored with the modified DeblurGAN, and recognition was performed for the testing data restored with the modified DeblurGAN, the recognition accuracy was the highest in scheme 4.

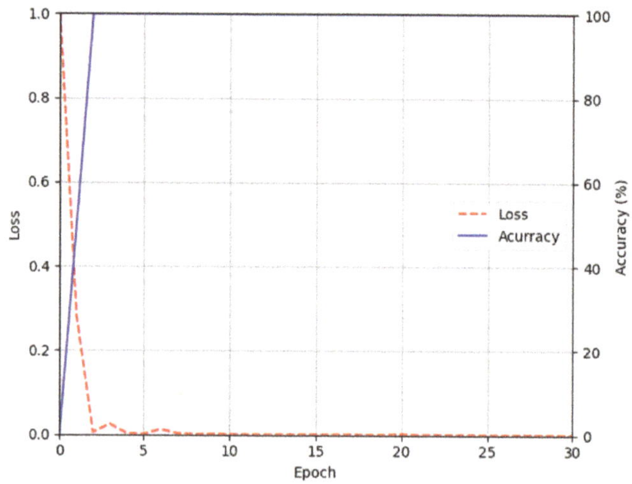

Figure 13. Training accuracy and loss graphs of DenseNet-161 using images restored by proposed modified DeblurGAN (SDU-DB).

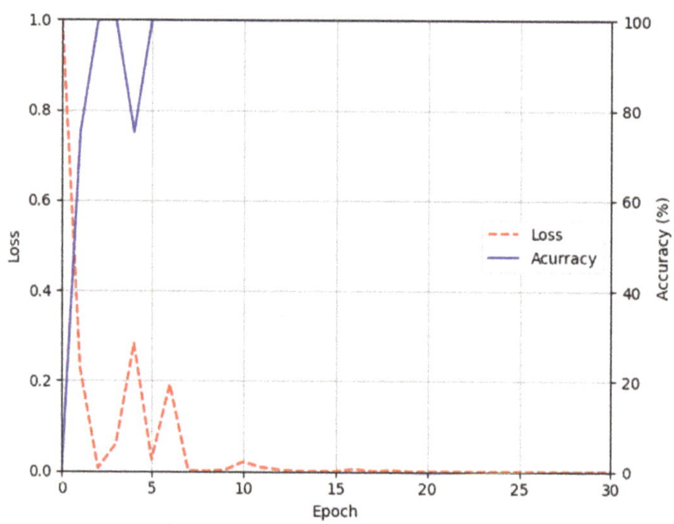

Figure 14. Training accuracy and loss graphs of DenseNet-161 using images restored by proposed modified DeblurGAN (PolyU-DB).

Table 5. Comparison of finger-vein recognition error (EER) with respect to the applicable of a motion blur with SDU-DB (unit: %).

Training & Testing with Original Images (Scheme 1)	Testing Blurred Images without Training (Scheme 2)	Training & Testing with Blurred Images (Scheme 3)	Training & Testing with Restored Images (Scheme 4) (Proposed Method)
2.932	14.618	6.420	5.270

Table 6. Comparison of finger-vein recognition error (EER) with respect to the applicable of a motion blur with PolyU-DB (unit: %).

Training & Testing with Original Images (Scheme 1)	Testing Blurred Images without Training (Scheme 2)	Training & Testing with Blurred Images (Scheme 3)	Training & Testing with Restored Images (Scheme 4) (Proposed Method)
1.534	18.303	5.886	4.536

Figures 15 and 16 show the receiver operating characteristics (ROC) curves for the recognition performance of schemes 1–4 of SDU-DB and PolyU-DB, respectively. Here, GAR is calculated as 100—FRR (%). As shown in Figures 15 and 16, in all cases, the recognition performance after restoration with the modified DeblurGAN proposed in this study (scheme 4) was higher than schemes 2 and 3.

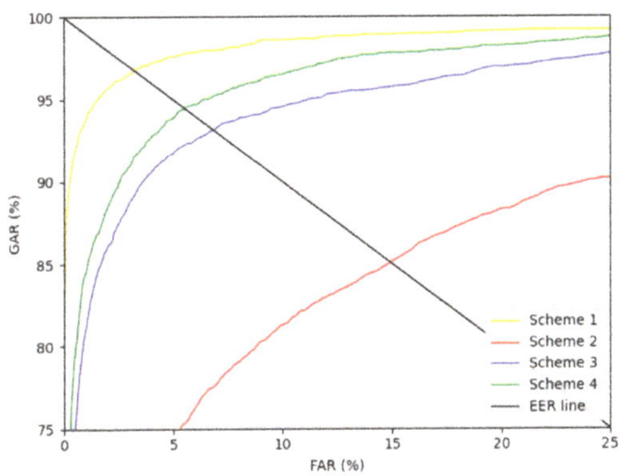

Figure 15. SDU-DB finger-vein recognition ROC curve for scheme 1–4.

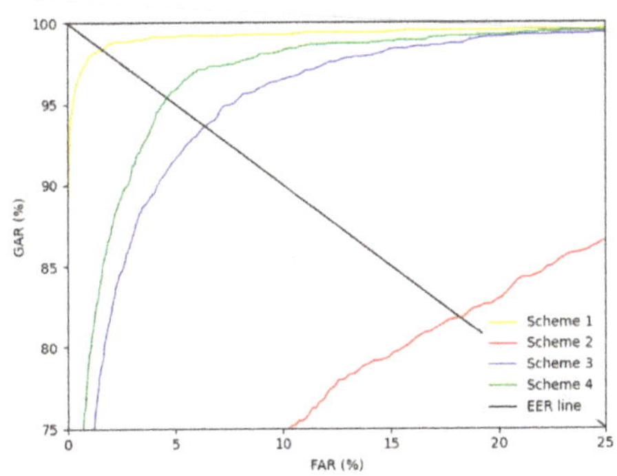

Figure 16. PolyU-DB finger-vein recognition ROC curve for schemes 1–4.

In Tables 7 and 8, the recognition performances of the modified DeblurGAN model were compared according to the changes in the perceptual loss based on the features extracted from the various CNN models and layers. For a fair performance evaluation, the same recognition model was used for all cases based on scheme 4 to measure the recognition accuracy. For VGG-19 (original DeblurGAN), features extracted from the third convolution layer before the third max-pooling were used. Moreover, features extracted from the first convolution layer before the third max-pooling were used for VGG-19 (conv3.1). This is a result of reflecting the features extracted from a layer prior to VGG-19 (original DeblurGAN) in the perceptual loss, indicating that VGG-19 (original DeblurGAN) showed better recognition performance. For ResNeXt-101 (conv2), better recognition performance was exhibited over VGG-19 (original DeblurGAN) and VGG-19 (conv3.1) for both experiments. Lastly, for ResNet-34 (conv2_x), the features extracted from the first residual block (conv2_x) were applied to a perceptual loss (proposed method), thus exhibiting the best performance in all cases with SDU-DB whereas VGG-19 (conv3.1) shows the better accuracies than other cases with PolyU-DB.

Table 7. Comparison of finger-vein recognition error (EER) of restored images in SDU-DB according to the perceptual loss based on the various CNN models and layers (unit: %).

VGG-19 [40] (Original DeblurGAN)	**VGG-19 [40] (Conv3.1)**	**ResNeXt-101 [51] (Conv2)**	**ResNet-34 [32] (Conv2_x)**
6.049	6.503	5.281	5.270

Table 8. Comparison of finger-vein recognition error (EER) of restored images in PolyU-DB according to the perceptual loss based on the various CNN models and layers (unit: %).

VGG-19 [40] (Original DeblurGAN)	**VGG-19 [40] (Conv3.1)**	**ResNeXt-101 [51] (Conv2)**	**ResNet-34 [32] (Conv2_x)**
4.777	4.536	4.764	4.983

4.6.2. Comparisons with the State-of-the-Art Methods

For the next experiment, the similarities between the images restored with the state-of-the-art methods and the proposed modified DeblurGAN and the original images were quantitatively evaluated. For a numerical comparison, a signal-to-noise ratio (SNR) [52], peak SNR (PSNR) [53], and SSIM [54] were measured. SNR and PSNR are evaluation metrics based on the MSE between two images. Equations (6)–(8) are mathematical equations of MSE, SNR, and PSNR, respectively.

$$\text{MSE} = \frac{1}{hw}\sum_{i=0}^{h-1}\sum_{j=0}^{w-1}[I_o(i,\, j) - I_r(i,\, j)]^2 \tag{6}$$

$$\text{SNR} = 10log_{10}\left(\frac{\frac{\sum_{i=0}^{h-1}\sum_{j=0}^{w-1}[I_o(i,\, j)]^2}{hw}}{MSE}\right) \tag{7}$$

$$\text{PSNR} = 10log_{10}\left(\frac{255^2}{MSE}\right) \tag{8}$$

where I_r is the restored image obtained from the state-of-the-art or proposed methods, and I_o is the original image. h and w are the height and width of an image, respectively. Equation (9) is the mathematical equation of SSIM:

$$\text{SSIM} = \frac{(2\mu_o\mu_r + C_1)(2\sigma_{or} + C_2)}{({\mu_o}^2 + {\mu_r}^2 + C_1)({\sigma_o}^2 + {\sigma_r}^2 + C_2)} \tag{9}$$

where μ_r and σ_r are the mean and standard deviation of the pixel values of the restored image, respectively. μ_o and σ_o are the mean and standard deviation of the pixel values of the original image, respectively. σ_{or} is the covariance of two images, and C_1 and C_2 are constants to prevent the denominator of each equation from becoming zero. Using the evaluation metrics of Equations (6)–(9), the enhancement quality of our proposed method and that of the state-of-the-art was numerically evaluated as shown in Tables 9 and 10. As shown in Tables 9 and 10, SRN-DeblurNet shows the higher values for PSNR, SNR, and SSIM compared to our modified DeblurGAN. That is, the qualities of restored images by SRN-DeblurNet are more similar to those of original ones than those by our method. However, the recognition accuracies by our method are higher than those by SRN-DeblurNet as shown in Tables 11 and 12. That is because the additional noises are included in the restored image and the features similar to the original features cannot be restored by SRN-DeblurNet, which causes the degradation of recognition accuracies although the qualities of restored images are similar to those of original ones.

Table 9. Comparisons of blur restoration by using the state-of-the-art methods and proposed modified DeblurGAN with PolyU-DB.

Methods	PSNR	SNR	SSIM
Original DeblurGAN [12]	28.98	21.45	0.90
DeblurGANv2 [14]	26.84	19.32	0.87
SRN-DeblurNet [15]	37.22	29.69	0.95
Modified DeblurGAN (proposed method) VGG-19 (conv3.1)	26.90	19.37	0.88
Modified DeblurGAN (proposed method) ResNet-34	27.70	20.17	0.90

Table 10. Comparisons of blur restoration by using the state-of-the-art methods and proposed modified DeblurGAN with SDU-DB.

Methods	PSNR	SNR	SSIM
Original DeblurGAN [12]	30.84	20.95	0.81
DeblurGANv2 [14]	29.63	19.73	0.82
SRN-DeblurNet [15]	39.17	29.28	0.90
Modified DeblurGAN (proposed method) VGG-19(conv3.1)	28.50	18.60	0.82
Modified DeblurGAN (proposed method) ResNet-34	32.64	22.75	0.85

Table 11. Comparisons of finger-vein recognition error (EER) by using the state-of-the-art restoration models and proposed methods with SDU-DB (unit: %).

Original DeblurGAN [12]	DeblurGANv2 [14]	SRN-DeblurNet [15]	Modified DeblurGAN
6.049	6.077	6.032	5.270

Table 12. Comparisons of finger-vein recognition error (EER) by using the state-of-the-art restoration models and proposed methods with PolyU-DB (unit: %).

Original DeblurGAN [12]	DeblurGANv2 [14]	SRN-DeblurNet [15]	Modified DeblurGAN
4.777	5.507	7.105	4.536

Figure 17 shows examples of the finger-vein images restored by state-of-the-art methods and the modified DeblurGAN. For the next experiment, finger-vein recognition performances were compared using the images restored by the modified DeblurGAN and those restored by the state-of-the-art restoration methods for SDU-DB and PolyU-DB, as shown in Tables 11 and 12. For the comparative experiment, the same recognition model was used for a fair performance evaluation to measure the recognition accuracy using the scheme 4 method of Tables 5 and 6. As shown in Tables 11 and 12, finger-vein recognition performance was higher than the existing state-of-the-art restoration methods, when the restoration was performed using the modified DeblurGAN method.

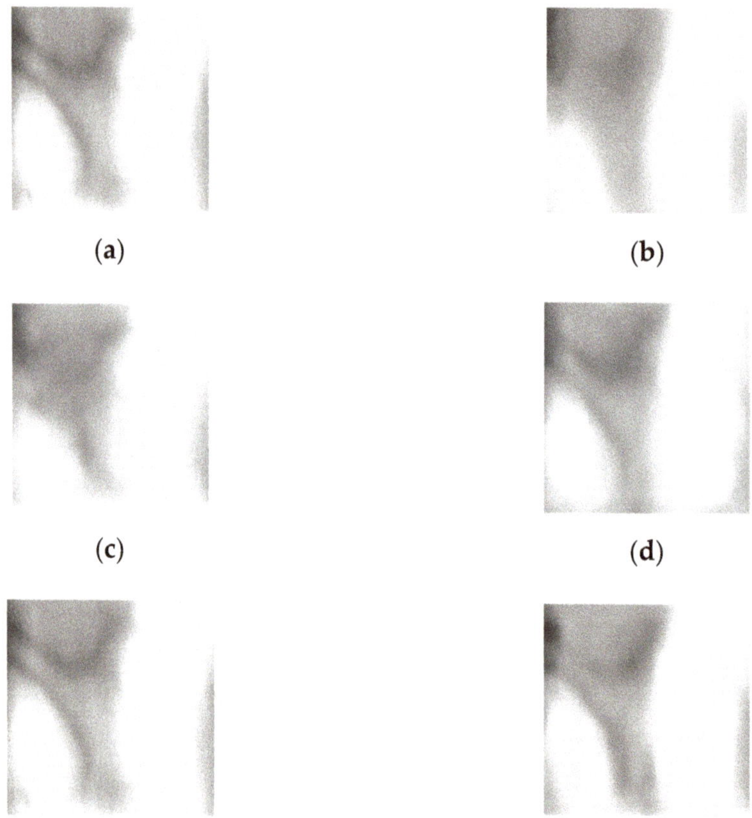

Figure 17. Examples of restored images using the state-of-the-art methods and the proposed modified DeblurGAN: (**a**) original images, (**b**) motion blurred images, and the restored images by (**c**) original DeblurGAN, (**d**) DeblurGANv2, (**e**) SRN-DeblurNet, and (**f**) proposed modified DeblurGAN.

Figure 18a,c are the result of authentic and imposter matching prior to restoration, which provide incorrect matching results caused by modified vein patterns and texture information due to motion blur. Authentic matching was falsely rejected as imposter matching, whereas imposter matching was falsely accepted as authentic, thus decreased the recognition performance. Figure 18b,d are the results of correct matching by restoring the incorrect matching problem in (a) and (c) by the modified DeblurGAN. Authentic matching was classified as correct acceptance, and imposer matching was classified as correct rejection.

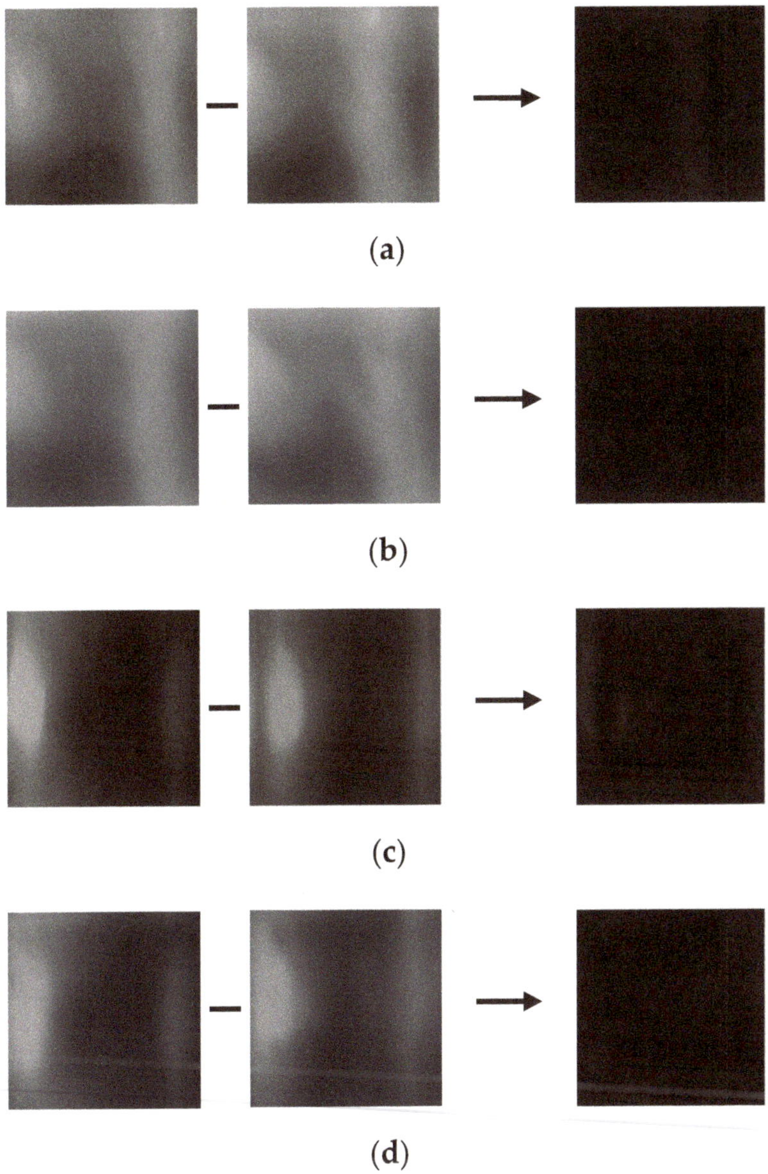

Figure 18. Correct recognition examples after restoring motion blur. (**a**) Incorrect genuine matching before restoring motion blur, (**b**) correct genuine matching after restoring motion blur, (**c**) incorrect imposter matching before restoring motion blur, and (**d**) correct imposter matching after restoring motion blur. From the left, examples in (**a–d**) present the registered, input, and difference images, respectively.

Figure 19 is an example of incorrect authentic matching and incorrect imposter matching despite the restoration method proposed in this study is applied. In the case of incorrect authentic matching, the difference in motion blur between the same classes is so severe that it is recognized as an imposter even after restoration, resulting in incorrect matching. In the case of incorrect imposter matching, the enrolled image and the input image appear similarly in dark shades, and the vein pattern is not clearly visible, so it recognized as authentic even after restoration, resulting in incorrect matching.

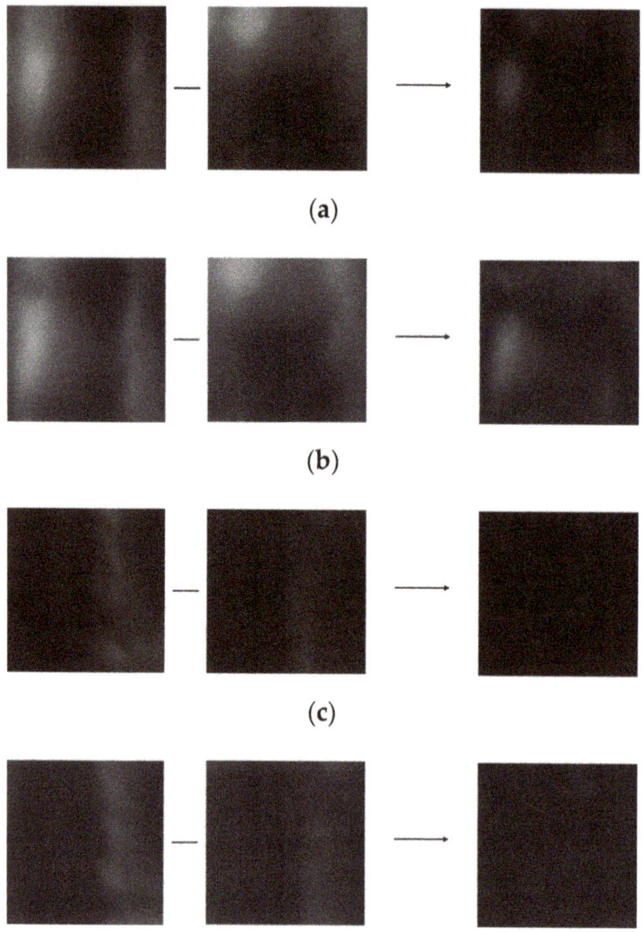

(a)

(b)

(c)

(d)

Figure 19. Incorrect recognition examples after restoring motion blur. (**a**) Incorrect genuine matching before restoring motion blur, (**b**) incorrect genuine matching after restoring motion blur, (**c**) incorrect imposter matching before restoring motion blur, and (**d**) incorrect imposter matching after restoring motion blur. From the left, examples in (**a**–**d**) present the registered, input, and difference images, respectively.

4.7. Processing Time of Proposed Method

For the next experiment, the inference time of the modified DeblurGAN proposed in this study and DenseNet-161 for the finger-vein recognition method was measured. The measurements were taken on the desktop described explained in Section 4.3 and the Jetson TX2 embedded system [55] shown in Figure 20. The reason for measuring using the embedded system is that on-board edge computing, which operates as an embedded system attached to the entrance door, is involved for most access-controlled type finger-vein recognition systems. Thus, it must be verified that on-board computing is feasible on the system proposed. Jetson TX2 has an NVIDIA Pascal^{TM}-family GPU

(256 CUDA cores), with 8-GB memory shared between the CPU and GPU, and 59.7-GB/s of memory bandwidth. It uses less than 7.5 watts of power. As presented in Table 13, in the case of the method proposed in this study, the recognition speed for one image was 16.2 ms on a desktop computer and 232.3 ms on the Jetson TX2 embedded system. This corresponds to 61.72 frames/s (1000/16.2) and 4.3 frames/s (1000/232.3), respectively. The processing time on the Jetson TX2 embedded system was longer than the desktop computer, due to limited computing resources. However, through the experiment, it was confirmed that the proposed method is applicable to an embedded system having limited computing resources.

Figure 20. Jetson TX2 embedded system.

Table 13. Comparisons of processing speed by proposed method on desktop computer and embedded system (unit: ms).

	Modified DeblurGAN for Restoration	DenseNet-161 for Finger-Vein Recognition	Total
Desktop computer	3.4	12.8	16.2
Jetson TX2	6.1	226.2	232.3

4.8. Analysis of Feature Map

4.8.1. Class Activation Map of Restored Image

Figure 21 shows the result of visualizing each class activation map [56] based on the original images and those restored by the proposed modified DeblurGAN in each layer of DenseNet-161. The location from which the class activation map is output is the 1st convolutional layer, the 1st transition layer, the 2nd transition layer, the 3rd transition layer, and the last dense block layer from top to bottom. Figure 21a,b show examples of authentic (genuine) and imposter matching. The left and middle images in (a) and (b) are the original and restored images, respectively. Important features are represented in red, whereas insignificant features are represented in blue in the class activation map. Therefore, if the red and blue regions of the two images appear to be similar, it generally indicates that the two images have similar characteristics. As shown in Figure 21a, in authentic matching, class activation occurs in a similar location of the original image and restored image. Accordingly, it was confirmed that the motion blurred finger-vein image was effectively restored and correct acceptance is possible. As shown in Figure 21b, with imposter matching, class activation occurs in different locations in the original and restored image, implying that correct rejection is possible.

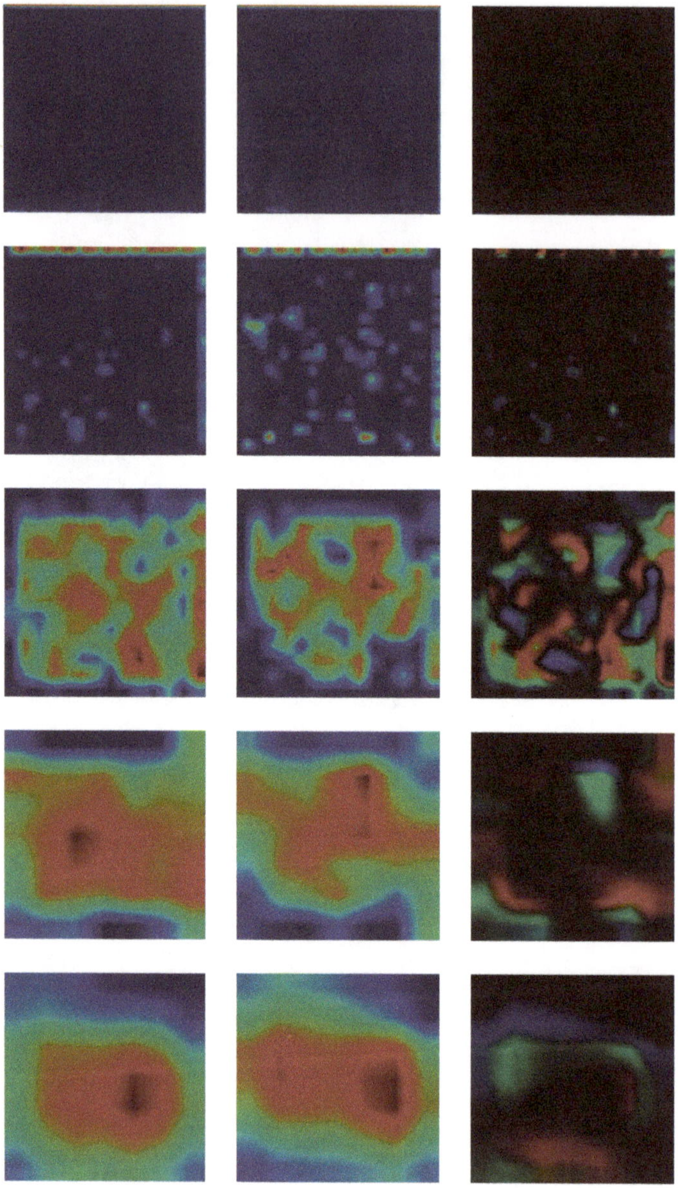

(a)

Figure 21. *Cont.*

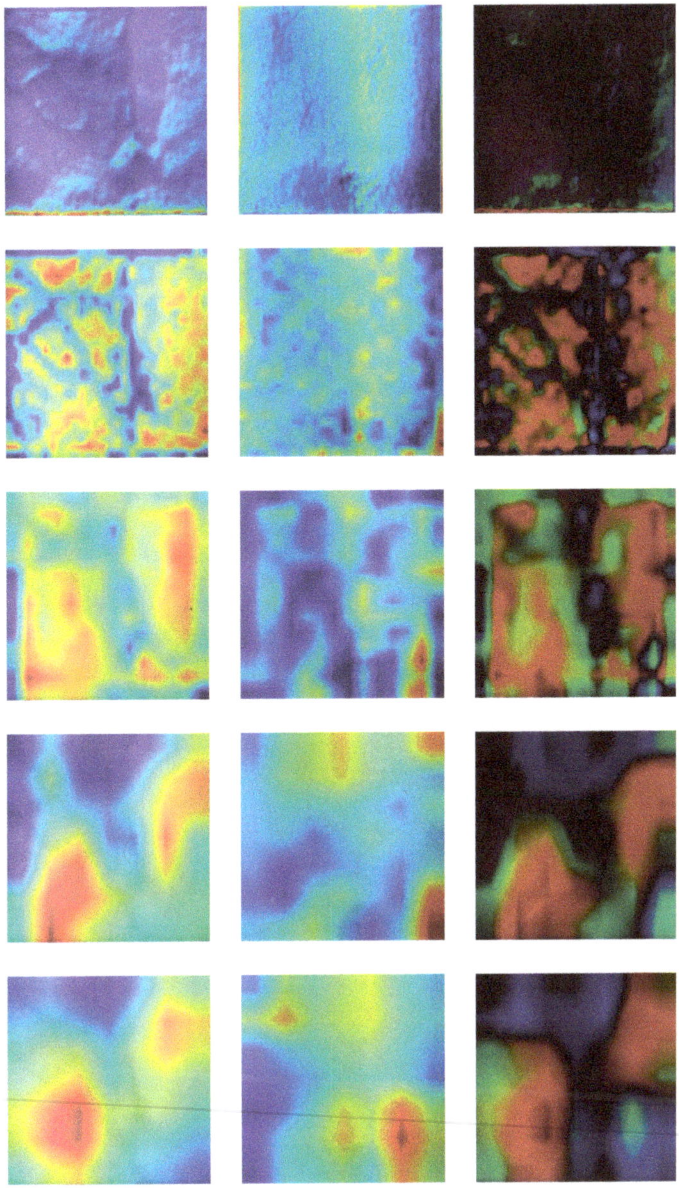

(b)

Figure 21. Comparisons of the class activation maps between the original and restored images. (**a**,**b**) are examples of authentic and imposter images, respectively. Images on the left of (**a**,**b**) are the original images, whereas those on the middle are the restored image by proposed modified DeblurGAN. In addition, the images on the right of (**a**,**b**) are the subtracted ones of the middle image from the left one. For both (**a**,**b**), the images from top to bottom are the class activation maps output from the 1st convolutional layer, the 1st transition layer, the 2nd transition layer, the 3rd transition layer, and the last dense block.

In addition, we included the subtracted CAM outputs of restored image from original (motion blurred) one in the right images of Figure 21a,b. The reasons of such differences in the subtracted CAM outputs are that the positions of important finger-vein features extracted are different in original and restored images. Nevertheless, the case of authentic matching (same class) shows the smaller differences as shown in the right image of the last row of Figure 21a compared to that of imposter matching (different classes) in the right image of the last row of Figure 21b. In addition, the reasons of such differences in the subtracted CAM outputs are that the important features of finger-vein can be newly extracted from the restored image (red color in the middle images of Figure 21a,b). However, they cannot be extracted from vein areas in original (motion blurred) image (red color in the left images of Figure 21a,b) due to the indistinctive vein patterns caused by motion blurring, but they are extracted from the other skin areas except for vein regions.

4.8.2. Feature Maps of Difference Image

Second, similar to Figure 21, the feature maps of DenseNet-161 were analyzed according to the layer depth in which the difference image between the restored enrolled and restored recognized image as input. The input of DenseNet-161 is the finger-vein image restored by the modified DeblurGAN. As the feature map dimension is too large, the feature maps presented in Figure 22 are each channel's output. Figure 22 presents the examples of the feature maps extracted from genuine and imposter matching images in several layers of DenseNet-161. Examples in Figure 22a–e are the feature maps extracted from the 1st convolutional layer, the 1st transition layer, the 2nd transition layer, the 3rd transition layer, and the last dense block, respectively. In addition, Figure 22f is the 3-dimensional feature map images created by averaging the feature map values of Figure 22e. The top and bottom images in Figure 22 show authentic and imposter matching, respectively.

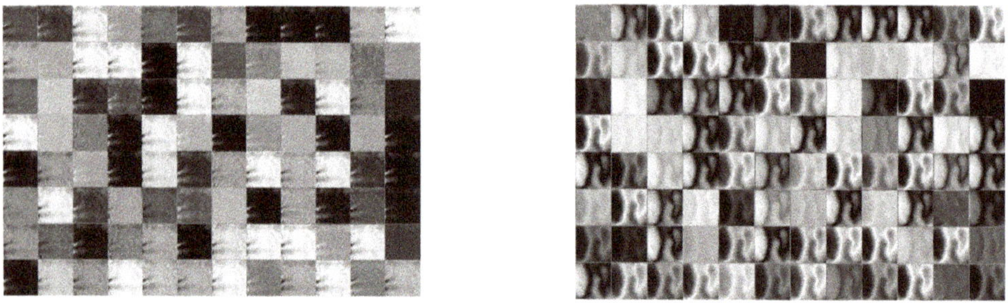

(**a**)

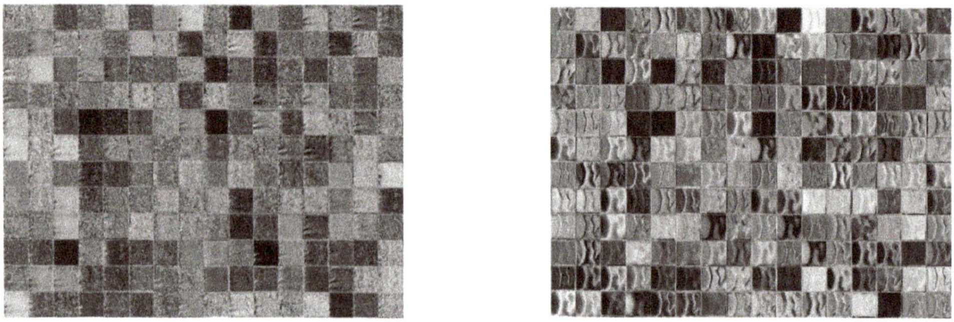

(**b**)

Figure 22. *Cont.*

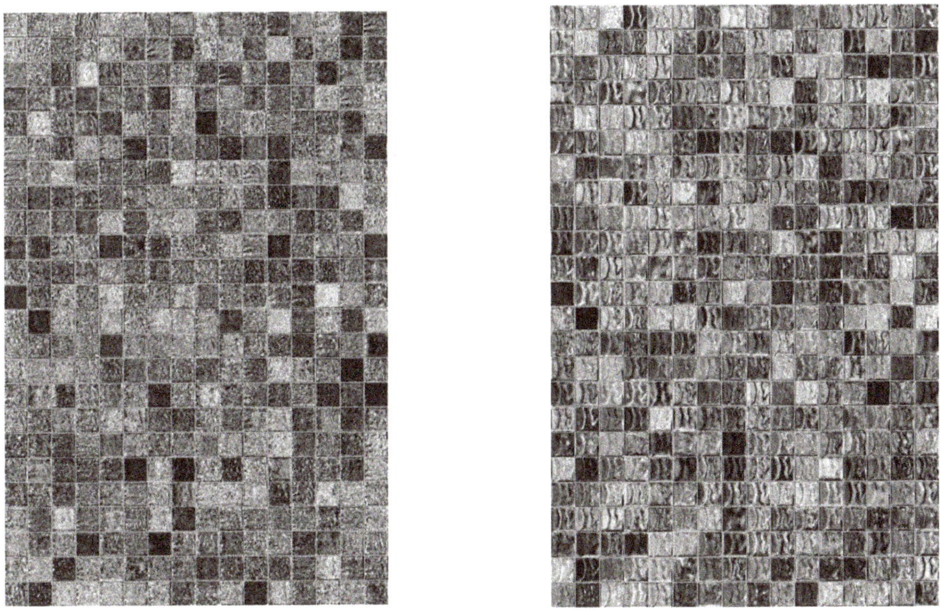

(c)

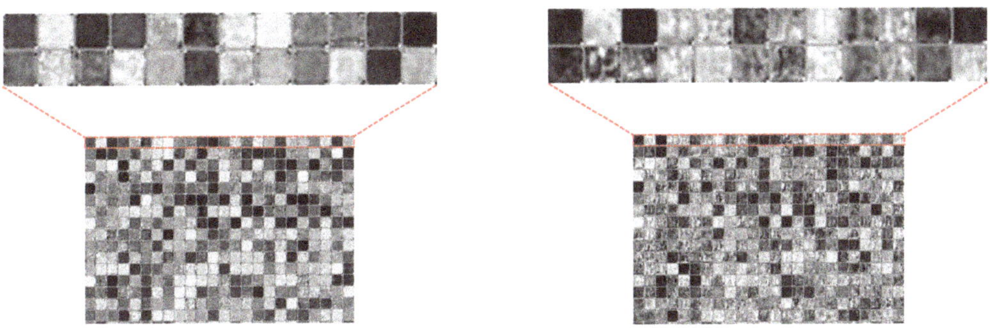

(d)

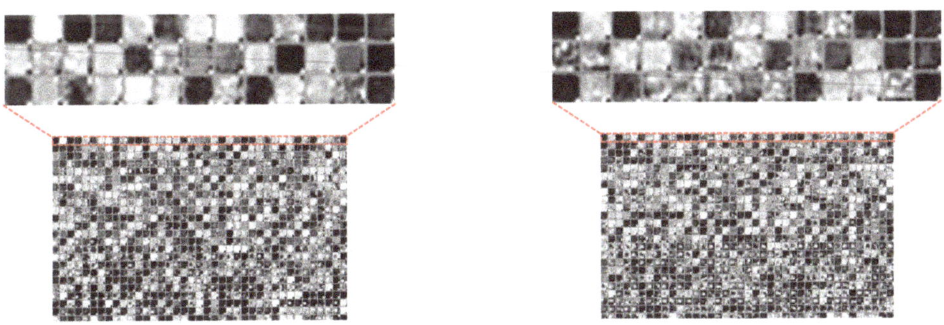

(e)

Figure 22. *Cont.*

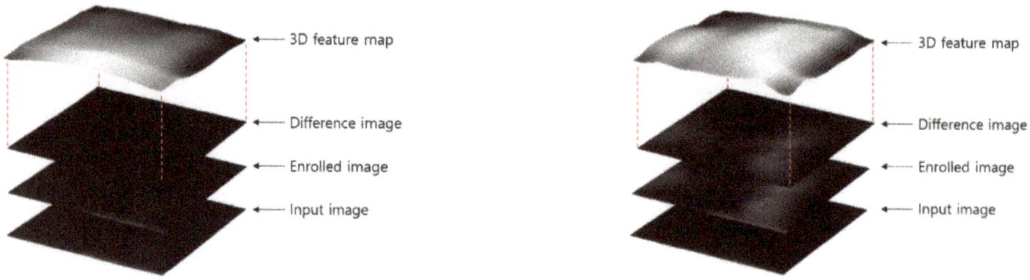

(f)

Figure 22. Feature maps extracted from a genuine matching image and an imposter matching image from several layers of the DenseNet-161. (**a**) Feature maps from the 1st convolutional layer; (**b**) feature maps from the 1st transition layer; (**c**) feature maps from the 2nd transition layer; (**d**) feature maps from the 3rd transition layer; (**e**) feature maps from the last dense block, and (**f**) 3D feature maps created by averaging feature map values of (**e**). Upper and lower examples in (**a–f**) represent genuine matching feature maps and imposter matching feature maps, respectively.

As shown in Figure 22, abstract features were extracted as the layer became deeper. For example, low-level features, such as lines and corners of the original image, were maintained in Figure 22a, whereas, in Figure 22e, only the abstracted features remained, and shape information mostly disappeared. As shown in Figure 22a–e, the feature maps of authentic and imposter matching do not seem to have a significant difference. However, as shown in Figure 22f, although the changes in the 3-dimensional feature map values drawn by calculating the average of feature map values for the authentic matching results from a step before the classification layer were mostly flat, the results of imposter matching showed that the changes in the feature-map values were greater than those of authentic matching. Therefore, the difference in the CNN feature maps of authentic and imposter matching by the proposed method was confirmed.

5. Conclusions

In this study, a motion blurred finger-vein image was restored to solve the problem of deterioration of finger-vein recognition performance due to motion blur, and a recognition method using deep CNN was studied to evaluate the performance of the restored image. A modified DeblurGAN was proposed by modifying the original DeblurGAN, which was a restoration model. Using two open databases, the recognition error rate was lower when recognition was performed using the restoration method proposed in this study than when images were not restored. Furthermore, based on the comparative experiments using various state-of-the-art restoration models, the proposed method was more effective in restoring an image from motion blur and had more improved recognition performance. Also, based on the analysis of class activation maps and feature maps, it was confirmed that the proposed modified DeblurGAN sufficiently maintained the effective characteristics for classifying authentic and imposter matching. However, as mentioned in Figure 19, it was confirmed that incorrect matching cases occurred despite the proposed restoration method. Therefore, in future studies, a method of increasing restoration and recognition performance by overcoming the extreme difference in motion blur in intra-class and reducing the degree of similarity between inter-classes will be studied. In our research, we used the previous methods [27,29,30] for the ROI detection of finger region, and just focused on the restoration of motion-blur by our proposed modified DeblurGAN and finger-vein recognition by our CNN with the selected ROI. That is because the performance analysis is difficult if both the ROI detection and feature extraction of finger-vein are affected by motion blurring. Therefore, we assume that the ROI without motion blurring is correctly detected by the previous methods [27,29,30], and we only consider that the detected ROI is motion blurred. We would research the motion blurring effect on the boundary detection of ROI in future work.

If the enrolled and recognized images are captured from different camera settings, the performance of finger-vein recognition based on image difference can be affected. However, the enrolled and recognized images are captured from the same capturing device including same camera setting in usual cases of actual finger-vein recognition system. In addition, in this case, the recognition based on image difference showed the better accuracies than those based on original image with the extracted feature vector [20]. Therefore, we use this scheme of image difference for recognition because we mainly focused on the restoration of motion blurring by proposed modified DeblurGAN. We would research the recognition method with the enrolled and recognized images captured from different camera settings in future work.

People usually put their finger on the device with some guiding bar in the actual finger-vein acquisition device (with fixed finger direction) [29,31]. Therefore, there exist only the limited variations of the horizontal and vertical translation and in-plane rotation in the captured finger-vein image. Our data augmentation method aims at covering these individual variations, and it can reduce the recognition error (false rejection case). However, horizontal and vertical mirroring does not happen in the case of a finger-vein image acquisition of the actual capturing device. Therefore, the mirroring generates the images of different classes, which increases the complexity of training data and difficulties of model training. As shown in [57,58], singular value decomposition (SVD) can generate the images of various styles, which can also produce the images of different classes, and it can also increase the complexity of training data and difficulties of model training. Therefore, we use our simple data augmentation method. In future work, we would research the various data augmentation method including SVD and mirroring.

Also, the application of the proposed motion blur restoration method to other biometric modalities, such as iris, face, and palm-vein recognition, will be examined. Moreover, a lighter model that can shorten the processing time will be studied. In future work, we would also research the method with the cases of two open databases combined. In addition, as a future work, we would introduce different types of blurring to the images and develop a generic solution.

Author Contributions: Methodology, J.C.; Conceptualization, J.S.H.; Validations, M.O., S.G.K.; Supervision, K.R.P.; Writing—original draft, J.C.; Writing—review and editing, K.R.P. All authors have read and agreed to the published version of the manuscript.

Funding: This research was supported in part by the National Research Foundation of Korea (NRF) funded by the Ministry of Science and ICT (MSIT) through the Basic Science Research Program (NRF-2021R1F1A1045587), in part by the NRF funded by the MSIT through the Basic Science Research Program (NRF-2020R1A2C1006179), and in part by the MSIT, Korea, under the ITRC (Information Technology Research Center) support program (IITP-2021-2020-0-01789) supervised by the IITP (Institute for Information & Communications Technology Planning & Evaluation).

Institutional Review Board Statement: Not applicable.

Informed Consent Statement: Not applicable.

Data Availability Statement: Not applicable.

Acknowledgments: Not applicable.

Conflicts of Interest: The authors declare no conflict of interest.

References

1. Liu, Z.; Yin, Y.; Wang, H.; Song, S.; Li, Q. Finger vein recognition with manifold learning. *J. Netw. Comput. Appl.* **2010**, *33*, 275–282. [CrossRef]
2. Lee, E.C.; Park, K.R. Restoration method of skin scattering blurred vein image for finger vein recognition. *Electron. Lett.* **2009**, *45*, 1074–1076. [CrossRef]
3. Yang, J.; Zhang, B.; Shi, Y. Scattering removal for finger-vein image restoration. *Sensors* **2012**, *12*, 3627–3640. [CrossRef] [PubMed]
4. Yang, J.; Zhang, B. Scattering removal for finger-vein image enhancement. In Proceedings of the International Conference on Hand-Based Biometrics (ICHB), Hong Kong, China, 17–18 November 2011; pp. 1–5.

5. Yang, J.; Shi, Y. Towards finger-vein image restoration and enhancement for finger-vein recognition. *Inf. Sci.* **2014**, *268*, 33–52. [CrossRef]
6. Shi, Y.; Yang, J.; Yang, J. A new algorithm for finger-vein image enhancement and segmentation. *Inf. Sci. Ind. Appl.* **2012**, *4*, 139–144.
7. Yang, J.; Bai, G. Finger-vein image restoration based on skin optical property. In Proceedings of the 11th International Conference on Signal Processing (ICSP), Beijing, China, 21–25 October 2012; pp. 749–752.
8. Yang, J.; Shi, Y.; Yang, J. Finger-vein image restoration based on a biological optical model. In *New Trends and Developments in Biometrics*; IntechOpen: London, UK, 2012; pp. 59–76.
9. You, W.; Zhou, W.; Huang, J.; Yang, F.; Liu, Y.; Chen, Z. A bilayer image restoration for finger vein recognition. *Neurocomputing* **2019**, *348*, 54–65. [CrossRef]
10. Lee, E.C.; Park, K.R. Image restoration of skin scattering and optical blurring for finger vein recognition. *Opt. Lasers Eng.* **2011**, *49*, 816–828. [CrossRef]
11. Choi, J.H.; Noh, K.J.; Cho, S.W.; Nam, S.H.; Owais, M.; Park, K.R. Modified conditional generative adversarial network-based optical blur restoration for finger-vein recognition. *IEEE Access* **2020**, *8*, 16281–16301. [CrossRef]
12. Kupyn, O.; Budzan, V.; Mykhailych, M.; Mishkin, D.; Matas, J. DeblurGAN: Blind motion deblurring using conditional adversarial networks. In Proceedings of the 2018 IEEE Conference on Computer Vision and Pattern Recognition (CVPR), Salt Lake City, UT, USA, 18–22 June 2018; pp. 8183–8192.
13. Szeliski, R. *Computer Vision: Algorithms and Applications*, 1st ed.; Springer: London, UK, 2010.
14. Kupyn, O.; Martyniuk, T.; Wu, J.; Wang, Z. Deblurgan-v2: Deblurring (orders-of-magnitude) faster and better. In Proceedings of the IEEE/CVF International Conference on Computer Vision (ICCV), Seoul, Korea, 27 October–2 November 2019; pp. 8878–8887.
15. Tao, X.; Gao, H.; Shen, X.; Wang, J.; Jia, J. Scale-recurrent network for deep image deblurring. In Proceedings of the 2018 IEEE Conference on Computer Vision and Pattern Recognition (CVPR), Salt Lake City, UT, USA, 18–22 June 2018; pp. 8174–8182.
16. Dongguk Modified DeblurGAN and CNN for Recognition of Blurred Finger-Vein Image with Motion Blurred Image Database. Available online: https://github.com/dongguk-dm/MDG_CNN (accessed on 29 June 2021).
17. Lee, E.C.; Lee, H.C.; Park, K.R. Finger vein recognition using minutia-based alignment and local binary pattern-based feature extraction. *Int. J. Imaging Syst. Technol.* **2009**, *19*, 179–186. [CrossRef]
18. Peng, J.; Wang, N.; El-Latif, A.A.A.; Li, Q.; Niu, X. Finger-vein verification using Gabor filter and SIFT feature matching. In Proceedings of the International Conference on Intelligent Information Hiding and Multimedia Signal Processing (IIHMSP), Piraeus, Greece, 18–20 July 2012; pp. 45–48.
19. Wu, J.-D.; Liu, C.-T. Finger-vein pattern identification using SVM and neural network technique. *Expert Syst. Appl.* **2011**, *38*, 14284–14289. [CrossRef]
20. Hong, H.G.; Lee, M.B.; Park, K.R. Convolutional neural network-based finger-vein recognition using NIR image sensors. *Sensors* **2017**, *17*, 1297. [CrossRef] [PubMed]
21. Kim, W.; Song, J.M.; Park, K.R. Multimodal biometric recognition based on convolutional neural network by the fusion of finger-vein and finger shape using near-infrared (NIR) camera sensor. *Sensors* **2018**, *18*, 2296. [CrossRef] [PubMed]
22. Qin, H.; El-Yacoubi, M.A. Deep representation-based feature extraction and recovering for finger-vein verification. *IEEE Trans. Inf. Forensic Secur.* **2017**, *12*, 1816–1829. [CrossRef]
23. Song, J.M.; Kim, W.; Park, K.R. Finger-vein recognition based on deep DenseNet using composite image. *IEEE Access* **2019**, *7*, 66845–66863. [CrossRef]
24. Noh, K.J.; Choi, J.; Hong, J.S.; Park, K.R. Finger-vein recognition based on densely connected convolutional network using score-level fusion with shape and texture images. *IEEE Access* **2020**, *8*, 96748–96766. [CrossRef]
25. Noh, K.J.; Choi, J.; Hong, J.S.; Park, K.R. Finger-vein recognition using heterogeneous databases by domain adaption based on a cycle-consistent adversarial network. *Sensors* **2021**, *21*, 524. [CrossRef]
26. Qin, H.; Wang, P. Finger-vein verification based on LSTM recurrent neural networks. *Appl. Sci.* **2019**, *9*, 1687. [CrossRef]
27. Gonzalez, R.C.; Woods, R.E. *Digital Image Processing*, 2nd ed.; Prentice-Hall: Upper Saddle River, NJ, USA, 2002.
28. Gonzalez, R.C.; Woods, R.E. *Digital Image Processing*, 3rd ed.; Prentice-Hall: Upper Saddle River, NJ, USA, 2010.
29. Kumar, A.; Zhou, Y. Human identification using finger images. *IEEE Trans. Image Process.* **2012**, *21*, 2228–2244. [CrossRef]
30. Kumar, A.; Zhang, D. Personal recognition using hand shape and texture. *IEEE Trans. Image Process.* **2006**, *15*, 2454–2461. [CrossRef]
31. Yin, Y.; Liu, L.; Sun, X. SDUMLA-HMT: A Multimodal Biometric Database. In Proceedings of the Chinese Conference on Biometric Recognition (CCBR), Beijing, China, 3–4 December 2011; pp. 260–268.
32. He, K.; Zhang, X.; Ren, S.; Sun, J. Deep residual learning for image recognition. In Proceedings of the 2016 IEEE Conference on Computer Vision and Pattern Recognition (CVPR), Las Vegas, NV, USA, 27–30 June 2016; pp. 770–778.
33. Nair, V.; Hinton, G.E. Rectified linear units improve restricted Boltzmann machines. In Proceedings of the 27th International Conference on International Conference on Machine Learning, Haifa, Israel, 21–24 June 2010; pp. 807–814.
34. Srivastava, N.; Hinton, G.; Krizhevsky, A.; Sutskever, I.; Salakhutdinov, R. Dropout: A simple way to prevent neural networks from overfitting. *J. Mach. Learn Res.* **2014**, *15*, 1929–1958.
35. Nah, S.; Kim, T.H.; Lee, K.M. Deep multi-scale convolutional neural network for dynamic scene deblurring. In Proceedings of the Computer Vision and Pattern Recognition (CVPR), Honolulu, HI, USA, 21–26 July 2017; pp. 3883–3891.

36. Köhler, R.; Hirsch, M.; Mohler, B.; Schölkopf, B.; Harmeling, S. Recording and playback of camera shake: Benchmarking blind deconvolution with a real-world database. In Proceedings of the Europe Conference Computer Vision (ECCV), Florence, Italy, 7–13 October 2012; pp. 27–40.
37. Gulrajani, I.; Ahmed, F.; Arjovsky, M.; Dumoulin, V.; Courville, A.C. Improved training of Wasserstein GANs. *arXiv* **2017**, arXiv:1704.00028.
38. Goodfellow, I.J.; Pouget-Abadie, J.; Mirza, M.; Xu, B.; Warde-Farley, D.; Ozair, S.; Courville, A.; Bengio, Y. Generative adversarial nets. In Proceedings of the 27th International Conference on Neural Information Processing Systems, Montreal, QC, Canada, 8–13 December 2014; pp. 2672–2680.
39. Salimans, T.; Goodfellow, I.; Zaremba, W.; Cheung, V.; Radford, A.; Chen, X. Improved techniques for training GANs. In Proceedings of the 30th International Conference on Neural Information Processing Systems, Barcelona, Spain, 5–10 December 2016; pp. 2234–2242.
40. Simonyan, K.; Zisserman, A. Very deep convolutional networks for large-scale image recognition. In Proceedings of the International Conference on Learning Representations, San Diego, CA, USA, 7–9 May 2015; pp. 1–14.
41. Zeiler, M.D.; Fergus, R. Visualizing and understanding convolutional networks. In Proceedings of the European Conference on Computer Vision (ECCV), Zurich, Switzerland, 6–12 September 2014; pp. 818–833.
42. Johnson, J.; Alahi, A.; Fei-Fei, L. Perceptual losses for real-time style transfer and super-resolution. In Proceedings of the European Conference on Computer Vision (ECCV), Amsterdam, The Netherlands, 8–16 October 2016; pp. 694–711.
43. Molchanov, D.; Ashukha, A.; Vetrov, D. Variational dropout sparsifies deep neural networks. *arXiv* **2017**, arXiv:1701.05369v3.
44. Image Differencing. Available online: https://en.wikipedia.org/wiki/Image_differencing (accessed on 20 December 2020).
45. Huang, G.; Liu, Z.; van der Maaten, L.; Weinberger, K.Q. Densely connected convolutional networks. In Proceedings of the IEEE Conference on Computer Society Conference on Computer Vision and Pattern Recognition (CVPR), Honolulu, HI, USA, 21–26 July 2017; pp. 2261–2269.
46. Ioffe, S.; Szegedy, C. Batch normalization: Accelerating deep network training by reducing internal covariate shift. In Proceedings of the International Conference on Machine Learning, Lille, France, 6–11 July 2015.
47. Deng, J.; Dong, W.; Socher, R.; Li, L.-J.; Li, K.; Fei-Fei, L. ImageNet: A large-scale hierarchical image database. In Proceedings of the IEEE Conference on Computer Vision and Pattern Recognition, Miami, FL, USA, 20–25 June 2009; pp. 248–255.
48. NVIDIA GeForce GTX 1070. Available online: https://www.geforce.com/hardware/desktop-gpus/geforce-gtx-1070/specifications (accessed on 27 December 2020).
49. Kingma, D.P.; Ba, J. Adam: A method for stochastic optimization. *arXiv* **2014**, arXiv:1412.6980.
50. Bottou, L. Large-scale machine learning with stochastic gradient descent. In Proceedings of the International Conference on Computational Statistics, Paris, France, 22–27 August 2010; pp. 177–186.
51. Xie, S.; Girshick, R.; Dollar, P.; Tu, Z.; He, K. Aggregated residual transformations for deep neural networks. In Proceedings of the IEEE Conference on Computer Society Conference on Computer Vision and Pattern Recognition (CVPR), Honolulu, HI, USA, 21–26 July 2017; pp. 5987–5995.
52. Stathaki, T. *Image Fusion: Algorithms and Applications*; Academic: Cambridge, MA, USA, 2008.
53. Salomon, D. *Data Compression: The Complete Reference*, 4th ed.; Springer: New York, NY, USA, 2006.
54. Wang, Z.; Bovik, A.C.; Sheikh, H.R.; Simoncelli, E.P. Image quality evaluation: From error visibility to structural similarity. *IEEE Trans. Image Process.* **2004**, *13*, 600–612. [CrossRef] [PubMed]
55. Jetson TX2 Module. Available online: https://developer.nvidia.com/embedded/jetson-tx2 (accessed on 2 January 2021).
56. Selvaraju, R.R.; Cogswell, M.; Das, A.; Vedantam, R.; Rarikh, D.; Batra, D. Grad-CAM: Visual explanations from deep networks through gradient-based localization. In Proceedings of the IEEE International Conference on Computer Vision (ICCV), Venice, Italy, 22–29 October 2017; pp. 618–626.
57. Zhu, Y.-C.; AlZoubi, A.; Jassim, S.; Jiang, Q.; Zhang, Y.; Wang, Y.-B.; Ye, X.-D.; DU, H. A generic deep learning framework to classify thyroid and breast lesions in ultrasound images. *Ultrasonics* **2021**, *110*, 1–8. [CrossRef] [PubMed]
58. Robb, E.; Chu, W.-S.; Kumar, A.; Huang, J.-B. Few-shot adaptation of generative adversarial networks. *arXiv* **2020**, arXiv:2010.11943v1.

MDPI
St. Alban-Anlage 66
4052 Basel
Switzerland
Tel. +41 61 683 77 34
Fax +41 61 302 89 18
www.mdpi.com

Sensors Editorial Office
E-mail: sensors@mdpi.com
www.mdpi.com/journal/sensors